D1081304

GCSE

Collins

Total Revision

GCSE Maths

WITHDRAWN

Paul Metcalf

Series Editor: Jayne de Courcy

Ridge Danyers - Marple Site
M0030790

Contents

HJ. June '02
M592
510.76 MET
30790

How this book will help you...

It doesn't matter whether you're heading for mocks in Year 11, or in the final run-up to your GCSE exam – **this book will help you to produce your very best.**

Whichever approach you decide to take to revision, this book will provide everything you need:

1. Total revision support
2. Quick revision check-ups
3. Exam practice

1 Total Revision Support

Everything you need to know

This book contains all the maths topics you'll have studied at school. **It covers all the Intermediate and Higher Tier topics set by all the Exam Boards.**

Lots of worked examples

If you feel you haven't understood something you have been taught in class, you can revise it using **the clear worked examples** in this book.

Short, easy-to-use sections

Each chapter is divided into a number of **short sections with clear headings.** Just look up what you want to know in the index and turn to that section in the book.

Higher Tier topics

Higher Tier topics are on purple tinted pages. You don't need to revise these if you are taking the Intermediate Tier exam.

...Turn over for QUICK REVISION CHECK-UPS and EXAM PRACTICE... →

2 Quick Revision Check-ups

Check yourself questions

It can be really hard knowing where to start when you're revising. Sitting down and wading through pages of maths isn't easy. You're probably asleep before the third page! This book makes it easy to stay awake – **because it makes revising ACTIVE.**

We came up with the idea of putting **'Check yourself' questions** at the end of each chapter. **The questions test your understanding of the maths in that chapter.** In this way, you can find out quickly and easily whether you do understand it. If you get all the questions right, you can move straight on to the next chapter. If you get several of the questions wrong, you know you need to work through the whole chapter carefully. **This really cuts down on revision time – and helps you focus on where you need to put most effort.**

Answers and Tutorials

If you want the 'Check yourself' questions to be a genuine test of how much you know, then you need to cover up the answers. But, if you'd rather, you can read through a question, then the answer and then the **'tutorial'**. This will still do you a lot of good – and doesn't require quite as much effort!

We've included 'tutorials', as well as answers, to give you even more help with your revision. The tutorials show how you should arrive at your answer.

Hint and note boxes

These boxes **highlight key points** that will make exams easier if you remember them.

3 Exam Practice

Exam technique

Knowing your maths is important. But **it's even more important to know how to use it to answer exam questions correctly.** The author sees hundreds of exam scripts a year and students very often lose marks not because they don't know their maths, but because **they haven't understood how to tackle exam questions**.

Questions and sample students' answers

It's often easiest to explain what to do and what not to do by looking at **actual examples of students' answers to exam questions**. This is why we've included sample answers in this book.

In the examiner's comments, the author runs through the answers, shows you pitfalls to avoid, and what you need to do to be certain of scoring full marks.

Questions to Answer

We've also included **lots of past exam questions from different Exam Boards for you to have a go at.** The answers are at the back of the book so it's easy not to cheat. Have a go at the questions yourself and then compare them with the answers. **We've provided comments on most of the answers to give you extra help** – and if you're still unsure you can go back to the relevant section in the book.

Three final tips:

1. **Work as consistently as you can during your whole GCSE Maths course.** If you don't understand something, ask your teacher straight away, or look it up in this book. You'll then find revision much easier.
2. **Plan your revision carefully and focus on the areas you know you find hard.** The 'Check yourself' questions in this book will help you do this.
3. **Try to do some exam questions as though you were in the actual exam.** Time yourself and don't cheat by looking at the answers until you've really had a good go at working out the answers.

About your GCSE Maths course

All GCSE Mathematics syllabuses have to conform to the requirements of the Mathematics National Curriculum. For this reason, the content and assessment of the different GCSE exams are very similar although there may be small differences in the way that coursework is assessed.

Tiers of entry

The Maths exam is offered at three tiers. The available grades are as follows:

Tier of entry

A*	A	B	C	Higher
B	C	D	E	Intermediate
D	E	F	G	Foundation

Your teacher will advise you on which tier is best for you.

Assessment

The assessment for all GCSE Maths exams is divided into four assessment objectives. Each assessment objective is weighted as follows:

Using and applying mathematics	20%
Number and algebra	40%
Shape, space and measures	20%
Handling data	20%

The first assessment objective of Using and applying mathematics is usually assessed through coursework involving practical and investigational work.

What this book covers

This book covers all the content of the assessment objectives which is tested on Intermediate and Higher tier written papers. It is divided into four sections covering:

Number chapters 1–27

Algebra chapters 28–51

Shape, space and measures chapters 52–78

Handling data chapters 79–96

Each section includes work covering the Intermediate tier and the Higher tier of entry. If you are entered for the Intermediate tier of entry, then you do not need to cover the work for the Higher tier which is collected at the end of each section and has a purple tint over the pages.

Written papers

The majority of exams are assessed on two written papers which include a calculator and non-calculator paper. The written papers assess the full range of grades for the tier of entry. The papers consist of questions of varying lengths and you must answer all the questions on the paper.

For the written papers you need the following equipment:

- **pens**
- **sharp pencil, pencil sharpener and eraser**
- **ruler, protractor and compasses**
- **a scientific calculator (for the calculator paper)**

CHAPTER 1

THE NUMBER SYSTEM

You need to know how different sets of numbers are defined.

natural numbers	**1, 2, 3, 4, 5, ...**	**also called counting numbers**
integers	**..., $^{-}5$, $^{-}4$, $^{-}3$, $^{-}2$, $^{-}1$, 0, 1, 2, 3, 4, 5, ...**	
positive integers	**$^{+}1$, $^{+}2$, $^{+}3$, $^{+}4$, $^{+}5$, ...**	**usually written 1, 2, 3, 4, 5, ...**
negative integers	**$^{-}1$, $^{-}2$, $^{-}3$, $^{-}4$, $^{-}5$, ...**	
even numbers	**2, 4, 6, 8, 10, ...**	
odd numbers	**1, 3, 5, 7, 9, ...**	
square numbers	**1, 4, 9, 16, 25, 36, ...**	
triangle numbers	**1, 3, 6, 10, 15, 21, ...**	

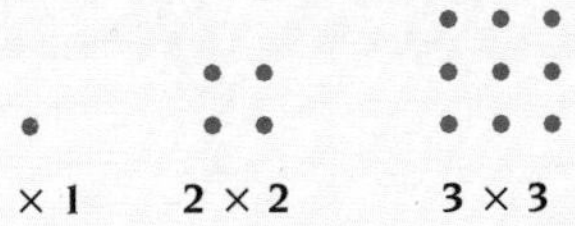

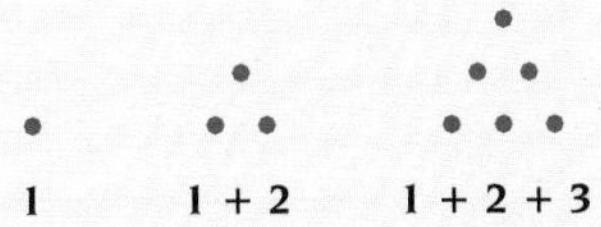

MULTIPLES

The multiples of a number are the products of the multiplication tables.

- Multiples of 5 are 5, 10, 15, 20, 25, 30, ...
- Multiples of 6 are 6, 12, 18, 24, 30, 36, ...

The **lowest common multiple** (LCM) of two or more numbers is the lowest multiple that is common to all of the given numbers.

Worked example

Find the lowest common multiple of 3 and 5.

Multiples of 3: 3 , 6, 9, 12, 15, 18, 21, 24, 27, 30, ...
Multiples of 5: 5, 10, 15, 20, 25, 30, 35, 40, 45, ...

Common multiples of 3 and 5: 15, 30, 45, 60, ...
The lowest common multiple of 3 and 5 is 15.

FACTORS

The factors of a number are the natural numbers that divide **exactly** into that number (i.e. without a remainder). The number 1 and the number itself are always factors of the given number.

- Factors of 8 are 1, 2, 4 and 8 because each of these divides exactly into 8.
- Factors of 12 are 1, 2, 3, 4, 6 and 12 because each of these divides exactly into 12.

The **highest common factor** (HCF) of two or more numbers is the highest factor which is common to all of the given numbers.

Worked example

Find the highest common factor of 16 and 24.

Factors of 16: 1, 2, 4, 8 and 16
Factors of 24: 1, 2, 3, 4, 6, 8, 12 and 24

Common factors of 16 and 24: 1, 2, 4 and 8
The highest common factor of 16 and 24 is 8.

NOTE

Factors of a number come in pairs.
For example:

$$24 = 1 \times 24 = 2 \times 12 = 3 \times 8 = 4 \times 6$$

Factors of 24 are:
1, 2, 3, 4, 6, 8, 12, 24.

NOTE

Square numbers such as 16 always have an odd number of factors as one of the factor pairs is made up of the same factor, repeated.

PRIME NUMBERS

A prime number is a natural number with exactly two factors.

Check yourself

QUESTIONS

From the numbers opposite write down:

1	2	3	4	5
6	7	8	9	10
11	12	13	14	15
16	17	18	19	20
21	22	23	24	25

(In the style of NEAB specimen paper 1998)

Q1 a multiple of 7

Q2 a factor of 10

Q3 a factor of 51

Q4 a square number bigger than 10

Q5 a prime number bigger than 16

Q6 a prime number that is even

Q7 a number that is a multiple of 3 and also a multiple of 7

REMEMBER! Cover the answers if you want to.

ANSWERS

A1 7, 14 or 21

A2 1, 2, 5 or 10

A3 1, 3 or 17

A4 16 or 25

A5 17, 19 or 23

A6 2

A7 21

TUTORIALS

T1 *The numbers 7, 14 and 21 are all multiples of 7 and appear in the table.*

T2 *The numbers 1, 2, 5 and 10 all divide exactly into 10 without a remainder.*

T3 *The numbers 1, 3 and 17 all divide exactly into 51 and appear in the table.*

T4 *The square numbers are 1, 4, 9, 16, 25, 36, 49, ... and the numbers 16 and 25 are both bigger than 10 and appear in the table.*

T5 *The numbers 2, 3, 5, 7, 11, 13, 17, 19, 23, 29, 31, 37, ... are all prime numbers (i.e. they each only have two factors) although only 17, 19 and 23 are bigger than 16 and appear in the table.*

T6 *The number 2 is the only prime number which is even. All other even numbers will have at least three factors i.e. 1, 2 and the number itself.*

T7 *The numbers 3, 6, 9, 12, 15, 18, 21, 24, 27, ... are multiples of 3 and the numbers 7, 14, 21, 28, 35, 42, ... are multiples of 7. Of the numbers that appear in the table, only 21 is a multiple of 3 and also a multiple of 7.*

PRIME FACTORS

A prime factor is a factor which is also a prime number.

All natural numbers can be written as a product of prime factors.

- The number 4 can be written as 2 × 2 where 2 is a prime factor.
- The number 15 can be written as 3 × 5 where 3 and 5 are prime factors.
- The number 40 can be written as 2 × 2 × 2 × 5 where 2 and 5 are prime factors.
- The number 90 can be written as 2 × 3 × 3 × 5 where 2, 3 and 5 are prime factors.

The prime factors of a number can be found by successively rewriting the number as a product of prime numbers in increasing order (i.e. 2, 3, 5, 7, 11, 13, 17, ... etc.)

Worked example

Write 756 as a product of its prime factors.

The number can be rewritten as follows.

756 = 2 × 378	2 is a factor of 756 and you can write 756 = 2 × 378 You now need to write 378 as a product of its prime factors.
756 = 2 × (2 × 189)	2 is a factor of 378 and you can write 378 = 2 × 189 You now need to write 189 as a product of its prime factors.
756 = 2 × 2 × (3 × 63)	2 is not a factor of 189, so try 3. 3 is a factor of 189 and you can write 189 = 3 × 63
756 = 2 × 2 × 3 × (3 × 21)	You now need to write 63 as a product of its prime factors. 3 is a factor of 63 and 63 = 3 × 21
756 = 2 × 2 × 3 × 3 × (3 × 7)	You now need to write 21 as a product its prime factors. 3 is a factor of 21 and 21 = 3 × 7
756 = 2 × 2 × 3 × 3 × 3 × 7	7 is also a prime number and cannot be divided any further.

Writing the number in full: 756 = 2 × 2 × 3 × 3 × 3 × 7

Writing the number using indices: $756 = 2^2 \times 3^3 \times 7$

NOTE

The number 1 is not a prime number as it has only one factor.

NOTE

See Chapter 6, Positive, negative and zero indices, for more information about indices.

Check yourself

QUESTIONS

Q1 Write 264 as a product of its prime factors.

Q2 Write 141 933 as a product of its prime factors.

ANSWERS

A1 $264 = 2 \times 2 \times 2 \times 3 \times 11$ or $2^3 \times 3 \times 11$

A2 $141\,933 = 3 \times 11 \times 11 \times 17 \times 23$ or $3 \times 11^2 \times 17 \times 23$

NOTE

See Chapter 6, Positive, negative and zero indices, for more information about indices.

TUTORIALS

T1 *The number 264 can be rewritten as a product of its prime factors as follows.*

$264 = 2 \times 132$
2 is a factor of 264 and $264 = 2 \times 132$

$264 = 2 \times (2 \times 66)$
2 is a factor of 132 and $132 = 2 \times 66$

$264 = 2 \times 2 \times (2 \times 33)$
2 is a factor of 66 and $66 = 2 \times 33$

$264 = 2 \times 2 \times 2 \times (3 \times 11)$
2 is not a factor of 33, so try 3.
3 is a factor of 33 and $33 = 3 \times 11$

$264 = 2 \times 2 \times 2 \times 3 \times 11$
11 is also a prime number and cannot be divided further.

So $264 = 2 \times 2 \times 2 \times 3 \times 11$
or $264 = 2^3 \times 3 \times 11$

T2 *The number 141 933 can be rewritten as a product of its prime factors as follows.*

$141\,933 = 3 \times 47\,311$
2 is not a factor of 141 933, so try 3.
3 is a factor and $141\,933 = 3 \times 47\,311$

$141\,933 = 3 \times (11 \times 4301)$
3 is not a factor of 47 311, so try 5.
5 is not a factor, so try 7.
7 is not a factor, so try 11.
11 is a factor and $47\,311 = 11 \times 4301$

$141\,933 = 3 \times 11 \times (11 \times 391)$
11 is a factor of 4301 and $4301 = 11 \times 391$

$141\,933 = 3 \times 11 \times 11 \times (17 \times 23)$
11 is not a factor of 391, so try 13.
13 is not a factor of 391, so try 17.
17 is a factor and $391 = 17 \times 23$

$141\,933 = 3 \times 11 \times 11 \times 17 \times 23$
23 is also a prime number and cannot be divided further.

So $141\,933 = 3 \times 11 \times 11 \times 17 \times 23$ *or*
$141\,933 = 3 \times 11^2 \times 17 \times 23$

CHAPTER 2

METRIC AND IMPERIAL UNITS

There are two systems of measurement used in the UK at present. Although metric units (based on a decimal system) are replacing the imperial system, you should be familiar with both.

IMPERIAL MEASURE

Length	12 inches (in)	= 1 foot (ft)
	3 feet	= 1 yard (yd)
	1760 yards	= 1 mile
Capacity	20 fluid ounces (fl oz)	= 1 pint (pt)
	8 pints	= 1 gallon (gall)
Weight	16 ounces (oz)	= 1 pound (lb)
	14 pounds	= 1 stone (st)
	8 stones	= 1 hundredweight (cwt)
	20 hundredweights	= 1 ton

METRIC (SI) MEASURE

Length	10 millimetres (mm)	= 1 centimetre (cm)
	100 centimetres	= 1 metre (m)
	1000 millimetres	= 1 metre
	1000 metres	= 1 kilometre (km)
Capacity	10 millilitres (ml)	= 1 centilitre (cl)
	1000 millilitres	= 1 litre
Weight	1000 milligrams	= 1 gram
	1000 grams	= 1 kilogram (kg)
	1000 kilograms	= 1 tonne (t)

CONVERSION FACTORS

You will need to know these approximate conversions which may be tested in the examination.

Imperial	Metric
1 inch	2.5 centimetres
1 foot	30 centimetres
5 miles	8 kilometres
1 litre	1.75 pints
1 gallon	4.5 litres
2.2 pounds	1 kilogram

Notation

When working with approximations, you may see the notation:

1 inch $\approx$ 2.5 cm

which is a short way of writing '1 inch is approximately equal to 2.5 cm'.

NOTE

The symbol $\approx$ simply means 'is approximately equal to'.

Worked example

How many millilitres are there in one pint?

1 gallon	$\approx$ 4.5 litres	
8 pints	$\approx$ 4.5 litres	As 1 gallon = 8 pints
8 pints	$\approx$ 4500 millilitres	As 1 litre = 1000 millilitres
1 pint	$\approx 4500 \div 8$	Dividing by 8 to find 1 pint.
	= 562.5 millilitres	

So there are 560 millilitres (to the nearest 10 ml) in one pint.

HINT

It is not reasonable to present the answer too accurately as the conversions are only approximate.

Check yourself

QUESTIONS

Q1 How many yards are there in one kilometre?

Q2 How many metres are there in 250 yards?

Q3 How many kilograms are there in one stone?

REMEMBER! Cover the answers if you want to.

ANSWERS

A1 1100 yards

A2 225 m

A3 6.4 kg or 6 kg

TUTORIALS

T1 *To answer the question you need to write kilometres in terms of yards, using*

8 kilometres ≈ 5 miles

= 5 × 1760 yards (1 mile = 1760 yards)

= 8800 yards

1 kilometre = $\frac{8800}{8}$ *Dividing by 8 to find 1 kilometre.*

= 1100 yards

There are approximately 1100 yards in one kilometre.

T2 *Writing 250 yards = 750 feet (3 feet = 1 yard)*

≈ 750 × 30 cm (1 foot ≈ 30 cm)

= 22 500 cm

= $\frac{22\,500}{100}$ *m Dividing by 100 to convert to metres.*

= 225 m

There are approximately 225 metres in 250 yards.

T3 *You know that 1 stone = 14 pounds and*

2.2 pounds ≈ 1 kilogram

So 1 pound ≈ $\frac{1}{2.2}$ *kg. Dividing by 2.2 to find 1 pound.*

1 stone = 14 pounds

≈ 14 × $\frac{1}{2.2}$ *kg (1 pound ≈* $\frac{1}{2.2}$ *kg)*

= 6.363 636... kg

= 6.4 kg or 6 kg Rounding to an appropriate degree of accuracy.

There are approximately 6.4 kg or 6 kg in 1 stone.

CHAPTER 3

DIRECTED NUMBERS

A directed number is one which has a + or – sign in front of it.

An example of the use of directed numbers is in temperature scales, where negative numbers imply temperatures below freezing.

ADDITION AND SUBTRACTION OF DIRECTED NUMBERS

To add or subtract directed numbers you find your starting position, then move up or down the number line.

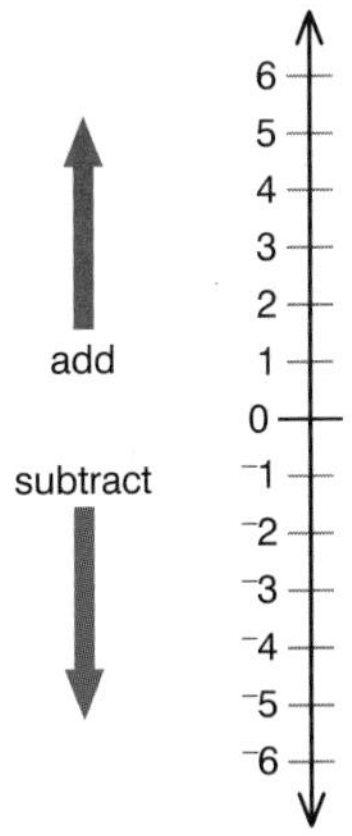

Worked example

$2 + 3 = {}^{+}5$ Start at $^{+}2$ and move up 3 places to $^{+}5$.

$4 - 7 = {}^{-}3$ Start at $^{+}4$ and move down 7 places to $^{-}3$.

${}^{-}3 + 6 = {}^{+}3$ Start at $^{-}3$ and move up 6 places to $^{+}3$.

${}^{-}1 - 4 = {}^{-}5$ Start at $^{-}1$ and move down 4 places to $^{-}5$.

When two signs appear together (e.g. $5 - {}^{-}4$) they can be replaced by one sign using the following rules.

+ + gives +

+ – gives –

– + gives –

– – gives +

$({}^{-}1) + ({}^{-}2)$ is the same as ${}^{-}1 - 2$ Since + – gives –.

$({}^{+}2) - ({}^{-}3)$ is the same as ${}^{+}2 + 3$ Since – – gives +.

MULTIPLICATION AND DIVISION OF DIRECTED NUMBERS

To multiply or divide directed numbers you multiply or divide the numbers and attach the sign according to the following rules.

- If the signs are the same then the answer will be positive.
- If the signs are opposite then the answer will be negative.

So:

$+ \times + = +$ $- \times - = +$

$+ \times - = -$ $- \times + = -$

and:

$+ \div + = +$ $- \div - = +$

$+ \div - = -$ $- \div + = -$

Worked example

${}^{+}3 \times {}^{-}4 = {}^{-}12$ As $+ \times - = -$.

${}^{-}6 \div {}^{+}2 = {}^{-}3$ As $- \div + = -$.

${}^{-}4 \div {}^{-}8 = {}^{+}\frac{1}{2}$ As $- \div - = +$ and $\frac{4}{8} = \frac{1}{2}$.

Check yourself

QUESTIONS

Q1 Work these out.

(a) $3 + 4$

(b) $4 - 6$

(c) $^{-}2 + 8$

(d) $^{-}9 + 3$

(e) $^{-}2 - 5$

(f) $^{+}3 + {}^{-}2$

(g) $^{+}7 - {}^{+}2$

(h) $^{-}3 - {}^{-}4$

(i) $^{-}11 + {}^{-}2$

(j) $^{-}10 - {}^{-}4$

Q2 Work these out.

(a) $^{+}4 \times {}^{+}6$

(b) $^{-}3 \times {}^{+}2$

(c) $^{+}7 \times {}^{-}3$

(d) $^{-}6 \times {}^{-}5$

(e) $^{+}12 \div {}^{+}3$

(f) $^{+}16 \div {}^{-}2$

(g) $^{-}10 \div {}^{+}4$

(h) $^{-}4 \times {}^{+}3 \times {}^{-}1$

REMEMBER! Cover the answers if you want to.

ANSWERS

A1

(a) $^{+}7$ or 7

(b) $^{-}2$

(c) $^{+}6$ or 6

(d) $^{-}6$

(e) $^{-}7$

(f) $^{+}1$ or 1

(g) $^{+}5$ or 5

(h) $^{+}1$ or 1

(i) $^{-}13$

(j) $^{-}6$

A2

(a) $^{+}24$ or 24

(b) $^{-}6$

(c) $^{-}21$

(d) $^{+}30$ or 30

(e) $^{+}4$ or 4

(f) $^{-}8$

(g) $^{-}2.5$ or $^{-}2\frac{1}{2}$

(h) $^{+}12$ or 12

TUTORIALS

T1

(a) $3 + 4 = 7$ *Start at* $^{+}3$ *and go up 4.*

(b) $4 - 6 = {}^{-}2$ *Start at* $^{+}4$ *and go down 6.*

(c) $^{-}2 + 8 = 6$ *Start at* $^{-}2$ *and go up 8.*

(d) $^{-}9 + 3 = {}^{-}6$ *Start at* $^{-}9$ *and go up 3.*

(e) $^{-}2 - 5 = {}^{-}7$ *Start at* $^{-}2$ *and go down 5.*

(f) $^{+}3 + {}^{-}2 = {}^{+}3 - 2 = 1$ $+ - = -$ *so rewrite as* $^{+}3 - 2$

(g) $^{+}7 - {}^{+}2 = {}^{+}7 - 2 = 5$ $- + = -$ *so rewrite as* $^{+}7 - 2$

(h) $^{-}3 - {}^{-}4 = {}^{-}3 + 4 = 1$ $- - = +$ *so rewrite as* $^{-}3 + 4$

(i) $^{-}11 + {}^{-}2 = {}^{-}11 - 2 = {}^{-}13$ $+ - = -$

(j) $^{-}10 - {}^{-}4 = {}^{-}10 + 4 = {}^{-}6$ $- - = +$

T2

(a) $^{+}4 \times {}^{+}6 = {}^{+}24$ *or* 24 $+ \times + = +$

(b) $^{-}3 \times {}^{+}2 = {}^{-}6$ $- \times + = -$

(c) $^{+}7 \times {}^{-}3 = {}^{-}21$ $+ \times - = -$

(d) $^{-}6 \times {}^{-}5 = {}^{+}30$ *or* 30 $- \times - = +$

(e) $^{+}12 \div {}^{+}3 = {}^{+}4$ *or* 4 $+ \div + = +$

(f) $^{+}16 \div {}^{-}2 = {}^{-}8$ $+ \div - = -$

(g) $^{-}10 \div {}^{+}4 = {}^{-}2.5$ $10 \div 4 = 2.5$ *and the sign is* $- \div + = -$

(h) $^{-}4 \times {}^{+}3 \times {}^{-}1 = 12$ $^{-}4 \times {}^{+}3 = {}^{-}12$ *and* $^{-}12 \times {}^{-}1 = {}^{+}12$ *or* 12

CHAPTER 4

ROUNDING – SIGNIFICANT FIGURES AND DECIMAL PLACES

Approximate answers are quite acceptable in many instances and rounding is a way that allows you to approximate answers.

In everyday applications, rounding is very common, for example:

- **populations are often expressed to the nearest million**
- **the number of people attending a pop concert may be expressed to the nearest thousand**
- **inflation may be expressed to the nearest whole number, or the nearest tenth.**

SIGNIFICANT FIGURES

Any number can be rounded to a given number of significant figures (written s.f.) using the following rules.

- Count along the digits to the required number of significant figures.
- Look at the next digit (to the right) in the number.
  If it is less than 5 leave the digit before it as it is.
  If it is 5 or more, increase the digit before it by 1.
- Replace all the digits to the right, but before the decimal point, by zeros, to keep the number at its correct size. (Digits to the right after the decimal point can just be removed.)

Worked example

Round 7638.462 to the number of significant figures shown.

6 s.f.	7638.462 = 7638.46 (6 s.f.)	
5 s.f.	7638.462 = 7638.5 (5 s.f.)	
4 s.f.	7638.462 = 7638 (4 s.f.)	
3 s.f.	7638.462 = 7640 (3 s.f.)	Fill with 0s to keep the number at its correct size.
2 s.f.	7638.462 = 7600 (2 s.f.)	Fill with 0s to keep the number at its correct size.
1 s.f.	7638.462 = 8000 (1 s.f.)	Fill with 0s to keep the number at its correct size.

DECIMAL PLACES

Any number can be rounded to a given number of decimal places (written d.p.) using the following rules.

- Count along the digits to the required number of decimal places.
- Look at the next digit (to the right) in the number.
  If it is less than 5 leave the digit before it as it is.
  If it is 5 or more, increase the digit before it by 1.
- Remove all the digits to the right.

Worked example

Round 6.427 509 3 to the number of decimal places shown.

6 d.p.	6.427 509 3 = 6.427 509 (6 d.p.)
5 d.p.	6.427 509 3 = 6.427 51 (5 d.p.)
4 d.p.	6.427 509 3 = 6.4275 (4 d.p.)
3 d.p.	6.427 509 3 = 6.428 (3 d.p.)
2 d.p.	6.427 509 3 = 6.43 (2 d.p.)
1 d.p.	6.427 509 3 = 6.4 (1 d.p.)

NOTE

The numbers to the left of the decimal point are not affected by this rounding process.

Check yourself

QUESTIONS

Write each of the following correct to 3 s.f., 2 s.f. and 1 s.f.

Q1 174.9

Q2 0.008 067 2

Q3 699.06

Write each of the following correct to 3 d.p., 2 d.p. and 1 d.p.

Q4 215.847 16

Q5 0.8006

REMEMBER! Cover the answers if you want to.

ANSWERS

A1

174.9 = 175 (3 s.f.)
174.9 = 170 (2 s.f.)
174.9 = 200 (1 s.f.)

TUTORIALS

T1

174.9 = 175 (3 s.f.)
174 are the first 3 s.f. and the next most significant figure is 9.
As 9 is bigger than 5, add 1 to the previous digit, giving 5.

174.9 = 170 (2 s.f.)
17 are the first 2 s.f. and the next most significant figure is 4.
As 4 is less than 5, leave the previous digit alone.
Fill with 0s to keep the number at its correct size.

174.9 = 200 (1 s.f.)
1 is the first s.f. and the next most significant figure is 7.
As 7 is bigger than 5, add 1 to the previous digit, giving 2.
Fill with 0s to keep the number at its correct size.

HINT

The number 175 = 180 (2 s.f.) but you must round the original 174.9.

A2

0.008 067 2 = 0.008 07 (3 s.f.)
0.008 067 2 = 0.0081 (2 s.f.)
0.008 067 2 = 0.008 (1 s.f.)

T2

0.008 067 2 = 0.008 07 (3 s.f.)
806 are the first 3 s.f. and the next most significant figure is 7.
As 7 is bigger than 5, add 1 to the previous digit, giving 807.
Check the decimal point and keep the number at its correct size.

0.008 067 2 = 0.0081 (2 s.f.)
80 are the first 2 s.f. and the next most significant figure is 6.
As 6 is bigger than 5, add 1 to the previous digit, giving 81.
Check the decimal point and keep the number at its correct size.

0.008 067 2 = 0.008 (1 s.f.)
8 is the first s.f. and the next most significant figure is 0.
As 0 is less than 5, leave the previous digit alone.
Check the decimal point and keep the number at its correct size.

ANSWERS

A3

699.06 = 699 (3 s.f.)
699.06 = 700 (2 s.f.)
699.06 = 700 (1 s.f.)

NOTE

Here the second 0 is being included as a significant figure.

A4

215.847 16 = 215.847 (3 d.p.)
215.847 16 = 215.85 (2 d.p.)
215.847 16 = 215.8 (1 d.p.)

NOTE

The number 215.85 = 215.9 (1 d.p.) but you must round the original number.

A5

0.8006 = 0.801 (3 d.p.)
0.8006 = 0.80 (2 d.p.)
0.8006 = 0.8 (1 d.p.)

NOTE

Here the second 0 is being included as a 'decimal place holder'.

TUTORIALS

T3

699.06 = 699 (3 *s.f.*)
699 are the first 3 s.f. and the next most significant figure is 0.
As 0 is less than 5, leave the previous digit alone.

699.06 = 700 (2 *s.f.*)
69 are the first 2 s.f. and the next most significant figure is 9.
As 9 is bigger than 5, add 1 to the previous digit, giving 70.
Fill with 0s to keep the number at its correct size.

699.06 = 700 (1 *s.f.*)
6 is the first s.f. and the next most significant figure is 9.
As 9 is bigger than 5, add 1 to the previous digit, giving 7.
Fill with 0s to keep the number at its correct size.

T4

215.847 16 = 215.847 (3 *d.p.*)
847 are in the first 3 d.p. and the next digit is 1. As 1 is less than 5, leave the previous digit alone. Rewrite the number, complete with decimal point, omitting all the digits after the third decimal place.

215.847 16 = 215.85 (2 *d.p.*)
84 are in the first 2 d.p. and the next digit is 7. As 7 is bigger than 5, add 1 to the previous digit, giving 85. Rewrite the number, complete with decimal point, replacing the 4 in the second d.p. by 5 and omitting all the digits after it.

215.847 16 = 215.8 (1 *d.p.*)
8 is in the first d.p. and the next digit is 4. As 4 is less than 5, leave the previous digit alone. Rewrite the number, complete with decimal point, omitting all the digits after the first decimal place.

T5

0.8006 = 0.801 (3 *d.p.*)
800 are in the first 3 d.p. and the next digit is 6.
As 6 is bigger than 5, add 1 to the previous digit, giving 801.
Rewrite the number, complete with decimal point, replacing the 0 in the third d.p. by 1 and omitting all the digits after it.

0.8006 = 0.80 (2 *d.p.*)
80 are in the first 2 d.p. and the next digit is 0.
As 0 is less than 5, leave the previous digit alone.
Rewrite the number, complete with decimal point, omitting all the digits after the second decimal place.

0.8006 = 0.8 (1 *d.p.*)
8 is in the first d.p. and the next digit is 0.
As 0 is less than 5, leave the previous digit alone.
Rewrite the number, complete with decimal point, omitting all the digits after the first decimal place.

CHAPTER 5

POWERS, ROOTS AND RECIPROCALS

When a number is multiplied by itself one or more times, the resulting number is a power of the first number. The base number is called a root of the power.

$5 \times 5 = 5^2 = 25$ **5^2 is 5 raised to the *power* of 2**
5 is the *square root* of 25

SQUARES

A **square number** is formed when another number is multiplied by itself.

The square of 8 is $8 \times 8 = 64$ so 64 is a square number.

CUBES

A **cube number** is formed when another number is multiplied by itself and then multiplied by itself again.

The cube of 5 is $5 \times 5 \times 5 = 125$ so 125 is a cube number.

SQUARE ROOTS

The **square root** of a number such as 36 is the number which, when squared, gives the first number (36). The square root of 36 is 6, because $6 \times 6 = 36$.

The sign $\sqrt[2]{\ }$ or $\sqrt{\ }$ is used to denote the square root so $\sqrt{36} = 6$.

Always take care with square roots.

$6 \times 6 = 36$

$^-6 \times {}^-6 = 36$

The square root of a number may be positive or negative.

You can find the square root of a number using the $\sqrt{\ }$ key on your calculator, but it will only give the positive square root.

NOTE

See Chapter 16, Using a calculator, for more information about using calculators.

CUBE ROOTS

The **cube root** of a number such as 27 is the number which, when cubed, gives the first number (27). The cube root of 27 is 3 because $3 \times 3 \times 3 = 27$.

The sign $\sqrt[3]{\ }$ is used to denote the cube root so $\sqrt[3]{27} = 3$.

You can find a cube root on a calculator if you have a key marked $\sqrt[3]{\ }$.

NOTE

See Chapter 16, Using a calculator, for more information about using calculators.

RECIPROCALS

The **reciprocal** of a number is found by dividing 1 by that number. The reciprocal of any non-zero number can be found by converting the number to a fraction and turning the fraction upside-down. The reciprocal of $\frac{2}{3}$ is $\frac{3}{2}$ and the reciprocal of 10 is $\frac{1}{10}$.

With a calculator you can find the reciprocal of a number by using the 1/x or x^{-1} key. You may need to use the INV or 2ndF key with it.

NOTE

See Chapter 16, Using a calculator, for more information about using calculators.

Check yourself

QUESTIONS

Q1 Write down the first six square numbers.

Q2 Write down the first six cube numbers.

Q3 Work out $\sqrt{49}$, $\sqrt{121}$, $\sqrt{256}$, $\sqrt{10}$.

Q4 Work out $\sqrt[3]{4096}$, $\sqrt[3]{10}$.

Q5 Work out the reciprocal of $\frac{3}{4}$, 15, $1\frac{1}{5}$.

REMEMBER! Cover the answers if you want to.

ANSWERS

A1 The first six square numbers are 1, 4, 9, 16, 25 and 36.

A2 The first six cube numbers are 1, 8, 27, 64, 125, 216.

A3 $\sqrt{49} = 7$, $\sqrt{121} = 11$, $\sqrt{256} = 16$, $\sqrt{10} = 3.162\,277\,7$

A4 $\sqrt[3]{4096} = 16$, $\sqrt[3]{10} = 2.154\,434\,7$

A5 $\frac{4}{3}, \frac{1}{15}, \frac{5}{6}$

TUTORIALS

T1 *Square numbers are found by multiplying numbers by themselves. The first six square numbers are found from* $1 \times 1 = 1$, $2 \times 2 = 4$, $3 \times 3 = 9$, $4 \times 4 = 16$, $5 \times 5 = 25$, $6 \times 6 = 36$.

T2 *Cube numbers are found by multiplying numbers by themselves and then by themselves again. The first six cube numbers are found from* $1 \times 1 \times 1 = 1$, $2 \times 2 \times 2 = 8$, $3 \times 3 \times 3 = 27$, $4 \times 4 \times 4 = 64$, $5 \times 5 \times 5 = 125$, $6 \times 6 \times 6 = 216$.

T3 *The square root of a number is the number which when squared gives that number.*

$\sqrt{49} = 7$ *as* $7 \times 7 = 49$, $\sqrt{121} = 11$ *as* $11 \times 11 = 121$, $\sqrt{256} = 16$ *as* $16 \times 16 = 256$.

$\sqrt{10}$ *is not an exact number but lies between 3 and 4 (as* $3 \times 3 = 9$ *and* $4 \times 4 = 16$*). Using the* [$\sqrt{\ }$] *key on your calculator gives* $\sqrt{10} = 3.162\,277\,7$.

T4 *The cube root of a number is the number which, when cubed gives that number.*

$\sqrt[3]{4096} = 16$ *as* $16 \times 16 \times 16 = 4096$.

$\sqrt[3]{10}$ *is not an exact number but lies between 2 and 3 (as* $2 \times 2 \times 2 = 8$ *and* $3 \times 3 \times 3 = 27$*).*

Using the [$\sqrt[3]{\ }$] *key on your calculator gives* $\sqrt[3]{10} = 2.154\,434\,7$.

T5 *The reciprocal of* $\frac{3}{4}$ *is found by turning the fraction upside-down to give* $\frac{4}{3}$. *The number 15 can be written as* $\frac{15}{1}$ *and the reciprocal of* $\frac{15}{1}$ *is* $\frac{1}{15}$. *Similarly the mixed number* $1\frac{1}{5}$ *can be written as* $\frac{6}{5}$ *(as a top-heavy or an improper fraction – see Chapter 9, Fractions) and the reciprocal of* $\frac{6}{5}$ *is* $\frac{5}{6}$.

CHAPTER 6

POSITIVE, NEGATIVE AND ZERO INDICES

When multiplying a number by itself you can use the following shorthand.

$5 \times 5 = 5^2$ You say '5 to the power 2 (or 5 squared).'

$5 \times 5 \times 5 = 5^3$ You say '5 to the power 3 (or 5 cubed).'

In general, this shorthand is written like this.

5^4 ← power (the 4); ← base (the 5)

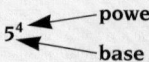

You say it as 'five to the power four'.

The power (or index) tells you how many times the base number is to be multiplied.

5^4 tells you to multiply together 4 (the power or index) 'lots' of 5 (the base number).

$5^4 = 5 \times 5 \times 5 \times 5$

Similarly 4^7 tells you to multiply together 7 'lots' of 4.

$4^7 = 4 \times 4 \times 4 \times 4 \times 4 \times 4 \times 4$

NOTE

You are multiplying the base number by itself, not by the index.

MULTIPLYING NUMBERS WITH INDICES

You can multiply numbers with indices like this.

$3^4 = 3 \times 3 \times 3 \times 3 \qquad 3^5 = 3 \times 3 \times 3 \times 3 \times 3$

$$\begin{aligned} \text{So } 3^4 \times 3^5 &= (3 \times 3 \times 3 \times 3) \times (3 \times 3 \times 3 \times 3 \times 3) \\ &= 3 \times 3 \times 3 \times 3 \times 3 \times 3 \times 3 \times 3 \times 3 \\ &= 3^9 \end{aligned}$$

Also

$12^4 = 12 \times 12 \times 12 \times 12 \qquad 12^6 = 12 \times 12 \times 12 \times 12 \times 12 \times 12$

$$\begin{aligned} \text{So } 12^4 \times 12^6 &= (12 \times 12 \times 12 \times 12) \times (12 \times 12 \times 12 \times 12 \times 12 \times 12) \\ &= 12 \times 12 \times 12 \times 12 \times 12 \times 12 \times 12 \times 12 \times 12 \times 12 \\ &= 12^{10} \end{aligned}$$

To multiply two numbers with indices **when their bases are the same** you just add their indices.

$3^4 \times 3^5 = 3^{4+5} = 3^9 \qquad 12^4 \times 12^6 = 12^{4+6} = 12^{10}$

NOTE

In general: $a^m \times a^n = a^{m+n}$

DIVIDING NUMBERS WITH INDICES

You can divide numbers with indices like this.

$$6^7 \div 6^4 = \frac{6 \times 6 \times 6 \times \cancel{6} \times \cancel{6} \times \cancel{6} \times \cancel{6}}{\cancel{6} \times \cancel{6} \times \cancel{6} \times \cancel{6}} = 6 \times 6 \times 6 = 6^3$$

$$15^9 \div 15^3 = \frac{15 \times 15 \times 15 \times 15 \times 15 \times 15 \times \cancel{15} \times \cancel{15} \times \cancel{15}}{\cancel{15} \times \cancel{15} \times \cancel{15}}$$

$$= 15 \times 15 \times 15 \times 15 \times 15 \times 15 = 15^6$$

To divide two numbers with indices **when their bases are the same** you just subtract their indices.

$6^7 \div 6^4 = 6^{7-4} = 6^3 \qquad 15^9 \div 15^3 = 15^{9-3} = 15^6$

NOTE

In general: $a^m \div a^n = a^{m-n}$

NEGATIVE POWERS

From the above it follows that:

$$7^4 \div 7^6 = 7^{4-6} = 7^{-2} \quad \text{and} \quad 7^4 \div 7^6 = \frac{\not{7} \times \not{7} \times \not{7} \times \not{7}}{7 \times 7 \times \not{7} \times \not{7} \times \not{7} \times \not{7}} = \frac{1}{7^2}$$

So $$7^{-2} = \frac{1}{7^2}$$

ZERO POWERS

Using the same ideas as before:

$$5^4 \div 5^4 = 5^{4-4} = 5^0 \quad \text{and} \quad 5^4 \div 5^4 = \frac{\not{5} \times \not{5} \times \not{5} \times \not{5}}{\not{5} \times \not{5} \times \not{5} \times \not{5}} = 1$$

$5^0 = 1, 6^0 = 1, 100^0 = 1, 645.321^0 = 1$ and so on.

NOTE

In general: $a^{-m} = \frac{1}{a^m}$

Any number raised to the power zero is equal to 1.

In general: $a^0 = 1$

Any number raised to the power 1 is equal to that number.

$5^1 = 5, 6^1 = 6, 100^1 = 100$

Check yourself

QUESTIONS

Q1 Find the value of the following.

(a) 9^3 (b) 4^{-2} (c) 6^1

Q2 Calculate the following, giving your answers in index form where possible.

(a) $3^{11} \times 3^{12}$ (b) $8^6 \div 8^4$

(c) $13^4 \div 13^4$ (d) $4^3 \times 5^2$

REMEMBER! Cover the answers if you want to.

ANSWERS

A1 (a) 729 (b) $\frac{1}{16}$ (c) 6

A2 (a) 3^{23} (b) 8^2 (c) 1 (d) 1600

NOTE

Check the rules above if you need to.

TUTORIALS

T1 (a) *An answer of 27 is a common mistake, caused by multiplying 9 × 3.*
9^3 tells you that three lots of 9 (the base number) are to be multiplied together (3 is the power or index).
So $9^3 = 9 \times 9 \times 9 = 729$

(b) *4^{-2} has a negative power.*

So $4^{-2} = \frac{1}{4^2} = \frac{1}{16}$ as $4^2 = 4 \times 4 = 16$

(c) *$6^1 = 6$ as any number raised to the power one is always equal to that number.*

T2 (a) $3^{11} \times 3^{12} = 3^{11+12} = 3^{23}$

(b) $8^6 \div 8^4 = 8^{6-4} = 8^2$

(c) $13^4 \div 13^4 = 13^{4-4} = 13^0 = 1$

(d) *As the base numbers are not the same you cannot use the rules of indices on this question and must work out the answer by evaluating 4^3 and 5^2.*

$4^3 = 4 \times 4 \times 4 = 64$
$5^2 = 5 \times 5 = 25$

So $4^3 \times 5^2 = 64 \times 25 = 1600$

CHAPTER 7

EXAMINATION QUESTIONS 1

EXAM PRACTICE

Sample Student's Answers & Examiner's Comments

1 Find the highest common factor (HCF) and the lowest common multiple (LCM) of the numbers 8 and 12.

Factors of 8 are 1, 8, 2, 4
Factors of 12 are 1, 12, 2, 6, 3, 4
Common factors are 1, 2, 4
HCF = 4
Multiples of 8 are 8, 16, 24, 32, 40, 48, 56, 64, 72, ...
Multiples of 12 are 12, 24, 36, 48, 60, 72, 84, 96, ...
Common multiples are 24, 48, 72, ...
LCM = 24

2 How many gallons are there in 14 litres?

$$1 \text{ litre} \approx 1.75 \text{ pints}$$
$$14 \text{ litres} \approx 14 \times 1.75 \text{ pints}$$
$$= 24.5 \text{ pints}$$
$$= \frac{24.5}{8} \text{ gallons}$$
$$= 3.0625 \text{ gallons}$$

EXAMINER'S COMMENTS

1 *The candidate should not produce the factors randomly because they might not find all of them. A system such as checking all the numbers in order or else finding factor pairs is recommended, because if the factors are written in order then it is easier to see common factors.*

The candidate does not need to continue with the multiples this far as the lowest common multiple is already visible.

2 *The fact that 1 litre ≈ 1.75 pints needs to be remembered, and also the alternative, 1 gallon ≈ 4.5 litres.*

The candidate has divided by 8 to convert the number of pints to the number of gallons. She should now round this figure to an appropriate degree of accuracy.

An answer of 3.1 gallons or 3 gallons would be more appropriate as the conversions are only approximate.

3 Calculate the difference between the positive square root of 64 and the cube root of 64.

$\sqrt{64} = 8$
$\sqrt[3]{64} = 4$
Difference $= 8 - 4 = 4$

4 Which is bigger 5^4 or 4^5?

$5^4 = 5 \times 5 \times 5 \times 5 = 625$
$4^5 = 4 \times 4 \times 4 \times 4 \times 4 = 1024$
4^5 is bigger than 5^4

EXAMINER'S COMMENTS

3 *The candidate has used the* $\sqrt{\ }$ *and* $\sqrt[3]{\ }$ *keys on the calculator to find* $\sqrt{64}$ *and* $\sqrt[3]{64}$.

4 *The candidate has remembered that* $5^4 = 5 \times 5 \times 5 \times 5$ *and not* 5×4 *which is a common mistake. He has also remembered to show his working which is important if method marks are to be awarded.*

It is important that this final statement is made in order to complete the answer.

Questions to Answer

1 Write the number 3960 as a product of its prime factors.

2 How many inches in one metre?

3 How many centimetres in 1 mile?

4 Find $\sqrt{81}$, $\sqrt{50}$.

5 Find $\sqrt[3]{512}$, $\sqrt[3]{50}$.

6 Find the reciprocal of 0.25.

7 Simplify $12^7 \times 12^{11}$.

8 Simplify $9^3 \div 9^5$.

9 Simplify $7^4 \times 7^2 \times 7^8$.

10 Simplify $4^3 \times 6^2$.

NUMBER

CHAPTER 8

STANDARD FORM – INVOLVING POSITIVE AND NEGATIVE INDICES

Standard form is a shorthand way of writing very large and very small numbers. Standard form numbers are always written in the form:

$A \times 10^n$

where A lies between 1 and 10 and n is a natural number.

VERY LARGE NUMBERS

Worked example

Write 35 000 in standard form.

$A = 3.5$ and

$35\,000 = 3.5 \times 10\,000$

$= 3.5 \times 10^4$

A QUICKER WAY

Write 35 000 in standard form.

Place the decimal point so A lies between 1 and 10 and find n.

3.5 0 0 0 so $A = 3.5$ and $n = 4$.

$35\,000 = 3.5 \times 10^4$

HINT

In standard form, n is positive for large numbers (e.g. $3.5 \times 10^4 = 35\,000$) and n is negative for small numbers (e.g. $4.78 \times 10^{-7} = 0.000\,000\,478$).

VERY SMALL NUMBERS

Worked example

Write 0.000 000 478 in standard form.

Place the decimal point so A lies between 1 and 10 and find n.

0 0 0 0 0 0 0 4.78 so $A = 4.78$ and $n = {}^-7$.

$0.000\,000\,478 = 4.78 \times 10^{-7}$

Check yourself

QUESTIONS

Q1 The distance from the Earth to the Moon is 250 000 miles. Write this number in standard form.

Q2 The distance from the Earth to the Sun is 9.3×10^7 miles. Write this as an ordinary number.

Q3 The weight of a hydrogen atom is given as 0.000 000 000 000 000 000 001 67 milligrams. Write this number in standard form.

Q4 Express 6.05×10^{-4} as an ordinary number.

REMEMBER! Cover the answers if you want to.

ANSWERS

A1 2.5×10^5 miles

A2 93 000 000 miles

A3 1.67×10^{-21} milligrams

A4 0.000 605

TUTORIALS

T1 *A = 2.5 and n = 5.*

T2 *You should check your answer by working backwards.*

T3 *A = 1.67 and n = ⁻21. The negative value of n tells you that the number is very small.*

T4 *You should check your answer by working backwards.*

ADDING AND SUBTRACTING NUMBERS IN STANDARD FORM

To add (or subtract) numbers in standard form when the powers are the same you can proceed as shown in the next example.

Worked example

Work out $(4.18 \times 10^{11}) + (3.22 \times 10^{11})$.

$$(4.18 \times 10^{11}) + (3.22 \times 10^{11}) = (4.18 + 3.22) \times 10^{11}$$
$$= 7.4 \times 10^{11}$$

To add (or subtract) numbers in standard form when the powers are not the same you need to convert the numbers to ordinary form.

Worked example

Work out $(8.42 \times 10^6) + (6 \times 10^7)$.

$$(8.42 \times 10^6) + (6 \times 10^7) = 8\,420\,000 + 60\,000\,000$$ Converting to ordinary form.

$$= 68\,420\,000$$
$$= 6.842 \times 10^7$$ Converting back to standard form.

HINT

Your calculator will deal with numbers in standard form, if you use the EXP or EE key. See Chapter 16, Using a calculator, for more information about using calculators.

Check yourself

QUESTIONS

Work these out.

Q1 $(2.69 \times 10^5) - (1.5 \times 10^5)$

Q2 $(4.31 \times 10^{-4}) + (3.5 \times 10^{-4})$

Q3 $(9.27 \times 10^{11}) + (2.631 \times 10^{11})$

Q4 $(4.11 \times 10^2) + (3.6 \times 10^1)$

ANSWERS

A1 1.19×10^5

A2 7.81×10^{-4}

A3 1.1901×10^{12}

A4 447 or 4.47×10^2

TUTORIALS

T1 $2.69 \times 10^5 - 1.5 \times 10^5 = (2.69 - 1.5) \times 10^5$
$= 1.19 \times 10^5$

T2 $4.31 \times 10^{-4} + 3.5 \times 10^{-4} = (4.31 + 3.5) \times 10^{-4}$
$= 7.81 \times 10^{-4}$

T3 $9.27 \times 10^{11} + 2.631 \times 10^{11} = (9.27 + 2.631) \times 10^{11}$
$= 11.901 \times 10^{11}$
$= 1.1901 \times 10^{12}$

Converting back to standard form where $11.901 = 1.1901 \times 10$ *or* 1.1901×10^1.

T4 *Since the powers are not the same you need to convert the numbers to ordinary form as*

$411 + 36 = 447$.

$447 = 4.47 \times 10^2$ *in standard form if required.*

MULTIPLYING AND DIVIDING NUMBERS IN STANDARD FORM

To multiply (or divide) numbers in standard form you need to use the rules of indices. See Chapter 6, Positive, negative and zero indices for more information about these rules.

Worked example

Work out $(8.5 \times 10^3) \times (4.2 \times 10^7)$.

$(8.5 \times 10^3) \times (4.2 \times 10^7) = (8.5 \times 4.2) \times (10^3 \times 10^7)$

Collecting powers of 10 together.

$= 35.7 \times 10^{3+7}$

Using the rules of indices where $a^m \times a^n = a^{m+n}$

$= 35.7 \times 10^{10}$

$= 3.57 \times 10^{11}$

Writing 35.7 as 3.57×10^1 to get the final answer in standard form.

Worked example

Work out $(6.3 \times 10^5) \div (2.1 \times 10^8)$.

$(6.3 \times 10^5) \div (2.1 \times 10^8) = (6.3 \div 2.1) \times (10^5 \div 10^8)$

Collecting powers of 10 together.

$= 3 \times 10^{5-8}$

Using the rules of indices where $a^m \div a^n = a^{m-n}$.

$= 3 \times 10^{-3}$

Check yourself

QUESTIONS

Q1 Work out $(6.5 \times 10^3) \times (4.2 \times 10^{-2})$.

Q2 Light travels at 2.998×10^8 m/s.
How far does it travel in one year?

Give your answer in metres using standard index form.

(MEG syllabus B, specimen paper 1998)

Q3 Work out $(3.12 \times 10^{11}) \div (6.5 \times 10^4)$.

Q4 In 1993 there were 5.51×10^9 people in the world. The gross national product of the world was $\$2.19 \times 10^{13}$. What is the gross national product per person? Give your answer to an appropriate degree of accuracy.

(SEG modular syllabus, specimen paper 1998)

REMEMBER! Cover the answers if you want to.

ANSWERS

A1 2.73×10^2

A2 9.45×10^{15} metres
(to an appropriate degree of accuracy)

A3 4.8×10^6

A4 \$3970
(to an appropriate degree of accuracy)

NOTE

You can also do these calculations using the EXP or EE key on your calculator. See Chapter 16, Using a calculator, for information on using calculators.

TUTORIALS

T1 $(6.5 \times 10^3) \times (4.2 \times 10^{-2})$
$= (6.5 \times 4.2) \times 10^3 \times 10^{-2}$
$= 27.3 \times 10^{3-2} = 27.3 \times 10^1$
$= 2.73 \times 10^2$
Rewriting in standard form where
$27.3 = 2.73 \times 10^1$.

T2 *If light travels* 2.998×10^8 *metres in one second then it travels*
$2.998 \times 10^8 \times 60$ *metres in one minute,*
$2.998 \times 10^8 \times 60 \times 60$ *metres in one hour,*
$2.998 \times 10^8 \times 60 \times 60 \times 24$ *metres in one day and*
$2.998 \times 10^8 \times 60 \times 60 \times 24 \times 365$ *metres in one year.*
$2.998 \times 10^8 \times 60 \times 60 \times 24 \times 365$
$= 9.45\,449 \times 10^{15}$ *metres*
$= 9.45 \times 10^{15}$ *metres*
(to an appropriate degree of accuracy)

T3 $(3.12 \times 10^{11}) \div (6.5 \times 10^4)$
$= (3.12 \div 6.5) \times (10^{11} \div 10^4)$
$= 0.48 \times 10^{11-4} = 0.48 \times 10^7$
$= 4.8 \times 10^6$
Rewriting in standard form where
$0.48 = 4.8 \times 10^{-1}$.

T4 *The gross national product per person*

$$= \frac{\text{gross national product}}{\text{number of people}} = \$\frac{2.19 \times 10^{13}}{5.51 \times 10^9}$$

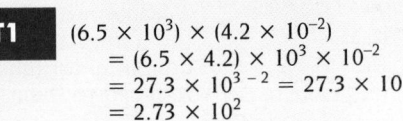

$= \$0.397\,459\,17 \times 10^4$
$= \$3974.5917$
$= \$3970$ *(to an appropriate degree of accuracy)*

CHAPTER 9

FRACTIONS

The top line of a fraction is called the **numerator** and the bottom line of a fraction is called the **denominator**.

EQUIVALENT FRACTIONS

Equivalent fractions can be found by multiplying or dividing the numerator and denominator by the same number.

$$\frac{1}{2} = \frac{2}{4} \ (\times 2) \qquad \frac{3}{4} = \frac{30}{40} \ (\times 10) \qquad \frac{4}{7} = \frac{12}{21} \ (\times 3) \qquad \frac{160}{200} = \frac{80}{100} \ (\div 2) \qquad \frac{9}{12} = \frac{3}{4} \ (\div 3)$$

NOTE

See Chapter 1, The number system, for more information about integers.

CANCELLING FRACTIONS

A fraction can be expressed in its **lowest terms** or **simplest form** by making the numerator and the denominator as small as possible. Both numerator and denominator must be integers. The process by which the fractions are reduced to their lowest terms is called **cancelling down** or **simplifying**.

ONE NUMBER AS A FRACTION OF ANOTHER

To find one number as a fraction of another, you write the numbers as a fraction, with the first number on the top and the second on the bottom.

Worked example

Write 55p as a fraction of 80p.

55p as a fraction of 80p $= \frac{55}{80} = \frac{11}{16}$ so 55p is $\frac{11}{16}$ of 80p.

Worked example

Write 4 mm as a fraction of 8 cm.

You must first ensure that the units are the same. 8 cm = 80 mm

4 mm as a fraction of 80 mm $= \frac{4}{80} = \frac{1}{20}$ so 4 mm is $\frac{1}{20}$ of 8 cm.

HINT

To find the common denominator of two numbers you find the lowest common multiple or LCM of the two denominators. The LCM of 8 and 5 is 40 (see Chapter 1, The number system).

NOTE

$$\frac{7}{8} = \frac{35}{40} \ (\times 5) \quad \text{and} \quad \frac{1}{5} = \frac{8}{40} \ (\times 8)$$

ADDITION AND SUBTRACTION

To add (or subtract) fractions, ensure they have the same denominators.

Worked example

(a) Add $\frac{2}{7} + \frac{4}{7}$.

$$\frac{2}{7} + \frac{4}{7} = \frac{6}{7}$$

(b) Subtract $\frac{7}{8} - \frac{1}{5}$.

$$\frac{7}{8} - \frac{1}{5} = \frac{35}{40} - \frac{8}{40} = \frac{27}{40}$$

Writing both fractions with a denominator of 40.

MIXED NUMBERS

A mixed number is one with a whole number part and a fractional part such as $1\frac{1}{5}$ or $5\frac{13}{20}$.

Any mixed number can be converted to a **top-heavy fraction** (or **improper fraction**).

$$1\tfrac{1}{5} = 1 + \tfrac{1}{5} = \tfrac{5}{5} + \tfrac{1}{5} = \tfrac{6}{5} \qquad \text{where } 1 = \tfrac{5}{5}$$

$$5\tfrac{13}{20} = 5 + \tfrac{13}{20} = \tfrac{100}{20} + \tfrac{13}{20} = \tfrac{113}{20} \qquad \text{where } 5 = \tfrac{100}{20}$$

Worked example

Add $1\frac{1}{5} + 5\frac{15}{20}$.

$$1\tfrac{1}{5} + 5\tfrac{13}{20} = \frac{6}{5} + \frac{113}{20} = \frac{24}{20} + \frac{113}{20}$$

Converting to top-heavy fractions and writing both fractions with a denominator of 20.

$$= \frac{137}{20} = 6\tfrac{17}{20}$$

Rewriting as a mixed number.

$$\frac{6}{5} \xrightarrow{\times 4} \frac{24}{20}$$

Alternatively you can add these mixed numbers by dealing with the whole numbers and the fractions separately.

$$1\tfrac{1}{5} + 5\tfrac{13}{20} = 1 + 5 + \frac{1}{5} + \frac{13}{20}$$

Splitting up the whole number part and the fraction part.

$$= 6 + \frac{4}{20} + \frac{13}{20}$$

Writing both fractions with a denominator of 20.

$$= 6 + \frac{17}{20} = 6\tfrac{17}{20}$$

$$\frac{1}{5} \xrightarrow{\times 4} \frac{4}{20}$$

Check yourself

QUESTIONS

Q1 Work out $3\frac{1}{4} - 1\frac{1}{5}$.

Q2 Work out $4\frac{1}{2} - 2\frac{3}{7} + 1\frac{1}{14}$.

REMEMBER! Cover the answers if you want to.

ANSWERS

A1 $2\frac{1}{20}$

A2 $3\frac{1}{7}$

TUTORIALS

T1 $3\frac{1}{4} - 1\frac{1}{5}$

$= \frac{13}{4} - \frac{6}{5}$ *Converting to top-heavy fractions.*

$= \frac{65}{20} - \frac{24}{20}$ *Writing both fractions with a denominator of 20.*

$= \frac{41}{20} = 2\frac{1}{20}$ *Rewriting as a mixed number.*

T2 $4\frac{1}{2} - 2\frac{3}{7} + 1\frac{1}{14}$

$= \frac{9}{2} - \frac{17}{7} + \frac{15}{14}$ *Converting to top-heavy fractions.*

$= \frac{63}{14} - \frac{34}{14} + \frac{15}{14}$ *Writing all fractions with a denominator of 14.*

$= \frac{44}{14} = \frac{22}{7} = 3\frac{1}{7}$ *Rewriting as a mixed number $3\frac{1}{7}$.*

Multiplication of fractions

To multiply fractions, multiply the numerators and multiply the denominators.

Worked example

Work out $\frac{4}{7} \times \frac{2}{11}$.

$$\frac{4}{7} \times \frac{2}{11} = \frac{4 \times 2}{7 \times 11}$$ Multiply the numerators and multiply the denominators.

$$= \frac{8}{77}$$

When working with mixed numbers you *must* convert to top-heavy fractions first.

NOTE

Alternatively, you can cancel the fractions.

$$\frac{{}^{2}\cancel{6} \times \cancel{20}^{4}}{{}_{1}\cancel{5} \times \cancel{3}_{1}} = \frac{2 \times 4}{1 \times 1} = 8$$

Worked example

Work out $1\frac{1}{5} \times 6\frac{2}{3}$.

$$1\tfrac{1}{5} \times 6\tfrac{2}{3} = \tfrac{6}{5} \times \tfrac{20}{3}$$ Converting to top-heavy fractions.

$$= \frac{6 \times 20}{5 \times 3}$$ Multiply the numerators and multiply the denominators.

$$= \frac{120}{15} = 8$$

NOTE

Alternatively, you can cancel the fractions.

$$\frac{2 \times \cancel{100}^{20}}{{}_{1}\cancel{5} \times 1} = \frac{2 \times 20}{1 \times 1} = \frac{40}{1} = 40$$

Worked example

Find $\frac{2}{5}$ of 100.

$$\tfrac{2}{5} \text{ of } 100 = \tfrac{2}{5} \times \tfrac{100}{1}$$ Writing 100 as a top-heavy fraction.

$$= \frac{\cancel{200}^{40}}{\cancel{5}_{1}}$$ Cancelling.

$$= 40$$

Division of fractions

To divide one fraction by another you multiply the first fraction by the **reciprocal** of the second fraction.

Worked example

Work out $\frac{3}{7} \div \frac{1}{7}$.

$$\frac{3}{7} \div \frac{1}{7} = \frac{3}{\cancel{7}_{1}} \times \frac{\cancel{7}^{1}}{1}$$ Multiplying by the reciprocal and cancelling the fractions.

$$= 3$$

Check yourself

QUESTIONS

Work these out.

Q1 $\frac{3}{4} \times \frac{2}{5}$

Q2 $4\frac{4}{5} \div 1\frac{1}{15}$

Q3 $\frac{3}{5}$ of 25

REMEMBER! Cover the answers if you want to.

ANSWERS		TUTORIAL	
A1	$\frac{3}{10}$	**T1** $\frac{3}{\cancel{4}_2} \times \frac{\cancel{2}^1}{5} = \frac{3 \times 1}{2 \times 5}$ $= \frac{3}{10}$	*Cancelling fractions.*
A2	$4\frac{1}{2}$	**T2** $4\frac{4}{5} \div 1\frac{1}{15} = \frac{24}{5} \div \frac{16}{15}$	*Converting to top-heavy fractions.*
		$= \frac{{}^3\cancel{24}}{{}_1\cancel{5}} \times \frac{\cancel{15}^3}{\cancel{16}_2}$	*Multiplying by reciprocal and cancelling fractions.*
		$= \frac{9}{2}$	
		$= 4\frac{1}{2}$	*Rewriting as a mixed number.*
A3	15	**T3** $\frac{3}{5}$ *of* $25 = \frac{3}{5} \times 25$	*'Of' means 'multiply'.*
		$= \frac{3 \times \cancel{25}^5}{{}_1\cancel{5} \times 1}$	*Cancelling fractions.*
		$= 15$	

Changing Fractions to Decimals

A fraction can be changed to a decimal by carrying out the division.

Worked example

Change $\frac{1}{4}$ to a decimal.

$$\frac{1}{4} = 1 \div 4$$
$$= 0.25$$

Worked example

Change $\frac{4}{15}$ to a decimal.

$$\frac{4}{15} = 4 \div 15$$
$$= 0.266\,666\,6\ldots$$

Notice that the decimal in the example above carries on infinitely. It is called a **recurring decimal**.

You write the decimal as $0.2\dot{6}$. The dot over the 6 tells you that the number carries on infinitely.

If a group of numbers repeats infinitely then two dots can be used to show the repeating numbers.

$$0.\dot{3}\dot{5} = 0.353\,535\,35\ldots$$
$$6.4\dot{1}\dot{7} = 6.417\,171\,717\ldots$$
$$3.\dot{2}0\dot{1} = 3.201\,201\,201\ldots$$
$$11.60\dot{2}5\dot{3} = 11.602\,532\,532\,53\ldots$$

CHANGING DECIMALS TO FRACTIONS

A decimal can be changed to a fraction by considering place value as follows.

Worked example

Change 0.459 to a fraction.

0.459 = 0 units and 4 tenths and 5 hundredths and 9 thousandths

$$0.459 = 0 + \frac{4}{10} + \frac{5}{100} + \frac{9}{1000}$$

$$= \frac{400}{1000} + \frac{50}{1000} + \frac{9}{1000}$$

Writing the fractions with a common denominator of 1000.

$$= \frac{459}{1000}$$

Check yourself

QUESTIONS

Change the following decimals to fractions.

Q1 0.25 **Q2** 0.48 **Q3** 0.162 **Q4** 0.4709

REMEMBER! Cover the answers if you want to.

ANSWERS

A1 $\frac{1}{4}$

A2 $\frac{12}{25}$

A3 $\frac{81}{500}$

A4 $\frac{4709}{10\,000}$

TUTORIALS

T1 $0.25 = \frac{2}{10} + \frac{5}{100} = \frac{20}{100} + \frac{5}{100} = \frac{25}{100}$

$= \frac{1}{4}$ *Cancelling down to the lowest terms.*

T2 $0.48 = \frac{4}{10} + \frac{8}{100} = \frac{40}{100} + \frac{8}{100} = \frac{48}{100}$

$= \frac{12}{25}$ *Cancelling down to the lowest terms.*

T3 $0.162 = \frac{1}{10} + \frac{6}{100} + \frac{2}{1000}$

$= \frac{100}{1000} + \frac{60}{1000} + \frac{2}{1000}$

$= \frac{162}{1000}$

$= \frac{81}{500}$ *Cancelling down.*

T4 $0.4709 = \frac{4}{10} + \frac{7}{100} + \frac{0}{1000} + \frac{9}{10\,000}$

$= \frac{4000}{10\,000} + \frac{700}{10\,000} + \frac{9}{10\,000}$

$= \frac{4709}{10\,000}$

HINT

By looking at these answers before cancelling you should be able to see a quick way to convert decimals to fractions.

CHAPTER 10 PERCENTAGES

Percentages are fractions with a denominator of 100.

50% means 50 out of 100 or $\frac{50}{100}$ (= $\frac{1}{2}$ in its lowest terms)

75% means 75 out of 100 or $\frac{75}{100}$ (= $\frac{3}{4}$ in its lowest terms)

CHANGING BETWEEN PERCENTAGES, FRACTIONS AND DECIMALS

CHANGING PERCENTAGES TO FRACTIONS

To change a percentage to a fraction, you divide by 100.

Worked example

Change 68% to a fraction.

$$68\% = \frac{68}{100}$$

$$= \frac{17}{25}$$ Cancelling down to the lowest terms.

Worked example

Change $45\frac{1}{2}\%$ to a fraction.

$$45\tfrac{1}{2}\% = \frac{45\frac{1}{2}}{100}$$

$$= \frac{91}{200}$$ Multiplying top and bottom by 2 to give integers on the top and bottom.

CHANGING PERCENTAGES TO DECIMALS

To change a percentage to a decimal you divide by 100.

Worked example

Change 68% to a decimal.

$$68\% = 68 \div 100$$

$$= 0.68$$

Worked example

Change $72\frac{3}{4}\%$ to a decimal.

$$72\tfrac{3}{4}\% = \frac{72\frac{3}{4}}{100}$$

$$= \frac{291}{400}$$ Multiplying top and bottom by 4 to remove the fraction on the top.

$$= 0.7275$$

NOTE

As $72\frac{3}{4} = 72.75$, you could just divide 72.75 by 100 to get the same result.

Changing fractions to percentages

To change a fraction to a percentage you multiply by 100.

Worked example

Change $\frac{1}{4}$ to a percentage.

$\frac{1}{4} = \frac{1}{4} \times 100\%$

$= 25\%$

Changing decimals to percentages

To change a decimal to a percentage you multiply by 100.

Worked example

Convert 0.2 to a percentage.

$0.2 = 0.2 \times 100\%$

$= 20\%$

Worked example

Change 0.005 to a percentage.

$0.005 = 0.005 \times 100\%$

$= 0.5\%$

Comparing percentages, fractions and decimals

To compare and order percentages, fractions and decimals you change them all into percentages.

Worked example

Place the following in order of size, starting with the smallest.

65%, $\frac{3}{5}$, 0.62, 63.5%, $\frac{3}{4}$, 0.7

65%	65%
$\frac{3}{5} = \frac{3}{5} \times 100\% = 60\%$	60%
$0.62 = 0.62 \times 100\% = 62\%$	62%
63.5%	63.5%
$\frac{3}{4} = \frac{3}{4} \times 100\% = 75\%$	75%
$0.7 = 0.7 \times 100\% = 70\%$	70%

So the order is $\frac{3}{5}$, 0.62, 63.5%, 65%, 0.7, $\frac{3}{4}$ (smallest to highest).

Check yourself

QUESTIONS

Q1 Complete this table.

Fraction	Decimal	Percentage
$\frac{1}{2}$	0.5	50%
	0.75	
		20%
$\frac{1}{8}$		

(In the style of NEAB specimen paper 1998)

Q2 Place the following in order of size, starting with the highest.

$\frac{38}{50}$, 81%, 0.85, $\frac{4}{5}$, 80.5%, 0.7888

REMEMBER! Cover the answers if you want to.

ANSWERS

A1

Fraction	Decimal	Percentage
$\frac{1}{2}$	0.5	50%
$\frac{3}{4}$	0.75	75%
$\frac{1}{5}$	0.2	20%
$\frac{1}{8}$	0.125	$12\frac{1}{2}$% or 12.5%

A2 0.85, 81%, 80.5%, $\frac{4}{5}$, 0.7888, $\frac{38}{50}$

TUTORIALS

T1 *Look at notes on converting between fractions, decimals and percentages for information on how to complete the table.*

T2 *Convert the fractions and decimals to percentages by multiplying by 100.*

$\frac{38}{50} = \frac{38}{50} \times 100\% = 76\%$	76%
81%	81%
$0.85 = 0.85 \times 100\% = 85\%$	85%
$\frac{4}{5} = \frac{4}{5} \times 100\% = 80\%$	80%
80.5%	80.5%
$0.7888 = 0.7888 \times 100\% = 78.88\%$	78.88%

So the order is 0.85, 81%, 80.5%, $\frac{4}{5}$, 0.7888, $\frac{38}{50}$ *(highest to smallest).*

Expressing one number as a percentage of another

To express one number as a percentage of another, write the first number as a fraction of the second and convert the fraction to a percentage.

Worked example

Write 55p as a percentage of 88p.

55p as a fraction of 88p $= \frac{55}{88}$.

$\frac{55}{88} = \frac{55}{88} \times 100\%$ Converting the fraction to a percentage.

$= 62.5\%$

Worked example

Write 2 feet as a percentage of 5 yards.

5 yards = 15 feet As 1 yard = 3 feet

So the problem becomes write '2 feet as a percentage of 15 feet'.

2 feet as a fraction of 15 feet $= \frac{2}{15}$.

$\frac{2}{15} = \frac{2}{15} \times 100\%$ Converting the fraction to a percentage.

$= 13.333\ 333...\%$

$= 13.3\%$ (3 s.f.)

Finding a percentage of an amount

To find the percentage of an amount, find 1% of the amount and then multiply to get the required amount.

Worked example

Calculate 50% of £72.

1% of £72 $= £\frac{72}{100}$

$= £0.72$

50% of £72 $= 50 \times £0.72$

$= £36$

Worked example

An investment valued at £2000 shows an increase of 6% one year. What is the new value of the investment?

To find 6% of £2000, first find 1%.

1% of £2000 = £20

So 6% of £2000 $= 6 \times £20$

$= £120$

The new value of the investment is £2000 + £120 = £2120

An alternative method uses the fact that after a 6% increase, the new amount will be 100% of the original amount + 6% of the original amount.

100% + 6% = 106% of the original amount

The new value of the investment is 106% of £2000.

1% of £2000 = £20

106% of £2000 = 106 × £20 = £2120 (as before)

NOTE

You can answer the question by finding 9% of the value and subtracting or else use the method introduced in the previous example.

Worked example

A caravan valued at £8000 depreciates by 9% each year. What is the value of the caravan after:

(a) one year (b) two years?

After a depreciation of 9%, the new amount

= 100% of the original amount – 9% of the original amount

= 91% of the original amount

(a) After one year, the value of the caravan is 91% of £8000.
1% of £8000 = £80
91% of £8000 = 91 × £80 = £7280

(b) After the second year, the value of the caravan is 91% of £7280.
1% of £7280 = £72.80
91% of £7280 = 91 × £72.80 = £6624.80

Check yourself

QUESTIONS

Q1 The highest percentage of people unemployed in Great Britain was in 1933. Out of 13 million people available for work, 3 million were unemployed.

What percentage were unemployed? Give your answer to the nearest whole number.

(MEG syllabus A, specimen paper 1998)

Q2 The price of a bathroom suite is advertised as £2800. A discount of 5.5% is agreed for a speedy sale. What is the final cost of the bathroom suite?

REMEMBER! Cover the answers if you want to.

ANSWERS

A1 23%

A2 £2646

TUTORIALS

T1 *Fraction unemployed* $= \frac{3 \text{ million}}{13 \text{ million}} = \frac{3}{13}$ *(as a fraction in its lowest terms).*

To convert the fraction to a percentage, multiply by 100 so the percentage $= \frac{3}{13} \times 100\% = 23.076923\%$

= 23% (to the nearest whole number).

T2 *The final cost = 94.5% (100% – 5.5%) of £2800.*
1% of £2800 = £28
94.5% of £2800 = 94.5 × £28 = £2646

Alternatively, you could find the discount as 5.5% of the price, and subtract.

Percentage change

To work out the percentage change, work out the increase or decrease and add or subtract it from the original amount.

$$\text{Percentage change} = \frac{\text{change}}{\text{original amount}} \times 100\%$$

where the change might be an increase, decrease, profit, loss, error, etc.

Worked example

A company produces 78 000 parts one year and 79 950 the following year. Calculate the percentage increase.

$$\text{Percentage increase} = \frac{\text{increase}}{\text{original amount}} \times 100\%$$

$$\text{Increase} = 79\,950 - 78\,000 = 1950$$

$$\text{Percentage increase} = \frac{1950}{78\,000} \times 100\% = 2.5\%$$

Worked example

A car valued at £3200 is sold for £2950. What is the percentage loss?

$$\text{Percentage loss} = \frac{\text{loss}}{\text{original amount}} \times 100\%$$

$$\text{Loss} = £3200 - £2950 = £250$$

$$\text{Percentage loss} = \frac{250}{3200} \times 100\% = 7.8125\%$$

$$= 7.81\% \text{ or } 7.8\% \text{ to an appropriate degree of accuracy.}$$

Check yourself

QUESTIONS

Q1 A retailer buys shirts at £12.72 each and sells them for £16.96. What is the percentage profit?

Q2 A weighing machine records a weight of 6 kg when the actual weight is 6.005 kg. What is the percentage error on the actual weight?

REMEMBER! Cover the answers if you want to.

ANSWERS

A1 33.3%

A2 0.08%

TUTORIALS

T1 *Percentage profit* $= \frac{\textit{profit}}{\textit{original amount}} \times 100\%$

Profit $= £16.96 - £12.72 = £4.24$

Percentage profit $= \frac{4.24}{12.72} \times 100\% = 33.3\%$

to an appropriate degree of accuracy.

T2 *Percentage error* $= \frac{\textit{error}}{\textit{actual amount}} \times 100\%$

Error $= 6.005 - 6 = 0.005$ *kg*

Percentage error $= \frac{0.005}{6.005} \times 100\% = 0.08\%$

to an appropriate degree of accuracy.

CHAPTER 11

REVERSE PERCENTAGES

To find the original amount after a percentage change you use reverse percentages.

USING REVERSE PERCENTAGES

Worked example

The value of a picture is given as £127 200 after a 6% increase. What was the original value of the picture?

£127 200 represents 106% of the original value (100% + 6%).

So 106% of the original value = £127 200

1% of the original value = $£\frac{127\,200}{106}$ = £1200

100% of the original value = 100 × £1200 = £120 000

The original value of the picture was £120 000 (100% of the original value).

Worked example

A washing machine is advertised at £335.75 after a price reduction of 15%. What was the original price of the washing machine?

£335.75 represents 85% of the original price (100% – 15%).

So 85% of the original price = £335.75

1% of the original price = $£\frac{335.75}{85}$ = £3.95

100% of the original price = 100 × £3.95 = £395

The original price of the washing machine was £395.

Check yourself

QUESTIONS

Q1 The price of a holiday is reduced by 5% to £361. What was the original cost of the holiday?

Q2 A car is sold for £5225 after a depreciation of 45% of the original purchase price. Calculate the original purchase price of the car.
(SEG modular syllabus, specimen paper 1998)

REMEMBER! Cover the answers if you want to.

ANSWERS

A1 £380

A2 £9500

TUTORIALS

T1 *After a reduction of 5%, the price represents 95% (100% – 5%) of the original cost of the holiday.*

95% of the original cost of the holiday = £361

1% of the original cost of the holiday = $£\frac{361}{95}$ = £3.80

100% of the original cost = 100 × £3.80 = £380

T2 *After a depreciation of 45%, the price represents 55% (100% – 45%) of the original purchase price.*

55% of original purchase price = £5225

1% of original purchase price = $£\frac{5225}{55}$ = £95

100% of original price = 100 × £95 = £9500

CHAPTER 12

EXAMINATION QUESTIONS 2

EXAM PRACTICE

Sample Student's Answers & Examiner's Comments

1 It is given that $m = 3 \times 10^{-2}$ and $n = 4 \times 10^{-3}$.
Calculate the value of mn giving your answer in standard form.
Calculate the value of $m + n$.

(In the style of MEG syllabus A, specimen paper 1998)

$mn = (3 \times 10^{-2}) \times (4 \times 10^{-3})$
$mn = 3 \times 4 \times 10^{-2} \times 10^{-3}$
$mn = 12 \times 10^{-2+-3}$
$mn = 12 \times 10^{-5}$

$mn = 1.2 \times 10^{1} \times 10^{-5}$
$mn = 1.2 \times 10^{1+-5}$
$mn = 1.2 \times 10^{-4}$

$m + n = 3 \times 10^{-2} + 4 \times 10^{-3}$
$m + n = 0.03 + 0.004$

$m + n = 0.034$

2 Ian and Kim share £400 between them. Ian gets $\frac{3}{5}$ of the money.
(a) How much does each of them get?
Kim puts £20 into her building society account.
(b) What fraction is this of her share?

(a) Ian gets $\frac{3}{5}$ of £400 $= \frac{3}{5} \times$ £400 = £240

Ian gets £240.

If Ian gets $\frac{3}{5}$ then Kim gets $\frac{2}{5}$.

$\frac{2}{5}$ of £400 $= \frac{2}{5} \times$ £400 = £160

Kim gets £160.

(b) Fraction $= \frac{20}{160}$

$\frac{20}{160} = \frac{1}{8}$

EXAMINER'S COMMENTS

1 *The candidate has used the laws of indices ($a^m \times a^n = a^{m+n}$) to multiply $10^{-2} \times 10^{-3}$ and converted 12×10^{-5} back to standard form.*

$12 = 1.2 \times 10^{1}$ in standard form

The final answer is given in standard form as required by the question.

To calculate $m + n$, the candidate has remembered that standard form numbers can only be added or subtracted if the power of 10 is the same. To find the sum, the numbers are converted to ordinary form first.

The answer can be given in standard form as 3.4×10^{-2} but this is not required by the question.

2 **a)**
It would be easier for the candidate to subtract Ian's £240 from £400 to find out how much Kim gets.

b)
Remembering that Kim's share is £160, the fraction is $\frac{20}{160}$.

This has been simplified by cancelling down to the lowest terms.

3 At the end of 1995, the population of Bridgeshire was 3000.
It is predicted that the population will decrease by 15% each year.
What will the population be at the end of 1998?
By the end of which year will the population fall below half the number recorded at the end of 1995?

At the end of 1996, population is
$\frac{85}{100} \times 3000 = 2550.$
At the end of 1997, population is
$\frac{85}{100} \times 2550 = 2167.5.$
At the end of 1998, population is
$\frac{85}{100} \times 2167.5 = 1842.375.$
The population at the end of 1998 is predicted to be 1842.
At the end of 1999, population is
$\frac{85}{100} \times 1842.375 = 1566.0188.$
At the end of 2000, the population is
$\frac{85}{100} \times 1566.0188 = 1331.1159.$
By the end of the year 2000, the population will fall below half the level recorded at the end of 1995.

4 The value of shares increases by 16% to £464.
What was the original value of the shares?

116% = £464
1% = £4
100% = £400
Original value = £400

EXAMINER'S COMMENTS

3 *A population decrease of 15% is equivalent to a multiplier of 85% or* $\frac{85}{100}$ *– in this example the candidate would be better advised to use 0.85 rather than* $\frac{85}{100}$ *on the calculator.*

The candidate has rounded off each answer to a realistic number and has remembered to use the original figures rather than the rounded figures when continuing with this work.

4 *The candidate has recognised this as a reverse percentage question and realised that after an increase of 16%, £464 represents 116% (100% + 16%) of the original value of the shares.*
£464 ÷ 116 = £4

100% represents the original value.

Questions to Answer

1 The circulation figure for a certain magazine is $12\frac{1}{2}$ million.

Write this number in standard form.

2 The distance from the Earth to the Moon is approximately 400 000 km and the distance from the Earth to the Sun is approximately 1.5×10^8 km.

Calculate the value of the expression:

$$\frac{\text{distance from the Earth to the Moon}}{\text{distance from the Earth to the Sun}}$$

giving your answer in standard form.

3 Light travels at an average speed of 299 800 km per second and the distance of the Sun to the Earth is approximately 1.496×10^8 km.

How long will it take light to travel from the Sun to the Earth, to the nearest minute?

4 List the following in order of size, starting with the smallest.

0.65 $\frac{3}{5}$ 62% $\frac{6}{11}$ $\frac{2}{3}$ 59% 0.666

5 Work out $5\frac{1}{5} - 1\frac{2}{3}$.

6 How many $\frac{3}{4}$-litre bottles can be filled from a container holding 12 litres?

7 A bank charges $2\frac{1}{2}$% interest on overdrafts. What interest will be charged on an overdraft of £220?

8 The number of casualties handled by a hospital increases by 12% each year. If the number of casualties this year is 500 how many casualties will there be after:

(a) one year (b) two years?

9 Chocolates are priced at £2.40. In a sale the chocolates cost £1.70. What percentage discount is this? Give your answer to an appropriate degree of accuracy.

10 If 20% of a certain number is 420, what is the original number?

11 A house is valued at £576 720 which represents an 8% increase on the original value. What was the original value?

CHAPTER 13

RATIOS AND PROPORTIONAL DIVISION

A ratio allows one quantity to be compared to another quantity in a similar way to fractions.

WRITING RATIOS

Worked example

In a box there are 12 lemons and 16 oranges. Write this as a ratio comparing the number of lemons to the number of oranges.

The ratio of the number of lemons to the number of oranges is 12 to 16.

You write this as $12:16$.

The order is important in ratios.

The ratio of the number of oranges to the number of lemons is 16 to 12 or $16:12$.

EQUIVALENT RATIOS

Equivalent ratios are ratios that are equal to each other.

The following ratios are all equivalent to $2:5$.

$2:5 = 4:10 = 6:15 = 8:20 = \dots$

Equivalent ratios can be found by multiplying or dividing both sides of the ratio by the same number. You can use this method to obtain the ratio in a form where both sides are integers.

- $1:2 = 2:4$
- $3:7 = 15:35$
- $16:20 = 4:5$
- $0.5:5 = 1:10$
- $\frac{3}{11}:\frac{4}{11} = 3:4$
- $12.5:15 = 5:6$

CANCELLING RATIOS

A ratio can be expressed in its **simplest form** or **lowest terms** by making both sides of the ratio as small as possible. Remember that both sides of the ratio must be integers.

Worked example

Write the following ratios in their simplest form.

(a) $10:15$ (b) $121:44$ (c) $4:\frac{1}{4}$ (d) $2.5:0.5$

(a) $10:15 = 2:3$ Dividing both sides by 5.

(b) $121:44 = 11:4$ Dividing both sides by 11.

(c) $4:\frac{1}{4} = 16:1$ Multiplying both sides by 4 to get integer values.

(d) $2.5:0.5 = 5:1$ Multiplying both sides by 2 to get integer values.

Worked example

Express the ratio 40p to £2 in its simplest form.

You must ensure that the units are the same.

£2 = 200p

Then the ratio is $40:200 = 1:5$ in its simplest form.

Worked example

Two lengths are in the ratio 4 : 5. If the first length is 60 cm, what is the second length?

The ratio is 4 : 5.

$4:5 = 1:1.25$ Dividing both sides by 4 to make an equivalent ratio, in the form 1 : *n*.

$= 1 \times 60 : 1.25 \times 60$ Multiplying both sides by 60 to find an equivalent ratio in the form 60 : *m*.

$= 60\text{ cm} : 75\text{ cm}$ As $1.25 \times 60 = 75$

So the second length is 75 cm.

Check yourself

QUESTIONS

Q1 Express the ratio of 5 km to 600 m in its simplest form.

Q2 Express the ratio 2 : 5 in the form 1 : *n*.

Q3 Express the ratio $\frac{1}{3} : \frac{1}{4}$ in its simplest form.

Q4 A plan of a college is drawn on a scale of 1 : 50. On the plan, a pathway is 22 cm long. How long is the actual pathway, in metres?

REMEMBER! Cover the answers if you want to.

ANSWERS

A1 25 : 3

A2 1 : 2.5 or $1 : 2\frac{1}{2}$

A3 4 : 3

A4 11 m

TUTORIALS

T1 *To find the ratio of 5 km to 600 m, both parts need to be expressed in the same units. 5 km = 5000 m so the ratio is 5000 : 600 = 25 : 3 in its simplest form.*

T2 $2:5 = 1:2.5$ *Dividing both sides by 2 to make an equivalent ratio in the form 1 : n.*

T3 $\frac{1}{3} : \frac{1}{4} = 4:3$ *This is found by multiplying both sides by 12 to make an equivalent ratio.*

T4 $1:50 = 1 \times 22 : 50 \times 22 = 22 : 1100$ *as an equivalent ratio. The pathway is 1100 cm = 11 metres.*

PROPORTIONAL PARTS

To share an amount into proportional parts, add up the individual parts and divide the amount by this number to find the value of one part.

Worked example

Divide £50 between two sisters in the ratio 3 : 2. How much does each get?

Number of parts = 3 + 2 = 5

Value of each part = £50 ÷ 5 = £10

The sisters receive £30 (3 parts at £10 each) and £20 (2 parts at £10 each).

HINT

It is useful to check that the separate amounts add up to the original amount (i.e. £30 + £20 = £50).

Check yourself

QUESTIONS

Q1 Three children raise money for a charity. The amounts they each raise are in the ratio 2 : 3 : 7. The total amount raised is £72. How much does each child raise?

Q2 Two villages with populations of 450 and 260 receive a grant for £3550. The councils agree to share the money in proportion to the population. How much does each village get?

REMEMBER! Cover the answers if you want to.

ANSWERS

A1 £12, £18 and £42

A2 £2250 and £1300

TUTORIALS

T1 *For the ratio 2 : 3 : 7, the number of parts = 2 + 3 + 7 = 12.*
The value of each part = £72 ÷ 12 = £6.
The children raise £12 (2 parts at £6), £18 (3 parts at £6) and £42 (7 parts at £6 each).
***Check** £12 + £18 + £42 = £72 as required.*

T2 *The grant is shared in the ratio of 450 to 260.*
Number of parts = 450 + 260 = 710
Value of each part = £3550 ÷ 710 = £5
The villages get £2250 (450 × £5) and £1300 (260 × £5).
***Check** £2250 + £1300 = £3550 as required.*

NUMBER

CHAPTER 14

ESTIMATION AND APPROXIMATION

Estimation and approximation are important parts of the examination syllabus. You will be required to give an estimation by rounding numbers to convenient approximations (usually 1 s.f .or 2 s.f.).

ROUNDING

Worked example

Estimate the value of $\frac{40.68 + 61.2}{9.96 \times 5.13}$.

Rounding these figures to 1 s.f. gives $\frac{40 + 60}{10 \times 5} = \frac{100}{50} = 2$

Using a calculator, the actual answer is 1.993 940 7... so the answer is a good approximation.

Check yourself

QUESTIONS

Estimate the value of each of these expressions.

Q1 $\frac{3.87 \times 5.07^3}{5.16 \times 19.87}$

Q2 $\frac{2.78 + \pi}{\sqrt{5.95 \times 6.32}}$

Q3 $\frac{59.96}{40.21 + 19.86} + \sqrt{80.652}$

REMEMBER! Cover the answers if you want to.

ANSWERS

A1 5

A2 1

A3 10

TUTORIALS

T1 $\frac{3.87 \times 5.07^3}{5.16 \times 19.87} \approx \frac{4 \times 125}{5 \times 20} = \frac{500}{100} = 5$

Take 5.07^3 *as approximately* $5^3 = 125$

T2 $\frac{2.78 + \pi}{\sqrt{5.95 \times 6.32}} \approx \frac{3 + 3}{\sqrt{6 \times 6}} = \frac{6}{6} = 1$

Take π *as approximately* 3.

T3 $\frac{59.96}{40.21 + 19.86} + \sqrt{80.652} \approx \frac{60}{40 + 20} + \sqrt{81}$

$= 1 + 9 = 10$

CHAPTER 15

MENTAL METHODS

In some questions you will not be allowed to use a calculator. For these questions you must show sufficient working to convince the examiner that you have not used a calculator.

NOTE

You can, of course, use a calculator to check your answers.

MULTIPLICATION

Worked example

Calculate 147 × 32.

$$\begin{aligned} 147 \times 32 &= 147 \times (30 + 2) \\ &= 147 \times 30 + 147 \times 2 \\ &= 4410 + 294 \\ &= 4704 \end{aligned}$$

$$\begin{array}{r} 147 \\ \times\ 30 \\ \hline 4410 \\ \hline \end{array} \qquad \begin{array}{r} 147 \\ \times\ 2 \\ \hline 294 \\ \hline \end{array}$$

It is more usual to set this multiplication out as follows.

$$\begin{array}{r} 147 \\ \times\ 32 \\ \hline 4410 \\ +\ 294 \\ \hline 4704 \\ \hline \end{array}$$

4410 Multiplying by 30.
294 Multiplying by 2.
4704 Adding.

MULTIPLICATION OF DECIMALS

To multiply two decimals without using a calculator:

- ignore the decimal points and multiply the numbers
- count the number of digits after the decimal point in each number and add them together, to find the number of digits after the decimal point in the answer
- place the decimal point in the answer to give the required number of digits after it.

HINT

This is more difficult to describe that it is to do. The examples should help.

Worked example

Find the product of 1.47 and 3.2.

147 × 32 = 4704 Ignore the decimal points and multiply the numbers, as in the previous example.

1.47 has two digits after the decimal point. 3.2 has one digit after the decimal point. The total number of digits after the decimal points is 2 + 1 = 3, so you need 3 decimal places in the answer.

1.47 × 3.2 = 4.704

Place the decimal point so that there are three digits after the decimal point.

HINT

It is always helpful to make a rough estimate of the answer to check that you haven't made a careless mistake.

Worked example

Find the product of 0.000 147 and 0.0032.

147 × 32 = 4704 Ignore the decimal points and multiply the numbers as in the earlier example.

The total number of digits after the decimal points = 6 + 4 = 10.

0.000 147 × 0.0032 = 0.000 000 470 4

Place the decimal point so that there are 10 digits after the decimal point in the answer.

HINT

If your answer after the first multiplication ends in 0, include this when you are counting the digits after the decimal point in your answer.

Check yourself

QUESTIONS

Work these out without using a calculator.

Q1 232×51

Q2 0.0614×3.5

REMEMBER! Cover the answers if you want to.

ANSWERS

A1 11 832

A2 0.2149

TUTORIALS

T1

$$\begin{array}{rrl} & 232 & \\ \times & 51 & \\ \hline & 11600 & \textit{Multiplying by 50.} \\ + & 232 & \textit{Multiplying by 1.} \\ \hline & 11832 & \textit{Adding.} \\ \hline \end{array}$$

T2 *Ignoring the decimal point and finding* 614×35:

$$\begin{array}{rrl} & 614 & \\ \times & 35 & \\ \hline & 18420 & \textit{Multiplying by 30.} \\ + & 3070 & \textit{Multiplying by 5.} \\ \hline & 21490 & \textit{Adding.} \\ \hline \end{array}$$

Count the number of digits after the decimal point in each number.
0.0614 *has 4 digits (4 d.p.)*
3.5 *has 1 digit (1 d.p.)*
The answer needs $4 + 1 = 5$ *decimal places.*
Place the decimal point to give 5 d.p. in the answer.
$0.0614 \times 3.5 = 0.21490$ *or* 0.2149

DIVISION

To divide by a two-digit number, proceed in exactly the same way as you would for any other division.

Worked example

Calculate $437 \div 19$.

$$\begin{array}{r} 23 \\ 19\overline{)437} \\ 38 \\ \hline 57 \\ 57 \\ \hline 0 \\ \hline \end{array}$$

Take the first two digits, $43 \div 19$ gives 2 with a remainder of 5.
Subtract $43 - 38$, bring down the next digit.
$57 \div 19$ gives 3 with no remainder.

So $437 \div 19 = 23$.

DIVISION OF DECIMALS

We can use the idea of equivalent fractions to deal with division of decimals.

Worked example

Calculate 43.7 ÷ 1.9.

43.7 ÷ 1.9 can be written as $\frac{43.7}{1.9} = \frac{437}{19}$ — Multiply the numerator and the denominator by 10 to obtain an equivalent fraction.

Work out 43.7 ÷ 1.9 as 437 ÷ 19 = 23 — As found in the previous example.

Worked example

Work out 0.003 08 ÷ 0.000 14.

0.003 08 ÷ 0.000 14 can be written as $\frac{0.003\,08}{0.000\,14} = \frac{308}{14}$ or 308 ÷ 14.

```
   22
14)308    Take the first two digits, 30 ÷ 14 gives 2 with a remainder of 2.
   28     Subtract 30 – 28, bring down the next digit.
   --
    28    28 ÷ 14 gives 2 with no remainder.
    28
    --
     0
```

So 0.003 08 ÷ 0.000 14 = 22.

Check yourself

QUESTIONS

Work these out without using a calculator.

Q1 459 ÷ 0.17

Q2 0.0936 ÷ 0.0013

REMEMBER! Cover the answers if you want to.

ANSWERS

A1 2700

A2 72

TUTORIALS

T1 459 ÷ 0.17

459 ÷ 0.17 *can be written as*

$\frac{459}{0.17} = \frac{45\,900}{17}$ *or* 45 900 ÷ 17

So 459 ÷ 0.17 = 2700.

```
   2700
17)45900
   34
   ---
   119
   119
   ---
     0
```

T2 0.0936 ÷ 0.0013

0.0936 ÷ 0.0013 *can be written as*

$\frac{0.0936}{0.0013} = \frac{936}{13}$ *or* 936 ÷ 13

So 0.0936 ÷ 0.0013 = 72.

```
   72
13)936
   91
   --
    26
    26
    --
     0
```

CHAPTER 16

USING A CALCULATOR

You should familiarise yourself with your calculator before the examination by reading the user manual. The following functions are to be found on most calculators, but your manual will provide further information.

You may need to use an INV or 2ndF key to access some of the functions.

Key	Explanation
C	Cancel – cancels only the last number entered.
AC	All cancel – cancels all of the data entered.
x^2	Calculates the square of the number.
x^3	Calculates the cube of the number.
$\sqrt{\ }$	Calculates the square root of the number.
$\sqrt[3]{\ }$	Calculates the cube root of the number.
$^1/x$ or x^{-1}	Calculates the reciprocal of the number.
+/–	Reverses the sign by changing positive numbers to negative numbers and negative numbers to positive numbers.
x^y	This is the power key. To enter 3^6 you key in 3 x^y 6
EXP or EE	This is the standard-form button. To enter 3.2×10^7 you key in 3.2 EXP 7 The display will show 3.2 ^{07} or 3.2 07
$a^b/_c$	This is the fraction key (not all calculators have this key). To enter $\frac{3}{4}$ key in 3 $a^b/_c$ 4. 3⌋4 in the display means $\frac{3}{4}$. To enter $1\frac{3}{4}$ key in 1 $a^b/_c$ 3 $a^b/_c$ 4. 1⌋3⌋4 in the display means $1\frac{3}{4}$.
Min or STO	Stores the displayed value in the memory.
MR or RCL	Recalls the value stored in the memory.
M+	Adds the displayed value to the number in the memory.
M–	Subtracts the displayed value from the number in the memory.
Mode	Gives the mode for calculations – refer to your user manual.
DRG	Gives the units for angles (degrees, radians or grads). Your calculator should normally be set to degrees.

Check yourself

QUESTIONS

Use your calculator to work these out.

Q1 5.1^2

Q2 $\sqrt{2.6}$

Q3 $\sqrt[3]{41.3}$

Q4 3^6

Q5 2^{-5}

Q6 $(3 \times 10^7) \times (2 \times 10^5)$

Q7 $(2.5 \times 10^3) \div 5$

Q8 $(2.1 \times 10^{-3}) + (4.62 \times 10^{-2})$

Q9 $\frac{3}{4} - \frac{1}{5}$

Q10 $\frac{4}{7} \div \frac{8}{9}$

Q11 $2\frac{1}{5} \times \frac{5}{9}$

REMEMBER! Cover the answers if you want to.

ANSWERS

A1 26.01

A2 1.612 451 6

A3 3.456 607 1

A4 729

A5 0.031 25

A6 6×10^{12}

A7 500

A8 0.0483

A9 $\frac{11}{20}$ or 0.55

A10 $\frac{9}{14}$ or 0.642 857 1

A11 $1\frac{2}{9}$ or 1.222 22

TUTORIAL

Refer to your user manual if you have any difficulties with this work.

CHAPTER 17

EXAMINATION QUESTIONS 3

EXAM PRACTICE

Sample Student's Answers & Examiner's Comments

1 The distance from the Earth to the Moon is approximately 384 400 km and the distance from the Earth to the Sun is approximately 1.496×10^8 km. Use these approximations to express the ratio:

distance from Earth to Moon : distance from Earth to Sun

in the form 1 : *n* where *n* is a whole number.

E to S = 149 600 000 km

The ratio is 384 400 : 149 600 000

$= 1 : \frac{149\,600\,000}{384\,400}$

= 1 : 389.17794

= 1 : 389

2 Ruth uses the formula:

$$t = \frac{2s}{u + v}$$

to calculate the value of *t* when *s* = 521.86, *u* = 21.19 and *v* = 58.66. Ruth estimates the value of *t* without using her calculator.

Write down the numbers Ruth should use in the formula to estimate the value of *t*.

Work out the estimate for the value of *t* given by your numbers.

Use your calculator to work out the actual value of *t*. Give your answer to an appropriate degree of accuracy.

(In the style of London specimen paper 1998)

s = 500, u = 20 and v = 60

$t = \frac{2s}{u + v} = \frac{2 \times 500}{20 + 60} = \frac{1000}{80} = 12.5$

$t = \frac{2 \times 521.86}{21.19 + 58.66}$

$t = \frac{1043.72}{79.85}$

t = 13.071 008...

t = 13.1

EXAMINER'S COMMENTS

1 *The candidate has not wasted time writing out the question again and is using a sensible shorthand. The standard form number has been converted to ordinary form although this is not necessary as a calculator should be used to divide 1.496×10^8 (using the* EXP *key) by 384 400.*

The candidate has made an equivalent ratio by dividing both sides by 384 400 to get the ratio in the form 1 : n.

She has remembered to round the number to a whole number as required.

2 *The candidate has rounded each of the values to 1 s.f.*

There is no need to write down all of the stages of this calculation. It would be better if he had used the memory facility on the calculator.

An answer to 3 s.f. or 2 s.f. is sufficient here as the original values contained 4 or 5 significant figures.

Questions to Answer

1 The speed of traffic on a three-lane stretch of road is in the ratio 2 : 3 : 5. If the speed of the traffic in the fastest lane is 60 miles per hour calculate the speed of traffic in the other two lanes.

(In the style of NEAB specimen paper 1998)

2 Bronze is a mixture of copper and tin. The bronze used for making nails is made from copper and tin in the ratio of 8 : 1 by weight. A large bronze nail has a weight of 18 g. What weight of copper does it contain?

(MEG syllabus B, specimen paper 1998)

3 Three friends pay £5, £7 and £8 into a sweepstake syndicate each week. They agree to split any winnings in proportion to their payments. How much will each get on a payout of £3556?

4 Give an approximate answer to:

$$\frac{28.65 \times 0.0852}{14.6 \times 3.22}$$

by rounding the numbers to a sensible degree of accuracy.

5 The formula:

$$s = \frac{v^2 - u^2}{2a}$$

gives the distance, in metres, travelled when the velocity changes from u m/s to v m/s due to a constant acceleration of a m/s^2.
Find an approximate value of s when $v = 19.7$, $u = 14.8$ and $a = 9.8$.
Use your calculator to find a more accurate value for s.

6 Use your calculator to work out the value of:

$$\frac{5.06 \times (10.32)^2}{281 + 217}$$

(a) Write down the full calculator display.

(b) Write down a calculation that could be carried out mentally to check this answer, using numbers rounded to one significant figure.

(c) Write down the answer to your calculation in part (b).

CHAPTER 18

COMPOUND MEASURES

Compound measures are those involving more than one unit such as speed (distance and time) or density (mass and volume).

SPEED

> **NOTE**
>
> The word 'per' means divide.

> **NOTE**
>
> The formula for speed can be rearranged as follows.
> distance = speed × time or
> $\text{time} = \frac{\text{distance}}{\text{speed}}$

> **NOTE**
>
> As the speed is measured in miles per hour, you need to express the distance in miles and the time in hours.

> **NOTE**
>
> You must remember that 0.45 hours is not 45 minutes as there are 60 minutes in one hour.
>
> To convert hours to minutes you multiply by 60.

The formula for speed is:

$$\text{speed} = \frac{\text{distance}}{\text{time}}$$ (expressed in units such as miles per hour)

Worked example

A taxi travels 16 miles in 20 minutes. What is the speed in miles per hour?

Distance = 16 miles Time = 20 minutes = $\frac{20}{60} = \frac{1}{3}$ hour

Using the formula speed = $\frac{\text{distance}}{\text{time}}$:

$$\text{speed} = \frac{16}{\frac{1}{3}} = 48 \text{ miles per hour (mph)}$$

Worked example

A cyclist travels 3.6 km at an average speed of 8 kilometres per hour. How long does the journey take?

Using the formula time = $\frac{\text{distance}}{\text{speed}}$:

$$\text{time} = \frac{3.6}{8} = 0.45 \text{ hours}$$

0.45 hours = 0.45 × 60 minutes = 27 minutes

The journey takes 27 minutes.

DENSITY

> **NOTE**
>
> The formula for density can be rearranged as follows.
> mass = density × volume or
> $\text{volume} = \frac{\text{mass}}{\text{density}}$

> **NOTE**
>
> As the density is measured in g/cm^3, you need to express the mass in grams and the volume in cm^3.

The formula for density is:

$$\text{density} = \frac{\text{mass}}{\text{volume}}$$ (expressed in units such as grams per cm^3)

Worked example

A piece of lead weighing 170 g has a volume of 15 cm^3. Give an estimate for the density of lead.

Using the formula density = $\frac{\text{mass}}{\text{volume}}$:

$$\text{density} = \frac{170}{15} = 11.3 \text{ g/cm}^3 \text{ (3 s.f.)}$$

Worked example

A sheet of metal measures 10 m by 6 m by 0.5 mm and has a density of 8.9 g/cm^3. What is the mass of the sheet of metal, in kilograms?

Volume = 1000 × 600 × 0.05 = 30 000 cm^3 Converting *all* lengths to cm.

Using the formula mass = density × volume:

mass = 8.9 × 30 000 g = 267 000 g = 267 kg Converting to kg where 1000 g = 1 kg.

Check yourself

QUESTIONS

Q1 A train travels 160 metres in 10 seconds. What is its speed in: (a) m/s (b) km/h?

Q2 A cyclist travels for 45 minutes at a speed of 14 miles per hour. What distance does the cyclist travel?

Q3 A lorry driver drives 14 km at an average speed of 40 kilometres per hour. How long does the journey take?

Q4 A silver cube of length 12 mm has a density of 10.5 g/cm^3. What is the mass of the cube?

REMEMBER! Cover the answers if you want to.

ANSWERS

A1 (a) 16 m/s
(b) 57.6 km/h

A2 10.5 miles

A3 21 minutes

A3 18.1 g (3 s.f.)

TUTORIALS

T1 (a) $\text{Speed} = \frac{\text{distance}}{\text{time}} = \frac{160}{10} = 16 \text{ m/s}$

(b) *For speed in km/h, you need to express the distance in kilometres and the time in hours.*

$160 \text{ m} = 0.16 \text{ km}$ *as 1 km = 1000 m*

$10 \text{ seconds} = \frac{1}{6} \text{ minutes}$ *as 60 seconds = 1 minute*

$= \frac{1}{360} \text{ hour}$ *as 60 minutes = 1 hour*

$$\text{Speed} = \frac{\text{distance}}{\text{time}} = \frac{0.16}{\frac{1}{360}} = 57.6 \text{ km/h}$$

T2 *As the speed is given in miles per hour, you need to express the time in hours.*

$\text{Time} = 45 \text{ minutes} = \frac{3}{4} \text{ hour}$ *or you could use 0.75 hours*

$\text{Distance} = \text{speed} \times \text{time} = 14 \times \frac{3}{4} = 10.5 \text{ miles}$

T3 $\text{Time} = \frac{\text{distance}}{\text{speed}} = \frac{14}{40} = 0.35 \text{ hours}$

To convert hours to minutes you multiply by 60.
$0.35 \text{ hours} = 0.35 \times 60 \text{ minutes} = 21 \text{ minutes}$

T4 *As the density is measured in g/cm^3, you need to express the mass in grams and the volume in cm^3.*

$\text{Volume} = 1.2 \times 1.2 \times 1.2 \text{ cm}^3$ *as each side of the cube is 12 mm and 12 mm = 1.2 cm*

$= 1.728 \text{ cm}^3$

Using mass = density × volume:

$\text{mass} = 10.5 \times 1.728$
$= 18.1 \text{ g (3 s.f.)}$

CHAPTER 19

SIMPLE AND COMPOUND INTEREST

With simple interest, the interest paid is not reinvested, whereas with compound interest, the amount of interest paid is reinvested and earns interest itself.

The formula for simple interest is:

$$A = P + \frac{PRT}{100}$$

where:

A = total amount

P = principal or original investment

R = rate (% per annum)

T = time (in years).

The formula for compound interest is:

$$A = P \times \left(1 + \frac{R}{100}\right)^T$$

where:

A = total amount

P = principal or original investment

R = rate (% per annum)

T = time (in years)

NOTE

The simple interest $I = \frac{PRT}{100}$

so $A = P + I$ or $A = P + \frac{PRT}{100}$

NOTE

The formula for compound interest questions can be worked out by repeatedly applying the simple interest formula.

SIMPLE INTEREST

Simple interest is calculated on a fixed principal, and can be calculated by the simple formula:

$$I = \frac{PRT}{100}$$

where:

I = interest

P = principal

R = rate (% per annum or per year)

T = time (in years)

Worked example

If £2000 is invested for 2 years at 6% per annum, calculate the simple interest and the total amount.

Using the formula $I = \frac{PRT}{100}$ where:

P = principal = £2000

R = rate = 6%

T = time = 2 years

$$I = \frac{2000 \times 6 \times 2}{100}$$

$= £240$

$A = P + I$

$= £2000 + £240$

$A = £2240$

The simple interest is £240 and the total amount is £2240.

COMPOUND INTEREST

In compound interest, the principal changes every year, as the previous year's interest is added into it.

Worked example

£1000 is invested at 6.2% p.a. compound interest. What is the amount after 3 years?

The compound interest can be worked out by repeatedly applying the simple interest formula:

$A = P + \frac{PRT}{100}$ where:

P = principal = £1000

R = rate = 6.2%

T = time = 1 year for each year

Year 1 $A = P + \frac{PRT}{100} = 1000 + \frac{1000 \times 6.2 \times 1}{100} = £1062$

Year 2 $A = P + \frac{PRT}{100}$ After 1 year, $P = £1062$.

$= 1062 + \frac{1062 \times 6.2 \times 1}{100} = £1127.844$

Year 3 $A = P + \frac{PRT}{100}$ After 2 years, $P = £1127.844$.

$= 1127.844 + \frac{1127.844 \times 6.2 \times 1}{100} = £1197.770\,328$

$= £1197.77$ (to the nearest penny)

Alternatively, using the compound interest formula:

$A = P \times \left(1 + \frac{R}{100}\right)^T$ where:

P = principal = £1000

R = rate = 6.2%

T = time = 3 years

$A = 1000\left(1 + \frac{6.2}{100}\right)^3 = £1197.770\,328$ (as before)

$A = £1197.77$ (to the nearest penny)

HINT

Do not round off until the final answer.

Check yourself

QUESTIONS

Q1 A sum of £5000 is invested at 7% p.a. simple interest. How long will it be before the amount equals £5875?

Q2 A sum of £250 is invested for 4 years. The simple interest paid is £52.50. What is the percentage rate per annum?

Q3 A sum of £2000 is invested for 2 years at 6% per annum. Calculate the total amount and the compound interest.

Q4 What is the compound interest on £10 000 invested over 2 years at 5.75% p.a?

REMEMBER! Cover the answers if you want to.

ANSWERS

A1 2.5 years or 2 years 6 months

A2 5.25% p.a.

A3 Total amount = £2247.20 and compound interest = £247.20

A4 £1183.06 (to the nearest penny)

TUTORIALS

T1 *The amount* $A = P + I$ *so the interest*
$I = £5875 - £5000 = £875$

$$I = \frac{PRT}{100}$$

$$875 = \frac{5000 \times 7 \times T}{100}$$ *Substituting* $I = £875$, $P = £5000$ *and* $R = 7\%$.

$$875 = 350 \times T$$

$$T = \frac{875}{350} = 2.5$$

The time is 2.5 years or 2 years and 6 months (12 months = 1 year).

T2
$$I = \frac{PRT}{100}$$

$$52.50 = \frac{250 \times R \times 4}{100}$$ *Substituting* $I = £52.50$, $P = £250$ *and* $T = 4$ *years.*

$$52.50 = 10 \times R$$

$$R = \frac{52.50}{10} = 5.25\%$$

The rate is 5.25% per annum.

T3 *Using the formula* $A = P \times \left(1 + \frac{R}{100}\right)^T$

$$A = 2000\left(1 + \frac{6}{100}\right)^2$$ *Substituting* $P = £2000$, $R = 6\%$ *and* $T = 2$ *years.*

$A = £2247.20$ *Remember to interpret 2247.2 on your calculator as £2247.20.*

$A = P + I$
$£2247.20 = £2000 + I$
$I = £247.20$

The total amount £2247.20 and the compound interest is £247.20.

T4
$$A = P \times \left(1 + \frac{R}{100}\right)^T$$

$$A = 10\,000\left(1 + \frac{5.75}{100}\right)^2$$ *Substituting* $P = £10\,000$, $R = 5.75\%$ *and* $T = 2$ *years.*

$A = £11\,183.06$ *(to the nearest penny)*

Compound interest $= £11\,183.06 - £10\,000$
$= £1183.06$ *(to the nearest penny)*

CHAPTER 20

PERSONAL AND HOUSEHOLD FINANCE

Questions on personal and household finance can be set under a variety of headings and contexts as illustrated in this chapter.

EARNINGS AND OVERTIME

A **wage** is the money earned for a week's work while a **salary** is the money earned in a year (which is usually shared into equal parts and paid monthly).

Overtime is money paid for working more than the agreed number of hours. Overtime is paid at different rates such as 'time and a half' or 'double time'.

Gross pay is the money earned before **deductions**.

Take home pay or **net pay** is the money left to 'take home' after deductions. Deductions might include income tax, national insurance, pension contributions etc.

Worked example

A courier is paid £171 for a basic 38-hour week and overtime is paid at time and a half. Calculate the courier's total wage for a week in which he works 43 hours.

Basic hourly rate $= £\frac{171}{38} = £4.50$ per hour

Overtime rate $= 1\frac{1}{2} \times £4.50 = £6.75$

Number of hours overtime $= 43 - 38 = 5$ hours

Overtime pay $= £6.75 \times 5 = £33.75$

Total wage $= £171 + £33.75$ basic + overtime
$= £204.75$

Worked example

A receptionist receives £182 for a basic 35-hour week. Overtime is paid at time and a half. How many hours are worked in a week when her total wage is £199.55?

Overtime pay $= £199.55 - £182 = £17.55$

Basic hourly rate $= £\frac{182}{35} = £5.20$ per hour

Overtime rate $= 1\frac{1}{2} \times £5.20 = £7.80$ per hour

Number of hours overtime $= \frac{\text{overtime pay}}{\text{overtime rate}}$

Number of hours overtime $= \frac{£17.55}{£7.80} = 2.25$ hours

Number of hours worked $= 35 + 2.25$ basic + overtime
$= 37.25$ hours or $37\frac{1}{4}$ hours

INCOME TAX

Income tax is a tax paid on income and is deducted under the Pay As You Earn (PAYE) system. The amount of income tax paid depends upon the money earned, the rate of tax and your **personal allowance**.

Your personal allowance is the amount of money that you can earn before you have to pay income tax. **Taxable income** is the income on which income tax is paid.

Worked example

An office worker has a monthly income of £860 and a personal tax allowance of £3765. He pays 20% on the first £3900 and 24% on the rest. How much tax does he pay during one year?

Annual income = £860 × 12 = £10 320

Taxable income = annual income – personal allowance
= £10 320 – £3765
= £6555

The office worker pays 20% on the first £3900 and 24% on the rest.

On £6555, he pays:

20% tax on the first £3900

24% tax on £2655 (£6555 – £3900 = £2655)

Amount of tax paid at 20% = 20% of £3900
= £780

Amount of tax paid at 24% = 24% of £2655
= £637.20

Total tax = £780 + £637.20
= £1417.20

Check yourself

QUESTIONS

Q1 Maneeta earns £5.70 per hour for a 38-hour week. Overtime is paid at time and a half, with double time on Sundays.
Calculate her total wages if she works $44\frac{1}{2}$ hours including $2\frac{1}{2}$ hours on a Sunday.

Q2 An insurance agent is paid a wage of £525 per month and receives commission of 0.05% on insurance policies sold. What is the insurance agent's total wage in a month in which she sells £32 600 of insurance?

Q3 Kathy earned £27 000 in 1995. She did not pay tax on the first £3525 of her income. She paid 20% tax on the first £3200 of her taxable income and 25% on the rest of her taxable income. Calculate the total amount of tax that she paid in 1995.

(SEG specimen paper 1998 – amended)

REMEMBER! Cover the answers if you want to.

ANSWERS

A1 £279.30

TUTORIALS

T1 *Basic wage* = 38 × £5.70 = £216.60

Maneeta works $2\frac{1}{2}$ *hours overtime on Sunday at double time and 4 hours at time and a half*

($44\frac{1}{2} - 38 - 2\frac{1}{2} = 4$)

Overtime:

double time = $2\frac{1}{2}$ *hours* × (2 × £5.70) = £28.50

time and a half = 4 *hours* × ($1\frac{1}{2}$ × £5.70) = £34.20

Total wage = £216.60 + £28.50 + £34.20 = £279.30

ANSWERS

A2 £541.30

A3 £5708.75

TUTORIALS

T2 Commission = 0.05% × £32 600

$$= \frac{0.05}{100} \times £32\,600$$

= £16.30

Total wage = wage + commission
= £525 + £16.30
= £541.30

T3 Taxable income = annual income – personal allowance
= £27 000 – £3525 = £23 475

Kathy paid 20% tax on the first £3200 of her taxable income and 25% on the rest.

On £23 475, she pays 20% tax on the first £3200 and 25% tax on £20 275 (£23 475 – £3200 = £20 275).

Amount of tax paid at 20% = 20% of £3200
= £640

Amount of tax paid at 25% = 25% of £20 275
= £5068.75

Total tax = £640 + £5068.75 = £5708.75

CREDIT AND HIRE PURCHASE

Hire purchase is a way of paying for goods or services over a period of time. You usually have to pay a **deposit**, plus a fixed amount of money paid at regular intervals until the total is paid off.

Worked example

A video camera costs £1790 or else it can be bought for a deposit of 20% plus 24 monthly payments of £86.50. Find:

(a) the deposit

(b) the credit price

(c) the amount saved by paying cash.

(a) The deposit = 20% of £1790

$$= \frac{20}{100} \times 1790$$

= £358

(b) The credit price = deposit + 24 monthly payments of £86.50
= £358 + 24 × £86.50
= £2434

(c) The amount saved by paying cash = credit price – cash price
= £2434 – £1790
= £644

VALUE ADDED TAX (VAT)

Value added tax (VAT) is a tax on goods or services.

Worked example

A dishwasher is advertised at £360 + VAT at $17\frac{1}{2}$%. Calculate the VAT and the total cost of the dishwasher.

VAT = $17\frac{1}{2}$% of £360 or 17.5% of £360

$$1\% \text{ of } £360 = £\frac{360}{100} = £3.60$$

$$17.5\% \text{ of } £360 = 17.5 \times £3.60 = £63$$

VAT = £63

$$\text{Total cost} = £360 + £63 = £423$$

Worked example

A CD system costs £282 including VAT at $17\frac{1}{2}$%. What is the cost of the CD system without the VAT?

£282 represents 117.5% (100% + 17.5%) of the cost of the CD system.

117.5% of the cost of the CD system = £282

$$1\% \text{ of the cost of the CD system} = £\frac{282}{117.5} = £2.40$$

$$100\% \text{ of the cost of the CD system} = 100 \times £2.40 = £240$$

The CD system costs £240 without the VAT.

BILLS, BILLS, BILLS

Electricity bills are worked out by costing the number of units of electricity used since the last reading. A quarterly charge is added to the bill and 8% VAT is applied to the total.

Gas bills are worked out by costing the number of units of gas used since the last reading. A standing charge is added to the bill and 8% VAT is applied to the total.

Telephone bills are worked out according to the number, length and distance of telephone calls made. A rental charge is added to the bill and $17\frac{1}{2}$% VAT is applied to the final amount.

Worked example

During one quarter a householder is charged £45.60 for telephone calls, along with a rental charge of £21.86. VAT at $17\frac{1}{2}$% is applied to the final amount. Calculate the cost of the telephone bill.

$$\text{Cost of bill} = \text{call charges} + \text{rental charge} = £45.60 + £21.86 = £67.46$$

Total with VAT = 117.5% × £67.46

Rate of VAT = $17\frac{1}{2}$% or 17.5%

100% + 17.5 % = 117.5%

= £79.2655
= £79.27 (to the nearest penny)

Check yourself

QUESTIONS

Q1 A store advertises a bedroom suite for £2300 + VAT at 17.5%. A mail order company advertises the same suite for £350 + 12 monthly payments of £198. Which method is the better buy? Give reasons for your answer.

Q2 A telephone bill includes £43.22 for local calls, £32.66 for long distance calls and a rental charge of £21.86. VAT at $17\frac{1}{2}$% is applied to the final amount.
Calculate the cost of the telephone bill.

Q3 A householder uses 926 units of electricity. The cost of electricity is 6.4p per unit. The quarterly charge is £10.96. VAT at 8% is charged on the final bill. What is the cost of the electricity bill?

Q4 During one quarter of 91 days, a householder uses 7650 kilowatt hours of gas. Gas is charged at 1.2p per kilowatt hour and the standing charge is 9.8p per day. VAT at 8% is charged on the final bill. What is the cost of the gas bill?

REMEMBER! Cover the answers if you want to.

ANSWERS

A1 Cost at the store = £2702.50 and by mail order = £2726.00 so the store is the better buy.

A2 £114.84 (to the nearest penny)

A3 £75.84 (to the nearest penny)

A4 £108.78 (to the nearest penny)

TUTORIALS

T1 *At the store: cost* = 117.5% × £2300 = £2702.50
By mail order: cost = £350 + 12 × £198 = £2726.00
The better buy is at the store as this is the lower price or, you might argue that mail order is the better buy as it allows you to pay the amount over a period of time (at a cost of £23.50*). You will gain credit for showing your working and arguing an appropriate response.*

T2 *Cost of bill* = *call charges* + *rental charge*
= £43.22 + £32.66 + £21.86
= £97.74

VAT = $17\frac{1}{2}$% *or* 17.5% *and* 100% + 17.5% = 117.5%

Total with VAT = 117.5% × £97.74
= £114.84 *(to the nearest penny)*

T3 *Cost of electricity* = 926 × 6.4*p*
= 5926.4*p* = £59.26
(to the nearest penny)

Cost of bill = *cost of electricity* + *quarterly charge*
= £59.26 + £10.96 = £70.22

VAT = 8% *and* 100% + 8% = 108%
Total with VAT = 108% × £70.22
= £75.84 *(to the nearest penny)*

T4 *Cost of gas* = 7650 × 1.2*p* = 9180*p* = £91.80

Standing charge
= *number of days* × *standing charge per day*
= 91 × 9.8*p* = 891.8*p* = £8.92 *(to the nearest penny)*

Cost of bill = *cost of gas* + *standing charge*
= £91.80 + £8.92 = £100.72

VAT = 8% *and* 100% + 8% = 108%
Total with VAT = 108% × £100.72
= £108.78 *(to the nearest penny)*

CHAPTER 21

EXAMINATION QUESTIONS 4

EXAM PRACTICE

Sample Student's Answers & Examiner's Comments

EXAMINER'S COMMENTS

1 *The candidate has realised that the average speed over the whole journey is not just the average of the two speeds given and each part of the journey needs to be treated separately.*

She finds the time using

$$time = \frac{distance}{speed}$$

and then the average speed using

$$speed = \frac{distance}{time}$$

and rounds off the answer to an appropriate degree of accuracy (3 s.f. or 2 s.f. would be most appropriate here).

The candidate would be well-advised to include the units in these time calculations.

2 *The candidate correctly uses the formulae for simple interest and compound interest (which is not usually given in the examination).*

An alternative would be to use the simple interest formula and add on the interest to the principal to work out the interest in the second year.

The compound interest is found by subtracting the principal from the amount.

He has failed to complete the question which asks for a calculation of the difference in interest earned. An answer of £370.80 – £360 = £10.80 would gain full marks.

1 A coach travels 45 miles at 60 mph on a motorway and 10 miles at 25 mph on country roads. What is the average speed over the whole of the journey?

Distance = 45 + 10 = 55 miles

45 miles at 60 mph = $\frac{45}{60}$ = 0.75

10 miles at 25 mph = $\frac{10}{25}$ = 0.4

Time = 0.75 + 0.4 = 1.15

Speed = $\frac{55}{1.15}$

= 47.82608696

= 47.8 mph (3 s.f.)

2 Calculate the difference in interest earned on £3000 over 2 years at:
(a) 6% p.a. simple interest (b) 6% p.a. compound interest.

(a) $I = \frac{3000 \times 6 \times 2}{100} = £360$

(b) $A = P \times \left(1 + \frac{R}{100}\right)^T$

$= 3000\left(1 + \frac{6}{100}\right)^2$

= £3370.80

I = £3370.80 – £3000

= £370.80

3 Leanne has an income of £318 per week. She has a tax allowance of £3765.

(a) What is her taxable income?

The rates of tax are: 20p in the £ on the first £3900 of taxable income
24p in the £ on any remaining taxable income.

(b) Find how much tax Leanne pays per month.

(In the style of SEG specimen paper 1998)

(a) Income = £318 X 52
= £16536

Taxable income = income − allowances
= £16536 − £3765
= £12771

(b) She pays 20p in the £ on the first £3900
= £780

She pays 24p in the £ on £8871
= £2129.04

Total tax = £780 + £2129.04
= £2909.04

EXAMINER'S COMMENTS

3 **a)**
The candidate has remembered to find the annual income.

The information given in the examination question should always be used even if this data is out of date due to budget changes.

b)
Again, the candidate has failed to complete the question which asks how much tax Leanne pays per month. An answer of £2909.04 ÷ 12 = £242.42 would gain full marks. It is always important to check that the question has been answered fully.

Questions to Answer

1 What is a speed of 100 kilometres per hour in metres per second?

2 The population of the Earth is approximately 5×10^9 and the Earth's surface area is approximately 4×10^{11} km^2. Calculate the approximate area, in km^2, per head of population. Give your answer in standard form.

3 A car gives a fuel consumption of 35 miles to the gallon on motorways and 26 miles to the gallon on other roads. A car user travels 80 miles on a motorway and 115 miles on other roads. Calculate the number of gallons needed for the journey.

4 Find the simple interest paid on £10 000 for 10 weeks at 6.85% p.a.

5 In how many years will an investment of £600 become £672.50 at 5% p.a. simple interest?

6 Ian invests £5000 at 7.4% interest per annum. Compound interest is paid every 6 months. What will be the value of his investment after one year?

CHAPTER 22

RATIONAL AND IRRATIONAL NUMBERS

NOTE

$\sqrt{a} \times \sqrt{b} = \sqrt{a \times b}$

$\frac{\sqrt{a}}{\sqrt{b}} = \sqrt{\frac{a}{b}}$

A rational number can be expressed in the form $\frac{p}{q}$ where p and q are integers.

RATIONAL AND IRRATIONAL NUMBERS

Examples of **rational** numbers include $\frac{1}{5}$, $0.\dot{3}$, 7, $\sqrt{9}$, $\sqrt[3]{64}$, etc.

Examples of **irrational** numbers include $\sqrt{2}$, $\sqrt{3}$, $\sqrt[3]{20}$, π, π^2, etc.

Irrational numbers involving square roots are also called **surds**. Surds can be multiplied and divided as follows.

Worked example

Work these out.

(a) $\sqrt{3} \times \sqrt{3}$ (b) $\sqrt{2} \times \sqrt{8}$ (c) $\frac{\sqrt{48}}{\sqrt{12}}$ (d) $\frac{\sqrt{30}}{\sqrt{6}}$

(a) $\sqrt{3} \times \sqrt{3} = \sqrt{3 \times 3} = \sqrt{9} = 3$ (b) $\sqrt{2} \times \sqrt{8} = \sqrt{2 \times 8} = \sqrt{16} = 4$

(c) $\frac{\sqrt{48}}{\sqrt{12}} = \sqrt{\frac{48}{12}} = \sqrt{4} = 2$ (d) $\frac{\sqrt{30}}{\sqrt{6}} = \sqrt{\frac{30}{6}} = \sqrt{5}$

Worked example

Simplify the following.

(a) $\sqrt{50}$ (b) $\sqrt{5} + \sqrt{45}$

(a) $\sqrt{50} = \sqrt{25 \times 2} = \sqrt{25} \times \sqrt{2} = 5 \times \sqrt{2}$
$= 5\sqrt{2}$ As $5 \times \sqrt{2}$ is usually written $5\sqrt{2}$.

(b) $\sqrt{5} + \sqrt{45} = \sqrt{5} + \sqrt{9 \times 5} = \sqrt{5} + \sqrt{9} \times \sqrt{5}$
$= \sqrt{5} + 3 \times \sqrt{5} = \sqrt{5} + 3\sqrt{5} = 4\sqrt{5}$

Check yourself

QUESTIONS

Q1 Which of these are rational numbers?

$3^{\frac{1}{2}}$ $(\sqrt{3})^2$ π^{-2} $\sqrt{5\frac{1}{4}}$ $\sqrt{6\frac{1}{4}}$

Write each of the rational numbers in the form $\frac{p}{q}$ where p and q are integers.

Q2 Simplify the following expressions leaving your answers in surd form.

(a) $\sqrt{5} \times \sqrt{15}$ (b) $\sqrt{5} + \sqrt{20}$

REMEMBER! Cover the answers if you want to.

ANSWERS

A1 $(\sqrt{3})^2$ and $\sqrt{6\frac{1}{4}}$ are rational numbers.

$(\sqrt{3})^2 = \frac{3}{1}$ and $\sqrt{6\frac{1}{4}} = \frac{5}{2}$

A2 (a) $5\sqrt{3}$ (b) $3\sqrt{5}$

TUTORIALS

T1 $(\sqrt{3})^2 = 3$ *since* $(\sqrt{3})^2 = 3 = \frac{3}{1}$ *in the form* $\frac{p}{q}$.

$\sqrt{6\frac{1}{4}} = \frac{5}{2}$ *since* $\sqrt{6\frac{1}{4}} = \sqrt{\frac{25}{4}} = \frac{\sqrt{25}}{\sqrt{4}} = \frac{5}{2}$

T2 *(a)* $\sqrt{5} \times \sqrt{15} = \sqrt{75} = \sqrt{25 \times 3} = \sqrt{25} \times \sqrt{3} = 5\sqrt{3}$

(b) $\sqrt{5} + \sqrt{20} = \sqrt{5} + \sqrt{4 \times 5} = \sqrt{5} + 2\sqrt{5} = 3\sqrt{5}$

CHAPTER 23

RECURRING DECIMALS

Recurring decimals are all rational numbers as they can be expressed as fractions.

RECURRING DECIMALS – NOTATION

0.166 666 666...	written $0.1\dot{6}$	$= \frac{1}{6}$
0.142 857 142 857...	written $0.\dot{1}4285\dot{7}$	$= \frac{1}{7}$
0.777 777 777...	written $0.\dot{7}$	$= \frac{7}{9}$
0.272 727 27...	written $0.\dot{2}\dot{7}$	$= \frac{3}{11}$

Worked example

Change $0.\dot{8}$ to a fraction.

Notice that	$10 \times 0.\dot{8} = 8.888\,888\,8...$	Multiplying both sides by 10.
and	$1 \times 0.\dot{8} = 0.888\,888\,8...$	
Subtracting:	$9 \times 0.\dot{8} = 8$	8.888 888 8... – 0.888 888 8...
and	$0.\dot{8} = \frac{8}{9}$	Dividing both sides by 9.

So $0.\dot{8} = \frac{8}{9}$

HINT

You should now check this by putting $\frac{8}{9}$ into your calculator.

Worked example

Convert $14.\dot{2}\dot{3}$ to a mixed number.

A mixed number consists of a whole number part and a fractional part. In this question you can split the number up and deal with the recurring decimal or else proceed as shown previously.

Notice that	$100 \times 14.\dot{2}\dot{3} = 1423.232\,323...$	Multiplying both sides by 100.
and	$1 \times 14.\dot{2}\dot{3} = 14.232\,323...$	
Subtracting:	$99 \times 14.\dot{2}\dot{3} = 1409$	1423.232 323... – 14.232 323...
and	$14.\dot{2}\dot{3} = \frac{1409}{99}$	Dividing both sides by 99.
	$= 14\frac{23}{99}$	Converting back to a mixed number.

So $14.\dot{2}\dot{3} = 14\frac{23}{99}$

Check yourself

QUESTIONS

Q1 Write a fraction which is equivalent to the recurring decimal $0.\dot{2}5\dot{3}$.

Q2 Change $0.8\dot{3}\dot{5}$ to a fraction.

Q3 Convert $0.1\dot{5}$ to a fraction.

(NEAB specimen paper 1998)

REMEMBER! Cover the answers if you want to.

ANSWERS

A1 $\dfrac{253}{999}$

A2 $\dfrac{827}{990}$

A3 $\dfrac{7}{45}$

TUTORIALS

T1 *Notice that* $1000 \times 0.\dot{2}5\dot{3} = 253.253\,253\ldots$ *Multiplying both sides by 1000.*

and $1 \times 0.\dot{2}5\dot{3} = 0.253\,253\ldots$

Subtracting: $999 \times 0.\dot{2}5\dot{3} = 253$

and $0.\dot{2}5\dot{3} = \dfrac{253}{999}$ *Dividing both sides by 999.*

T2 *Notice that* $100 \times 0.8\dot{3}\dot{5} = 83.535\,353\,5\ldots$ *Multiplying both sides by 100.*

and $1 \times 0.8\dot{3}\dot{5} = 0.835\,353\,5\ldots$

Subtracting: $99 \times 0.8\dot{3}\dot{5} = 82.7$

and $0.8\dot{3}\dot{5} = \dfrac{82.7}{99}$ *Dividing both sides by 99.*

$= \dfrac{827}{990}$ *Writing as a proper fraction.*

T3 *Notice that* $10 \times 0.1\dot{5} = 1.555\,555\,5\ldots$ *Multiplying both sides by 10.*

and $1 \times 0.1\dot{5} = 0.155\,555\,5\ldots$

Subtracting: $9 \times 0.1\dot{5} = 1.4$

and $0.1\dot{5} = \dfrac{1.4}{9}$ *Dividing both sides by 9.*

$= \dfrac{14}{90}$ *Writing as a proper fraction.*

$= \dfrac{7}{45}$ *In its lowest form.*

CHAPTER 24

FRACTIONAL INDICES

As was stated in Chapter 6, Positive, negative and zero indices, for any real numbers a, m and n:

$$a^m \times a^n = a^{m+n}$$

$$a^m \div a^n = a^{m-n}$$

$$a^{-m} = \frac{1}{a^m}$$

$$a^1 = a$$

$$a^0 = 1$$

Using these rules, you can also see that:

$$a^{\frac{1}{2}} \times a^{\frac{1}{2}} = a^{\frac{1}{2}+\frac{1}{2}} = a^1 = a$$

and $a^{\frac{1}{3}} \times a^{\frac{1}{3}} \times a^{\frac{1}{3}} = a^{\frac{1}{3}+\frac{1}{3}+\frac{1}{3}} = a^1 = a$

So that any number raised to the power $\frac{1}{2}$ means $\sqrt[2]{\ }$ or $\sqrt{\ }$ $\quad a^{\frac{1}{2}} = \sqrt{a}$

and any number raised to the power $\frac{1}{3}$ means $\sqrt[3]{\ }$ $\quad a^{\frac{1}{3}} = \sqrt[3]{a}$

Similarly $\quad a^{\frac{1}{n}} = \sqrt[n]{a}$

EVALUATING NUMBERS WITH FRACTIONAL INDICES

Worked example

Evaluate the following. (a) $49^{\frac{1}{2}}$ (b) $64^{\frac{1}{3}}$ (c) $3125^{\frac{1}{5}}$

(a) $49^{\frac{1}{2}} = \sqrt{49}$
$= 7$

(b) $64^{\frac{1}{3}} = \sqrt[3]{64}$
$= 4$

(c) $3125^{\frac{1}{5}} = \sqrt[5]{3125}$
$= 5$

HINT

Strictly, $\sqrt{49} = \pm 7$
as $7 \times 7 = 49$ and $^-7 \times {}^-7 = 49$
although it is usual to take the positive square root.

Worked example

Evaluate $81^{\frac{3}{4}}$.

$81^{\frac{3}{4}} = (81^{\frac{1}{4}})^3 = (3)^3$ As $81^{\frac{1}{4}} = 3$.
$= 27$

Alternatively, you can use:

$81^{\frac{3}{4}} = (81^3)^{\frac{1}{4}}$
$= (531\,441)^{\frac{1}{4}}$
$= 27$

although this method is rather longwinded.

Check yourself

QUESTIONS

Q1 Evaluate the following.

(a) $36^{\frac{1}{2}}$ (b) $343^{\frac{1}{3}}$ (c) $256^{\frac{1}{4}}$

Q2 Evaluate the following.

a) $125^{\frac{2}{3}}$ (b) $125^{-\frac{2}{3}}$

Q3 What is the value of n if $4^n = \frac{1}{2}$?

REMEMBER! Cover the answers if you want to.

ANSWERS

A1

(a) 6

(b) 7

(c) 4

A2

(a) 25

(b) $\frac{1}{25}$

A3 $n = -\frac{1}{2}$

TUTORIALS

T1

(*a*) $36^{\frac{1}{2}} = \sqrt{36} = 6$

(*b*) $343^{\frac{1}{3}} = \sqrt[3]{343} = 7$

(*c*) $256^{\frac{1}{4}} = \sqrt[4]{256} = 4$

T2

(*a*) $125^{\frac{2}{3}} = (125^{\frac{1}{3}})^2 = (5)^2 = 25$

(*b*) $125^{-\frac{2}{3}} = \frac{1}{125^{\frac{2}{3}}}$

$= \frac{1}{(125^{\frac{1}{3}})^2}$

$= \frac{1}{5^2}$

$= \frac{1}{25}$

T3 $4^{-\frac{1}{2}} = \frac{1}{2}$ so $n = -\frac{1}{2}$

CHAPTER 25

DIRECT AND INVERSE PROPORTION

With direct proportion as one variable increases, the other increases and as one decreases, the other one decreases.

With inverse proportion as one variable increases, the other decreases and as one decreases, the other one increases.

If y is proportional to x then you can write $y \propto x$ or else $y = kx$.

If y is inversely proportional to x then you can write $y \propto \frac{1}{x}$ or else $y = \frac{k}{x}$.

The value of k is a constant and is called the constant of proportionality.

WORKING WITH PROPORTION

HINT

The term 'varies with' is also used instead of 'is proportional to'.

Worked example

Write down the following statements using the $\propto$ sign and the constant of proportionality.

(a) y varies directly as x cubed.

(b) s is proportional to the square root of t.

(c) v varies inversely as the square of w.

(d) p is inversely proportional to the cube root of q.

(a) $y \propto x^3$ or
$y = kx^3$

(b) $s \propto \sqrt{t}$ or
$s = k\sqrt{t}$

(c) $v \propto \frac{1}{w^2}$ or
$v = \frac{k}{w^2}$

(d) $p \propto \frac{1}{\sqrt[3]{q}}$ or
$p = \frac{k}{\sqrt[3]{q}}$

Worked example

Given that a varies directly as the cube of b and $a = 4$ when $b = 2$, find the value of k and the value of a when $b = 3$.

If a varies directly as the cube of b then $a \propto b^3$ or $a = kb^3$.

Since we know that $a = 4$ when $b = 2$, then: $4 = k \times 2^3$

$4 = k \times 8$

so $k = \frac{1}{2}$

The equation is $a = \frac{1}{2}b^3$.

When $b = 3$ then $a = \frac{1}{2} \times 3^3$

$a = \frac{1}{2} \times 27$

$a = 13.5$

Check yourself

QUESTIONS

Q1 Given that T is proportional to the positive square root of W and $T = 36$ when $W = 16$:

(a) calculate T when W is 100
(b) calculate W when T is 18.

Q2 Given that V varies inversely as the cube of Y and $V = \frac{3}{8}$ when $Y = 2$, find the value of V when:

(a) $Y = 3$ (b) $Y = 10$.

REMEMBER! Cover the answers if you want to.

ANSWERS

A1
(a) $T = 90$
(b) $W = 4$

A2
(a) $V = \frac{1}{9}$
(b) $V = \frac{3}{1000}$ or 0.003 or 3×10^{-3}

TUTORIALS

T1 *If T is proportional to the positive square root of W then* $T \propto \sqrt{W}$ *and* $T = k\sqrt{W}$.
Since $T = 36$ *when* $W = 16$ *then* $36 = k\sqrt{16}$.
$36 = k \times 4$
i.e. $k = 9$
The equation is $T = 9\sqrt{W}$.

(a) *When* $W = 100$ *then:*
$$T = 9\sqrt{100} = 9 \times 10 = 90$$

(b) *When* $T = 18$ *then* $18 = 9\sqrt{W}$
$$\sqrt{W} = 2$$
$$W = 4 \text{ as } \sqrt{4} = 2$$

T2 *If V varies inversely as the cube of Y then* $V \propto \frac{1}{Y^3}$ *and* $V = \frac{k}{Y^3}$.

Since $V = \frac{3}{8}$ *when* $Y = 2$ *then:*

$$\frac{3}{8} = \frac{k}{2^8}$$

$$\frac{3}{8} = \frac{k}{8}$$

$k = 3$

The equation is $V = \frac{3}{Y^3}$.

(a) *When* $Y = 3$ *then:*

$$V = \frac{3}{3^3} = \frac{3}{27}$$

$$= \frac{1}{9}$$ *Cancelling down.*

(b) *When* $Y = 10$ *then:*

$$V = \frac{3}{10^3} = \frac{3}{1000}$$ *or* 0.003 *or* 3×10^{-3}

CHAPTER 26

UPPER AND LOWER BOUNDS

If a weight is given as 10 grams to the nearest gram, then the actual weight will lie in the interval 9.5 grams to 10.499 999... grams as all values in this interval will be rounded to 10 grams to the nearest gram. The weight 10.499 999... grams is usually written as 10.5 g although it is accepted that 10.5 g would be rounded to 11 g (to the nearest gram).

The value 9.5 g is called the lower bound as it is the lowest value which would be rounded to 10 g while 10.5 g is called the upper bound.

WORKING WITH BOUNDS

Worked example

A rectangle measures 10 cm by 6 cm, where each measurement is given to the nearest cm. Write down an interval approximation for the area of a rectangle.

The lower bound (minimum area) = 9.5 × 5.5 = 52.25 cm^2

The upper bound (maximum area) = 10.5 × 6.5 = 68.25 cm^2

The interval approximation = 52.25 cm^2 to 68.25 cm^2

Worked example

To the nearest whole number, the value of $p = 215$ and the value of $q = 5$. Calculate the maximum and minimum values of the following expressions.

(a) $p + q$ (b) $p - q$ (c) $p \times q$ (d) $p \div q$

$p_{min} = 214.5$ $p_{max} = 215.5$

$q_{min} = 4.5$ $q_{max} = 5.5$

(a) For $p + q$ maximum = 215.5 + 5.5 = 221
minimum = 214.5 + 4.5 = 219

(b) For $p - q$ maximum = 215.5 − 4.5 = 211
minimum = 214.5 − 5.5 = 209

(c) For $p \times q$ maximum = 215.5 × 5.5 = 1185.25
minimum = 214.5 × 4.5 = 965.25

(d) For $p \div q$ maximum = 215.5 ÷ 4.5 = 47.888 888 ...
minimum = 214.5 ÷ 5.5 = 39

HINT

To get the maximum value of $p - q$ we need to work out $p_{max} - q_{min}$ and to get the minimum value of $p - q$ we need to work out $p_{min} - q_{max}$.

HINT

To get the maximum value of $p \div q$ we need to work out $p_{max} \div q_{min}$ and to get the minimum value of $p \div q$ we need to work out $p_{min} \div q_{max}$.

Check yourself

QUESTIONS

Q1 Roger Bannister's time of 239.2 seconds is known to be correct to the nearest tenth of a second. What is the shortest time that it could actually be?

(SEG specimen paper 1998, part question)

Q2 A Ford Scorpio costs £18 700 and a Vauxhall Calibra costs £17 300, both prices being given to the nearest £100. What is the least possible difference in price between the two cars?

ANSWERS

A1 239.15 seconds

A2 £1300.01

NOTE

You must always take care when working with money or ages to identify lower and upper bounds.

TUTORIALS

T1 *If 239.2 seconds is correct to the nearest tenth of a second then the shortest time it could be is 239.15 seconds.*

T2 *As each price is given to the nearest £100 then:*

Ford Scorpio at £18 700

maximum = £18 749.99

Note that £18 750 *would be rounded up to* £18 800.

minimum = £18 650.00

Vauxhall Calibra at £17 300

maximum = £17 349.99

Note that £17 350 *would be rounded up to* £17 400.

minimum = £17 250.00

Least possible difference in price

$= S_{min} - C_{max}$

= £18 650 – £17 349.99

= £1300.01

CHAPTER 27

EXAMINATION QUESTIONS 5

EXAM PRACTICE

Sample Student's Answers & Examiner's Comments

1 (a) Which of the following numbers are rational?

$1 + \sqrt{2}$ π^2 $3^0 + 3^{-1} + 3^{-2}$

(b) When p and q are two different irrational numbers, $p \times q$ can be rational. Write down one example to show this.

(SEG specimen paper 1998)

(a) $1 + \sqrt{2}$ no

π^2 no

$3^0 + 3^{-1} + 3^{-2} = 1 + \frac{1}{3} + \frac{1}{9}$

$= \frac{9}{9} + \frac{3}{9} + \frac{1}{9}$

$= \frac{13}{9}$

yes

(b) $\sqrt{2} \times \sqrt{8} = 4$

$\sqrt{3} \times \sqrt{12} = 6$

2 Change $0.8\dot{3}\dot{5}$ to a fraction.

$100x = 83.535\,353\,5\ldots$

$1x = 0.835\,353\,5\ldots$

$99x = 82.7$

$x = \frac{82.7}{99}$

EXAMINER'S COMMENTS

1 **a)**

The candidate has appreciated that $1 + \sqrt{2}$ and π^2 are irrational numbers and has converted the third expression into a series of rational numbers.

It was not necessary to complete the addition as the sum of rational numbers is always rational. The candidate would also be advised to make her answers more clear so that the examiner is in no doubt which numbers they consider to be rational.

b)

There are a number of examples to illustrate this fact although it is important that p and q are different irrational numbers and that $p \times q$ is the square root of a square number. The candidate is advised to give only one example for the answer as any incorrect responses will mean no marks being awarded.

2 *The candidate has remembered how to deal with recurring decimals and has used the letter x to find the fraction.*

He has correctly divided both sides of the equation by 99 but has forgotten to write the result as a proper fraction.

The equivalent fraction $\frac{827}{990}$ will gain full marks.

EXAMINER'S COMMENTS

3 a)
The candidate would be better to work out $(1024^{\frac{1}{5}})^3$ although the correct answer will get full marks.

b)
She has realised that $a^{-n} = \frac{1}{a^n}$.

c)
She has used the laws of indices to solve the problem.

In general $(a^m)^n = a^{m \times n}$

4 *The candidate has used the fact that $b \propto s^2$ to write $b = ks^2$.*

To find k the candidate used the fact that $s = 50$ when $b = 125$.

He has correctly found the formula linking the braking distance and the speed.

a)
The inclusion of units serves as a useful check.

b)
Again, the inclusion of units serves as a useful check.

3 Write down the values of:

(a) $1024^{\frac{3}{5}}$ (b) $81^{-\frac{3}{4}}$ (c) $(2^{\frac{1}{3}})^6$.

(a) $1024^{\frac{3}{5}} = (1024^3)^{\frac{1}{5}}$

$= 107\,374\,182\,4^{\frac{1}{5}}$

$= 64$

(b) $81^{-\frac{3}{4}} = \dfrac{1}{81^{\frac{3}{4}}}$

$81^{\frac{3}{4}} = (81^{\frac{1}{4}})^3$

$= 3^3$

$= 27$

$81^{-\frac{3}{4}} = \dfrac{1}{81^{\frac{3}{4}}} = \dfrac{1}{27}$

(c) $(2^{\frac{1}{3}})^6 = 2^{\frac{1}{3}} \times 2^{\frac{1}{3}} \times 2^{\frac{1}{3}} \times 2^{\frac{1}{3}} \times 2^{\frac{1}{3}} \times 2^{\frac{1}{3}}$

$= 2^{\frac{1}{3}+\frac{1}{3}+\frac{1}{3}+\frac{1}{3}+\frac{1}{3}+\frac{1}{3}}$

$= 2^2$

$= 4$

4 The braking distance, b, of a car is proportional to the square of the speed, s. If the braking distance for a car travelling at 50 mph is 125 feet, find:

(a) the braking distance when the speed is 70 mph

(b) the speed when the braking distance is 45 feet.

$b = ks^2$

$125 = k \times 50^2$

$125 = k \times 2500$

$k = \dfrac{125}{2500}$

$k = 0.05$

So $b = 0.05s^2$

(a) When $s = 70$, $b = 0.05 \times 70^2$

$b = 245$

Braking distance = 245 feet

(b) When $b = 45$, $45 = 0.05 \times s^2$

$s^2 = 900$

$s = 30$

Speed = 30 mph

Questions to Answer

1 Simplify the following expressions, leaving your answers in surd form.

(a) $\sqrt{12} + \sqrt{3}$
(b) $\sqrt{12} \times \sqrt{6}$

(SEG modular syllabus, specimen paper 1998)

2 Show that $0.\dot{2}3\dot{4}$ is a rational number.

3 Write down a fraction which is equivalent to 0.033 636 363 ...

(SEG specimen paper 1998)

4 Write down the following in order of size with the smallest first.

$243^{\frac{3}{5}}$ $\quad 32^{\frac{4}{5}}$ $\quad 5^{-1}$ $\quad 20^{0}$ $\quad (^{-}2)^{3}$ $\quad 2^{-2}$

5 The value of a car, £v, is inversely proportional to its age, a years. After 1 year a car has a value of £7000. Find:
(a) the value of the car after $3\frac{1}{2}$ years
(b) the age of the car when it is worth £2500.

6 Given that R is inversely proportional to the square root of T and $R = 1$ when $T = 4$, find R when $T = 9$ and when $T = 5$.

7 Twenty video recorders are packed in a large container. Each video recorder weighs 12.6 kg to the nearest 0.1 kg.
(a) Calculate the lower bound for the total weight of the video recorders.
(b) Calculate the difference between the upper and lower bounds for the total weight of the video recorders.

(SEG modular syllabus, specimen paper 1998)

8 A train travels 42 miles (to the nearest mile) in a time of 35 minutes (to the nearest minute). Find the maximum and minimum speed of the train in miles per hour, correct to 3 significant figures.

9 A company makes rectangular sheets of tinplate for use in cans.

Dimensions of rectangular sheets		
Thickness	Length	Width
0.15 mm	830 mm	635 mm

The length and width are given to the nearest mm and the thickness is given to the nearest 0.01 mm.

Calculate the percentage saving in volume to the company if it produces sheets to the minimum dimensions rather than the maximum dimensions.

(SEG specimen paper 1998)

CHAPTER 28

PATTERNS AND SEQUENCES

A sequence is a set of numbers which follow a particular rule. The word 'term' is often used to describe the numbers in the sequence. In the following sequence 7 is called the first term, 9 is called the second term, 11 is called the third term, etc.

7, 9, 11, 13, 15, 17, ...

The nth term is often used to denote the value of any term in the sequence.

If you are given a sequence in which the nth term is $2n + 5$ then:

	first term (where $n = 1$) gives	$2 \times 1 + 5 = 7$
	second term (where $n = 2$) gives	$2 \times 2 + 5 = 9$
	third term (where $n = 3$) gives	$2 \times 3 + 5 = 11$
Similarly:	50th term (where $n = 50$) gives	$2 \times 50 + 5 = 105$
and	1000th term (where $n = 1000$) gives	$2 \times 1000 + 5 = 2005$

SEQUENCE RULES

Most number sequences involve adding/subtracting or multiplying/dividing in the rule for finding one term from the one before it. Once you have found the rule then you can use it to find subsequent terms.

Worked example

Find the sequence rule and find the next three terms of these sequences.

(a) 1, 6, 11, 16, 21, ... (b) 3, 6, 12, 24, 48, ...

(c) 243, 81, 27, 9, 3, ... (d) 1,4, 13, 40, 121, ...

(a) 1 (+5) 6 (+5) 11 (+5) 16 (+5) 21 ...

The rule for moving from term to term is + 5.
The next three terms are 26, 31 and 36

(b) 3 (×2) 6 (×2) 12 (×2) 24 (×2) 48 ...

The rule for moving from term to term is × 2.
The next three terms are 96, 192 and 384.

(c) 243 (÷ 3) 81 (÷ 3) 27 (÷ 3) 9 (÷ 3) 3 ...

The rule for moving from term to term is ÷ 3.
The next three terms are 1, $\frac{1}{3}$ and $\frac{1}{9}$.

(d) 1 (× 3 + 1) 4 (× 3 + 1) 13 (× 3 + 1) 40 (× 3 + 1) 121 ...

The rule for moving from term to term is × 3 + 1.
The next three terms are 364, 1093 and 3280

FINDING TERMS

You can use the method shown above for finding subsequent terms of a sequence but it would be rather time-consuming if you needed to find the 100th term or the 500th term etc. The following method can be used to find the nth term of a sequence where the difference between the terms is a constant or else the difference *between the differences* is a constant.

Worked example

Find the *n*th term of the following sequences.

(a) 3, 7, 11, 15, 19, ... (b) 2, 6, 12, 20, 30, ...

(a) 3 7 11 15 19 ...

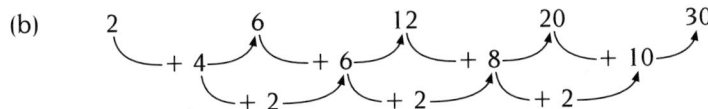

+4 +4 +4 +4 ... Difference between terms = 4

As the differences are constant then a linear rule can be applied to the terms to find the *n*th term. You can construct a table showing the sequence and term numbers.

Term number	1	2	3	4	5
Sequence	3	7	11	15	19
Try 4 × (term number)	4	8	12	16	20
Difference	1	1	1	1	1
The difference is always equal to 1.					
So try 4 × (term number) – 1	3	7	11	15	19
Sequence	3	7	11	15	19
Difference	0	0	0	0	0

So the *n*th term is $4 \times n - 1$ or $4n - 1$.

(b) 2 6 12 20 30 ...

+ 4 + 6 + 8 + 10 ...

+ 2 + 2 + 2 ... Difference of differences = 2

As the difference of differences is constant then a quadratic rule can be applied to the terms to find the *n*th term. You can construct a table showing the sequence and term numbers.

Term number	1	2	3	4	5
Sequence	2	6	12	20	30
Try (term number) × (term number)	1	4	9	16	25
Difference	1	2	3	4	5
The difference is always equal to the term number.					
So try (term number) × (term number) + (term number)	2	6	12	20	30
Sequence	2	6	12	20	30
Difference	0	0	0	0	0

So the *n*th term is $n \times n + n$ or $n^2 + n$.

SPECIAL SEQUENCES

You should be able to recognise the following special sequences of numbers.

1, 4, 9, 16, 25, ... square numbers

1, 3, 6, 10, 15, ... triangle numbers

1, 8, 27, 64, 125, ... cube numbers

2, 3, 5, 7, 11, 13, 17, ... prime numbers

NOTE

See Chapter 1, The number system, for further information on these sequences of numbers.

Another sequence which you should also know is the Fibonacci sequence
Each term after the second is found by adding up the two previous terms.

1, 1, 2, 3, 5, 8, 13, 21, ...

So 3rd term = 1st term + 2nd term
  4th term = 2nd term + 3rd term etc.

Check yourself

QUESTIONS

Q1 Write down the first five terms of a sequence where the nth term is given as:

(a) $2n + 1$ (b) $3n - 8$

(c) $n^2 - 3n$ (d) $\frac{n}{n+1}$

Q2 Find the sequence rule and write down the next three terms of each of these sequences.

(a) 4, 2.5, 1, $^{-}0.5$, $^{-}2$, $^{-}3.5$, ...
(b) 4, $^{-}8$, 16, $^{-}32$, 64, ...
(c) 2, 5, 11, 23, 47, ...
(d) 2, 3, 5, 9, 17, ...

Q3 Write down the nth term of each of the following sequences.

(a) 3, 8, 13, 18, 23, ...
(b) 2, 5, 10, 17, 26, ...

Q4 The nth term of a sequence is given by $\frac{4n}{n+1}$.

(a) Write down the value of the 10th term.
(b) Write down the value of the 100th term.
(c) What happens to the values of the sequence as n increases in size?

REMEMBER! Cover the answers if you want to.

ANSWERS

A1
(a) 3, 5, 7, 9, 11, ...
(b) $^{-}5$, $^{-}2$, 1, 4, 7, ...
(c) $^{-}2$, $^{-}2$, 0, 4, 10, ...
(d) $\frac{1}{2}, \frac{2}{3}, \frac{3}{4}, \frac{4}{5}, \frac{5}{6}, \ldots$

A2
(a) Subtract 1.5.
Next three terms: $^{-}5$, $^{-}6.5$, $^{-}8$
(b) Multiply by $^{-}2$.
Next three terms: $^{-}128$, 256, $^{-}512$
(c) Multiply by 2 and add 1.
Next three terms: 95, 191, 383
(d) Multiply by 2 and subtract 1.
Next three terms: 33, 65, 129

TUTORIAL

T1
(a) $2 \times 1 + 1 = 3$, $2 \times 2 + 1 = 5$, $2 \times 3 + 1 = 7$, *etc.*
(b) $3 \times 1 - 8 = {}^{-}5$, $3 \times 2 - 8 = {}^{-}2$, $3 \times 3 - 8 = 1$, *etc.*
(c) $1 \times 1 - 3 \times 1 = {}^{-}2$, $2 \times 2 - 3 \times 2 = {}^{-}2$, $3 \times 3 - 3 \times 3 = 0$, *etc.*
(d) $\frac{1}{1+1} = \frac{1}{2}$, $\frac{2}{2+1} = \frac{2}{3}$, $\frac{3}{3+1} = \frac{3}{4}$, *etc.*

T2
(a) 4 → 2.5 → 1 → $^{-}0.5$ → $^{-}2$ → $^{-}3.5$ ...
(each step -1.5)

(b) 4 → $^{-}8$ → 16 → $^{-}32$ → 64 ...
(each step $\times {}^{-}2$) ...

Any series alternating between positive and negative numbers is invariably being multiplied or divided by a negative number.

(c) 2 → 5 → 11 → 23 → 47 ...
(each step $\times 2 + 1$) ...

(d) 2 → 3 → 5 → 9 → 17 ...
(each step $\times 2 - 1$)

ANSWERS

A3

(a) nth term = $5n - 2$
(b) nth term = $n^2 + 1$

A4

(a) 10th term = $\frac{40}{11}$

(b) 100th term = $\frac{400}{101}$

(c) The value gets closer and closer to 4.

TUTORIAL

T3

(a) *As the differences are constant then a linear rule can be applied to the terms to find the nth term. Each term is the previous term + 5 so try 5n first.*

Term	1	2	3	4	5
Sequence	3	8	13	18	23
Try 5 × *(term number)*	5	10	15	20	25
Sequence	3	8	13	18	23
Difference	2	2	2	2	2
The difference is always 2.					
So try 5 × *(term number)* – 2	3	8	13	18	23
Sequence	3	8	13	18	23
Difference	0	0	0	0	0

So the nth term is $5 \times n - 2$ *or* $5n - 2$.

(b) *As the differences of differences are constant then a quadratic rule can be applied to the terms to find the nth term.*

Term	1	2	3	4	5
Sequence	2	5	10	17	26
Try (term number) × *(term number)*	1	4	9	16	25
Sequence	2	5	10	17	26
Difference	1	1	1	1	1
The difference is always 1.					
So try (term number) × *(term number)* + 1	2	5	10	17	26
Sequence	2	5	10	17	26
Difference	0	0	0	0	0

So the nth term is $n \times n + 1$ *or* $n^2 + 1$.

T4

(a) *10th term* $= \frac{4 \times 10}{10 + 1}$

$= \frac{40}{11}$

(b) *100th term* $= \frac{4 \times 100}{100 + 1}$

$= \frac{400}{101}$

(c) *Trying out values of n as n increases in size, you should notice that the value gets closer and closer to 4.*

CHAPTER 29

SUBSTITUTION

Substitution means that you replace the letters in a formula or expression by the given numbers.

NOTE

$3x$ means $3 \times x$, $4y$ means $4 \times y$ and $5z$ means $5 \times z$.

Worked example

Find the value of:

(a) $3x + 4y - 5z$ where $x = 3$, $y = 7$ and $z = {}^{-}4$

(b) $3ab + ac^2 - \frac{b}{c}$ where $a = 2$, $b = {}^{-}10$ and $c = 5$.

(a) Substituting $x = 3$, $y = 7$ and $z = {}^{-}4$ in $3x + 4y - 5z$:

$$\begin{aligned} 3x + 4y - 5z &= 3 \times 3 + 4 \times 7 - 5 \times {}^{-}4 \\ &= 9 + 28 - {}^{-}20 \quad \text{Remember that } -(-) = +. \\ &= 9 + 28 + 20 \\ &= 57 \end{aligned}$$

NOTE

$ac^2 = a \times c \times c$ – Remember that only the c is squared.

(b) Substituting $a = 2$, $b = {}^{-}10$ and $c = 5$ in $3ab + ac^2 - \frac{b}{c}$:

$$\begin{aligned} 3ab + ac^2 - \frac{b}{c} &= 3 \times 2 \times {}^{-}10 + 2 \times 5 \times 5 - \frac{{}^{-}10}{5} \\ &= {}^{-}60 + 50 - {}^{-}2 \quad \text{Remember that } -(-) = +. \\ &= {}^{-}60 + 50 + 2 \\ &= {}^{-}8 \end{aligned}$$

Check yourself

QUESTIONS

Q1 The formula for finding the angle sum of an n-sided polygon is $(2n - 4) \times 90°$.

Use the formula to find the angle sum of:

(a) a triangle (b) a quadrilateral
(c) a 15-sided polygon.

Q2 The surface area, A, of a closed cone is given by $A = \pi rl + \pi r^2$ where r is the radius of the base and l is the slant height.

Find the surface area of a cone with radius 3 cm and slant height 8 cm.

Q3 Given that $\frac{1}{u} = \frac{1}{f} - \frac{1}{v}$, find the value of u when:

(a) $f = 4$ and $v = 8$
(b) $f = 3$ and $v = 5$.

REMEMBER! Cover the answers if you want to.

ANSWERS

A1

(a) 180°
(b) 360°
(c) 2340°

A2

104 cm² (3 s.f.)

A3

(a) $u = 8$
(b) $u = 7.5$ or $7\frac{1}{2}$

TUTORIAL

T1

(*a*) *Substituting $n = 3$ (as a triangle has 3 sides) in the equation:*
angle sum $= (2 \times 3 - 4) \times 90° = 180°$.

(*b*) *Substituting $n = 4$ (as a quadrilateral has 4 sides) in the equation:*
angle sum $= (2 \times 4 - 4) \times 90° = 360°$.

(*c*) *Substituting $n = 15$ in the equation:*
angle sum $= (2 \times 15 - 4) \times 90° = 2340°$.

T2

Substituting $r = 3$ (as the radius is 3 cm) and $l = 8$ (as the slant height is 8 cm) in the equation:

$A = \pi rl + \pi r^2$

$A = \pi \times 3 \times 8 + \pi \times 3 \times 3$ *As $r^2 = r \times r$*

$A = \pi \times 24 + \pi \times 9$
Using the fact that
$\pi \times 24 + \pi \times 9 = \pi \times (24 + 9) = \pi \times 33$

$A = 103.672\,56$

$A = 104\,cm^2$ (3 s.f.)
Rounding to an appropriate degree of accuracy and remembering to include the units of area.

T3

(*a*) *Substituting $f = 4$ and $v = 8$ in the formula*
$\frac{1}{u} = \frac{1}{f} - \frac{1}{v}$:

$\frac{1}{u} = \frac{1}{4} - \frac{1}{8}$

$\frac{1}{u} = \frac{1}{8}$ As $\frac{1}{4} - \frac{1}{8} = \frac{1}{8}$

$u = 8$

(*b*) *Substituting $f = 3$ and $v = 5$ in the formula*
$\frac{1}{u} = \frac{1}{f} - \frac{1}{v}$:

$\frac{1}{u} = \frac{1}{3} - \frac{1}{5}$

$\frac{1}{u} = \frac{2}{15}$ As $\frac{1}{3} - \frac{1}{5} = \frac{5}{15} - \frac{3}{15} = \frac{2}{15}$
– see Chapter 9, Fractions.

$u = \frac{15}{2}$ *Reciprocating both sides.*

$u = 7.5$ *or* $7\frac{1}{2}$

ALGEBRA

CHAPTER 30

SIMPLIFYING EXPRESSIONS

An algebraic expression is a collection of algebraic terms along with their + and – signs.

LIKE TERMS

Like terms are numerical multiples of the same algebraic quantity. For example, $3x$, ^-5x, $\frac{1}{2}x$ and $0.55x$ are all like terms because they are all multiples of the same algebraic quantity i.e. x.

Worked example

Collect together the like terms from the following list.

x	$3x$	y	^-6x	xy^2
$3ab$	^-3y	xy	^-4xy	$\frac{1}{3}y$
x^2y	$7ab$	^-2ba	yx^2	$^-\frac{3}{4}xy^2$

Like terms are numerical multiples of the same algebraic quantity.

x, $3x$, ^-6x	are all terms in x	
y, ^-3y, $\frac{1}{3}y$	are all terms in y	
xy^2, $^-\frac{3}{4}xy^2$	are all terms in xy^2	
$3ab$, $7ab$, ^-2ba	are all terms in ab	where $^-2ba = {}^-2ab$
xy, ^-4xy	are all terms in xy	
x^2y, yx^2	are all terms in x^2y	where $yx^2 = x^2y$

NOTE

In algebraic terms involving more than one letter, it is useful to write them in alphabetical order, so that $3ba$ is written $3ab$ and $14zxy$ is written $14xyz$ etc.

ADDING, SUBTRACTING, MULTIPLYING AND DIVIDING

You can add or subtract like terms. The process of adding and subtracting like terms in an expression or equation is called **simplifying**.

Worked example

Simplify the following expressions.

(a) $3x + 2y - 7z + 4x - 3y$

(b) $4p + 3pq - 3p + 8qp - 2r$

(a) $3x + 2y - 7z + 4x - 3y$

$= 3x + 4x + 2y - 3y - 7z$ Putting like terms together, *along with the signs in front of them.*

$= 7x - 1y - 7z$ Adding and subtracting like terms.

$= 7x - y - 7z$ Rewriting $1y$ as y.

(b) $4p + 3pq - 3p + 8qp - 2r$

$= 4p - 3p + 3pq + 8qp - 2r$ Putting like terms together, *along with the signs in front of them.*

$= 4p - 3p + 3pq + 8pq - 2r$ Writing $8qp$ as $8pq$ since $p \times q$ is the same as $q \times p$.

$= p + 11pq - 2r$

You can also multiply or divide terms and this process is also called simplifying. When dividing terms, they can be simplified by cancelling.

Worked example

Simplify the following expressions.

(a) $3f \times 4g$ (b) $8pq^2 \div 2pq$

(a) $3f \times 4g = 3 \times f \times 4 \times g$

$= 12 \times f \times g$ Multiplying $3 \times 4 = 12$.

$= 12fg$ Rewriting without the × signs.

(b) $8pq^2 \div 2pq = \dfrac{8 \times p \times q \times q}{2 \times p \times q}$ Writing as a fraction with $q^2 = q \times q$.

$= \dfrac{{}^4\cancel{8} \times \cancel{p} \times q \times \cancel{q}}{{}^1\cancel{2} \times \cancel{p} \times \cancel{q}}$ Cancelling down.

$= 4q$ Rewriting without the × signs.

Check yourself

QUESTIONS

Q1 Simplify the following expressions.

(a) $3a + 6b - 2a - 5c$
(b) $5x + 7y - 3xy + 2x + 2yx$
(c) $x^3 + 3x^2 - 4x - 9x^2 + 7x - 2$

Q2 Simplify the following expressions.

(a) $3x \times 4y \times 2z$
(b) $5a \times 2a^2$
(c) $(4abc)^2 \times a^2b$
(d) $8mn^3 \div 4n$

REMEMBER! Cover the answers if you want to.

ANSWERS

A1

(a) $a + 6b - 5c$
(b) $7x + 7y - xy$
(c) $x^3 - 6x^2 + 3x - 2$

A2

(a) $24xyz$
(b) $10a^3$
(c) $16a^4b^3c^2$
(d) $2mn^2$

TUTORIAL

T1

(a) $3a + 6b - 2a - 5c = 3a - 2a + 6b - 5c$
Collecting like terms together along with their signs.
$= a + 6b - 5c$

(b) $5x + 7y - 3xy + 2x + 2yx$
$= 5x + 2x + 7y - 3xy + 2xy$
Collecting like terms and rewriting 2yx as 2xy.
$= 7x + 7y - xy$

(c) $x^3 + 3x^2 - 4x - 9x^2 + 7x - 2$
$= x^3 - 6x^2 + 3x - 2$ *Collecting like terms.*

T2

(a) $3x \times 4y \times 2z = 3 \times 4 \times 2 \times x \times y \times z$
$= 24xyz$

(b) $5a \times 2a^2 = 5 \times 2 \times a \times a^2$
$= 10 \times a \times a \times a$ As $a^2 = a \times a$
$= 10a^3$

(c) $(4abc)^2 \times a^2b = 4abc \times 4abc \times a^2b$
$= 16a^2b^2c^2 \times a^2b$
Multiplying $4abc \times 4abc$.
$= 16a^4b^3c^2$

(d) $8mn^3 \div 4n = \dfrac{8 \times m \times n \times n \times n}{4 \times n}$

$= \dfrac{{}^2\cancel{8} \times m \times \cancel{n} \times n \times n}{{}^1\cancel{4} \times \cancel{n}} = 2mn^2$

CHAPTER 31

EXPANDING AND FACTORISING

Brackets are used to group algebraic terms. The process of removing brackets from an expression (or an equation) is called expanding and the process of rewriting an expression (or an equation) so that it includes terms in brackets is called factorising.

EXPANDING BRACKETS

When expanding brackets in an expression, all the terms inside the brackets must be multiplied by the term just before the bracket.

Worked example

Expand the following expressions.

(a) $3(5a - 2b)$ (b) $^-7(a + 2b - 3c)$

(a) $3(5a - 2b) = 3 \times 5a + 3 \times {^-2b} = 15a - 6b$

(b) $^-7(a + 2b - 3c) = {^-7} \times a - 7 \times 2b - 7 \times {^-3c}$

$= {^-7a} - 14b + 21c$ Remembering that $^-7 \times {^-3c} = {^+21c}$.

Worked example

Expand and simplify the following expressions.

(a) $5(p + 2q) + 2(p - 5q)$ (b) $4a(b - c) - 3b(a - c)$

(a) $5(p + 2q) + 2(p - 5q) = 5 \times p + 5 \times 2q + 2 \times p + 2 \times -5q$

$= 5p + 10q + 2p - 10q = 7p + 0q = 7p$

(b) $4a(b - c) - 3b(a - c) = 4a \times b + 4a \times {^-c} - 3b \times a - 3b \times {^-c}$

The term outside the second bracket is ^-3b.

$= 4ab - 4ac - 3ab + 3bc$

Writing $3ba$ as $3ab$ and remembering that $^-3b \times {^-c} = {^+3bc}$.

$= 1ab - 4ac + 3bc$

$= ab - 4ac + 3bc$ As $1ab$ is usually written ab.

HINT

Care must be taken when expanding brackets to ensure that every term in the bracket is multiplied by the term outside the bracket.

FACTORISING EXPRESSIONS INTO BRACKETS

To factorise an expression, you need to look for terms that have **common factors**. Then you rewrite the expression with the factors outside brackets. Remember that common factors of two (or more) terms are factors that appear in both (or all) of the terms.

Worked example

Factorise the following expressions.

(a) $5a - 15$ (b) $9xy + 33yz$ (c) $pq^3 - p^2q$

To factorise these expressions, write them inside brackets, with common factors taken outside the brackets.

(a) $5a - 15 = 5(a - 3)$ 5 is a common factor, as $5a = 5 \times a$ and $15 = 5 \times 3$

(b) $9xy + 33yz = 3y(3x + 11z)$ $3y$ is a common factor as $9xy = 3y \times 3x$ and $33yz = 3y \times 11z$

(c) $pq^3 - p^2q = pq(q^2 - p)$ pq is a common factor as $pq^3 = pq \times q^2$ and $p^2q = pq \times q$

HINT

Always check the contents of the brackets to ensure that there are no further common factors. It is also a good idea to check your work by expanding the brackets to make sure you get back to the original expression.

Check yourself

QUESTIONS

Q1 Multiply out $3m(5 - 2m)$.

Q2 Expand and simplify these expressions.

(a) $5x - (3y + 4x)$
(b) $5(a + b - 2c) - 2(a - 2b + 3c)$
(c) $a(2a + b) - 2b(a - b^2)$

Q3 Factorise these expressions completely.

(a) $4x^2 - 6x$
(b) $2lw + 2wh + 2hl$
(c) $5x^2y - 10xy^2$

REMEMBER! Cover the answers if you want to.

ANSWERS

A1 $15m - 6m^2$

A2
(a) $x - 3y$
(b) $3a + 9b - 16c$
(c) $2a^2 - ab + 2b^3$

A3
(a) $2x(2x - 3)$
(b) $2(lw + wh + hl)$
(c) $5xy(x - 2y)$

TUTORIAL

T1 $3m(5 - 2m) = 3m \times 5 + 3m \times {}^{-}2m$
$= 15m - 6m^2$ *As* $m \times m = m^2$

T2
(a) $5x - (3y + 4x) = 5x - 1 \times 3y - 1 \times 4x$
The – outside the bracket is taken as $^{-}1$.
$= 5x - 3y - 4x$ *Multiplying out.*
$= x - 3y$ *Collecting like terms together.*

(b) $5(a + b - 2c) - 2(a - 2b + 3c)$
$= 5 \times a + 5 \times b + 5 \times {}^{-}2c - 2 \times a$
$- 2 \times {}^{-}2b - 2 \times 3c$
$= 5a + 5b - 10c - 2a + 4b - 6c$
$= 3a + 9b - 16c$ *Collecting like terms together.*

(c) $a(2a + b) - 2b(a - b^2)$
$= a \times 2a + a \times b - 2b \times a - 2b \times {}^{-}b^2$
$= 2a^2 + ab - 2ab + 2b^3$ $a \times a = a^2$
ba is the same as ab
$^{-}2b \times {}^{-}b^2 = 2b^3$
$= 2a^2 - ab + 2b^3$ *Collecting like terms together.*

T3
(a) $4x^2 - 6x = 2x(2x - 3)$ *2x is a common factor, as* $4x^2 = 2x \times 2x$ *and* $6x = 2x \times 3$.
(b) $2lw + 2wh + 2hl = 2(lw + wh + hl)$
2 is the only factor common to all three terms.
(c) $5x^2y - 10xy^2 = 5xy(x - 2y)$
5xy is a common factor as $5x^2y = 5xy \times x$ *and* $10xy^2 = 5xy \times 2y$.

BINOMIAL EXPRESSIONS

A binomial expression consists of two terms such as $(a + b)$ or $(5x - 2z)$.

To expand the product of two binomial expressions, you must multiply each term in the first expression by each term in the second expression.

$$(a + b)(c + d) = a \times (c + d) + b \times (c + d)$$
$$= a \times c + a \times d + b \times c + b \times d$$
$$= ac + ad + bc + bd$$

NOTE

The word FOIL can be used to remind you how to expand these binomial expressions where
F = First
O = Outsides
I = Insides
L = Last.

		Product
F = First	$(a + b)(c + d)$	$a \times c$
O = Outsides	$(a + b)(c + d)$	$a \times d$
I = Insides	$(a + b)(c + d)$	$b \times c$
L = Last	$(a + b)(c + d)$	$b \times d$

Worked example

Expand these expressions.

(a) $(x + 3)(x + 5)$ (b) $(3x - 1)(4x - 5)$

(a) $(x + 3)(x + 5) = x \times x + x \times 5 + 3 \times x + 3 \times 5$
$= x^2 + 5x + 3x + 15$
$= x^2 + 8x + 15$

(b) $(3x - 1)(4x - 5) = 3x \times 4x + 3x \times {}^-5 + {}^-1 \times 4x + {}^-1 \times {}^-5$
$= 12x^2 - 15x - 4x + 5$ Remembering that ${}^-1 \times {}^-5 = {}^+5$.
$= 12x^2 - 19x + 5$

You can use the reverse process to write a quadratic as a product of brackets, as shown in the following worked example.

Worked example

Factorise $x^2 + 5x + 6$.

You know that $x^2 + 5x + 6 = (x \quad)(x \quad)$ Since $x \times x = x^2$ as required.

Look for pairs of numbers that multiply together to give ${}^+6$.

Possibilities include	${}^+1 \times {}^+6$	$(x + 1)(x + 6)$	$= x^2 + 6x + 1x + 6 = x^2 + 7x + 6$
	${}^-1 \times {}^-6$	$(x - 1)(x - 6)$	$= x^2 - 6x - 1x + 6 = x^2 - 7x + 6$
	${}^+2 \times {}^+3$	$(x + 2)(x + 3)$	$= x^2 + 2x + 3x + 6 = x^2 + 5x + 6$
	${}^-2 \times {}^-3$	$(x - 2)(x - 3)$	$= x^2 - 2x - 3x + 6 = x^2 - 5x + 6$

and the correct solution is:

$(x + 2)(x + 3) = x^2 + 2x + 3x + 6 = x^2 + 5x + 6$.

An alternative method is to look at pairs of numbers that multiply together to give ${}^+6$ (i.e. with product ${}^+6$) and add to give ${}^+5$ (i.e. with sum ${}^+5$).

Numbers	Product	Sum	
${}^+1$ and ${}^+6$	${}^+6$	${}^+7$	✗
${}^-1$ and ${}^-6$	${}^+6$	${}^-7$	✗
${}^+2$ and ${}^+3$	${}^+6$	${}^+5$	✓
${}^-2$ and ${}^-3$	${}^+6$	${}^-5$	✗

From this, you can quickly see the numbers you need.

Again, the correct solution is $(x + 2)(x + 3)$.

Worked example

Factorise $x^2 + x - 12$.

You know that $x^2 + x - 12 = (x \quad)(x \quad)$ Since $x \times x = x^2$ as required.

You now need to look at pairs of numbers which multiply together to give $^-12$.

Possibilities include	$^-1 \times {}^+12$	$(x - 1)(x + 12)$	$= x^2 - 1x + 12x - 12$	$= x^2 + 11x - 12$
	$^+1 \times {}^-12$	$(x + 1)(x - 12)$	$= x^2 + 1x - 12x - 12$	$= x^2 - 11x - 12$
	$^-2 \times {}^+6$	$(x - 2)(x + 6)$	$= x^2 - 2x + 6x - 12$	$= x^2 + 4x - 12$
	$^+2 \times {}^-6$	$(x + 2)(x - 6)$	$= x^2 + 2x - 6x - 12$	$= x^2 - 4x - 12$
	$^-3 \times {}^+4$	$(x - 3)(x + 4)$	$= x^2 - 3x + 4x - 12$	$= x^2 + 1x - 12$
	$^+3 \times {}^-4$	$(x + 3)(x - 4)$	$= x^2 + 3x - 4x - 12$	$= x^2 - 1x - 12$

and the correct solution is:

$(x - 3)(x + 4) = x^2 - 3x + 4x - 12 = x^2 + x - 12$.

Again, using the alternative method, look for numbers with product $^-12$ and sum $^+1$.

Numbers	Product	Sum	
$^-1$ and $^+12$	$^-12$	$^+11$	✗
$^+1$ and $^-12$	$^-12$	$^-11$	✗
$^-2$ and $^+6$	$^-12$	$^+4$	✗
$^+2$ and $^-6$	$^-12$	$^-4$	✗
$^-3$ and $^+4$	$^-12$	$^+1$	✓
$^+3$ and $^-4$	$^-12$	$^-1$	✗

The correct solution is $(x - 3)(x + 4)$.

NOTE

$(x - 3)(x + 4)$ can be written as $(x + 4)(x - 3)$, which also gives $x^2 + x - 12$.

Check yourself

QUESTIONS

Q1 Expand and simplify these expressions.

(a) $(x + 1)(x + 4)$
(b) $(2a + 3)(2b + 1)$
(c) $(4y - 2)(y + 2)$
(d) $(3y - 5)(2y - 7)$
(e) $(3x + 1)^2$

Q2 Expand and simplify $(x + y)(y - z) - (x - z)(y + z)$.

Q3 Factorise these expressions completely.

(a) $x^2 + 6x + 8$
(b) $x^2 - x - 2$
(c) $x^2 - 7x - 18$

REMEMBER! Cover the answers if you want to.

ANSWERS

A1
(a) $x^2 + 5x + 4$
(b) $4ab + 2a + 6b + 3$
(c) $4y^2 + 6y - 4$
(d) $6y^2 - 31y + 35$
(e) $9x^2 + 6x + 1$

A2
$^-2xz + y^2 + z^2$

A3
(a) $(x + 2)(x + 4)$
(b) $(x + 1)(x - 2)$
(c) $(x + 2)(x - 9)$

TUTORIAL

T1

(a) $(x + 1)(x + 4)$
$= x \times x + x \times 4 + 1 \times x + 1 \times 4$
$= x^2 + 4x + 1x + 4$
$= x^2 + 5x + 4$ *Collecting like terms.*
$(x + 1)(x + 4) = x^2 + 5x + 4$

(b) $(2a + 3)(2b + 1)$
$= 2a \times 2b + 2a \times 1 + 3 \times 2b + 3 \times 1$
$= 4ab + 2a + 6b + 3$
$(2a + 3)(2b + 1) = 4ab + 2a + 6b + 3$

(c) $(4y - 2)(y + 2)$
$= 4y \times y + 4y \times 2 + {}^-2 \times y + {}^-2 \times 2$
$= 4y^2 + 8y - 2y - 4$
$= 4y^2 + 6y - 4$ *Collecting like terms.*
$(4y - 2)(y + 2) = 4y^2 + 6y - 4$

(d) $(3y - 5)(2y - 7)$
$= 3y \times 2y + 3y \times {}^-7 + {}^-5 \times 2y + {}^-5 \times {}^-7$
$= 6y^2 - 21y - 10y + 35$ As ${}^-5 \times {}^-7 = {}^+35$.
$= 6y^2 - 31y + 35$ *Collecting like terms.*
$(3y - 5)(2y - 7) = 6y^2 - 31y + 35$

(e) $(3x + 1)^2 = (3x + 1)(3x + 1)$
$= 3x \times 3x + 3x \times 1 + 1 \times 3x + 1 \times 1$
$= 9x^2 + 3x + 3x + 1$
$= 9x^2 + 6x + 1$ *Collecting like terms.*
$(3x + 1)^2 = 9x^2 + 6x + 1$

T2

$(x + y)(y - z) - (x - z)(y + z)$
$= \{xy - xz + y^2 - yz\} - \{xy + xz - yz - z^2\}$

$= xy - xz + y^2 - yz - xy - xz + yz + z^2$
Removing brackets and remembering the $^-1$ outside the second bracket.

$= {}^-2xz + y^2 + z^2$ *Collecting like terms.*

T3

(a) $x^2 + 6x + 8$
$= (x \quad)(x \quad)$ *Look for numbers that multiply together to give a product of $^+8$.*

Try ${}^+1 \times {}^+8 \quad {}^-1 \times {}^-8$
${}^+2 \times {}^+4 \quad {}^-2 \times {}^-4$

$= (x + 2)(x + 4)$ *Check by multiplying out.*

(b) $x^2 - x - 2$
$= (x \quad)(x \quad)$ *Look for numbers that multiply together to give a product of $^-2$.*

Try ${}^-1 \times {}^+2 \quad {}^+1 \times {}^-2$

$= (x + 1)(x - 2)$ *Check by multiplying out.*

(c) $x^2 - 7x - 18$
$= (x \quad)(x \quad)$ *Look for numbers that multiply together to give a product of $^-18$.*

Try ${}^-1 \times {}^+18 \quad {}^+1 \times {}^-18$
${}^-2 \times {}^+9 \quad {}^+2 \times {}^-9$
${}^-3 \times {}^+6 \quad {}^+3 \times {}^-6$

$= (x + 2)(x - 9)$

CHAPTER 32

EXAMINATION QUESTIONS 6

EXAM PRACTICE

Sample Student's Answers & Examiner's Comments

1 The *n*th term of a sequence is $3n^2 + 1$.

(a) Write down the first five terms of this sequence.

(b) Write down the first four differences between consecutive terms of the sequence in (a).

(c) What is the *n*th term of the sequence formed by these differences?

(MEG syllabus B, specimen paper 1998)

(a)

$3 \times 1^2 + 1 = 4$

$3 \times 2^2 + 1 = 13$

$3 \times 3^2 + 1 = 28$

$3 \times 4^2 + 1 = 49$

$3 \times 5^2 + 1 = 76$

(b)

Terms: 4 13 28 49 76

Differences: +9 +15 +21 +27

(c)

Terms: +9 +15 +21 +27

Differences: +6 +6 +6

Term	1	2	3	4
Sequence	9	15	21	27
6 X (term number)	6	12	18	24
Sequence	9	15	21	27
Difference	3	3	3	3

The nth term is $6n + 3$

EXAMINER'S COMMENTS

1 **a)**
The candidate has correctly substituted in the given formula and has remembered that $3n^2 = 3 \times n^2$ and not $3n \times 3n$ which is a common mistake.

b)
It is important to label your answer clearly so that the examiner can see what your answers are.

c)
The candidate has worked out the differences of this new sequence. The differences are constant which means that a linear rule can be applied to the terms to find the nth term.

She has noticed that the differences now are always 3 so the nth term is $6n + 3$. She should then check this on the given terms.

EXAMINER'S COMMENTS

2 The air temperature, T°C, outside an aircraft flying at a height of h feet is given by the formula:

$$T = 26 - \frac{h}{500}$$

The air temperature outside an aircraft is ⁻52 °C.
Calculate the height of the aircraft.

(London specimen paper 1998)

$$-52 = 26 - \frac{h}{500}$$

$$-52 - 26 = -\frac{h}{500}$$

$$-78 = -\frac{h}{500}$$

$$39000 = h$$

2 *The candidate has substituted $T = {}^{-}52$ into the given formula and 26 is subtracted from both sides. The candidate has remembered to include the – sign on the right-hand side.*

The candidate has multiplied both sides by ⁻1 to remove the – signs. He has multiplied both sides of the formula by 500 to obtain h.

3 A rectangle is $(2x + 5)$ cm long and $(x - 4)$ cm wide.
What is the perimeter of the rectangle in its simplest form?

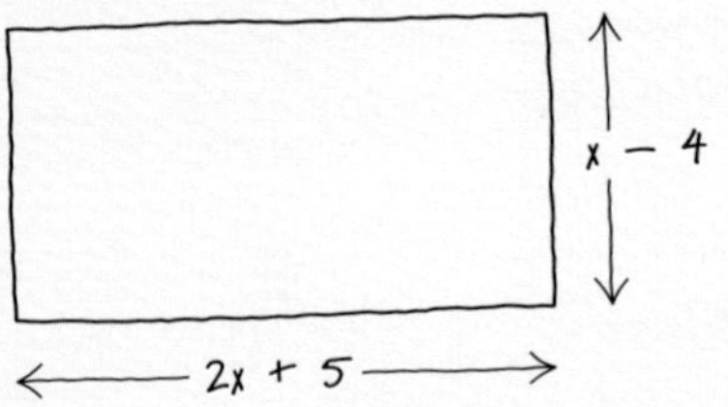

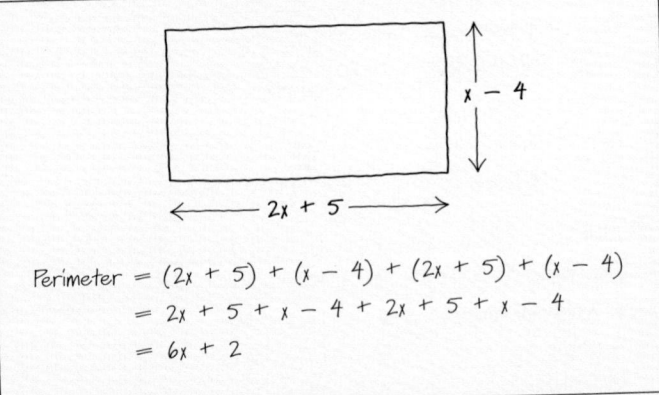

$$\text{Perimeter} = (2x + 5) + (x - 4) + (2x + 5) + (x - 4)$$
$$= 2x + 5 + x - 4 + 2x + 5 + x - 4$$
$$= 6x + 2$$

3 *The candidate has drawn a rectangle to help her with the question. The sides have been correctly labelled. She has collected together like terms and given the answer in its simplest form thus gaining maximum marks.*

4 Expand $(p + q)(p - q)$ and use this to work out $999^2 - 998^2$ without a calculator.

$$(p + q)(p - q) = p^2 - pq + pq - q^2$$
$$= p^2 - q^2$$

$$999^2 - 998^2 = 998\,001 - 996\,004$$
$$= 1997$$

4 *The candidate has successfully expanded the brackets and simplified the expression as appropriate.*

He has not used the relationship to find $999^2 - 998^2$ and will probably have used a calculator as well. In this case, although the answer is correct, no marks will be awarded as he has not used the method asked for in the question.

The candidate should use the fact that $p^2 - q^2 = (p + q)(p - q)$

so that $999^2 - 998^2$
$= (999 + 998)(999 - 998)$
$= 1997 \times 1$
$= 1997$

Questions to Answer

1 Write down the next three terms of the following sequence.

3, 8, 15, 24, 35, ...

Hence or otherwise find the nth term of the sequence and use this to find the difference between the 25th and 26th term.

2 A sequence of numbers is shown below.

Term	1st	2nd	3rd	4th	5th
Sequence	2	8	18	32	50

(a) Write down the next term of the sequence.
(b) Write down the nth term of the sequence in terms of n.
(c) Evaluate the 50th term of the sequence.

(SEG specimen paper 1998)

3 The square has the same perimeter as the triangle.
The triangle is right-angled and has an area of $84\,cm^2$.
Calculate the length, d, of the diagonal of the square.
Give your answer correct to one decimal place.

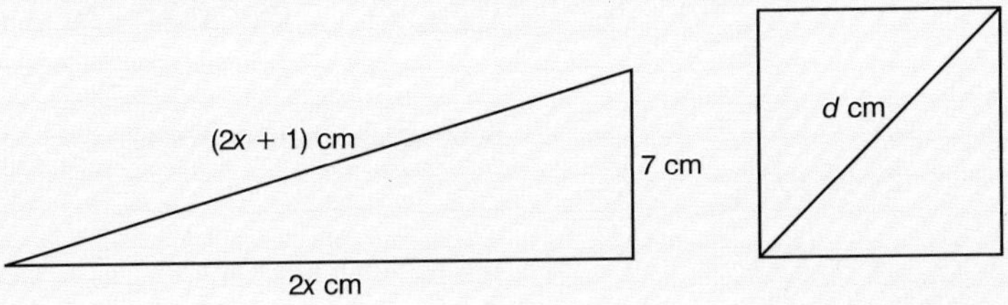

(SEG specimen paper 1998)

4 What is the value of $x^2 - 3x + 5$ when:
(a) $x = 2$ (b) $x = {}^{-}2$?

5 Factorise fully the expression $2\pi rh + 2\pi r^2$.

6 Given that $(2x - 1)(x + 3) = 2x^2 + ax + b$ find a and b.

7 Simplify the expression $(x + 5)^2 - (x - 5)^2$.

8 Factorise completely the expression $x^2 - 14x + 33$.

CHAPTER 33

SOLVING EQUATIONS

An algebraic equation is made up of two algebraic expressions separated by an equals sign. The equals sign provides a balance between the two algebraic expressions. To maintain the balance of an equation, you must make sure that whatever you do to one side you also do to the other side.

$x + 8 = 11$

$x + 8 - 8 = 11 - 8$ **Subtracting 8 from both sides.**

$x = 3$

$x - 2.5 = 3$

$x - 2.5 + 2.5 = 3 + 2.5$ **Adding 2.5 to both sides.**

$x = 5.5$

$4x = 12$

$\frac{4x}{4} = \frac{12}{4}$ **Dividing both sides by 4.**

$x = 3$

$\frac{x}{5} = 3.2$

$\frac{x}{5} \times 5 = 3.2 \times 5$ **Multiplying both sides by 5.**

$x = 16$

Worked example

Solve the following equations.

(a) $4x + 5 = 17$ (b) $9x - 1 = 7x + 14$ (c) $6(2x + 5) = 48$

(a) $4x + 5 = 17$

$4x = 17 - 5$ Subtracting 5 from both sides.

$4x = 12$

$x = \frac{12}{4}$ Dividing both sides by 4.

$x = 3$

(b) $9x - 1 = 7x + 14$ This equation has the unknown on both sides.

$9x = 7x + 14 + 1$ Adding 1 to both sides.

$9x = 7x + 15$

$9x - 7x = 15$ Subtracting $7x$ from both sides.

$2x = 15$

$x = \frac{15}{2}$ Dividing both sides by 2.

$x = 7\frac{1}{2}$ or 7.5

(c) $6(2x + 5) = 48$

$6 \times 2x + 6 \times 5 = 48$ Expanding the brackets.

$12x + 30 = 48$

$12x = 48 - 30$ Subtracting 30 from both sides.

$12x = 18$

$x = \frac{18}{12}$ Dividing both sides by 12.

$x = 1\frac{1}{2}$ or 1.5

You can apply the same method to solve equations in a variety of number problems.

Worked example

Find three consecutive numbers with a sum of 15.

To solve a problem like this, it is helpful to call one of the numbers x. As the numbers are consecutive then the other two numbers are $(x + 1)$ and $(x + 2)$.

If the sum is 15 then $x + (x + 1) + (x + 2) = 15$.

$x + x + 1 + x + 2 = 15$

$3x + 3 = 15$ Collecting like terms.

$3x = 15 - 3$ Subtracting 3 from both sides.

$3x = 12$

$x = \frac{12}{3}$ Dividing both sides by 3.

$x = 4$

The answer $x = 4$ needs to be interpreted. Since the consecutive numbers were x, $(x + 1)$ and $(x + 2)$ then the required numbers are 4, $(4 + 1)$ and $(4 + 2)$ or 4, 5 and 6.

HINT

You should now check your solutions to make sure that they are correct.

Worked example

A group of people shared a pay-out of £45 000 equally, and each of them received £5000. How many people shared the money?

You need to form an equation and solve it. Let the number of people be x.

$\frac{45\,000}{x} = 5000$

$45\,000 = 5000x$ Multiplying both sides by x.

$\frac{45\,000}{5000} = x$ Dividing both sides by 5000.

$9 = x$

$x = 9$ Reversing the answer to get x on the LHS.

Nine people shared the pay-out.

Check yourself

QUESTIONS

Q1 Solve the following equations.

(a) $3x + 4.5 = 15$ (b) $20x = 30$

(c) $\frac{27}{z} = 3$

(d) $3y + 5 = 25 - y$ (e) $4(2x - 3) = 44$

(f) $4x + 1 = 3(x + 2)$ (g) $\frac{y + 3}{10} = 3$

(h) $6(x - 2) - 2(2x + 1) = 0$

Q2 Find three consecutive numbers with a sum of 78.

Q3 Three sisters are x years, $(x + 2)$ years and $2x$ years old. The sum of their ages is 22. How old is each of the sisters?

Q4 The angles of a triangle are $x°$, $2x°$ and $(3x + 30)°$. Find the value of x.

ANSWERS

A1

(a) $x = 3.5$
(b) $x = 1.5$ or $1\frac{1}{2}$
(c) $z = 9$
(d) $y = 5$
(e) $x = 7$
(f) $x = 5$
(g) $y = 27$
(h) $x = 7$

A2 25, 26, and 27

A3 5, 7 and 10 years old

A4 $x = 25$

TUTORIAL

T1

(a) $3x = 10.5$ *Subtracting 4.5 from both sides.*
$x = 3.5$ *Dividing both sides by 3.*

(b) $x = \frac{30}{20} = 1.5$ *or* $1\frac{1}{2}$

(c) $27 = 3z$ *Multiplying both sides by z.*
$z = 9$ *Dividing both sides by 3.*

(d) $3y = 20 - y$ *Subtracting 5 from both sides.*
$4y = 20$ *Adding y to both sides.*
$y = 5$ *Dividing both sides by 4.*

(e) $8x - 12 = 44$ *Expanding the brackets.*
$8x = 56$ *Adding 12 to both sides.*
$x = 7$ *Dividing both sides by 8.*

(f) $4x + 1 = 3x + 6$ *Expanding the brackets.*
$4x = 3x + 5$ *Subtracting 1 from both sides.*
$x = 5$ *Subtracting 3x from both sides.*

(g) $y + 3 = 30$ *Multiplying both sides by 10.*
$y = 27$ *Subtracting 3 from both sides.*

(h) $6x - 12 - 4x - 2 = 0$
Remember the – sign outside of the second bracket.
$2x - 14 = 0$ *Collecting like terms.*
$2x = 14$ *Adding 14 to both sides.*
$x = 7$ *Dividing both sides by 2.*

T2 *Let the three consecutive numbers be* x, $x + 1$ *and* $x + 2$, *then* $x + (x + 1) + (x + 2) = 78$ *and* $x = 25$ *so the numbers are 25, 26 and 27.*

T3 *The sum of their ages is* $x + (x + 2) + 2x = 4x + 2$. *But the sum of their ages is 22 so:*

$4x + 2 = 22$
$4x = 20$ *Subtracting 2 from both sides.*
$x = 5$ *Dividing both sides by 4.*

The ages are x years, (x + 2) years and 2x years which makes the girls 5 years, 7 years and 10 years old.

T4 *The angles of a triangle add up to 180° so:*

$x° + 2x° + (3x + 30)° = 180°$ *and* $x = 25$

CHAPTER 34

REARRANGING FORMULAE

You can rearrange (or **transpose**) a formula in exactly the same way as you solve an equation. To maintain the balance of the formula you must make sure that whatever you do to one side of the formula you also do to the other side of the formula.

For the formula $S = \frac{D}{T}$, S is called the **subject** of the formula. The formula can be rearranged to make D or T the subject as follows.

$S = \frac{D}{T}$

$S \times T = D$ Multiplying both sides of the formula by T.

$D = S \times T$ or $D = ST$ Turning the formula round so that D is the subject of the formula.

Or, from $D = ST$:

$\frac{D}{S} = T$ Dividing both sides of the formula by S.

$T = \frac{D}{S}$ Turning the formula around so that T is the subject.

Worked example

Make x the subject of the formula $y = 3x + 2$.

$y = 3x + 2$

$y - 2 = 3x$ Subtracting 2 from both sides.

$\frac{y-2}{3} = x$ Dividing both sides of the formula by 3.

$x = \frac{y-2}{3}$ Turning the formula around so that x is the subject.

Check yourself

QUESTIONS

Q1 Rewrite the following with the letter indicated in brackets as the subject.

(a) $C = 2\pi r$ (r) (b) $v = u + at$ (u)

(c) $v = u + at$ (a) (d) $A = \pi r^2$ (r)

(e) $V = \pi r^2 h$ (h) (f) $V = \pi r^2 h$ (r)

(g) $I = \frac{PRT}{100}$ (T)

Q2 Rearrange the formula $a = \frac{m}{b^2}$ to make b the subject.

Q3 A formula states:

$$h = \frac{S}{6.3r}$$

(a) Find the value of h when $S = 50$ and $r = 3$, giving your answer correct to one decimal place.

(b) Rearrange the formula to make S the subject.

(c) Find the value of S when $h = 6$ and $r = 2$.

(d) Rearrange the formula to make r the subject.

(MEG *syllabus* B, *specimen paper* 1998)

ANSWERS

A1

(a) $r = \frac{C}{2\pi}$

(b) $u = v - at$

(c) $a = \frac{v-u}{t}$

(d) $r = \sqrt{\frac{A}{\pi}}$

(e) $h = \frac{V}{\pi r^2}$

(f) $r = \sqrt{\frac{V}{\pi h}}$

(g) $T = \frac{100I}{PR}$

A2

$b = \sqrt{\frac{m}{a}}$

A3

(a) $h = 2.6$ (correct to one decimal place)
(b) $S = 6.3hr$
(c) $S = 75.6$

(d) $r = \frac{S}{6.3h}$

TUTORIAL

T1

(a) $r = \frac{C}{2\pi}$ *Dividing both sides by 2π.*

(b) $u = v - at$ *Subtracting at from both sides.*

(c) $at = v - u$ *Subtracting u from both sides.*

$a = \frac{v-u}{t}$ *Dividing both sides by t.*

(d) $r^2 = \frac{A}{\pi}$ *Dividing both sides by π.*

$r = \sqrt{\frac{A}{\pi}}$ *Taking the square root of both sides.*

(e) $h = \frac{V}{\pi r^2}$ *Dividing both sides by πr^2.*

(f) $r^2 = \frac{V}{\pi h}$ *Dividing both sides by πh.*

$r = \sqrt{\frac{V}{\pi h}}$ *Taking the square root of both sides.*

(g) $100I = PRT$ *Multiplying both sides by 100.*

$T = \frac{100I}{PR}$ *Dividing both sides by PR.*

T2

$a = \frac{m}{b^2}$

$ab^2 = m$ *Multiplying both sides by b^2.*

$b^2 = \frac{m}{a}$ *Dividing both sides by a^2.*

$b = \sqrt{\frac{m}{a}}$ *Taking the square root of both sides.*

T3

(a) *When S = 50 and r = 3 then* $h = \frac{50}{6.3 \times 3}$

$h = \frac{50}{18.9} = 2.645\,502\,6$

$h = 2.6$ *(correct to one decimal place)*

(b) $h = \frac{S}{6.3r}$

$S = h \times 6.3r$ *Multiplying both sides by 6.3r.*

$S = 6.3hr$ *Rearranging the right-hand side.*

(c) *When h = 6 and r = 2 then:*

$S = 6.3 \times 6 \times 2 = 75.6$

(d) $S = 6.3hr$

$r = \frac{S}{6.3h}$ *Dividing both sides by 6.3h.*

CHAPTER 35

ALGEBRAIC INDICES

In Chapter 6, Positive, negative and zero indices, you saw that, in general:

$a^m \times a^n = a^{m+n}$ $\quad$ $a^m \div a^n = a^{m-n}$

$a^{-m} = \frac{1}{a^m}$ $\quad$ **and** $\quad$ $a^0 = 1$

Worked example

Simplify the following expressions.

(a) $a^2 \times a^5$ $\quad$ (b) $b^7 \div b^7$ $\quad$ (c) $c^4 \times c \times c^{11}$

(a) $a^2 \times a^5 = a^{2+5} = a^7$

(b) $b^7 \div b^7 = b^{7-7} = b^0 = 1$

(c) $c^4 \times c \times c^{11} = c^{4+1+11} = c^{16}$ $\quad$ As $c = c^1$.

REMEMBER

Any number to the power 0 = 1

Check yourself

QUESTIONS

Q1 Simplify the following expressions.

(a) $x^4 \times x$ $\quad$ (b) $(y^3)^2$

(c) $(2a^4)^3$ $\quad$ (d) $d^{12} \div d^9$

(e) $6x^7 \div 3x^4$ $\quad$ (f) $3x^6 \div 9x^8$

Q2 Find the values of the letters in the following equations.

(a) $3^x = 81$

(b) $2^{3y} = 64$

(c) $5^{2x+1} = 125$

REMEMBER! Cover the answers if you want to.

ANSWERS

A1

(a) x^5

(b) y^6

(c) $8a^{12}$

(d) d^3

(e) $2x^3$

(f) $\frac{1}{3}x^{-2}$ or $\frac{1}{3x^2}$

A2

(a) $x = 4$

(b) $y = 2$

(c) $x = 1$

TUTORIAL

T1

(a) $x^4 \times x = x^{4+1} = x^5$ $\quad$ As $x = x^1$.

(b) $(y^3)^2 = y^3 \times y^3 = y^{3+3} = y^6$

(c) $(2a^4)^3 = (2a^4) \times (2a^4) \times (2a^4)$

$= 2 \times a^4 \times 2 \times a^4 \times 2 \times a^4$ $\quad$ *Removing the brackets.*

$= 8 \times a^4 \times a^4 \times a^4$ $\quad$ As $2 \times 2 \times 2 = 8$.

$= 8 \times a^{4+4+4} = 8a^{12}$

(d) $d^{12} \div d^9 = d^{12-9} = d^3$

(e) $6x^7 \div 3x^4 = 2x^{7-4} = 2x^3$

(f) $3x^6 \div 9x^8 = \frac{1}{3}x^{6-8} = \frac{1}{3}x^{-2}$ $\quad$ As $\frac{3}{9} = \frac{1}{3}$.

$x^{-2} = \frac{1}{x^2}$ *so* $\frac{1}{3}x^{-2}$ *can be written as* $\frac{1}{3x^2}$.

T2

(a) $3^x = 81$ $\quad 3^4 = 81$ $\quad$ *so* $x = 4$

(b) $2^{3y} = 64$ $\quad 2^6 = 64$ $\quad$ *so* $3y = 6$ *or* $y = 2$

(c) $5^{2x+1} = 125$ $\quad 5^3 = 125$ $\quad$ *so* $2x + 1 = 3$

$2x = 2$ *or* $x = 1$

CHAPTER 36

EXAMINATION QUESTIONS 7

EXAM PRACTICE

Sample Student's Answers & Examiner's Comments

1 The length of a man's forearm (f cm) and his height (h cm) are approximately related by the formula:

$$h = 3f + 90$$

(a) Part of the skeleton of a man is found and the forearm is 20 cm long. Use the formula to estimate the man's height.

(b) A man's height is 162 cm. Use the formula to estimate the length of his forearm.

(c) George is 1 year old and he is 70 cm tall. Find the value of the formula for the length of his forearm and state why this value is impossible.

(d) Use the formula to find an expression for f in terms of h.

(MEG syllabus A, specimen paper 1998)

(a) $f = 20\text{cm}$
$h = 3f + 90$
$h = 3 \times 20 + 90$
$h = 150\text{cm}$

(b) $h = 162\text{cm}$
$h = 3f + 90$
$162 = 3f + 90$
$72 = 3f$
$f = 24\text{cm}$

(c) $h = 70\text{cm}$
$h = 3f + 90$
$70 = 3f + 90$
$^{-}20 = 3f$
$f = -\frac{20}{3}$
Length is negative.

(d) $h = 3f + 90$
$h - 90 = 3f$
$3f = h - 90$
$f = \frac{h - 90}{3}$

EXAMINER'S COMMENTS

1 **a)**
The candidate has successfully substituted the value into the equation. It is always a good idea to include the units of measurement.

b)
He has successfully substituted the value into the equation.

c)
The candidate loses marks here by not explaining why such a value is impossible. Some suggestion that a negative length is impossible would gain the marks. A possible reason for this might be that the equation is not valid for maturing infants or only applicable to the length of a grown man's forearm and his height.

d)
He has written down an expression for f in terms of h.

2 A rocket is fired vertically upwards with velocity u metres per second. After t seconds the rocket's velocity, v metres per second, is given by the formula $v = u + gt$.

(a) Find the value of v when $u = 100$, $g = {}^{-}9.8$ and $t = 5$.

(b) Rewrite the formula in terms of t.

(c) Find t when $u = 93.5$, $g = {}^{-}9.8$ and $v = 20$.

(a) $u = 100,\ g = {}^{-}9.8,\ t = 5$

$v = u + gt$

$v = 100 + {}^{-}9.8 \times 5$

$v = 51$

(b) $v = u + gt$

$v - u = gt$

$\frac{v - u}{g} = t$

(c) $u = 93.5,\ g = {}^{-}9.8,\ v = 20$

$t = \frac{v - u}{g} = \frac{20 - 93.5}{-9.8}$

$t = \frac{-73.5}{-9.8} = 7.5$

EXAMINER'S COMMENTS

2 a)
The candidate has successfully substituted the values into the equation.

b)
It is usual to write the formula as $t = \frac{v - u}{g}$ *where t is more easily recognised as the subject of the formula.*

c)
The candidate has successfully substituted the values into the equation but must be careful with the negative signs as a negative divided by a negative equals a positive.

She might be suspicious if the answer gives a negative value for time.

Questions to Answer

1 Find three consecutive even numbers with sum 84.

2 A room is 4 metres longer than its width. If the perimeter of the room is 72 metres, what is the length of the room?

3 Rewrite the following with a as the subject of the formula.

$$s = ut + \tfrac{1}{2}at^2$$

4 The formula for the surface area of a sphere, with area A cm^2 and radius r cm, is given as $A = 4\pi r^2$.

(a) Find A when $r = 20$ cm.

(b) Rearrange the formula to make r the subject.

(c) Use this formula to find r when $A = 100$ cm^2.

5 Find the value of the letters in the following equations.

(a) $5^{2x} = 625$ (b) $10^y = 1$ (c) $2^{3x-1} = 256$

CHAPTER 37

INTERPRETING GRAPHS

You need to be familiar with interpreting graphs which can be presented in many different contexts.

Worked example

Draw a conversion graph to show the relationship between miles and kilometres, given that 5 miles is approximately 8 kilometres.

Use your graph to find:

(a) how many kilometres there are in 15 miles

(b) how many kilometres there are in 8 miles

(c) how many miles there are in $32\frac{1}{2}$ kilometres.

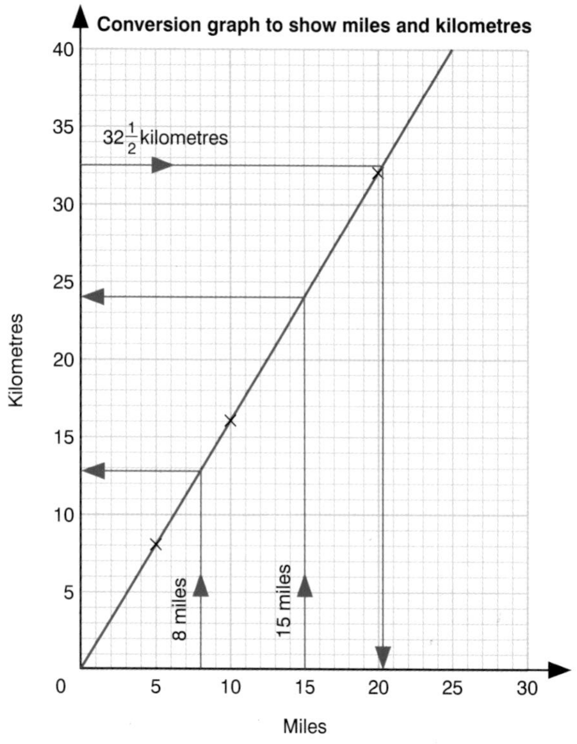

The first requirement is to draw the graph, using the fact that 5 miles is approximately 8 kilometres. This means that:

5 miles is approximately 8 kilometres

10 miles is approximately 16 kilometres

20 miles is approximately 32 kilometres etc.

As this is a straight-line graph, three points will be sufficient (two points for the line and one point as a check) to draw the graph. It is also helpful to note that 0 miles is equal to 0 kilometres so the straight line should pass through the origin.

The question asks you to 'use your graph' to find answers to the questions asked and so it is important to show the examiner how you arrived at your solutions and convince them that you did not just calculate the answers.

From the graph:

(a) there are approximately 24 kilometres in 15 miles

(b) there are approximately 12.8 kilometres in 8 miles (Remember that each small square is 1 kilometre on this scale.)

(c) there are approximately 20.3 miles in $32\frac{1}{2}$ kilometres. (Remember that each small square is 1 mile on this scale.)

Worked example

A salesperson leaves home at 1030 hours and their distance from home is shown on the graph on page 97.

Use the graph to answer the following questions.

(a) How many kilometres are travelled before the first stop?

(b) How long does it take to reach the first stop?

(c) How far is the salesperson away from home at 1600 hours?

(d) What time does the salesperson arrive back at home?

(e) How far does the salesperson travel between 1200 hours and 1600 hours?

(f) What is the average speed of the salesperson between 1030 hours and 1130 hours?

(g) What is the average speed of the salesperson between 1300 hours and 1330 hours?

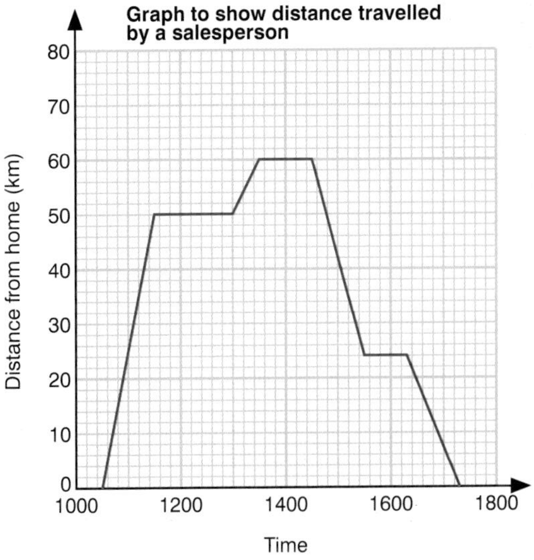

(a) The salesperson travels 50 kilometres before the first stop.
(b) It takes 1 hour (from 1030 hours to 1130 hours) to reach the first stop.
(c) The salesperson is 24 kilometres away from home at 1600 hours.
(d) The salesperson arrives back home at 1718 hours (Remember that each small square represents $\frac{1}{5}$ hour or 12 minutes.)
(e) The salesperson travels 10 km + 36 km = 46 km
(f) Between 1030 hours and 1130 hours:
distance travelled = 50 km
time taken = 1 hour
so speed = 50 kilometres per hour
(speed = distance ÷ time)
(g) Between 1300 hours and 1330 hours:
distance travelled = 10 km
time taken = $\frac{1}{2}$ hour
so speed = 20 kilometres per hour (speed = distance ÷ time)

Check yourself

QUESTIONS

Q1 The cooking instructions for a piece of meat are given as:
40 minutes per kilogram plus 25 minutes.
Draw a graph and use it to find:
(a) how long a piece of meat weighing 2 kilograms would take to cook
(b) how long a piece of meat weighing 1.4 kilograms would take to cook
(c) the weight of a piece of meat which takes $1\frac{1}{2}$ hours to cook.

Q2 The following graph shows the journeys of two cyclists travelling between two places that are 25 miles apart.

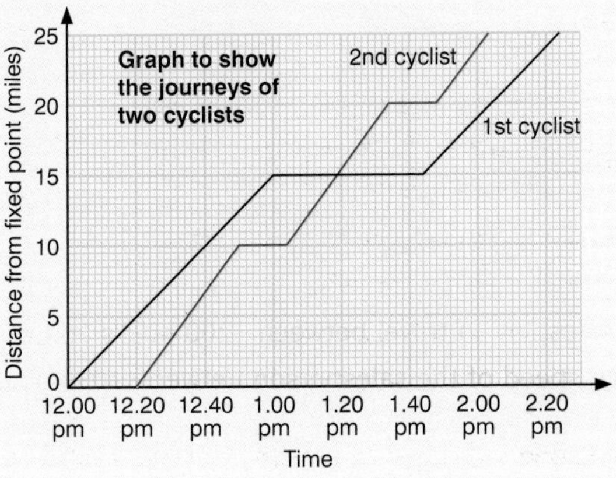

Use the graph to answer the following questions.
(a) How many miles does the first cyclist travel before the first stop?
(b) What is the average speed of the first cyclist over this first part of the journey?
(c) How long does this first cyclist stop?
(d) At what time does the second cyclist overtake the first cyclist?
(e) What is the average speed of the second cyclist at this time?
(f) What time does the second cyclist arrive at the destination?
(g) What is the greatest distance between the two cyclists?

ANSWERS

A1

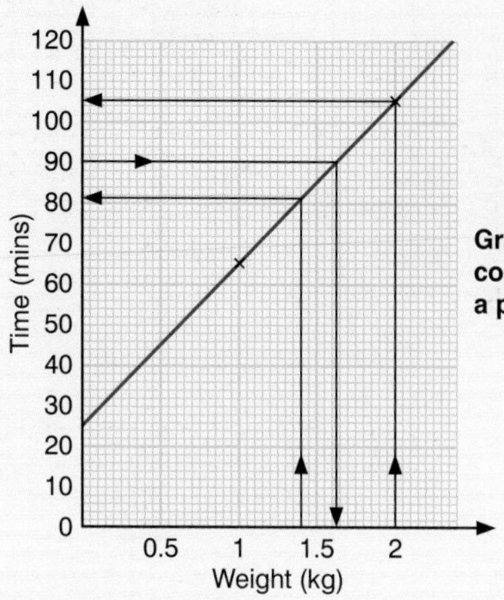

(a) 105 minutes
(b) 81 minutes
(c) 1.62 or 1.63 kg

A2

(a) 15 miles
(b) 15 miles per hour
(c) 44 minutes
(d) 1.19 p.m.
(e) 20 miles per hour
(f) 2.03 p.m.
(g) 5.2 or 5.3 miles

TUTORIAL

T1

Reading from the graph:

(a) *A piece of meat weighing 2 kilograms would take 105 minutes to cook.*
(b) *A piece of meat weighing 1.4 kilograms would take 81 minutes to cook.*
(c) *A piece of meat which takes $1\frac{1}{2}$ hours to cook would weigh 1.62 or 1.63 kilograms (converting $1\frac{1}{2}$ hours to 90 minutes).*

T2

Reading from the given graph:

(a) *The first cyclist travels 15 miles before the first stop.*
(b) *Distance travelled = 15 miles*

Time taken = 1 hour

So speed = 15 miles per hour
(speed = distance ÷ time)

(c) *The first cyclist stops for 44 minutes.*
(Each small square represents 2 minutes.)
(d) *The second cyclist overtakes the first cyclist at 1.19 p.m.*
(e) *Distance travelled = 10 miles*

Time taken = 30 minutes

So speed = distance ÷ time

$= 10 \div \frac{1}{2}$ *(as 30 minutes = $\frac{1}{2}$ hour)*

= 20 miles per hour

(f) *The second cyclist arrives at the destination at 2.03 p.m.*
(g) *The greatest distance between the two cyclists is 5.2 miles to 5.3 miles (when the second cyclist arrives at the destination).*

CHAPTER 38

LINEAR GRAPHS AND COORDINATES

Coordinates are used to locate points on a graph. A **linear graph** is one in which the points can be joined to give a straight line.

COORDINATES IN FOUR QUADRANTS

Negative coordinates can be used by extending the *x*-axis and *y*-axis in the negative direction dividing the graph into four quadrants.

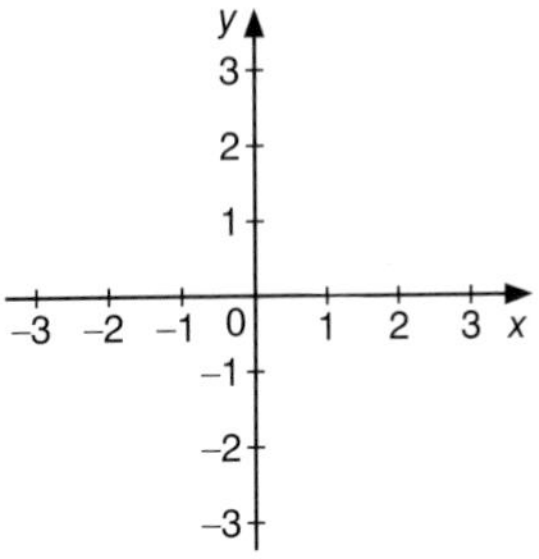

Check yourself

QUESTIONS

For each question, draw a set of axes, choosing suitable scales. Plot the points and join them up in order.

Q1 $(30, 10), (30, 20), (0, 5), (^-30, ^-20), (^-30, 10), (^-10, 0), (^-30, ^-10), (^-30, ^-20), (0, ^-5), (30, ^-20), (30, ^-10), (10, 0), (30, 10)$

Q2 $(0, 0), (0, 13), (15, 13), (15, 8), (5, 8), (5, 3), (15, 3), (15, ^-2), (5, ^-2), (5, ^-12), (0, ^-12), (0, 0), (^-5, 0), (^-5, 3), (^-15, 3), (^-15, 8), (^-5, 8), (^-5, 13), (^-20, 13), (^-20, ^-12), (^-5, ^-12), (^-5, ^-7), (^-15, ^-7), (^-15, ^-2), (^-5, ^-2), (^-5, 0)$

REMEMBER! Cover the answers if you want to.

ANSWERS

A1

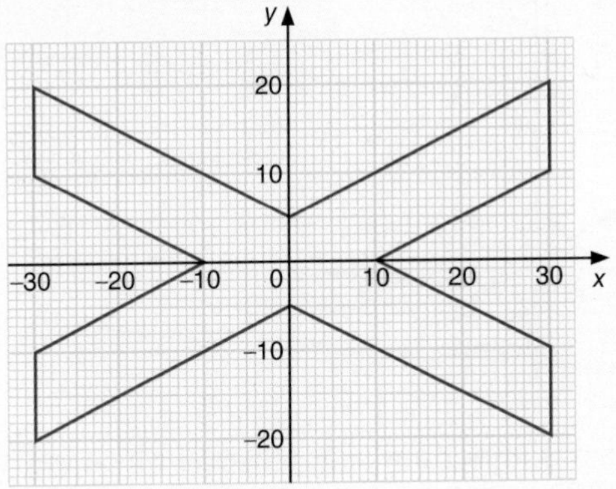

A2

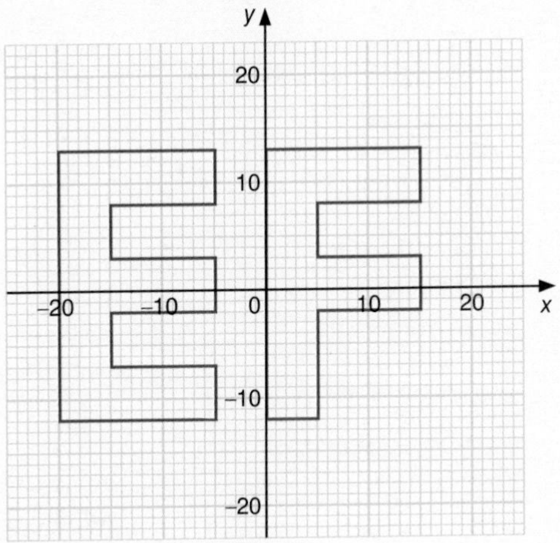

LINEAR GRAPHS

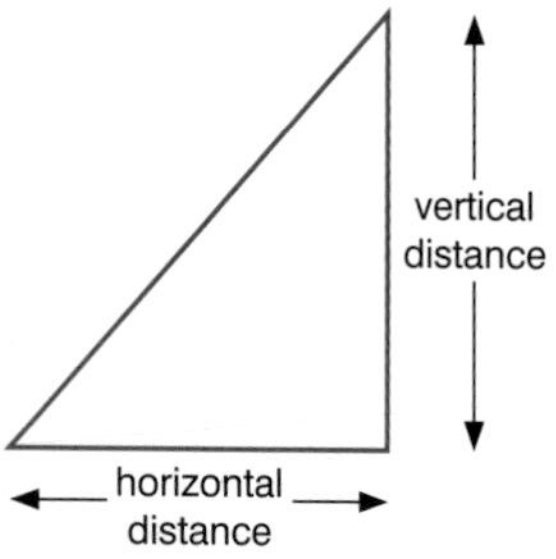

The word linear means 'straight line'.

The gradient of a line is defined as: $\dfrac{\text{vertical distance}}{\text{horizontal distance}}$

Gradients can be either positive or negative depending on their direction of slope.

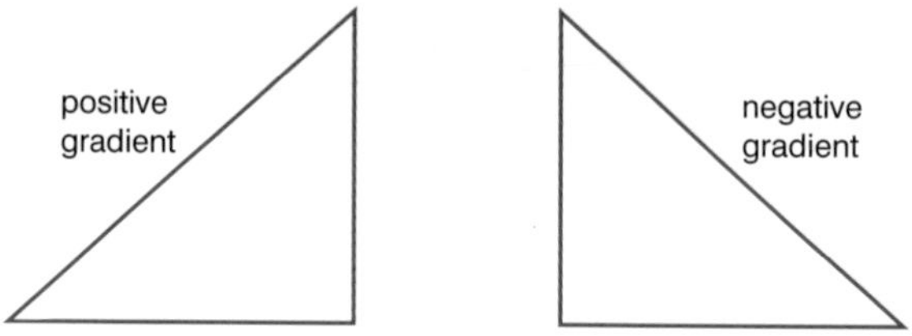

NOTE

Parallel lines have the same gradient, and lines with the same gradient are parallel.

Worked example

Calculate the gradients of the lines joining the points:

(a) (3, 2) and (7, 10)

(b) ($^{-}8$, 5) and (2, $^{-}1$).

(a) Sketching the two points (3, 2) and (7, 10) on a graph:

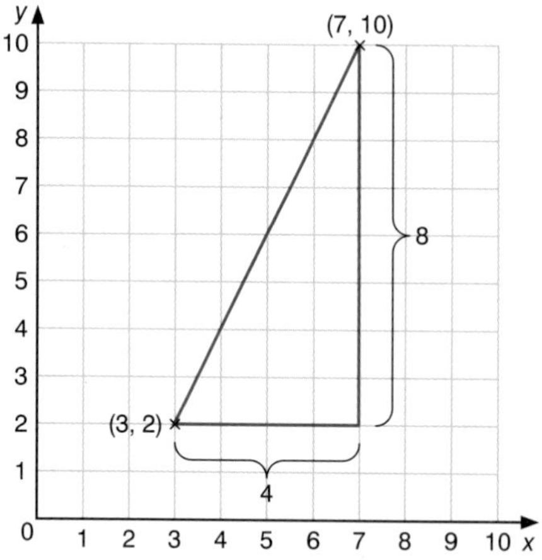

$$\text{Gradient} = \frac{\text{vertical distance}}{\text{horizontal distance}} = \frac{8}{4} = 2$$

(b) Sketching the two points ($^{-}8$, 5) and (2, $^{-}1$) on a graph:

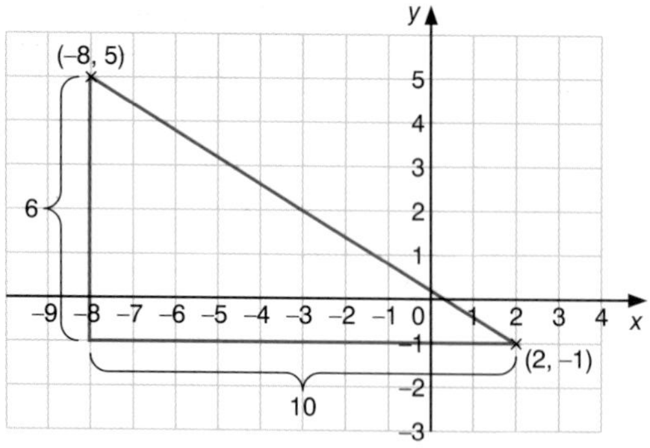

$$\text{Gradient} = \frac{\text{vertical distance}}{\text{horizontal distance}} = -\frac{6}{10} = -\frac{3}{5}$$

The negative sign is important as it tells you which way the graph is sloping. This graph is sloping down, from left to right.

All linear graphs can be written in the form:

$$y = mx + c$$

where m is the **gradient** of the line and c is the cut-off on the y-axis. This is where the line **intersects** the y-axis (also called the **y-intercept**).

Worked example

Sketch the following straight-line graphs.

(a) $y = 2x - 1$ (b) $y + 3x = 3$

The graphs can easily be sketched by comparing their equations with the form $y = mx + c$ where m is the gradient and c is the cut-off on the y-axis (i.e. the y-intercept).

(a) For $y = 2x - 1$:

$m = 2$ and $c = {}^{-}1$

So the gradient is 2 and the cut-off on the y-axis is $^{-}1$.

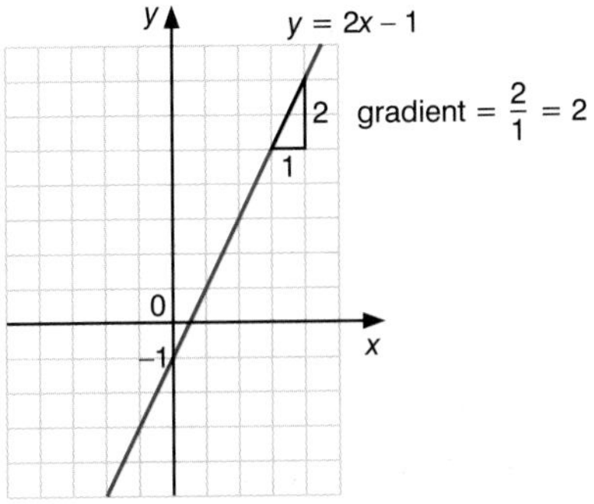

(b) $y + 3x = 3$ Rearranging the formula to get it in the form $y = mx + c$.

$y = 3 - 3x$ Subtracting $3x$ from both sides.

$y = {}^{-}3x + 3$ Rewriting in the required form.

$m = {}^{-}3$ and $c = 3$

So the gradient is $^{-}3$ and the cut-off on the y-axis is 3.

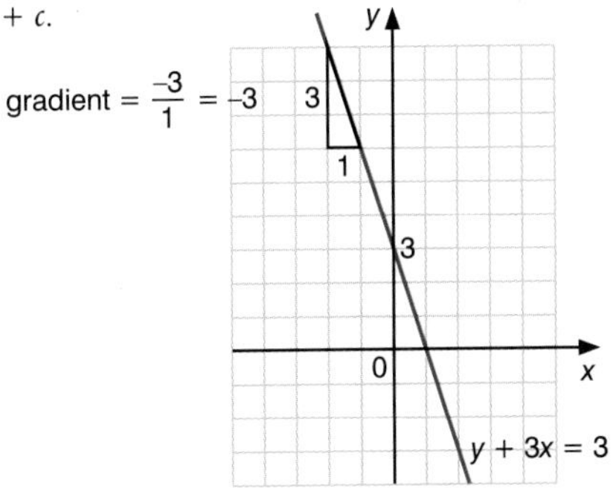

Check yourself

QUESTIONS

Q1 On the same set of axes, sketch the following lines.

(a) $y = \frac{1}{2}x + 5$

(b) $y = \frac{1}{2}x + 3$

(c) $y = \frac{1}{2}x - 2$

What can you say about the lines?

Q2 On the same set of axes, sketch the following graphs.

(a) $y + 3x = 5$
(b) $x = 2y + 6$

Q3 Write down the equations of the following straight-line graphs.

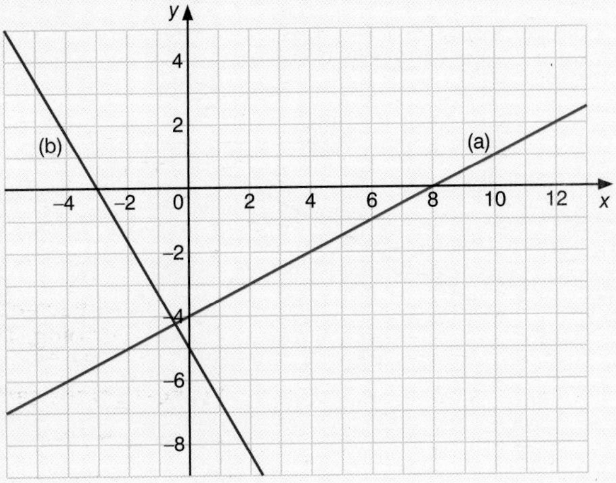

ANSWERS

A1

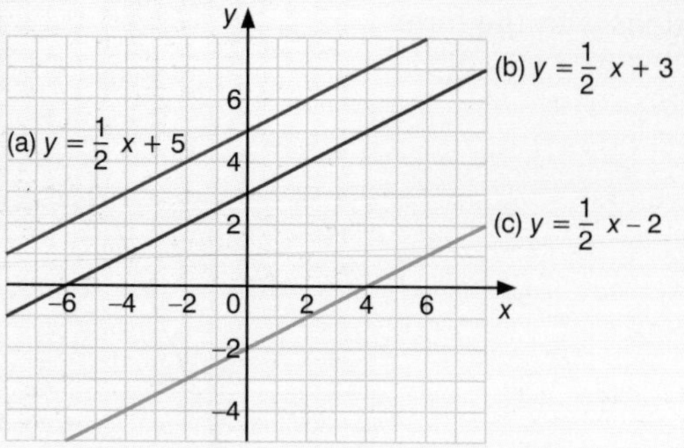

The lines are all parallel and have the same gradient.

A2 (a) $y + 3x = 5$ (b) $x = 2y + 6$

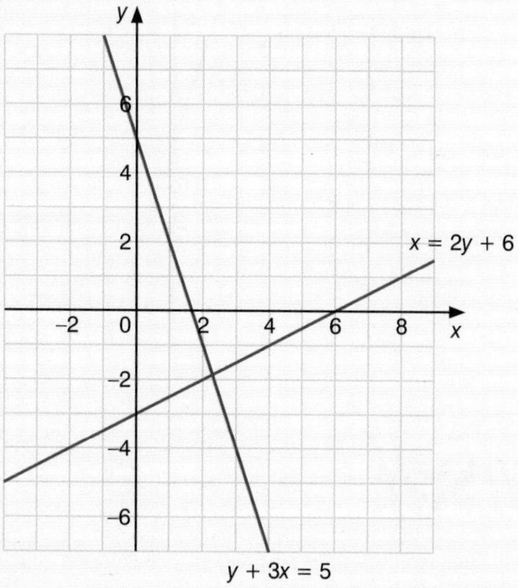

A3 (a) $y = \frac{1}{2}x - 4$ (b) $y = -\frac{5}{3}x - 5$

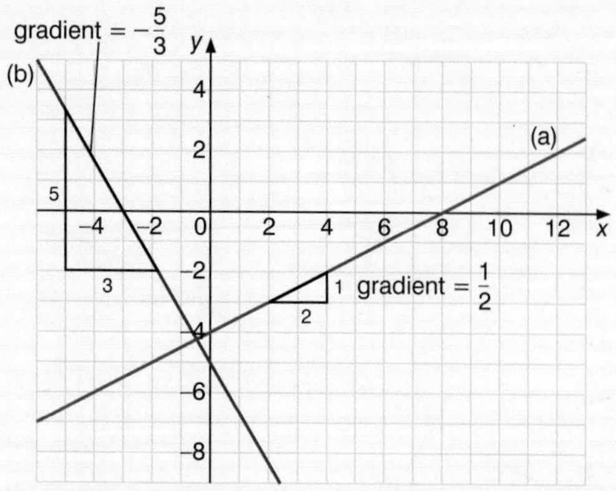

TUTORIAL

T1 *Parallel lines have the same gradient and lines with the same gradient are parallel.*

In this case the gradient, $m = \frac{1}{2}$.

T2 *To sketch the graphs it is helpful to find the value of m (the gradient) and c (the cut-off on the y-axis, or y-intercept).*

(a) $y + 3x = 5$ *Rearranging the formula to get it in the form* $y = mx + c$.

$y = {}^{-}3x + 5$ *Subtracting 3x from both sides.*

$m = {}^{-}3$ *and* $c = 5$

(b) $x = 2y + 6$ *Rearranging the formula to get it in the form* $y = mx + c$.

$x - 6 = 2y$ *Subtracting 6 from both sides.*

$2y = x - 6$ *Turning the formula around.*

$y = \frac{1}{2}(x - 6)$ *Multiplying both sides by* $\frac{1}{2}$ *(or dividing both sides by 2).*

$y = \frac{1}{2}x - 3$ *Multiplying out the bracket.*

$m = \frac{1}{2}$ *and* $c = {}^{-}3$.

T3 (a) *From the graph you can see that* $m = \frac{1}{2}$ *and* $c = {}^{-}4$.

so $y = mx + c$

$y = \frac{1}{2}x - 4$

(b) *From the graph you can see that* $m = \frac{^{-}5}{3}$ *and* $c = {}^{-}5$.

so $y = mx + c$

$y = \frac{^{-}5}{3}x - 5$

CHAPTER 39

QUADRATIC, CUBIC AND RECIPROCAL GRAPHS

Quadratic, cubic and reciprocal graphs all have the same basic shapes as illustrated in the sketches below.

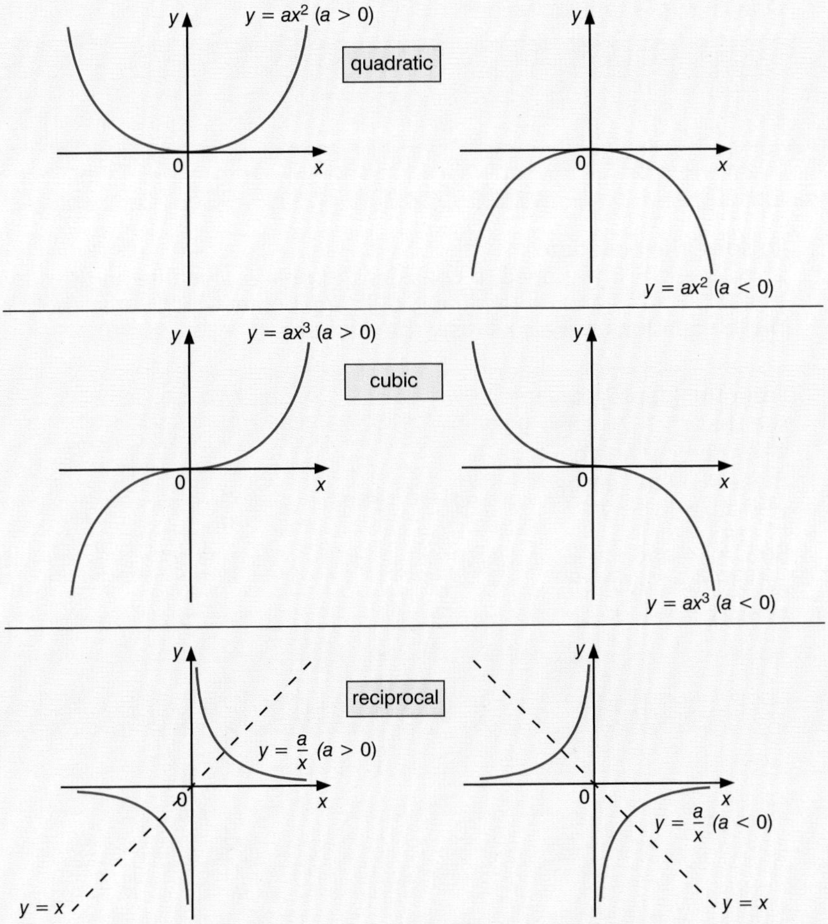

QUADRATIC GRAPHS

Quadratic graphs can be written in the form $y = ax^2 + bx + c$ (where a is non-zero). They all have approximately the same basic shape as shown in the examples below.

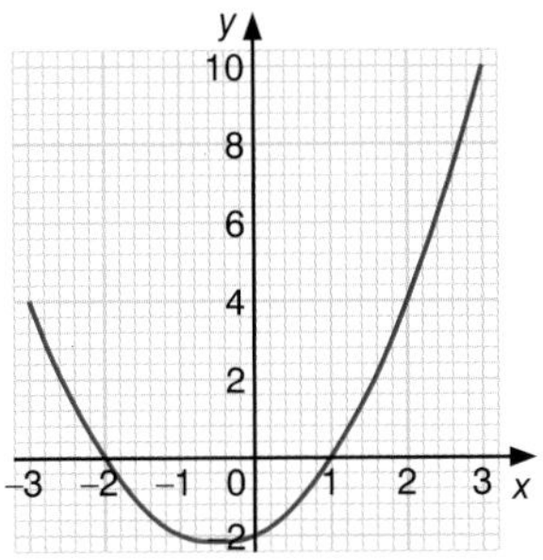

$y = x^2 + x - 2$

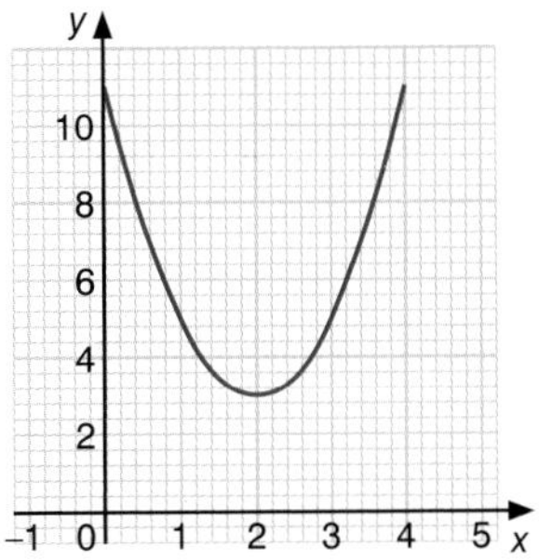

$y = 2x^2 - 8x + 11$

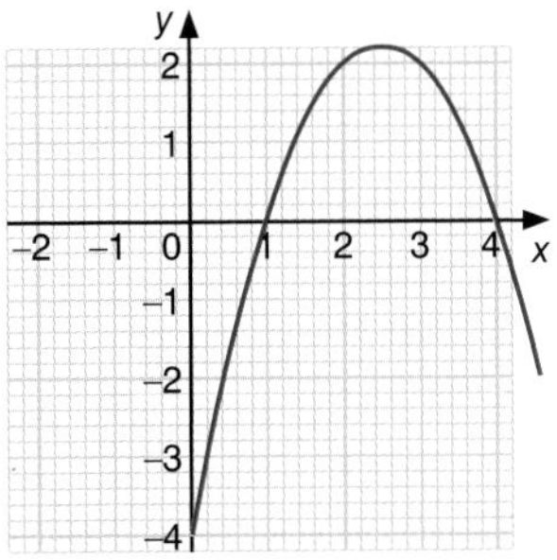

$y = -x^2 + 5x - 4$

NOTE

When drawing such graphs it is important that you join the points up with a smooth curve rather than a series of straight lines.

Worked example

Draw the graph of $y = x^2 + 2x - 8$ and use it to solve the equation $x^2 + 2x - 8 = 0$.

Drawing up a table of values:

x	$^-5$	$^-4$	$^-3$	$^-2$	$^-1$	0	1	2	3
$y = x^2 + 2x - 8$	7	0	$^-5$	$^-8$	$^-9$	$^-8$	$^-5$	0	7
Coordinates	($^-5$, 7)	($^-4$, 0)	($^-3$, $^-5$)	($^-2$, $^-8$)	($^-1$, $^-9$)	(0, $^-8$)	(1, $^-5$)	(2, 0)	(3, 7)

To solve the equation:
$x^2 + 2x - 8 = 0$ you need to consider the points which lie on the curve $y = x^2 + 2x - 8$ and on the line $y = 0$. Any points which satisfy both of these equations will also satisfy the equation $x^2 + 2x - 8 = 0$.

HINT

You should check your answers by substituting them into the equation to see if they are correct.

From the graph you can see that $x^2 + 2x - 8 = 0$ when the curve crosses the $y = 0$ line, giving $x = {}^-4$ and $x = 2$.

So $x = {}^-4$ and $x = 2$ satisfy the equation $x^2 + 2x - 8 = 0$.

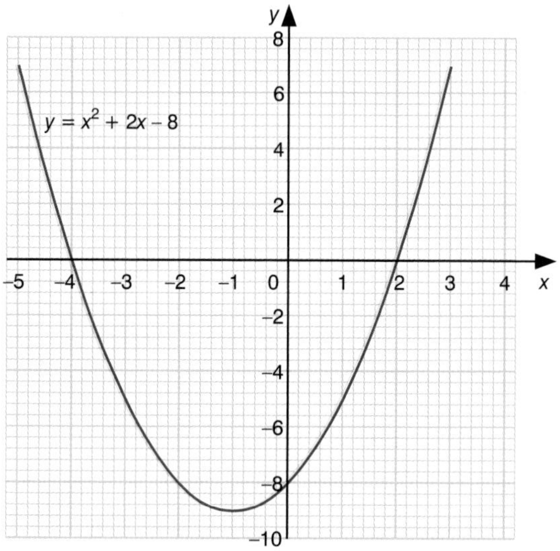

Worked example

Draw the graph $y = 2x^2$ and use it to solve the equation $2x^2 - 5x - 3 = 0$.

Drawing up a table of values:

x	$^-2$	$^-1$	0	1	2	3	4
$y = 2x^2$	8	2	0	2	8	18	32
Coordinates	($^-2$, 8)	($^-1$, 2)	(0, 0)	(1, 2)	(2, 8)	(3, 18)	(4, 32)

To solve the equation $2x^2 - 5x - 3 = 0$ using $y = 2x^2$, rewrite the given equation as follows:

$2x^2 - 5x - 3 = 0$

$2x^2 = 5x + 3$ Adding $5x + 3$ to both sides.

To solve the equation $2x^2 - 5x - 3 = 0$ you need to consider the points which lie on the curve $y = 2x^2$ and on the line $y = 5x + 3$. Any points which satisfy both of these equations will also satisfy the equation $2x^2 - 5x - 3 = 0$.

Draw the graph of $y = 2x^2$ from the table, and then the line $y = 5x + 3$.

From the graph you can see that where the two graphs $y = 2x^2$ and $y = 5x + 3$ cross, $x = {}^-\frac{1}{2}$ and $x = 3$, so these values satisfy the equation $2x^2 - 5x - 3 = 0$.

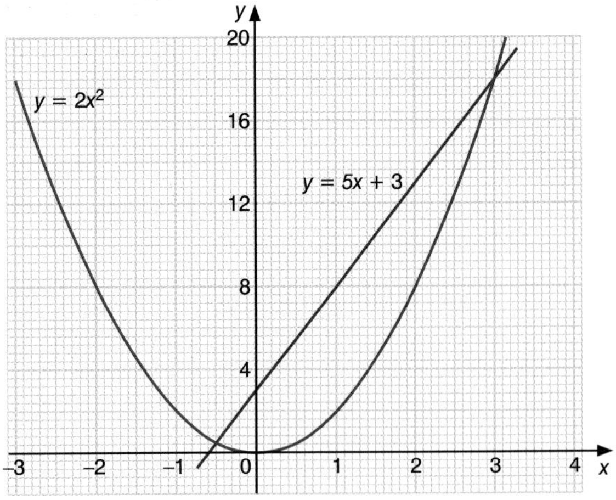

Worked example

Draw the graph of $y = 3x - x^2$ for values of x from $x = {}^-1$ to $x = 4$.

Use your graph to find:

(a) the value of x when $3x - x^2$ is as large as possible

(b) the values of x for which $3x - x^2$ is greater than 1.5.

To draw the graph of $y = 3x - x^2$ for $x = -1$ to $x = 4$, first draw up a table of values:

x	$^-1$	0	1	2	3	4
$y = 3x - x^2$	$^-4$	0	2	2	0	$^-4$
Coordinates	$(^-1, {}^-4)$	(0, 0)	(1, 2)	(2, 2)	(3, 0)	$(4, {}^-4)$

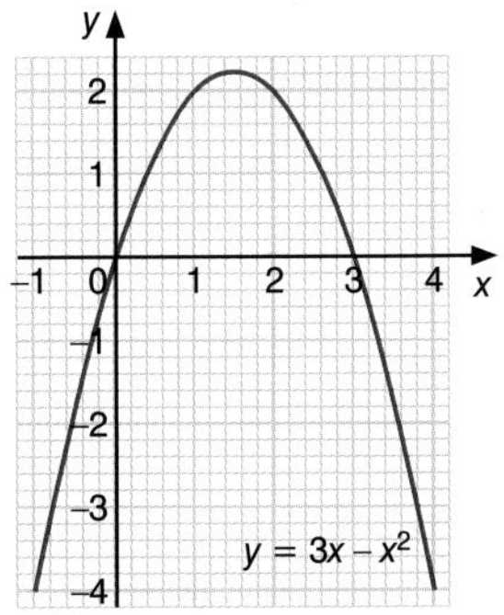

From the graph you can see that:

(a) $3x - x^2$ is as large as possible at $x = 1.5$

(b) $3x - x^2 = 1.5$ when $x = 0.6$ and when $x = 2.4$.

The values of x for which $3x - x^2$ is greater than 1.5 are $0.6 < x < 2.4$.

Check yourself

QUESTIONS

Q1 On the same set of axes, draw and label the following graphs.

(a) $y = x^2$ (b) $y = x^2 + 5$
(c) $y = x^2 - 2$

Q2 Draw the graph of $y = x^2 - 6x + 5$.

Use your graph to find:

(a) the coordinates of the minimum value of $x^2 - 6x + 5$

(b) the values of x when
(i) $x^2 - 6x + 5 = 0$
(ii) $x^2 - 6x + 5 = 5$.

Q3 The height reached by an object thrown into the air is given by the formula:

$h = 20t - 5t^2$

where h is the height in metres and t is the time in seconds.

Plot the graph of h against t for $0 \leqslant t \leqslant 4$ and use your graph to find the maximum height reached by the object.

ANSWERS

A1

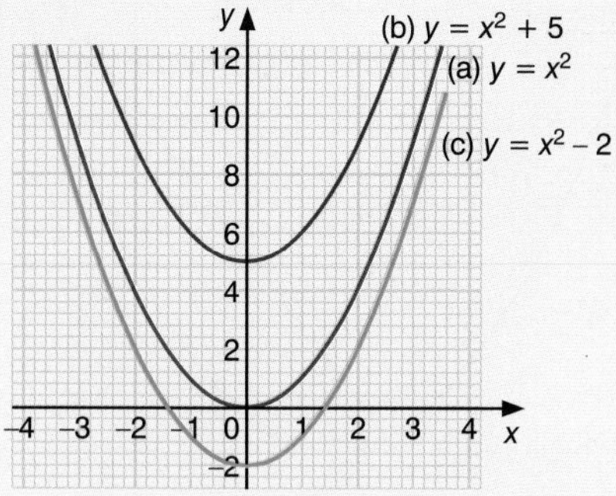

A2

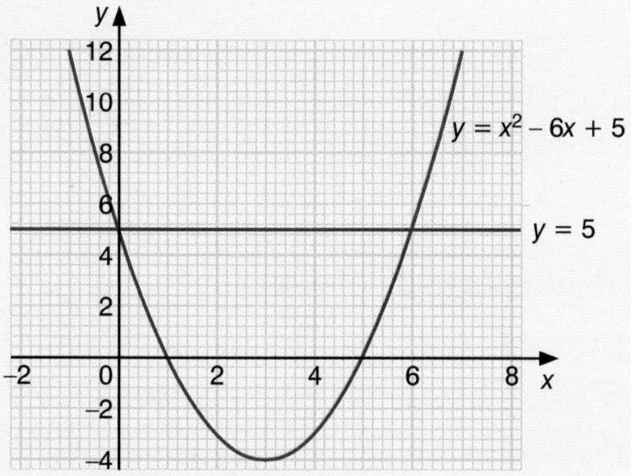

(a) Minimum = (3, ⁻4)
(b) (i) $x = 1$ and $x = 5$
(ii) $x = 0$ and $x = 6$

A3

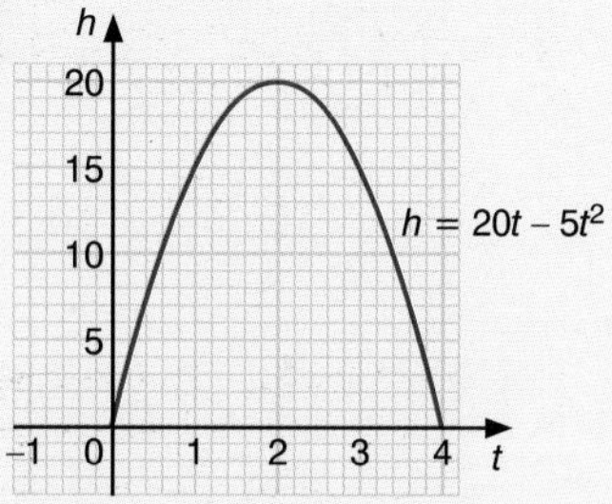

Maximum height = 20 metres

TUTORIAL

T2

From the graph you can see that:

(*a*) *the coordinates of the minimum value of $x^2 - 6x + 5$ are (3, ⁻4)*

(*b*) (*i*) *the values of x when $x^2 - 6x + 5 = 0$ will lie on $y = x^2 - 6x + 5$ and $y = 0$, so the values $x = 1$ and $x = 5$ satisfy the equation $x^2 - 6x + 5 = 0$*

(*ii*) *the values of x when $x^2 - 6x + 5 = 5$ will lie on $y = x^2 - 6x + 5$ and $y = 5$, so the values $x = 0$ and $x = 6$ satisfy the equation $x^2 - 6x + 5 = 5$.*

T3

From the graph you can see that the maximum height occurs at the point (2, 20) so the maximum height is 20 metres (which occurs when the time is 2 seconds).

CUBIC GRAPHS

Cubic graphs can be written in the form $y = ax^3 + bx^2 + cx + d$ (where a is non-zero). They all have approximately the same basic shape as shown in the examples below.

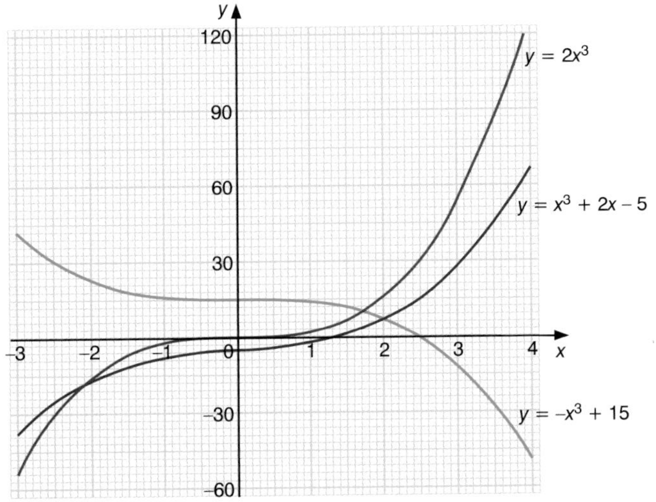

Worked example

Draw the graph of $y = (x - 1)(x - 2)(x - 4)$ for $0 \leqslant x \leqslant 5$ and use it to solve the equation $(x - 1)(x - 2)(x - 4) = 0$.

Drawing up a table of values:

x	0	0.5	1	1.5	2	2.5	3	3.5	4	4.5	5
y	$^-8$	$^-2.625$	0	0.625	0	$^-1.125$	$^-2$	$^-1.875$	0	4.375	12
Coordinates	(0, $^-8$)	(0.5, $^-2.625$)	(1, 0)	(1.5, 0.625)	(2, 0)	(2.5, $^-1.125$)	(3, $^-2$)	(3.5, $^-1.875$)	(4, 0)	(4.5, 4.375)	(5, 12)

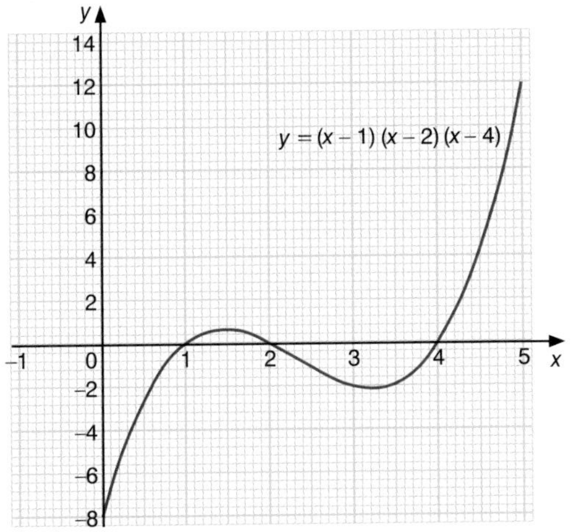

NOTE

You need to use non-integer values to find out sufficient detail about the behaviour of the curve and to be able to draw its graph.

From the graph you can see that $(x - 1)(x - 2)(x - 4) = 0$ when the curve crosses the line $y = 0$, giving $x = 1$, $x = 2$ and $x = 4$ so that $x = 1$, $x = 2$ and $x = 4$ satisfy the equation $(x - 1)(x - 2)(x - 4) = 0$.

RECIPROCAL GRAPHS

Reciprocal graphs can be written in the form $y = \frac{a}{x}$ or $xy = a$ (where a is a constant). They all have approximately the same basic shape as shown in the examples below.

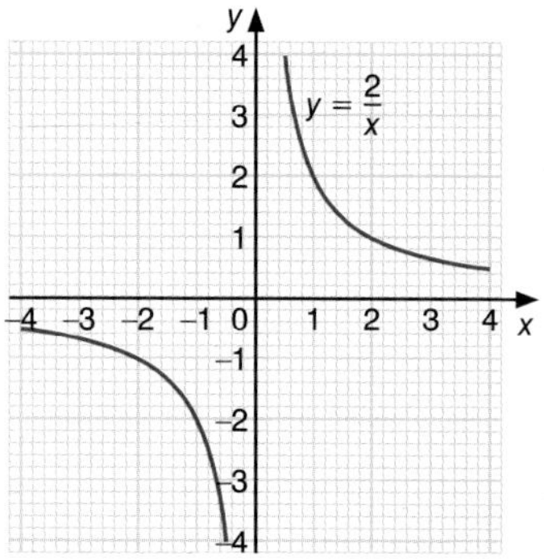

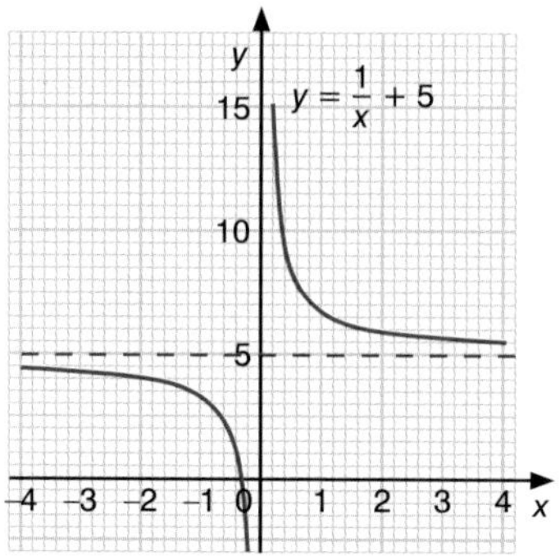

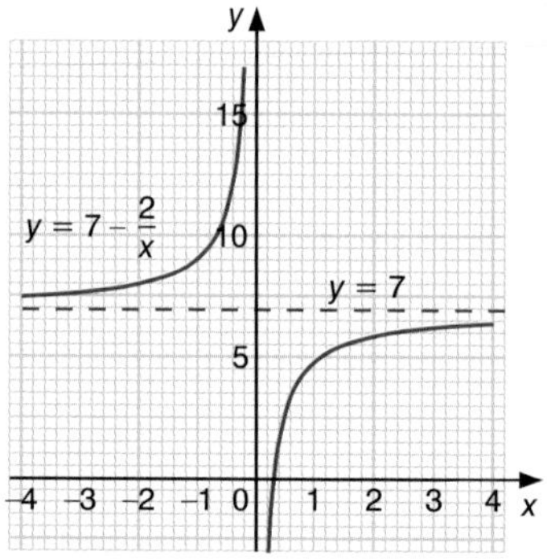

Worked example

On the same set of axes, draw the graph of $xy = 12$ and $y = x$. Use your graphs to solve the equation $x^2 = 12$.

The graph of $xy = 12$ is the same as $y = \frac{12}{x}$ (or $x = \frac{12}{y}$).

Drawing up a table of values:

x	$^{-}4$	$^{-}3$	$^{-}2$	$^{-}1$	0	1	2	3	4
$y = x$	$^{-}4$	$^{-}3$	$^{-}2$	$^{-}1$	0	1	2	3	4
$y = \frac{12}{x}$	$^{-}3$	$^{-}4$	$^{-}6$	$^{-}12$		12	6	4	3

The reciprocal curve $y = \frac{12}{x}$ is not defined at $x = 0$. You will need to use non-integer values, between 0 and 1, and 0 and $^{-}1$, to find out sufficient detail about the behaviour of the curve and to be able to draw its graph.

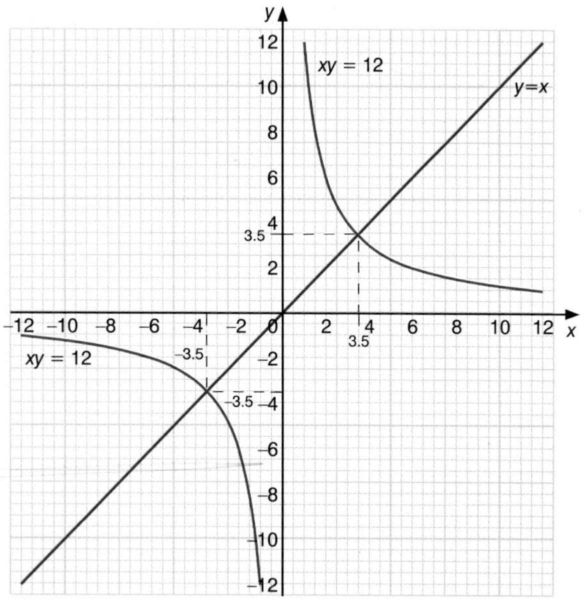

From the graph you can see that the curve $y = \frac{12}{x}$ crosses the line $y = x$ when $x \approx {}^{-}3.5$ and $x \approx 3.5$ correct to 1 d.p.

So $x = {}^{-}3.5$ and $x = 3.5$ satisfy the equation $\frac{12}{x} = x$.

i.e. $12 = x \times x$ Multiplying both sides by x.

or $x^2 = 12$ As required.

NOTE

Since $xy = 12$ and $y = x$, at the point of intersection $x \times x = 12$ which gives $x^2 = 12$, $x = \sqrt{12} \approx 3.5$.

Check yourself

QUESTIONS

Q1 Draw the graph $y = 2x^3 - 3x^2 - 11x + 6$ for $^-2 \leqslant x \leqslant 4$.

Q2 By drawing the graphs of $y = \frac{1}{x}$ and $y = x^2 - 1$ on the same axes, solve the equation $\frac{1}{x} = x^2 - 1$.

REMEMBER! Cover the answers if you want to.

ANSWERS

A1

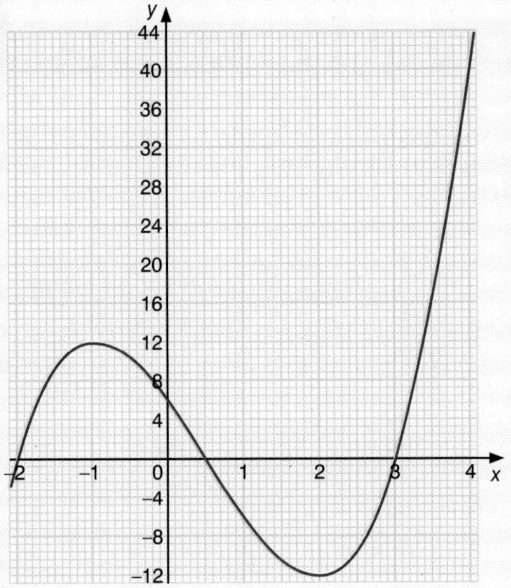

A2

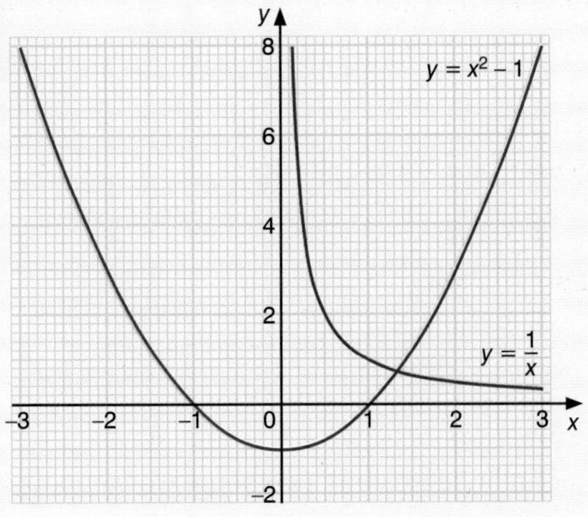

$x \approx 1.3$ satisfies the equation $\frac{1}{x} = x^2 - 1$.

TUTORIAL

T1 *A table of values in the given range should be used to draw the graph, bearing in mind the general shape of a cubic curve.*

x	$^-2$	$^-1$	0	1	2	3	4
y	0	12	6	$^-6$	$^-12$	0	42
Coordinates	($^-2$, 0)	($^-1$, 12)	(0, 6)	(1, $^-6$)	(2, $^-12$)	(3, 0)	(4, 42)

T2 *The reciprocal curve, $y = \frac{1}{x}$ is not defined at $x = 0$.*

You will need to use non-integer values, between 0 and 1, to find out sufficient detail about the behaviour of the curve to be able to draw its graph.

From the graph you can see that the curve $y = \frac{1}{x}$ crosses the curve $y = x^2 - 1$ when $x \approx 1.3$ so that $x \approx 1.3$ satisfies the equation $\frac{1}{x} = x^2 - 1$.

CHAPTER 40

INEQUALITIES AND GRAPHS

The solution of the inequality $x < 5$ can take many values such as 4, π, $\sqrt{2}$, $2\frac{1}{2}$, 0, $^-3\frac{1}{4}$, $^-100\,000$ etc. although such inequalities can be best shown on a number line as follows.

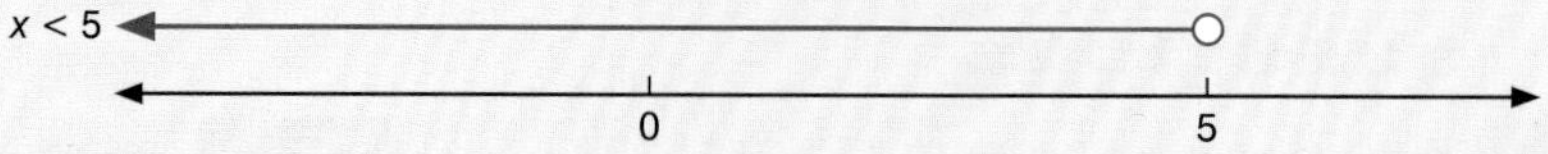

The open circle, ○, at the end of the line shows that the value 5 is not included as $x < 5$.

The solid circle, ●, is used to show that the number is included as shown below for $x \leqslant 5$.

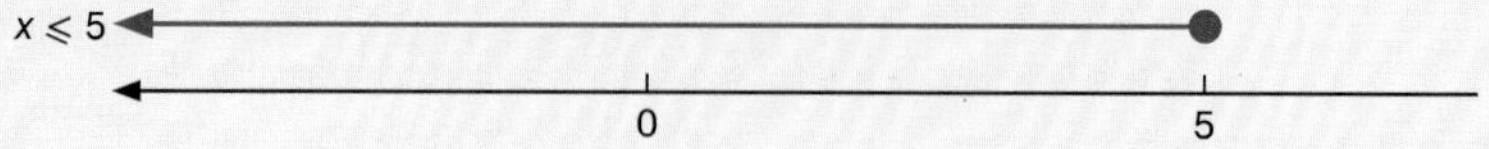

SOLVING INEQUALITIES

Inequalities can be solved in exactly the same way as equalities (i.e. equations) except that **when you multiply or divide by a negative number you must reverse the inequality sign**.

Worked example

Solve the following inequalities and show their solutions on a number line.

(a) $3y + 5 < 17$ (b) $^-10c > 5$ (c) $5 - 8m \leqslant 13$

(d) $4x < 5x + 2$ (e) $2 \leqslant \frac{2}{3}(x + 5) \leqslant 6$

(a) $3y + 5 < 17$

$3y < 12$ Subtracting 5 from both sides.

$y < 4$ Dividing both sides by 3.

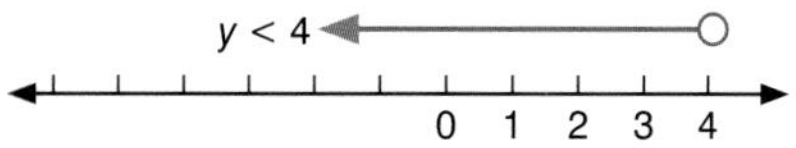

(b) $^-10c > 5$

$c < \frac{5}{^-10}$ Dividing both sides by $^-10$ and reversing the sign.

$c < -\frac{1}{2}$ Cancelling and taking the sign to the front of the expression.

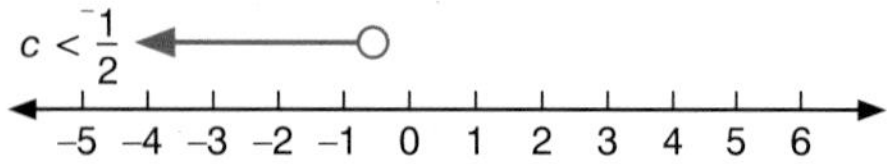

(c) $5 - 8m \leqslant 13$

$^-8m \leqslant 8$ Subtracting 5 from both sides.

$m \geqslant {}^-1$ Dividing both sides by $^-8$ and reversing the sign.

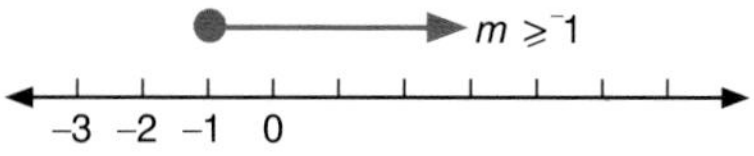

(d) $4x < 5x + 2$

$^-x < 2$ Subtracting $5x$ from both sides.

$x > {}^-2$ Multiplying both sides by $^-1$ and reversing the sign.

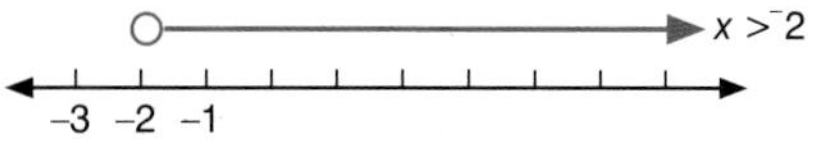

(e) $2 \leqslant \frac{2}{3}(x + 5) \leqslant 6$

This inequality actually represents two separate inequalities.

i.e. $2 \leqslant \frac{2}{3}(x + 5)$ and $\frac{2}{3}(x + 5) \leqslant 6$

$2 \leqslant \frac{2}{3}(x + 5) \leqslant 6$

$6 \leqslant 2(x + 5) \leqslant 18$ Multiplying both sides (of both inequalities) by 3.

$3 \leqslant x + 5 \leqslant 9$ Dividing both sides (of both inequalities) by 2.

$^-2 \leqslant x \leqslant 4$ Subtracting 5 from both sides (of both inequalities).

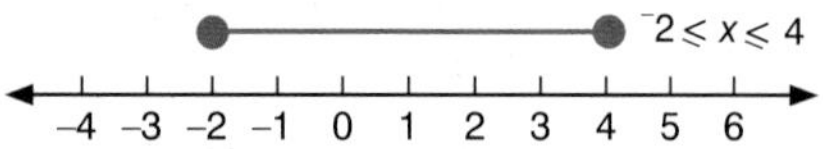

Check yourself

QUESTIONS

Q1 Show the following inequalities on a number line.

(a) $x > 3$
(b) $x \geqslant {}^{-}2$
(c) ${}^{-}2 < x \leqslant 7$

Q2 List the integer values of $18 \leqslant 2x - 10 < 22$.

Q3 Write down the whole numbers in the range ${}^{-}5 < x < 5$ which satisfy the inequality $2(x - 5) > {}^{-}4$.

Q4 Solve these inequalities.

(a) $4x + 2 \leqslant 17 - x$
(b) $18 - 6x > 3 - 3x$

REMEMBER! Cover the answers if you want to.

ANSWERS / TUTORIAL

A1

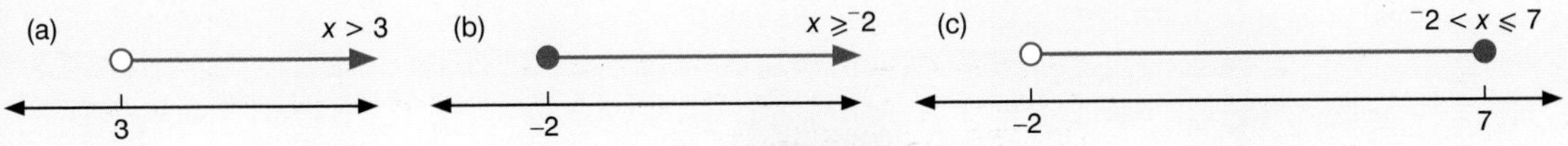

A2 14 and 15

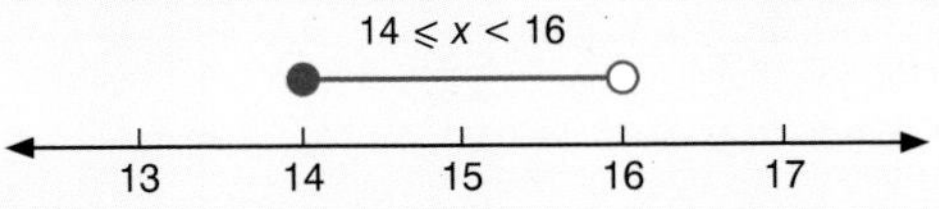

T2 $18 \leqslant 2x - 10 < 22$

$28 \leqslant 2x < 32$ *Adding 10 to both sides (of both inequalities).*

$14 \leqslant x < 16$ *Dividing both sides (of both inequalities) by 2.*

The number line can be used to show the required integer values more clearly.

A3 4

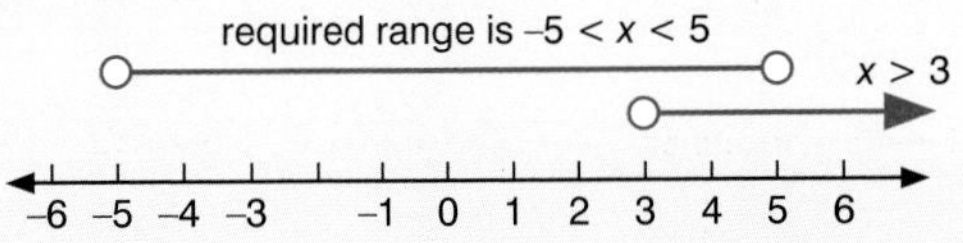

T3 $2(x - 5) > {}^{-}4$

$2x - 10 > {}^{-}4$ *Multiplying out the bracket.*

$2x > 6$ *Adding 10 to both sides of the inequality.*

$x > 3$ *Dividing both sides by 2.*

The number line can be used to show the required whole numbers more clearly.

A4 (a) $x \leqslant 3$
(b) $x < 5$

T4 *(a)* $4x + 2 \leqslant 17 - x$

$4x \leqslant 15 - x$ *Subtracting 2 from both sides.*

$5x \leqslant 15$ *Adding x to both sides.*

$x \leqslant 3$ *Dividing both sides by 5.*

(b) $18 - 6x > 3 - 3x$

$18 > 3 + 3x$ *Adding 6x to both sides.*

$15 > 3x$ *Subtracting 3 from both sides.*

$5 > x$ *Dividing both sides by 3.*

$x < 5$ *Rewriting the inequality to take x to the left-hand side.*

GRAPHING INEQUALITIES

Inequalities can easily be shown on a graph by replacing the inequality sign by an equals (=) sign and drawing the line. This will divide the graph into two regions. You need to decide which of these regions is required.

It is usual to shade out the region which is not required, although some examination questions ask you to shade the required region. You must make it clear to the examiner which is your required region, by labelling it as appropriate.

You also need to make clear whether the line is included (i.e. the inequality is $\leqslant$ or $\geqslant$), or excluded (i.e.the inequality is $<$ or $>$). To do this, use a solid line if the line is included, or a dotted line if it is not included.

Worked example

Draw graphs of these lines.

$x = 2 \qquad y = 4 \qquad y = 6 - x$

On your graph, label the region where the points (x, y) satisfy the inequalities:

$x \geqslant 2 \qquad y < 4 \qquad y \leqslant 6 - x$

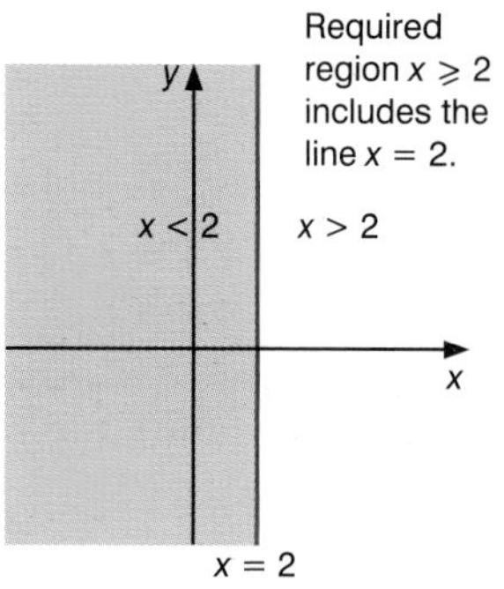

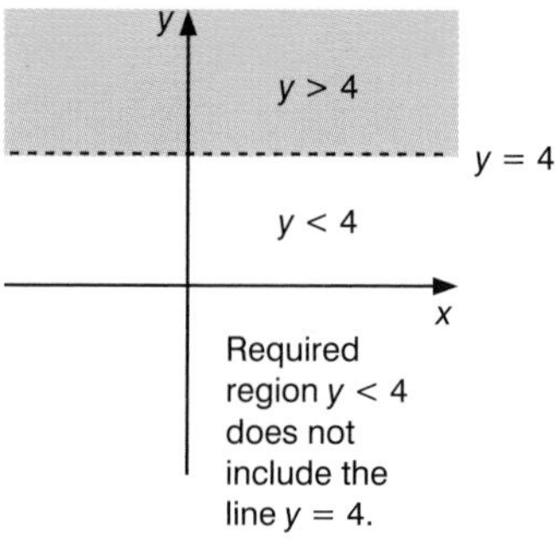

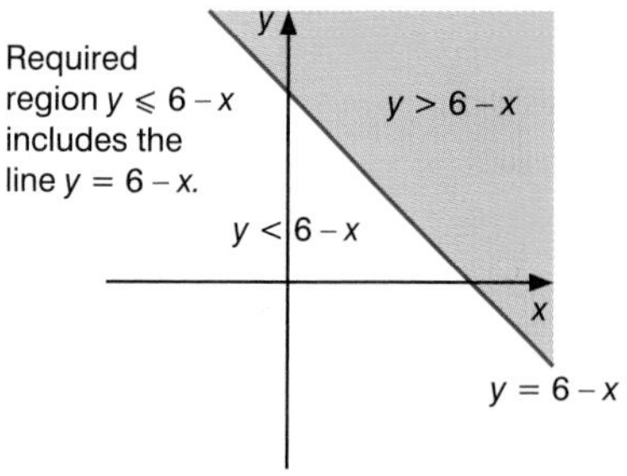

Now combining the graphs:

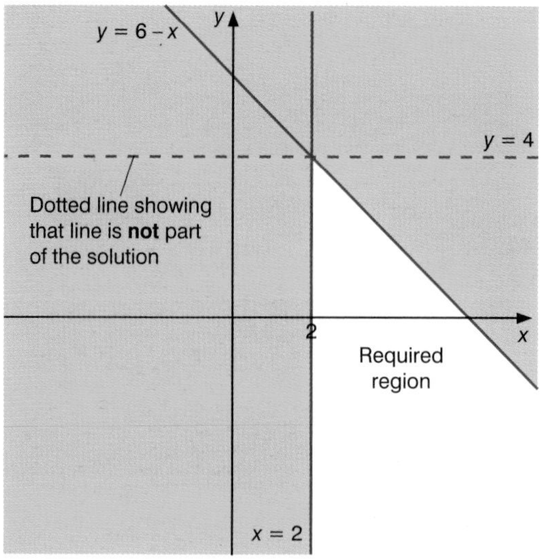

Check yourself

QUESTIONS

Q1 Draw individual graphs to show each of the following inequalities.

(a) $x \geqslant 3$ (b) $y \leqslant 5$
(c) $x + y < 8$ (d) $y > 2x + 5$
(e) $y \geqslant x^2$

Q2 Write down the inequalities illustrated in the unshaded parts of the diagrams below.

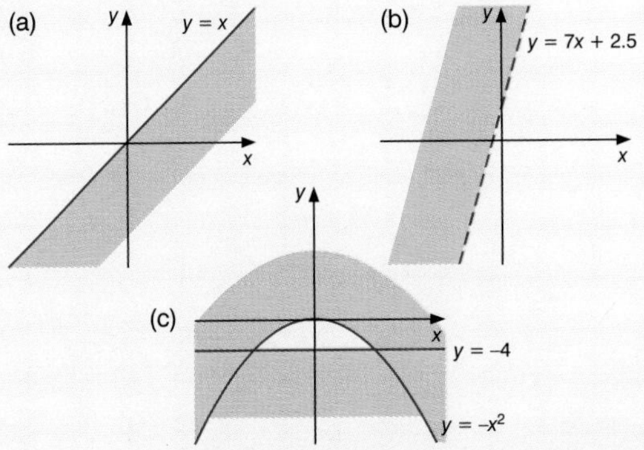

Q3 Draw a set of axes, with each axis labelled from 0 to 10. Label the region represented by these inequalities.

$x \geqslant 2$ $y \geqslant 1$ $x + y \leqslant 9$

What is the maximum value of $x + y$ which satisfies all of these conditions?

Q4 Draw a set of axes, with each axis labelled from $^-2$ to 6. Shade and label the region represented by the inequalities $y \leqslant x + 2$, $x + y < 6$ and $y \geqslant 0$.

REMEMBER! Cover the answers if you want to.

ANSWERS

A1

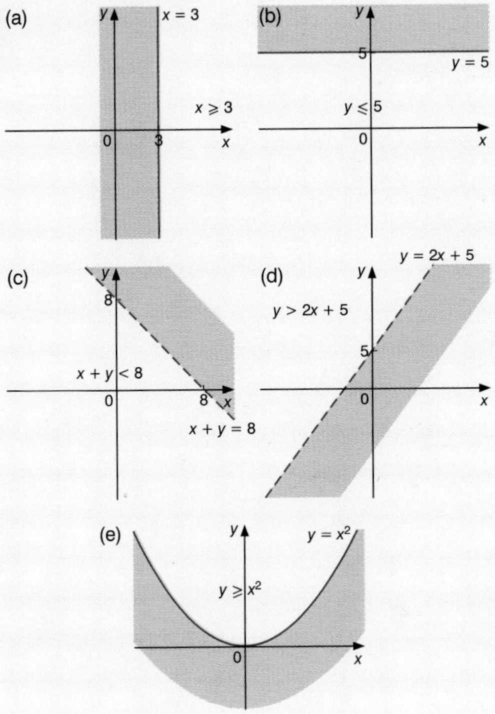

TUTORIAL

T1 *Remember to use dotted lines to show the inequalities for 'less than' (<) or 'greater than' (>), and solid lines for 'less than or equal to' (≤) and 'greater than or equal to' (≥).*

ANSWERS

A2

(a) $y \leqslant x$
(b) $y < 7x + 2.5$
(c) $y \leqslant {}^-x^2$ and $y \geqslant {}^-4$

TUTORIAL

T2

(a) $y \leqslant x$
Values can be taken either side of the given line to ascertain the required region.

(b) $y < 7x + 2.5$
The dotted line shows that the line is ***not*** *included in the required region.*

(c) $y \leqslant {}^-x^2$ *and* $y \geqslant {}^-4$
Both inequalities must be satisfied to give the required region.

A3

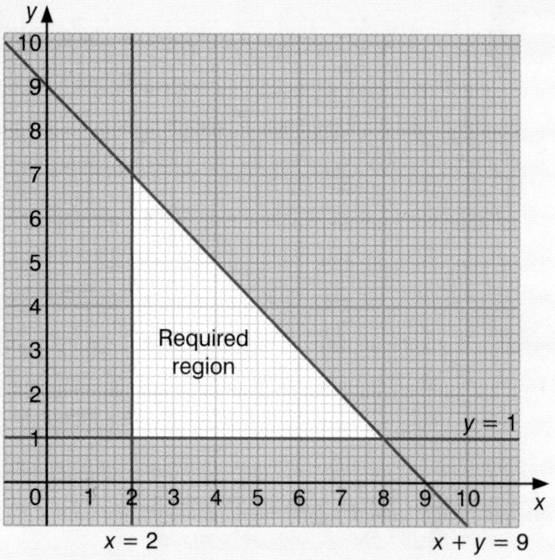

The maximum value for $x + y$ which satisfies all of these conditions is 9 at the points (2, 7) and (8, 1) on the graph.

A4

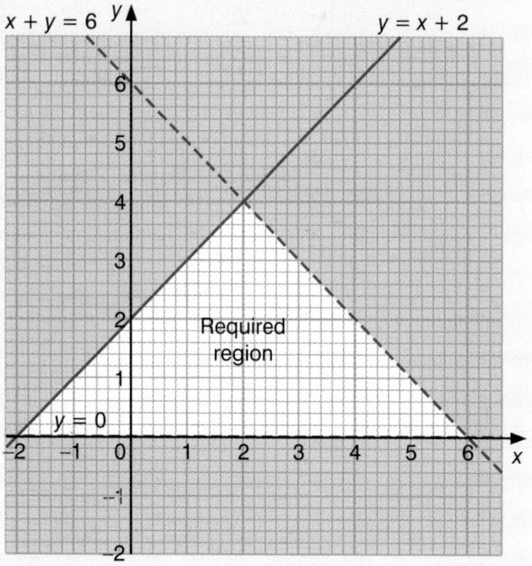

EXAM PRACTICE

Sample Student's Answers & Examiner's Comments

1 The diagram shows the graphs of $y = \frac{1}{2}x + 1$, $5x + 6y = 30$ and $x = 2$.

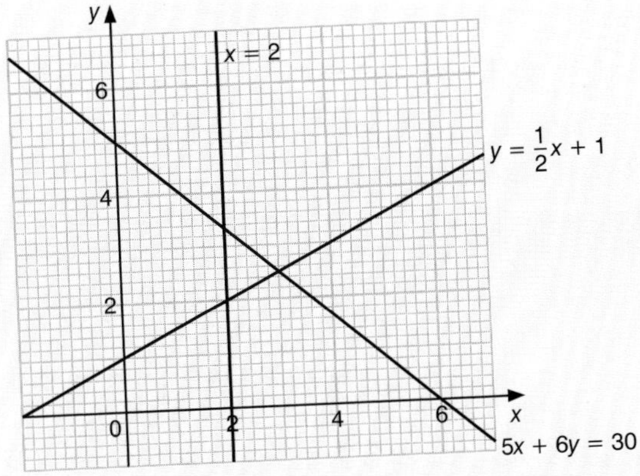

(a) On the diagram, shade and label, with the letter R, the region for which the points (x, y) satisfy the three inequalities:

$y \leqslant \frac{1}{2}x + 1$

$5x + 6y \leqslant 30$ and

$x \geqslant 2$

(b) Solve the inequality $\frac{1}{2}x + 1 < 3$.

(c) Represent your answer to part (b) on the number line.

(MEG syllabus A, specimen paper 1998)

(a)

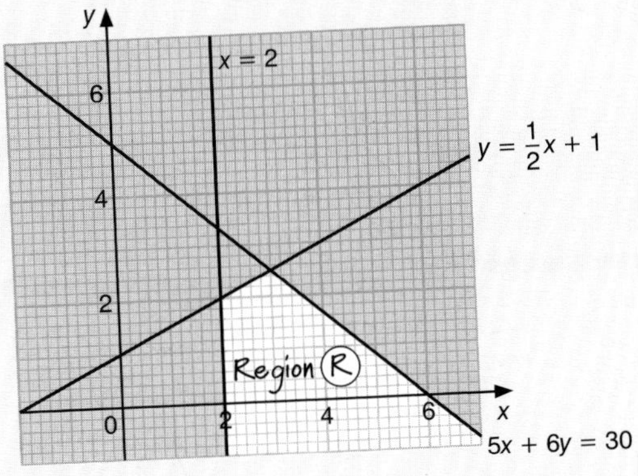

(b) $\frac{1}{2}x + 1 < 3$

$\frac{1}{2}x < 3 - 1$

$\frac{1}{2}x < 2$

$x < 4$

(c)

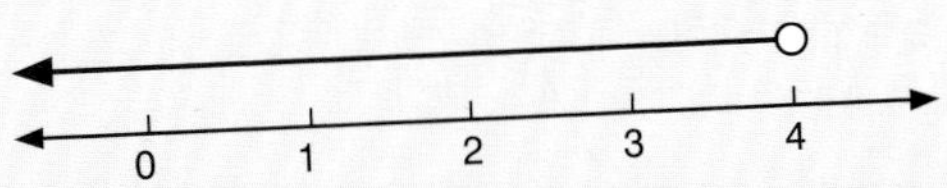

EXAMINER'S COMMENTS

1 *The candidate has identified the given lines and marked these on the diagram. The question asks the candidate to shade and label, with the letter R, the required region rather than the region not required. He has, however, successfully found the region which is suitably highlighted.*

EXAMINER'S COMMENTS

2 (a) Draw the graph of $y = x^2 - 3x$ for $^-2 \leqslant x \leqslant 4$.

(b) By drawing a suitable straight line on your graph estimate, correct to one decimal place, the two solutions of the equation $x^2 - 3x + 1 = 0$.

(c) By using your graph, or otherwise, solve the inequality $x^2 - 3x < 0$.

$x^2 - 3x + 1 = 0$

$x^2 - 3x = {}^-1$

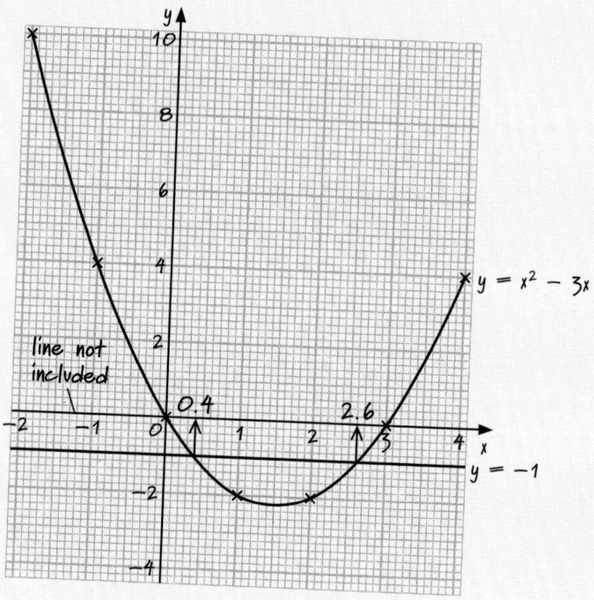

~~$x^2 - 3x < 0$~~

~~$x^2 < 3x$~~

~~$x < 3$~~

$0 < x < 3$

2 a)
The candidate has drawn the curve $y = x^2 - 3x$ for $^-2 \leqslant x \leqslant 4$ and has drawn the line $y = {}^-1$ to solve $x^2 - 3x = {}^-1$ which is the same as $x^2 - 3x + 1 = 0$.

b)
The values obtained from the graph are 0.4 and 2.6 although she has not stated these explicitly which might result in some marks being lost.

c)
The candidate did attempt to solve the inequality without reference to the graph but this attempt was unsuccessful and was subsequently crossed out.

By referring to the graphs he has correctly identified the required values of x and appreciated that the points $x = 0$ and $x = 3$ are not included as they do not satisfy $x^2 - 3x < 0$.

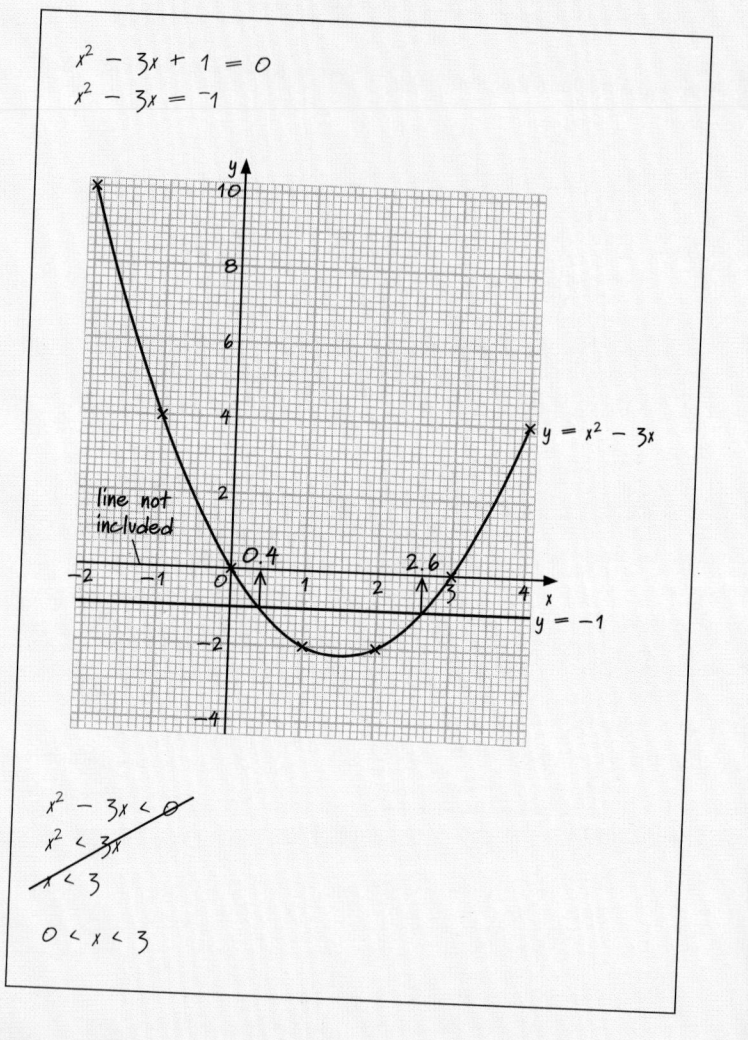

Questions to Answer

1 Gas is charged at 10 pence per therm and a standing charge of £9.80 is applied. Show this information on a graph and use this graph to find:

(a) the cost of 25 therms

(b) the cost of 66 therms

(c) the number of therms used if the bill is £14.60.

2 The travel graph shows a train journey between two towns A and C stopping at B.

Use the graph to find:

(a) the average speed between towns A and B

(b) the average speed between towns A and C.

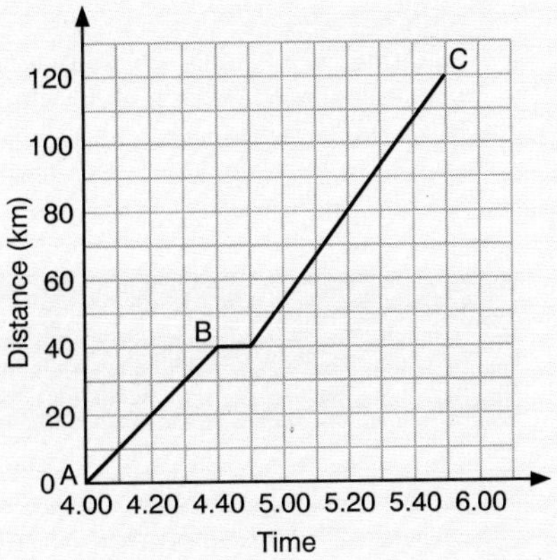

3 Draw the graph of $y = 2x^2$ for $^-3 \leqslant x \leqslant 3$ and use your graph to find approximate solutions to the following equations.

(a) $2x^2 = 2$ (b) $2x^2 = 6x$ (c) $2x^2 - x - 6 = 0$

4 On the same axes, draw the graphs of $xy = 6$ and $y = x + 1$.
Use these graphs to solve the equation $x^2 + x - 6 = 0$.

5 Solve the inequality $2(3 + 2x) > 4$ and show the solutions on a number line.

6 Given that $8 - 2x \leqslant 12$ what is the least value of x?

7 Write down the three inequalities which define the triangular region ABC.

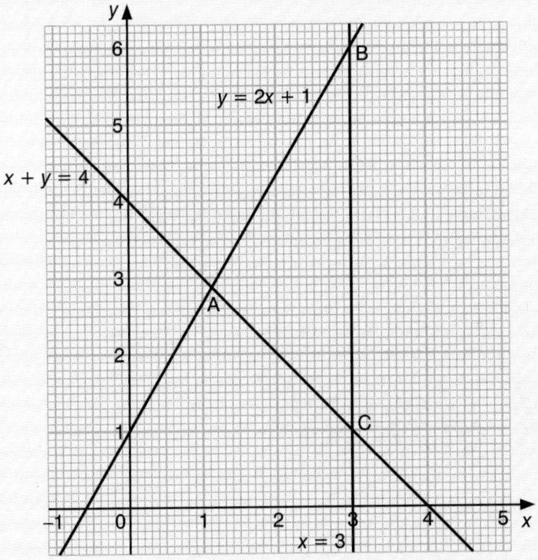

(MEG syllabus A, specimen paper 1998)

CHAPTER 42

SIMULTANEOUS EQUATIONS

A pair of simultaneous equations is a pair of equations in two unknowns. Both equations are correct at the same time, or simultaneously. Simultaneous equations are usually solved by graphical or algebraic methods. The examination question should make it clear which method you should use or else any method will be acceptable.

GRAPHICAL SOLUTION OF SIMULTANEOUS EQUATIONS

Simultaneous equations can be solved by using graphs to plot the two equations. The coordinates of the point of intersection gives the solution of the simultaneous equations.

Worked example

Solve these simultaneous equations.

$y = x + 4$

$y = 2x + 3$

The simultaneous equations $y = x + 4$ and $y = 2x + 3$ can be plotted as a pair of straight lines, like this.

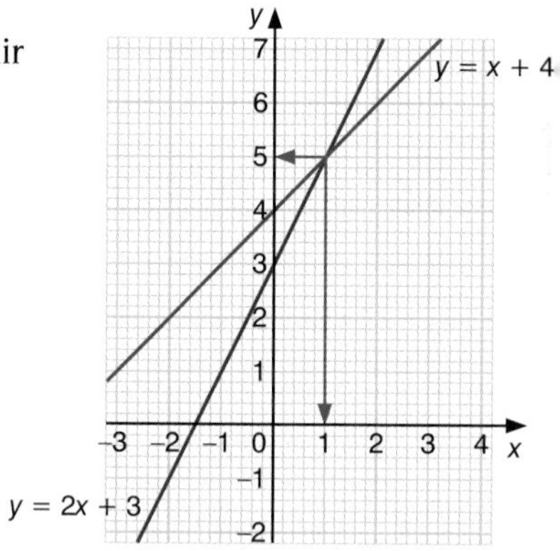

The coordinates of the point of intersection, (1, 5) give the solution of the simultaneous equations. At the point of intersection (1, 5) , $x = 1$ and $y = 5$. So the solution is $x = 1$ and $y = 5$.

HINT

You should now check this solution by substituting the values in the original equations.

Check yourself

QUESTIONS

By drawing suitable graphs, solve the following pairs of simultaneous equations.

Q1 $y = x + 4$
$y = 4x - 2$

Q2 $x - y = 1$
$2x + y + 7 = 0$

ANSWERS

A1 $x = 2$ and $y = 6$

A2 $x = {}^{-}2$ and $y = {}^{-}3$

TUTORIAL

T1 *The graphs of the two simultaneous equations can easily be sketched if you know the gradient and the cut-off on the y-axis (y-intercept).*

$y = x + 4$ *$m = 1$ and $c = 4$ (where m is the gradient and c is the cut-off on the y-axis or the y-intercept)*

$y = 4x - 2$ *$m = 4$ and $c = {}^{-}2$*

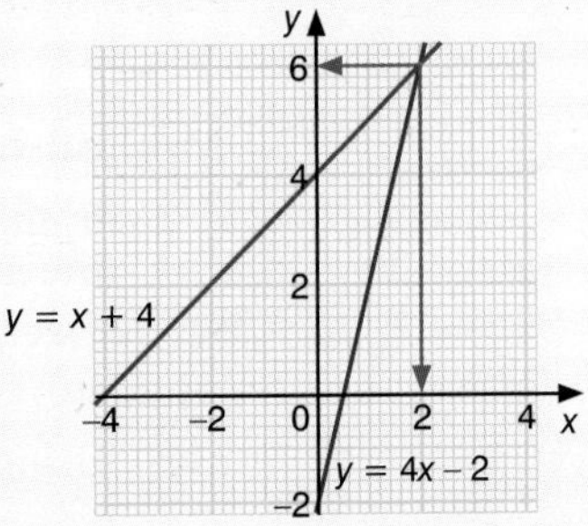

The coordinates of the point of intersection, (2, 6) give the solution of the simultaneous equations, $x = 2$ and $y = 6$.

T2 $x - y = 1$
$2x + y + 7 = 0$

Rearranging the above formulae to get them in the form $y = mx + c$:

$x - y = 1$		
$x = 1 + y$	*Adding y to both sides.*
$x - 1 = y$	*Subtracting 1 from both sides.*
$y = x - 1$	*Putting y on the left-hand side.*

$m = 1$ and $c = {}^{-}1$

$2x + y + 7 = 0$	
$y + 7 = {}^{-}2x$	*Subtracting 2x from both sides.*
$y = {}^{-}2x - 7$	*Subtracting 7 from both sides.*

$m = {}^{-}2$ and $c = {}^{-}7$

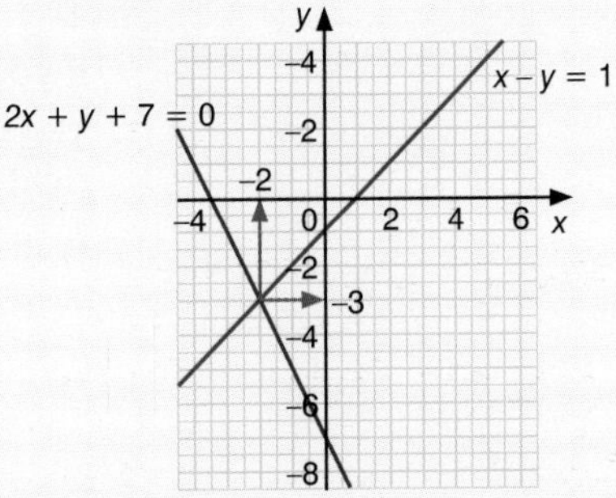

The coordinates of the point of intersection, $({}^{-}2, {}^{-}3)$ give the solution of the simultaneous equations, $x = {}^{-}2$ and $y = {}^{-}3$.

ALGEBRAIC SOLUTION OF SIMULTANEOUS EQUATIONS

Two methods are commonly employed when solving simultaneous equations as shown in the following worked example.

Worked example

Solve these simultaneous equations.

$x + 2y = 5$

$3x - 2y = 7$

Method of substitution

In this method, one equation is rewritten to make one of the unknowns the subject of the equation.

This is then substituted into the second equation which can then be solved as follows.

Using $x + 2y = 5$ you can write:

$x = 5 - 2y$ Making x the subject of the equation.

Now substitute this value of x into the second equation.

$3x - 2y = 7$

$3(5 - 2y) - 2y = 7$ Substituting $x = 5 - 2y$.

$15 - 6y - 2y = 7$ Expanding the brackets.

$8 = 8y$ Collecting like terms on each side.

$y = 1$

You can now use $x = 5 - 2y$ with $y = 1$ to find x.

$x = 5 - 2y$

$x = 5 - 2 \times 1$ As $y = 1$.

$x = 3$

Method of elimination

In this method, the equations, or multiples of the equations, are added or subtracted in order to eliminate one of the unknowns. The resulting equation can then be solved as follows.

$x + 2y = 5$

$3x - 2y = 7$

If you add the left-hand sides of the equations then y will be eliminated.

The sum of the two left-hand sides must equal the sum of the two right-hand sides.

$(x + 2y) + (3x - 2y) = 5 + 7$

$x + 2y + 3x - 2y = 12$

$4x = 12$

$x = 3$

or

$$\begin{array}{rl} x + 2y &= 5 \\ + \; 3x - 2y &= 7 \\ \hline 4x \quad\quad &= 12 \\ \hline x &= 3 \end{array}$$

Substituting for y in the first equation:

$x + 2y = 5$

$3 + 2y = 5$ As $x = 3$.

$2y = 2$ Subtracting 3 from both sides.

$y = 1$ Dividing both sides by 2.

Check yourself

QUESTIONS

Q1 Solve the following simultaneous equations by:

(i) the graphical method
(ii) the method of substitution
(iii) the method of elimination.

(a) $x + 3y = 10$
$2x - 3y = 2$

(b) $x = 5y - 3$
$3x - 8y = 12$

(c) $x - 1 = y$
$3x + 4y = 6$

Q2 Calculate the coordinates of the point of intersection of these lines.

$2x + 3y = 7$

$5x - 3y = 21$

Q3 Find two numbers with a sum of 36 and a difference of 4.

Q4 The cost of two ties and a shirt is £32.50 while the cost of one tie and two shirts is £41.00. What is the cost of a tie and a shirt?

REMEMBER! Cover the answers if you want to.

ANSWERS

A1 (a) $x = 4$ and $y = 2$

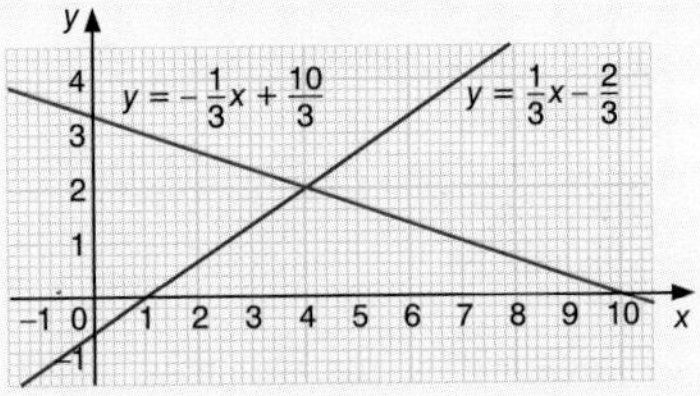

(b) $x = 12$ and $y = 3$

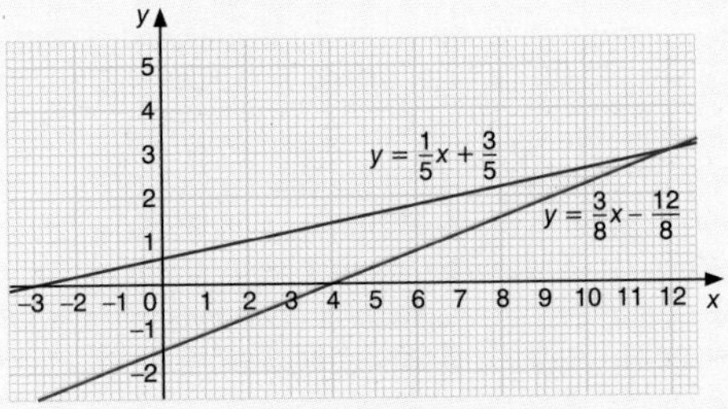

(c) $x = \frac{10}{7}$ or $1\frac{3}{7}$ and $y = \frac{3}{7}$

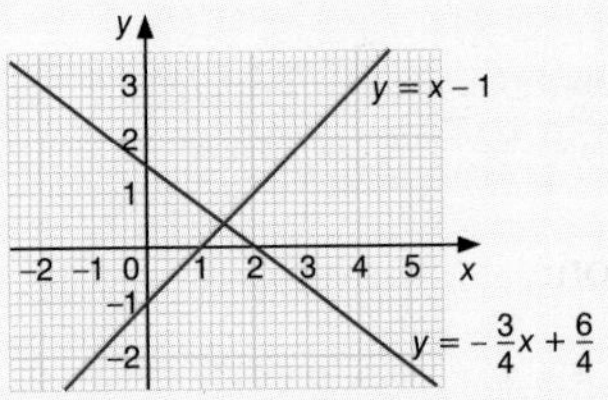

TUTORIAL

T1 *(a)* $x + 3y = 10$
$2x - 3y = 2$

(i) *Graphical method* — *Plot* $y = -\frac{1}{3}x + \frac{10}{3}$ *and* $y = \frac{2}{3}x - \frac{2}{3}$. *Intersect at (4, 2)*

(ii) *Method of substitution* — *Substitute* $x = 10 - 3y$ *in the second equation.*

(iii) *Method of elimination* — *Add the two equations to eliminate the term in* y.

(b) $x = 5y - 3$
$3x - 8y = 12$

(i) *Graphical method* — *Plot* $y = \frac{1}{5}x + \frac{3}{5}$ *and* $y = \frac{3}{8}x - \frac{12}{8}$. *Intersect at (12, 3).*

(ii) *Method of substitution* — *Substitute* $x = 5y - 3$ *in the second equation.*

(iii) *Method of elimination* — *Rewrite the first equation as* $x - 5y = {}^{-}3$ *and then* $3x - 15y = {}^{-}9$ *(multiplying by 3). Subtract the two equations to eliminate the term in* x.

(c) $x - 1 = y$
$3x + 4y = 6$

(i) *Graphical method* — *Plot* $y = x - 1$ *and* $y = -\frac{3}{4}x + \frac{6}{4}$. *Intersect at* $(\frac{10}{7}, \frac{3}{7})$.

(ii) *Method of substitution* — *Substitute* $y = x - 1$ *in the second equation.*

(iii) *Method of elimination* — *Rewrite the first equation as* $x - y = 1$ *and then* $4x - 4y = 4$ *(multiplying by 4). Add the two equations to eliminate the term in* y.

ANSWERS

A2 $(4, {}^-\frac{1}{3})$

A3 16 and 20

A4 The cost of a tie is £8.00 and the cost of a shirt is £16.50. The cost of a tie and a shirt is £24.50.

TUTORIAL

T2

$2x + 3y = 7$
$5x - 3y = 21$

The question says 'calculate' so solving these simultaneous equations by the method of elimination and adding the two equations:

$$(2x + 3y) + (5x - 3y) = 7 + 21$$
$$7x = 28$$
$$x = 4$$

Substituting in the first equation:

$2x + 3y = 7$
$8 + 3y = 7$ As $x = 4$.
$3y = {}^-1$
$y = {}^-\frac{1}{3}$

So the coordinates of the point of intersection of the lines are $(4, {}^-\frac{1}{3})$.

T3

Let the two numbers be x and y (say).

$x + y = 36$ *As their sum is 36.*
$x - y = 4$ *As their difference is 4.*

Solving these simultaneous equations by the method of elimination and adding the two equations:

$$(x + y) + (x - y) = 36 + 4$$
$$2x = 40$$
$$x = 20$$

Substituting in the first equation:

$x + y = 36$
$20 + y = 36$ As $x = 20$.
$y = 16$

So the two numbers are 20 and 16.

T4

Let £t be the cost of one tie and £s be the cost of one shirt.

$2t + s = 32.50$ *As the cost of two ties and a shirt is £32.50.*
$t + 2s = 41.00$ *As the cost of one tie and two shirts is £41.00.*

Solving these simultaneous equations by the method of substitution:
Using $2t + s = 32.50$ *you can write* $s = 32.50 - 2t$.

Substituting this value of s into the second equation:

$t + 2s = 41$ *41 is the same as 41.00*
$t + 2(32.50 - 2t) = 41$ *Substituting* $s = 32.50 - 2t$.
$t + 65 - 4t = 41$ *Expanding the brackets.*
${}^-3t = {}^-24$ *Collecting like terms on each side.*
$t = 8$ *Dividing both sides by* ${}^-3$.

You can now use $s = 32.50 - 2t$ *with* $t = 8$ *to find s.*

$s = 32.50 - 2t$
$s = 32.50 - 2 \times 8$ As $t = 8$.
$s = 16.50$

The cost of a tie is £8.00 and the cost of a shirt is £16.50.
A tie and a shirt together cost £24.50.

CHAPTER 43

QUADRATIC EQUATIONS

Quadratic equations are equations of the form $ax^2 + bx + c = 0$ where $a \neq 0$. Quadratic equations can be solved in a number of ways but at this level they are usually solved by graphical or algebraic methods. The examination question should make it clear which method you should use. If it does not, any method will be acceptable.

See Chapter 39, Quadratic, cubic and reciprocal graphs for further information on how to solve quadratic equations by graphical methods.

SOLUTION BY FACTORS

If the product of two numbers is zero then one or both of them must be zero.

If $ab = 0$ then either $a = 0$ or $b = 0$ or both $a = 0$ and $b = 0$.

This fact can also be applied to the solution of quadratic equations.

Worked example

Solve the quadratic equation $(x - 3)(x + 1) = 0$.

Since the product of the two factors is zero then one or both of them must be zero so:

either $(x - 3) = 0$ which implies that $x = 3$

or $(x + 1) = 0$ which implies that $x = {}^-1$.

The solutions of the equation $(x - 3)(x + 1) = 0$ are $x = 3$ and $x = {}^-1$.

Worked example

Solve the quadratic equation $x^2 + 4x - 21 = 0$.

Factorising the left-hand side of the equation:

$x^2 + 4x - 21 = (x \quad)(x \quad)$
$= (x - 3)(x + 7)$

You need to look for numbers which, when multiplied together, give ⁻21 and which, when added, give 4.

Try ⁺1 and ⁻21 ⁻1 and ⁺21

⁺3 and ⁻7 ⁻3 and ⁺7

The quadratic equation can be written $(x - 3)(x + 7) = 0$.

Since the product of the two factors is zero then one or both of them must be zero so:

either $(x - 3) = 0$ which implies that $x = 3$

or $(x + 7) = 0$ which implies that $x = {}^-7$.

The solutions of the equation $x^2 + 4x - 21 = 0$ are $x = 3$ and $x = {}^-7$.

NOTE

See Chapter 31, Expanding and factorising for further information on factorising quadratic equations.

Check yourself

QUESTIONS

Q1 Solve the following quadratic equations.

(a) $(x - 5)(x - 7) = 0$
(b) $(x - 6)(2x + 1) = 0$
(c) $x^2 + 4x - 5 = 0$

Q2 Write down quadratic equations with solutions as follows.

(a) $x = 2$ and $x = 5$ (b) $x = {}^-3$ and $x = \frac{1}{5}$

Q3 Solve $x^2 = 6x$.

Q4 Solve $x^2 - 5x + 2 = 8$.

Q5 The length of a rectangle is 4 centimetres longer than the width. The area of the rectangle is 96 square centimetres. What are the length and the width of the rectangle?

REMEMBER! Cover the answers if you want to.

ANSWERS

A1
(a) $x = 5$ and $x = 7$
(b) $x = 6$ and $x = {}^{-}\frac{1}{2}$
(c) $x = 1$ and $x = {}^{-}5$

A2
(a) $(x - 2)(x - 5) = 0$ or $x^2 - 7x + 10 = 0$
(b) $(x + 3)(5x - 1) = 0$ or $5x^2 + 14x - 3 = 0$

A3
$x = 0$ and $x = 6$

A4
$x = 6$ and $x = {}^{-}1$

A5
width = 8 metres, length = 12 metres

TUTORIAL

T1
(a) $(x - 5)(x - 7) = 0$ *Either* $(x - 5) = 0$ *or* $(x - 7) = 0$
The solutions are $x = 5$ *and* $x = 7$.

(b) $(x - 6)(2x + 1) = 0$ *Either* $(x - 6) = 0$ *or* $(2x + 1) = 0$
The solutions are $x = 6$ *and* $x = {}^{-}\frac{1}{2}$.

(c) $x^2 + 4x + 5 = 0$
$(x - 1)(x + 5) = 0$ *Either* $(x - 1) = 0$ *or* $(x + 5) = 0$
The solutions are $x = 1$ *and* $x = {}^{-}5$.

T2
Reverse the above process to find the factors.

(a) $x = 2$ *and* $x = 5$
Either $(x - 2) = 0$ *or* $(x - 5) = 0$
$(x - 2)(x - 5) = 0$ *or* $x^2 - 7x + 10 = 0$

(b) $x = {}^{-}3$ *and* $x = \frac{1}{5}$
Either $(x + 3) = 0$ *or* $(x - \frac{1}{5}) = 0$
$x - \frac{1}{5} = 0$ *gives* $5x - 1 = 0$ *Multiplying both sides by 5.*
$(x + 3)(5x - 1) = 0$ *or* $5x^2 + 14x - 3 = 0$

T3
$x^2 = 6x$
$x^2 - 6x = 0$ *Writing in the form* $ax^2 + bx + c = 0$ *and factorising.*
$x(x - 6) = 0$ *Either* $x = 0$ *or* $(x - 6) = 0$
The solutions are $x = 0$ *and* $x = 6$.

T4
$x^2 - 5x + 2 = 8$
$x^2 - 5x - 6 = 0$ *Writing in the form* $ax^2 + bx + c = 0$ *and factorising.*
$(x - 6)(x + 1) = 0$ *Either* $(x - 6) = 0$ *or* $(x + 1) = 0$
The solutions are $x = 6$ *and* $x = {}^{-}1$.

T5
Let the width be x *centimetres, then the length is* $(x + 4)$ *centimetres.*
Area $= x(x + 4) = 96$
$x^2 + 4x = 96$ *Expanding the brackets.*
$x^2 + 4x - 96 = 0$ *Writing in the form* $ax^2 + bx + c = 0$ *and factorising.*
$(x - 8)(x + 12) = 0$ *Either* $x = 8$ *or* $x = {}^{-}12$.
But a width of ${}^{-}12$ *centimetres is impossible so you ignore this value. Since the length is 4 cm greater than the width, this gives width = 8 cm and length = 12 cm.*

CHAPTER 44

TRIAL AND IMPROVEMENT METHODS

Trial and improvement can be used to provide successively better approximations to the solution of a problem. A first approximation is repeatedly refined to provide an improved solution until the desired accuracy is obtained.

Worked example

The length of a rectangle is 2 cm greater than the width. The area of the rectangle is 30 cm^2. Use trial and improvement to obtain the length and width, to the nearest millimetre.

A useful way to solve this problem is to draw up a table.

width	length (width + 2)	area	comments	
4	6	24	too small	
5	7	35	too large	width must lie between 4 and 5
4.5	6.5	29.25	too small	width must lie between 4.5 and 5
4.7	6.7	31.49	too large	width must lie between 4.5 and 4.7
4.6	6.6	30.36	too large	width must lie between 4.5 and 4.6
4.55	6.55	29.8025	too small	width must lie between 4.55 and 4.6

Since 4.55 and 4.6 are both equal to 4.6 (correct to the nearest one-tenth of a centimetre, or millimetre) then you can stop and say that the solution is 4.6 (correct to the nearest millimetre).

Alternatively, notice that:

using 4.5 gives 29.25, which is $^-0.75$ from 30

using 4.6 gives 30.36, which is 0.36 from 30.

So 4.6 is closer.

Worked example

A solution of the equation $x^3 - 3x = 25$ lies between 3 and 4. Use a trial and improvement method to find this solution, giving your answer correct to 1 d.p.

When $x = 3$ $\quad x^3 - 3x = 3^3 - 3 \times 3 = 18$

When $x = 4$ $\quad x^3 - 3x = 4^3 - 3 \times 4 = 52$

So the solution lies between 3 and 4 (and seems to be closer to $x = 3$).

Try $x = 3.5$ $\quad x^3 - 3x = 3.5^3 - 3 \times 3.5 = 32.375$

So the solution lies between 3 and 3.5 (and seems to be closer to $x = 3.5$).

Try $x = 3.3$ $\quad x^3 - 3x = 3.3^3 - 3 \times 3.3 = 26.037$

So the solution lies between 3 and 3.3 (and seems to be closer to $x = 3.3$).

Try $x = 3.2$ $\quad x^3 - 3x = 3.2^3 - 3 \times 3.2 = 23.168$

So the solution lies between 3.2 and 3.3.

Try $x = 3.25$ $\quad x^3 - 3x = 3.25^3 - 3 \times 3.25 = 24.578\,125$

So the solution lies between 3.25 and 3.3.

Since 3.25 and 3.3 are both equal to 3.3 (correct to 1 decimal place) then you can stop and say that the solution = 3.3 (correct to 1 decimal place).

Alternatively:

using 3.2 gives 23.168 which is $^-1.832$ from 25

using 3.3 gives 26.037 which is 1.037 from 25.

So $x = 3.3$ is the better solution.

HINT

An initial approximation can always be obtained by drawing a graph of the function although this initial value will usually be given in the question. Alternatively, you could try a few calculations, using whole numbers, in your head.

Check yourself

QUESTIONS

Q1 A solution of the equation $x^3 + x = 100$ lies in the range $4 \leqslant x \leqslant 5$. Use trial and improvement to find the solution correct to 1 decimal place.

Q2 Using the method of trial and improvement, solve the equation:

$x^3 - x = 5$

correct to 1 decimal place.

REMEMBER! Cover the answers if you want to.

ANSWERS

A1 4.6 (1 d.p.)

A2 1.9 (1 d.p.)

TUTORIAL

T1 *We know the solution lies between 4 and 5, so try 4.5 first.*

When $x = 4.5$ $x^3 + x = 95.625$ *(too small)*
When $x = 4.6$ $x^3 + x = 101.936$ *(too large)*
When $x = 4.55$ $x^3 + x = 98.746375$

So x lies between 4.55 and 4.6 which are both equivalent to 4.6 (1 d.p.).

T2 *To find an approximate solution of the equation it is helpful to draw a graph of the curve* $y = x^3 - x$ *and the line* $y = 5$ *and note the solution where they cross. Alternatively consider a few different values to find an appropriate starting point.*

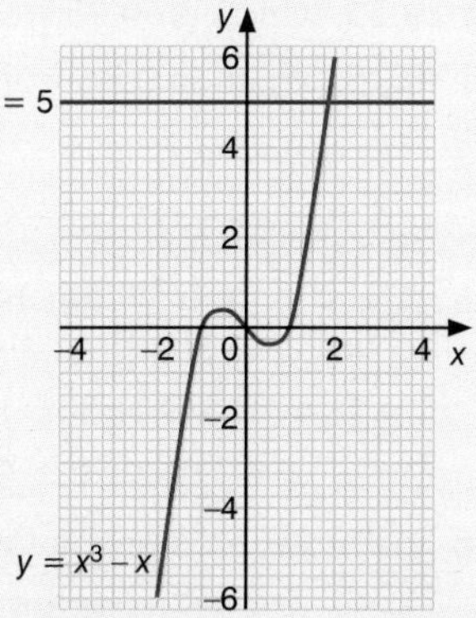

From the graph you can see that the solution lies between 1.8 and 2.

When $x = 1.8$ $x^3 - x = 4.032$ *(Substituting* $x = 1.8$*.)*
When $x = 2$ $x^3 - x = 6$ *(Substituting* $x = 2$*.)*
When $x = 1.9$ $x^3 - x = 4.959$ *(So x is between 1.9 and 2.)*
When $x = 1.95$ $x^3 - x = 5.464875$ *(So x is between 1.9 and 1.95.)*
When $x = 1.92$ $x^3 - x = 5.157888$

So x lies between 1.9 and 1.92 which are both equal to 1.9 (1 d.p.).

EXAMINATION QUESTIONS 9

EXAM PRACTICE

Sample Student's Answers & Examiner's Comments

1 Greg sold 40 tickets for a concert. He sold x tickets at £2 each and y tickets at £3.50 each. He collected £92.00. Write down two equations connecting x and y.

Solve these simultaneous equations to find out how many of each kind of ticket he sold.

(NEAB specimen paper 1998)

$x + y = 40$

$2x + 3.50y = 92.00$

$x = 40 - y$

$2(40 - y) + 3.5y = 92.00$

$80 - 2y + 3.5y = 92.00$

$80 + 1.5y = 92.00$

$1.5y = 12.00$

$y = 8$

$x + 8 = 40$

$x = 32$

$2 \times 32 + 3.50 \times 8 = 64 + 28 = 92$ ✓

2 A rectangle has a length $(x + 5)$ cm and a width $(x - 2)$ cm.

(a) If the perimeter of the rectangle is 24 cm, what is the value of x?

(b) If the area of the rectangle is 60 cm^2, show that $x^2 + 3x - 70 = 0$.

(c) Find the value of x when the area of the rectangle is 60 cm^2.

(NEAB specimen paper 1998)

(a) Perimeter $= (x + 5) + (x - 2) + (x + 5) + (x - 2)$

$= 4x + 6 = 24$

$4x = 18$

$x = 4.5$

(b) Area $= (x + 5) \times (x - 2)$

$= x^2 - 2x + 5x - 10$

$= x^2 + 3x - 10 = 60$

$= x^2 + 3x - 10 - 60 = 0$

$= x^2 + 3x - 70 = 0$

(c) $x^2 + 3x - 70 = 0$

$(x - 7)(x + 10) = 0$

$x = 7$ or $x = {}^-10$

EXAMINER'S COMMENTS

1 *The candidate has correctly identified the fact that $x + y = 40$ and appreciated that the amount of money collected is $x \times$ £2 plus $y \times$ £3.50.*

This equation might be better written as $2x + 3.5y = 92$ or else (doubled) as $4x + 7y = 184$.

The candidate has found the value of y and substituted this in the first equation to find x.

He has checked his answers by substituting them in the second equation.

2 **a)**
The working is untidy but the candidate equates the perimeter to the given value of 24 cm. This is used to find the value of x.

b)
Again, the working is untidy but the candidate has equated the area to the given value of 60 cm^2 and shown that $x^2 + 3x - 70 = 0$.

c)
She has correctly factorised the quadratic and correctly given answers of $x = 7$ or $x = {}^-10$ but has not interpreted the results.

The answer $x = {}^-10$ needs to be rejected as it is impossible to get a negative length.

EXAMINER'S COMMENTS

3 (a) Complete the following table for the equation $y = x^3 - 2x + 1$.

x	$^-3$	$^-2$	$^-1$	0	1	2	3
y							

(b) Draw a graph of the equation from $x = {}^-3$ to $x = 3$.

(c) By drawing a suitable straight line on the graph solve the equation $x^3 - 2x + 1 = 3$.

(d) Using the method of trial and improvement, or otherwise, solve the equation $x^3 - 2x + 1 = 3$ correct to one decimal place.

(a) $y = x^3 - 2x + 1$

x	$^-3$	$^-2$	$^-1$	0	1	2	3
y	$^-20$	$^-3$	2	1	0	5	22

(b)

(c)

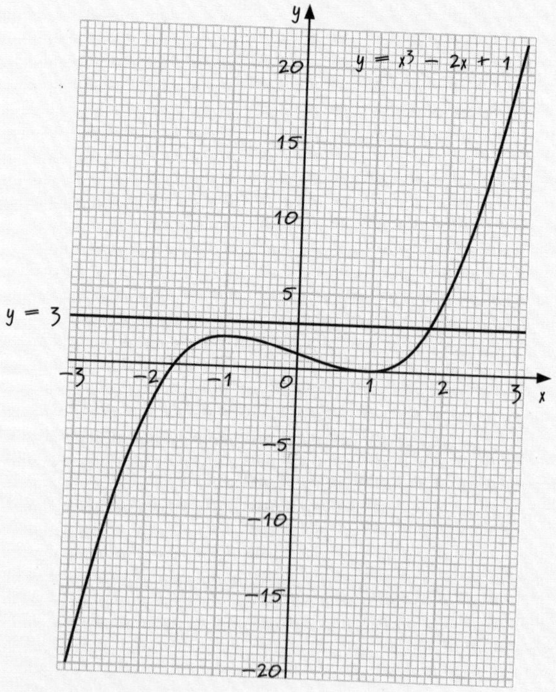

(d)

1.5	1.375
2	5
1.7	2.513
1.8	3.232
1.75	2.859375

$x = 1.8$ (1 d.p.)

3 **a)**
The candidate has successfully completed the table and used this to draw the graph of the equation.

b) and c)
The candidate has also drawn the line $y = 3$ on the graph which allows him to find an approximation for $x^3 - 2x + 1 = 3$.

d)
He has used his graph to find an initial approximation for the value of x which lies between $x = 1.5$ and $x = 2$. The candidate has drawn up a table for the values of x and the values of $x^3 - 2x + 1$ although this is not very clear from the presentation. The value of x lies between 1.75 and 1.8 so $x = 1.8$ (1 d.p.) as required.

Questions to Answer

1 Solve these simultaneous equations.

$4x - y = 13$

$3x + y = 15$

2 The equation of the straight line $y = mx + c$ is satisfied at the points (2, 3) and (1, $^{-}2$). What is the equation of the straight line?

3 Two numbers have a sum of 12 and a difference of 2. What are the two numbers?

4 Solve the following quadratic equations.

(a) $s^2 - 6s - 55 = 0$

(b) $n^2 + 10 = 7n$

5 Express $y^2 - 25$ as a product of two factors (brackets) and hence solve $y^2 - 25 = 0$.

6 Find the number which, when added to its square, gives a total of 12.

7 A solution of the equation $x^3 + x = 12$ lies between 2 and 3. Use the method of trial and improvement to find this solution of the equation. Give your answer to 1 decimal place.

(NEAB specimen paper 1998)

8 (a) Find the value of $x^3 + 5x$ when $x = 3.7$.

(b) Use trial and improvement to solve $x^3 + 5x = 60$. Give the value of x correct to two decimal places.

(MEG syllabus B, specimen paper 1998)

CHAPTER 46
AREA UNDER A GRAPH – TRAPEZIUM RULE

The area under a graph can be found by various methods, including counting squares. A better approximation can be found by dividing the area into small trapezia, and finding the sum of the areas of these trapezia.

Function notation is a useful means of dealing with the difficult equations and formulae used in this chapter.

FUNCTION NOTATION

Function notation is another useful tool for describing the relationship between two variables. In function notation the relationship $y = ...$ is written as $f(x) = ...$ so that the relationship $y = ax^2 + bx + c$ is written $f(x) = ax^2 + bx + c$ where $y = f(x)$. It is important to appreciate that $f(x)$ does not mean f multiplied by x but it is shorthand for 'the function of x'.

If $f(x) = x^2 - 4x + 3$ then $f(5)$ is the value of the function when $x = 5$ and $f(^-2)$ is the value of the function when $x = {^-2}$. In this case:

$$f(5) = 5^2 - 4 \times 5 + 3 = 8 \text{ and}$$
$$f(^-2) = (^-2)^2 - 4 \times {^-2} + 3 = 15$$

etc.

TRAPEZIUM RULE

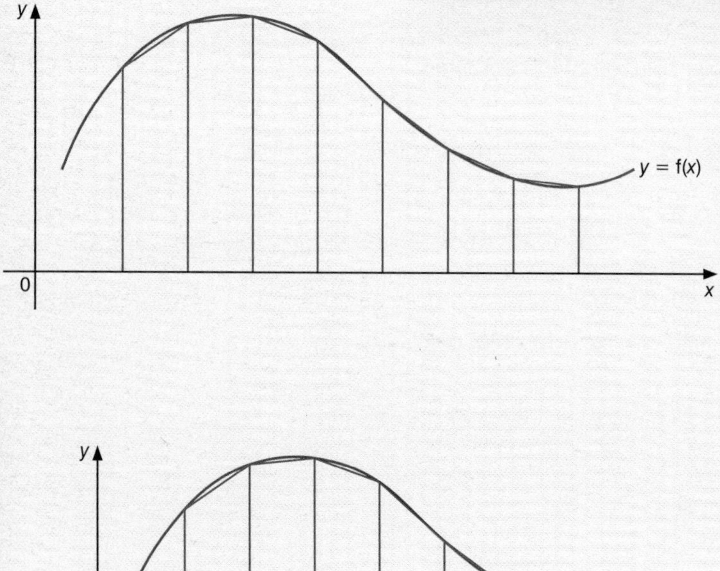

One method to find the area under a graph is to count squares, although this method can be rather cumbersome. A better method is to divide the area under the graph into a series of strips and approximate each strip to a trapezium to find an approximation for the area.

The area of a trapezium $= \frac{1}{2}h(a + b)$ where a and b are the lengths of the parallel sides and h is the perpendicular distance between them.

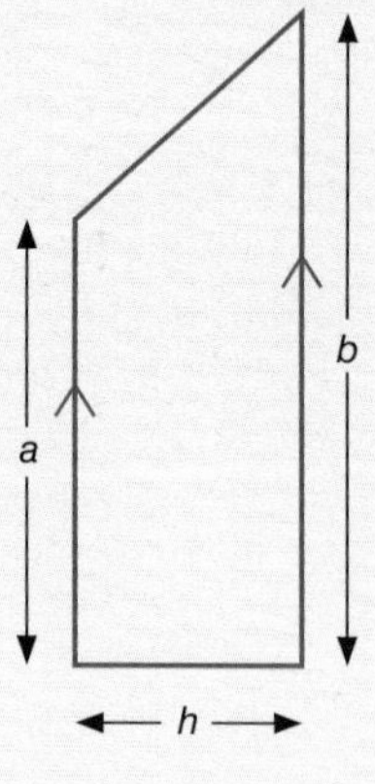

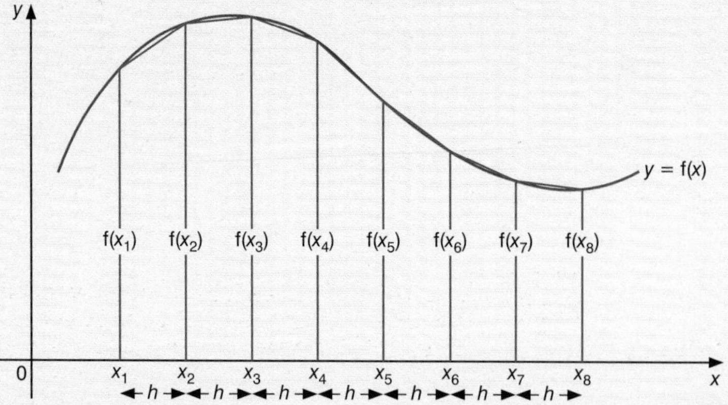

An approximation for the area under the graph can be found by adding the individual trapezia.

$$\text{Area} = \tfrac{1}{2}h\{f(x_1) + f(x_2)\} + \tfrac{1}{2}h\{f(x_2) + f(x_3)\} + \tfrac{1}{2}h\{f(x_3) + f(x_4)\} + \tfrac{1}{2}h\{f(x_4) + f(x_5)\}$$
$$+ \tfrac{1}{2}h\{f(x_5) + f(x_6)\} + \tfrac{1}{2}h\{f(x_6) + f(x_7)\} + \tfrac{1}{2}h\{f(x_7) + f(x_8)\}$$

$= \frac{1}{2}h\{f(x_1) + f(x_2) + f(x_2) + f(x_3) + f(x_3) + f(x_4) + f(x_4) + f(x_5) + f(x_5)$

$+ f(x_6) + f(x_6) + f(x_7) + f(x_7) + f(x_8)\}$

$= \frac{1}{2}h\{f(x_1) + 2f(x_2) + 2f(x_3) + 2f(x_4) + 2f(x_5) + 2f(x_6) + 2f(x_7) + f(x_8)\}$

NOTE

It is sometimes helpful to remember:

Area $= \frac{1}{2}h\{\text{ends} + 2 \times \text{middles}\}$

Worked example

By dividing the area into four equal intervals, find an approximation for the area under the curve $f(x) = x^2 + 1$ between $x = 1$ and $x = 5$.

Start by drawing a sketch to see the required area.

Area $= \frac{1}{2}h\{\text{ends} + 2 \times \text{middles}\}$

$= \frac{1}{2}\{f(1) + f(5) + 2\{f(2) + f(3) + f(4)\}\}$ As $h = 1$.

$= \frac{1}{2}\{2 + 26 + 2\{5 + 10 + 17\}\}$

$= \frac{1}{2}\{92\}$

$= 46$ square units

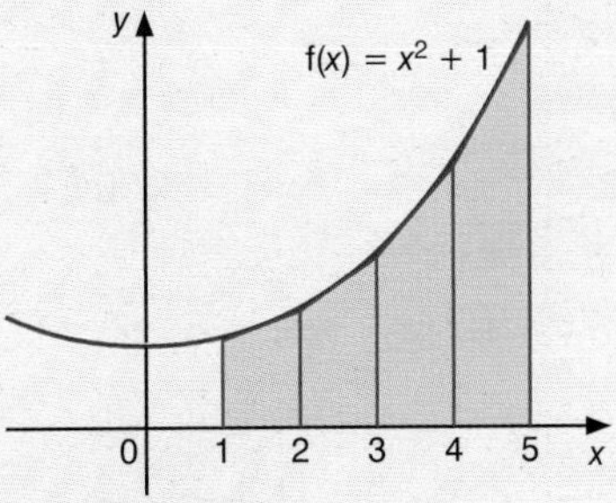

THE MEANING OF THE AREA UNDER A GRAPH

The area under a graph can be expressed as:

units along the y-axis × units along the x-axis

so that if the y-axis is speed (in miles/hour) and the x-axis is time (in hours) then:

the area under the graph = units along the y-axis × units along the x-axis

$= \frac{\text{miles}}{\text{hour}} \times \text{hour} = \text{miles}$

As miles are units of distance then the area under the graph measures the total distance in miles.

Similarly if the y-axis is rate of flow (in gallons/second) and the x-axis is time (in seconds) then:

the area under the graph = units along the y-axis × units along the x-axis

$= \frac{\text{gallons}}{\text{second}} \times \text{second} = \text{gallons}$

As gallons are a measurement of flow then the area under the graph measures the total flow in gallons.

Check yourself

QUESTIONS

Q1 By dividing the area into four equal intervals, find an approximation for the area under the curve $f(x) = 2x + 3$ between $x = 1$ and $x = 5$.

Q2 By dividing the area into three equal intervals, find an approximation for the area under the curve $f(x) = x^3$ between $x = 1$ and $x = 4$.

Q3 By dividing the area into six equal intervals, find an approximation for the area under the curve $f(x) = x^3$ between $x = 1$ and $x = 4$.

ANSWERS

A1 36 square units

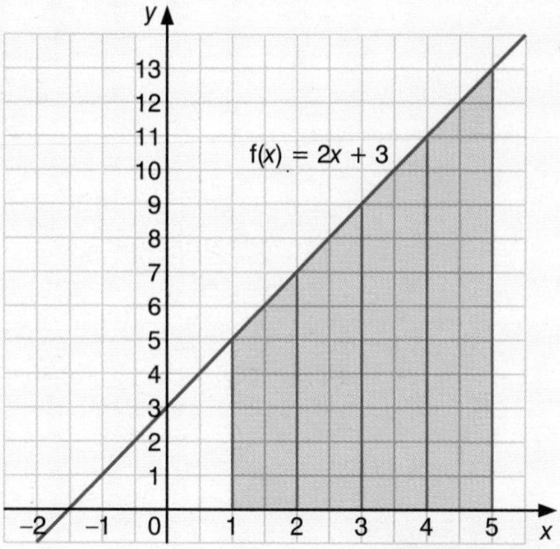

A2 67.5 square units

A3 64.7 square units (3 s.f.)

TUTORIAL

T1 $Area = \frac{1}{2}h\{ends + 2 \times middles\}$

$= \frac{1}{2}\{f(1) + f(5) + 2\{f(2) + f(3) + f(4)\}\}$

$= \frac{1}{2}\{5 + 13 + 2\{7 + 9 + 11\}\}$

$= \frac{1}{2}\{18 + 2 \times 27\}$

$= 36$ *square units*

Since f(x) is a straight line it is easy to check this answer by reference to the graph.

T2 $Area = \frac{1}{2}h\{ends + 2 \times middles\}$

$= \frac{1}{2}\{f(1) + f(4) + 2\{f(2) + f(3)\}\}$

$= \frac{1}{2}\{1 + 64 + 2\{8 + 27\}\}$

$= \frac{1}{2}\{135\}$

$= 67.5$ *square units*

T3 $Area = \frac{1}{2}h\{ends + 2 \times middles\}$

$= \frac{1}{2} \times \frac{1}{2}\{f(1) + f(4) + 2\{f(1.5) + f(2)$
$+ f(2.5) + f(3) + f(3.5)\}\}$

This time $h = 0.5$ of a unit.

$= \frac{1}{4}\{1 + 64 + 2\{3.375 + 8 + 15.625$
$+ 27 + 42.875\}\}$

$= \frac{1}{4}\{258.75\}$

$= 64.6875$ *square units*

This answer should be approximated as it is unreasonable to suggest such accuracy when approximating the area under the graph using trapezia.

CHAPTER 47 GRADIENT AND TANGENTS

Unlike a straight line, the gradient of a curve will change depending upon where you want to find the gradient. The gradient at a point is found by drawing a tangent at that point and working out the gradient as before.

To draw a tangent to a curve you place your ruler on the curve at the required point so that the angles produced at either side are approximately equal.

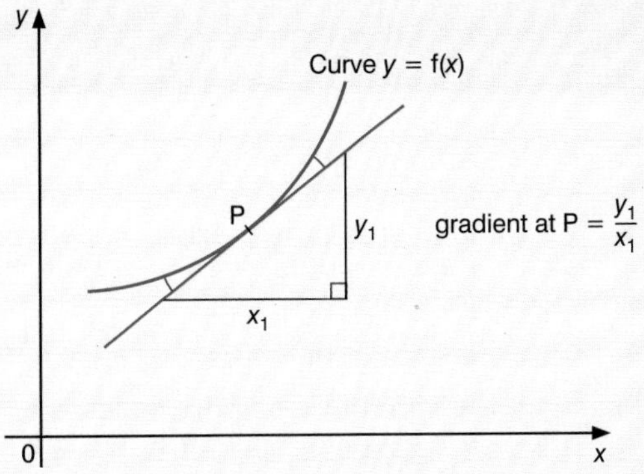

Worked example

Find the gradient at the following points for the curve $y = 10x - x^2$.

(a) $x = 3$ (b) $x = 9$

Write down the coordinates of the point where the gradient = 0.

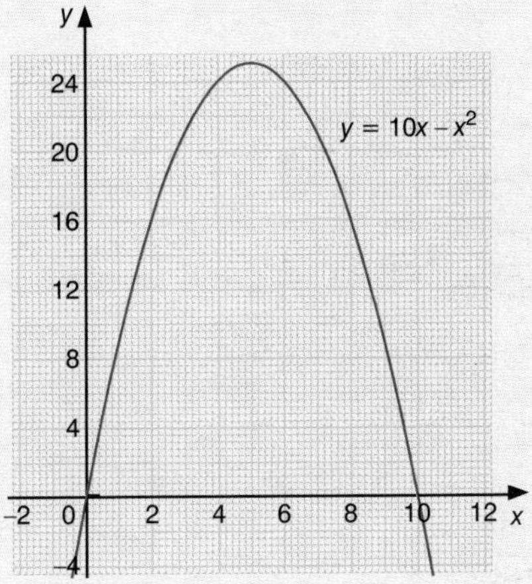

(a)

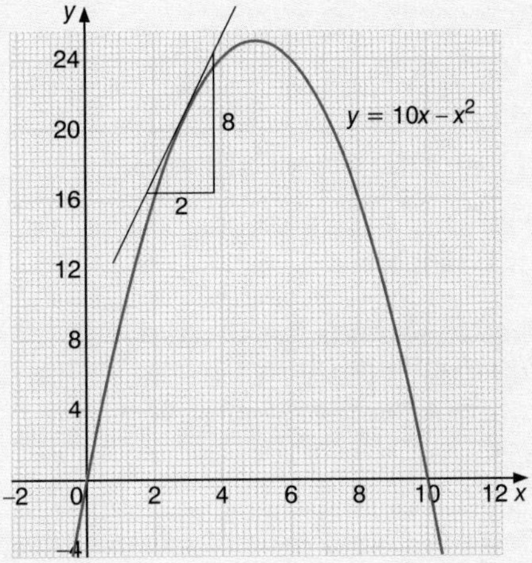

(b)

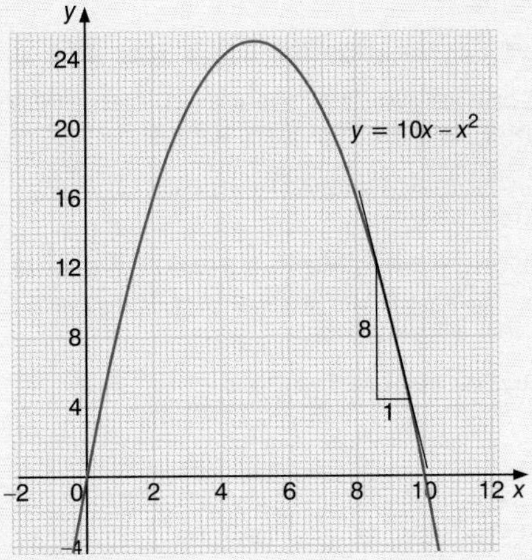

(a) To find the gradient at $x = 3$ you need to draw a tangent at the point and work out the gradient.

$$\text{Gradient} = \frac{\text{vertical distance}}{\text{horizontal distance}} = \frac{8}{2} = 4$$

(b) To find the gradient at $x = 9$ you need to draw a tangent at the point and work out the gradient.

$$\text{Gradient} = \frac{\text{vertical distance}}{\text{horizontal distance}} = -\frac{8}{1} = {}^{-}8$$

As gradient is negative.

From the graph you can see that the gradient is 0 when $x = 5$. The y-value at this point can be found by substituting $x = 5$ in the equation $y = 10x - x^2$.

$y = 10x - x^2 = 10 \times 5 - 5^2 = 25$

So the coordinates when the gradient is 0 are (5, 25).

HINT

The point at which the gradient = 0 is sometimes called the turning point and represents the position of a maximum or minimum value for the function. In this case the maximum value of $y = 10x - x^2$ is 25.

MEANING OF THE GRADIENT OF A GRAPH

The gradient of a graph can be expressed as $\frac{\text{distance along the } y\text{-axis}}{\text{distance along the } x\text{-axis}}$

so if the variable along the y-axis is distance (in miles) and the variable along the x-axis is time (in hours) then the gradient of the graph

$$= \frac{\text{distance along the } y\text{-axis}}{\text{distance along the } x\text{-axis}} = \frac{\text{miles}}{\text{hours}} \text{ or miles per hour (i.e. speed).}$$

Similarly if the variable along the y-axis is mass (in grams) and the variable along the x-axis is volume (in cm^3) then the gradient of the graph

$$= \frac{\text{distance along the } y\text{-axis}}{\text{distance along the } x\text{-axis}} = \frac{\text{grams}}{\text{cm}^3} \text{ or grams per cubic centimetre (i.e. density).}$$

Worked example

The graph shows the speed of an object over a period of 1 minute.

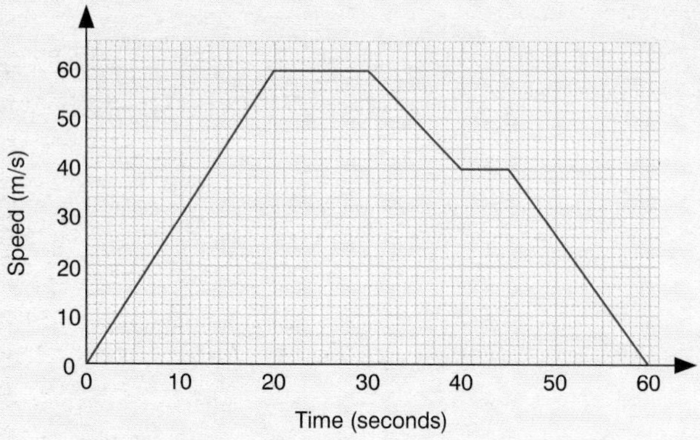

Describe, fully, the motion of the object:

(a) during the first 20 seconds

(b) between 30 and 40 seconds

(c) between 40 and 45 seconds

Use your graph to find the total distance travelled:

(d) in the first 30 seconds

(e) over the period of 1 minute.

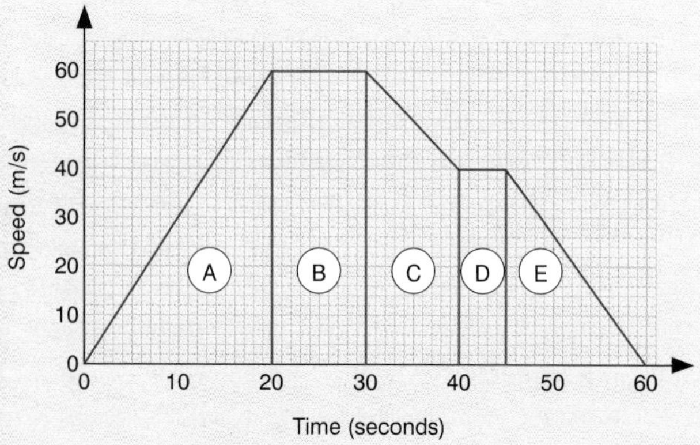

(a) During the first 20 seconds the speed of the object increases uniformly from 0 m/s to 60 m/s.

The gradient of this line gives the acceleration $= \frac{60}{20}$ or 3 m/s^2.

(b) Between 30 and 40 seconds the speed of the object decreases uniformly from 60 m/s to 40 m/s.

The gradient of this line gives the acceleration $= \frac{^-20}{10}$ or $^-2 \text{ m/s}^2$.

This negative acceleration implies a deceleration.

(c) Between 40 and 45 seconds the object is continuing at a constant speed (in this case 40 m/s).

To find the total distance travelled you need to find out the area under the graph.

(c) The total distance travelled in the first 30 seconds is found by adding up the area under the graph between 0 and 30 seconds.

Total distance = area A + area B
= 600 + 600 = 1200 metres

(e) The total distance travelled over the period of 1 minute is found by adding up the area under the graph between 0 and 60 seconds.

Total distance = area A + area B + area C + area D + area E
= 600 + 600 + 500 + 200 + 300
= 2200 metres

NOTE

See Chapter 46, Area under a graph – trapezium rule, for more information about the meaning of the area under a graph.

Check yourself

QUESTIONS

Q1 Find the gradient of the curve $y = x^2$ at the following points.

(a) $x = 2$ (b) $x = {}^{-}2$
(c) $x = 5$ (d) $x = 0$

Q2 Find the gradient of the curve $y = x^3 + 2$ at the following points.

(a) $x = 2$ (b) $x = 4$
(c) $x = {}^{-}1$ (d) $x = {}^{-}3$

REMEMBER! Cover the answers if you want to.

ANSWERS

A1
(a) gradient = 4
(b) gradient = ${}^{-}4$
(c) gradient = 10
(d) gradient = 0

A2
(a) gradient = 12
(b) gradient = 48
(c) gradient = 3
(d) gradient = 27

TUTORIAL

T1

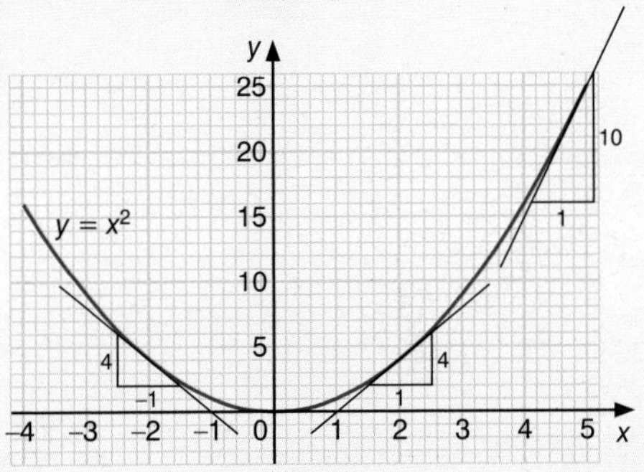

T2

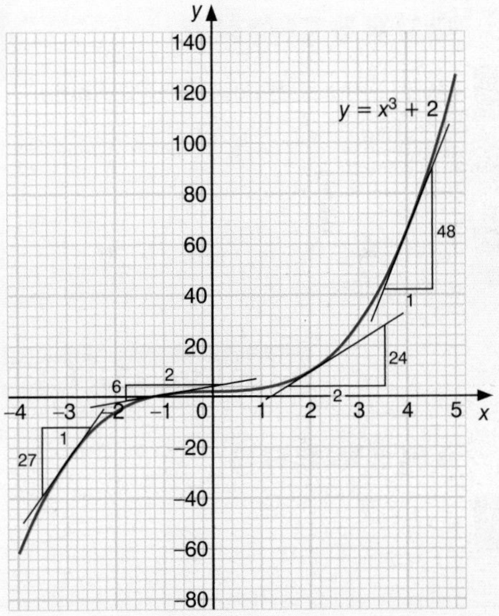

CHAPTER 48

FURTHER REARRANGEMENT OF FORMULAE

At the Higher level you are required to rearrange formulae where the subject exists in more than one term. Once again, a formula can be rearranged (or transposed) as before but to maintain the balance you must make sure that whatever you do to one side you do to the other side.

Worked example

Make x the subject of the formula $p = \dfrac{xy}{x - y}$.

$p = \dfrac{xy}{x-y}$	
$p(x - y) = xy$	Multiplying both sides by $(x - y)$.
$px - py = xy$	Expanding the brackets.
$px = xy + py$	Adding py to both sides and attempting to isolate terms in x.
$px - xy = py$	Collecting terms in x on one side only.
$x(p - y) = py$	Factorising the left-hand side with x outside the bracket.
$x = \dfrac{py}{p-y}$	Dividing both sides by $(p - y)$ to make x the subject.

Check yourself

QUESTIONS

Q1 Make y the subject of the formula $p = \dfrac{xy}{x - y}$.

Q2 Rearrange the formula $r = \sqrt{\dfrac{s}{s + t}}$ to make s the subject.

Q3 A formula states $x = \dfrac{1 - t^2}{1 + t^2}$.

(a) Find the value of x when $t = \frac{1}{2}$.

(b) Rearrange the formula to make t the subject.

REMEMBER! Cover the answers if you want to.

ANSWERS

A1 $y = \dfrac{px}{x + p}$

TUTORIAL

T1

$p = \dfrac{xy}{x-y}$	
$p(x - y) = xy$	*Multiplying both sides by $(x - y)$.*
$px - py = xy$	*Expanding the brackets.*
$px = xy + py$	*Collecting together terms in y on one side only.*
$y(x + p) = px$	*Factorising the terms in y.*
$y = \dfrac{px}{x+p}$	*Dividing both sides by $(x + p)$ to make y the subject.*

ANSWERS

A2 $s = \dfrac{r^2t}{1 - r^2}$

A3 (a) $x = \frac{3}{5}$ or 0.6

(b) $t = \sqrt{\dfrac{1 - x}{1 + x}}$

TUTORIAL

T2

$r = \sqrt{\dfrac{s}{s + t}}$

$r^2 = \dfrac{s}{s + t}$ *Squaring both sides.*

$r^2(s + t) = s$ *Multiplying both sides by (s + t).*

$r^2s + r^2t = s$ *Expanding the brackets.*

$s - r^2s = r^2t$ *Collecting together terms in s on one side.*

$s(1 - r^2) = r^2t$ *Factorising the terms in s.*

$s = \dfrac{r^2t}{1 - r^2}$ *Dividing both sides by $(1 - r^2)$ to make s the subject.*

T3 (*a*) $x = \dfrac{1 - t^2}{1 + t^2}$

$x = \dfrac{1 - \left(\frac{1}{2}\right)^2}{1 + \left(\frac{1}{2}\right)^2}$ *Substituting $\frac{1}{2}$ for t.*

$x = \dfrac{\frac{3}{4}}{\frac{5}{4}}$

$x = \frac{3}{5}$

(*b*) $x = \dfrac{1 - t^2}{1 + t^2}$

$x(1 + t^2) = 1 - t^2$ *Multiplying both sides by $(1 + t^2)$.*

$x + xt^2 = 1 - t^2$ *Expanding the brackets.*

$xt^2 + t^2 = 1 - x$ *Collecting together terms in t on one side only.*

$t^2(x + 1) = 1 - x$ *Factorising the terms in t^2.*

$t^2 = \dfrac{1 - x}{1 + x}$ *Dividing both sides by $x + 1$ to make t^2 the subject.*

$t = \sqrt{\dfrac{1 - x}{1 + x}}$ *Taking the square root of both sides to make t the subject.*

ALGEBRA

CHAPTER 49

FURTHER QUADRATIC EQUATIONS

At the Higher level you will be expected to solve quadratic equations by a variety of methods including using the formula and iteration techniques. The examination question should make it clear which method you should use or else any method will be acceptable

See Chapter 39, Quadratic, cubic and reciprocal graphs for further information on how to solve quadratic equations by graphical methods and see Chapter 43, Quadratic equations for further information on other algebraic methods

QUADRATIC EQUATIONS – SOLUTION BY FORMULA

NOTE

This formula for solving quadratic equations will be given in the examination so you do not need to memorise it but you do need to know how to use it correctly.

Where a quadratic does not factorise then you can use the formula:

$$x = \frac{-b \pm \sqrt{b^2 - 4ac}}{2a}$$

to solve any quadratic of the form

$$ax^2 + bx + c = 0$$

as shown in the following worked example.

Worked example

Solve the equation $x^2 + 3x - 2 = 0$.

Comparing the given quadratic with the general form $ax^2 + bx + c = 0$ then:

$a = 1, b = 3, c = {}^{-}2$.

Substituting these values in the formula:

$$x = \frac{-b \pm \sqrt{b^2 - 4ac}}{2a}$$

$$x = \frac{{}^{-}3 \pm \sqrt{3^2 - 4 \times 1 \times {}^{-}2}}{2 \times 1}$$

$$x = \frac{{}^{-}3 \pm \sqrt{9 - {}^{-}8}}{2}$$

$$x = \frac{{}^{-}3 \pm \sqrt{17}}{2}$$

$$x = \frac{{}^{-}3 \pm 4.123\,105\,626}{2}$$

$$x = \frac{{}^{-}3 + 4.123\,105\,626}{2} \quad \text{or} \quad x = \frac{{}^{-}3 - 4.123\,105\,626}{2}$$

$$x = \frac{1.123\,105\,626}{2} \quad \text{or} \quad x = \frac{{}^{-}7.123\,105\,626}{2}$$

$x = 0.561\,552\,813$ or $x = {}^{-}3.561\,552\,813$

$x = 0.562$ (3 s.f.) or $x = {}^{-}3.56$ (3 s.f.)

NOTE

You should always round your answers here to an appropriate degree of accuracy.

Check yourself

QUESTIONS

Q1 Solve the following quadratic equations by using the formula $x = \frac{-b \pm \sqrt{b^2 - 4ac}}{2a}$.

(a) $x^2 + 2x - 1 = 0$
(b) $2x^2 = 10x - 4$

Q2 The length of a room is 4 metres longer than its width. Find the dimensions of the room if the area is 32 square metres.

REMEMBER! Cover the answers if you want to.

ANSWERS

A1 (a) $x = 0.414$ or $x = {}^-2.414$ (3 s.f.)
(b) $x = 4.56$ or $x = 0.438$ (3 s.f.)

TUTORIAL

T1 (*a*) *Comparing* $x^2 + 2x - 1 = 0$ *with the general form* $ax^2 + bx + c = 0$:
$a = 1, b = 2, c = {}^-1.$

$$x = \frac{{}^-2 \pm \sqrt{2^2 - 4 \times 1 \times {}^-1}}{2 \times 1}$$

$$x = \frac{{}^-2 \pm \sqrt{8}}{2}$$

$$x = \frac{{}^-2 + 2.828\,427\,125}{2}$$

$$\text{or } x = \frac{{}^-2 - 2.828\,427\,125}{2}$$

$x = 0.414$ *or* $x = {}^-2.414$ (3 *s.f.*)

(*b*) *The equation* $2x^2 = 10x - 4$ *needs to be rearranged in order to compare it with the general form* $ax^2 + bx + c = 0$.

$2x^2 = 10x - 4$

$2x^2 - 10x + 4 = 0$

Comparing $2x^2 - 10x + 4 = 0$ *with the general form* $ax^2 + bx + c = 0$:
$a = 2, b = {}^-10, c = 4.$

$$x = \frac{-{}^-10 \pm \sqrt{({}^-10)^2 - 4 \times 2 \times 4}}{2 \times 2}$$

$$x = \frac{10 \pm \sqrt{68}}{4}$$

Remember that $- \times - = +$ *and* $({}^-10)^2 = 100$.

$$x = \frac{10 + 8.246\,211\,251}{4}$$

$$\text{or } x = \frac{10 - 8.246\,211\,251}{4}$$

$x = 4.56$ *or* $x = 0.438$ (3 *s.f.*)

ANSWERS

A2 width = 4 metres, length = 8 metres

HINT

The fact that the square root worked out to be an exact number ($\sqrt{144} = 12$) suggests that the quadratic equation $x^2 + 4x - 32 = 0$ could have been factorised which would have made the problem a lot easier and quicker. As a rule, you should always use factorising rather than the formula, if you can.

TUTORIAL

T2 *Let the width be x metres, then the length is $(x + 4)$ metres.*

$\text{Area} = x(x + 4) = 32$

$x^2 + 4x = 32$

$x^2 + 4x - 32 = 0$

Comparing $x^2 + 4x - 32 = 0$ with the general form $ax^2 + bx + c = 0$:

$a = 1, b = 4, c = {}^-32.$

$$x = \frac{{}^-4 \pm \sqrt{4^2 - 4 \times 1 \times {}^-32}}{2 \times 1}$$

$$x = \frac{{}^-4 \pm \sqrt{144}}{2}$$

$$x = \frac{{}^-4 + 12}{2} \quad \text{or } x = \frac{{}^-4 - 12}{2}$$

$x = 4$ *or* $x = {}^-8$

But a width of ⁻8 metres is impossible so you ignore this value. Since the length of the room is 4 m greater than the width, this gives:
width = 4 m, length = 8 m.

QUADRATIC EQUATIONS – SOLUTION BY ITERATION

Iteration involves using an initial value (usually denoted by x_1) to find successive solutions ($x_2, x_3, x_4, \ldots$). Each solution is based on the previous solution to improve the accuracy, so that x_{n+1} is calculated from x_n, using the given iterative formula.

Iterative formula can be used as an alternative method for solving quadratic equations. The iterative formula can be found by rearranging the equation.

Worked example

Solve the quadratic equation $x^2 - 5x + 1 = 0$.

$x^2 - 5x + 1 = 0$

$x^2 + 1 = 5x$ Adding $5x$ to both sides.

$5x = x^2 + 1$ Turning the equation around.

$x = \frac{1}{5}(x^2 + 1)$

Now, writing this as an iterative formula:

$x_{n+1} = \frac{1}{5}(x_n{}^2 + 1)$

Any equation can give rise to a number of different iterative formulae, although they are not always of any use in solving the quadratic. In the above example $x_{n+1} = \frac{1}{5}(x_n{}^2 + 1)$ provides a solution to the quadratic whereas $x_{n+1} = 5 - \frac{1}{x_n}$ will not, as successive iterations will diverge.

Check yourself

QUESTIONS

Q1 Use the iteration $x_{n+1} = 10 + \frac{1}{x_n}$ with $x_1 = 5$ to find a root of the equation $x^2 - 10x - 1 = 0$ correct to four decimal places.

Q2 A sequence is given by $x_{n+1} = \frac{6}{x_n + 5}$.

The first term, x_1, of the sequence is 3.

(a) Find the next three terms.

(b) What do you think is the value of x_n as n becomes very large? Write down this value.

(c) Show that the quadratic equation which the sequence above is intended to solve is $x^2 + 5x - 6 = 0$.

(d) Solve this quadratic equation.

REMEMBER! Cover the answers if you want to.

ANSWERS

A1 $x = 10.0990$ (4 d.p.)

A2 (a) $x_2 = 0.75$

$x_3 = 1.043\,478\,3$

$x_4 = 0.992\,805\,7$

(b) The value of x_n would seem to be tending towards an answer of 1.

(c) $x = \frac{6}{x + 5}$

$x(x + 5) = 6$

$x^2 + 5x = 6$

$x^2 + 5x - 6 = 0$

(d) $x = 1$ or $x = {}^{-}6$

TUTORIAL

T1 *Using $x_{n+1} = 10 + \frac{1}{x_n}$ and $x_1 = 5$:*

$$x_2 = 10 + \frac{1}{x_1} = 10 + \frac{1}{5} = 10.2$$

$$x_3 = 10 + \frac{1}{x_2} = 10 + \frac{1}{10.2} = 10.098\,039\,22$$

$$x_4 = 10 + \frac{1}{x_3} = 10 + \frac{1}{10.098\,039\,22}$$

$$= 10.099\,029\,13$$

$$x_5 = 10 + \frac{1}{x_4} = 10 + \frac{1}{10.099\,029\,13}$$

$$= 10.099\,019\,42$$

Since $x_5 = x_4$ to 4 d.p. then a root of the equation is 10.0990 (4 d.p.).

T2 *(a)* *Using $x_{n+1} = \frac{6}{x_n + 5}$ and $x_1 = 3$:*

$x_2 = 0.75$ $x_3 = 1.043\,478\,3$ $x_4 = 0.992\,805\,7$

(b) *The value of x_n would seem to be tending towards an answer of 1 as n becomes very large.*

(c) *The quadratic equation is found by rearranging:*

$x = \frac{6}{x + 5}$

$x(x + 5) = 6$ *Multiplying both sides by $(x + 5)$.*

$x^2 + 5x - 6 = 0$ *Expanding and rearranging.*

(d) *Factorising the left-hand side of the equation:*

$x^2 + 5x - 6 = (x - 1)(x + 6)$

giving $(x - 1)(x + 6) = 0$

So the solutions of the quadratic equation $x^2 + 5x - 6 = 0$ are $x = 1$ and $x = {}^{-}6$.

ALGEBRA

CHAPTER 50

FURTHER FUNCTIONS AND GRAPHS

There are four different graph transformations with which you need to be familiar. When using these transformations it is helpful to try out a few points to check that you have the correct idea. The four transformations take the form:

$y = k\mathrm{f}(x)$ $\quad y = \mathrm{f}(x) + a$ $\quad y = \mathrm{f}(kx)$ $\quad y = \mathrm{f}(x + a)$

Each of these transformations is shown on the function $\mathrm{f}(x) = x^3$.

$y = k\mathrm{f}(x)$

Under this transformation the graph of the function is stretched (or shrunk if $k < 1$) along the y-axis.

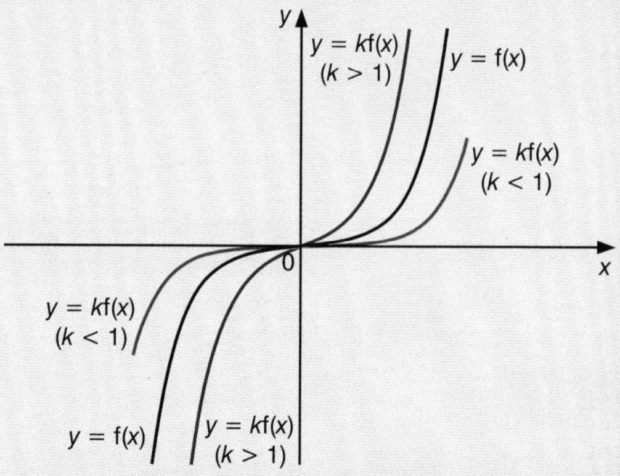

$y = \mathrm{f}(x) + a$

Under this transformation the function is translated along the y-axis. If $a > 0$ then the graph of the function moves up (positive direction) and if $a < 0$ then the graph of the function moves down (negative direction).

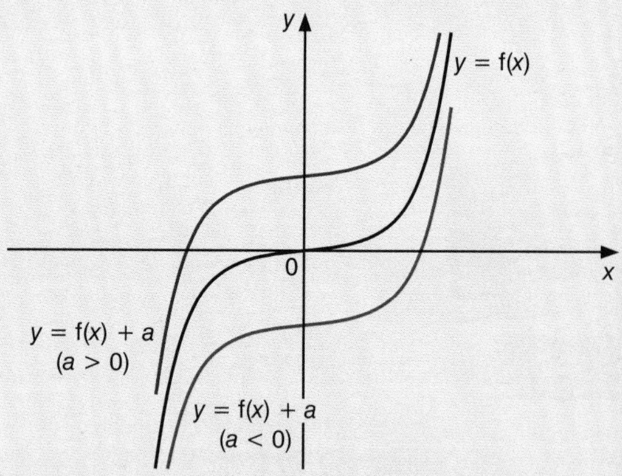

$y = \mathrm{f}(kx)$

Under this transformation the graph of the function is shrunk (or stretched if $k < 1$) along the x-axis.

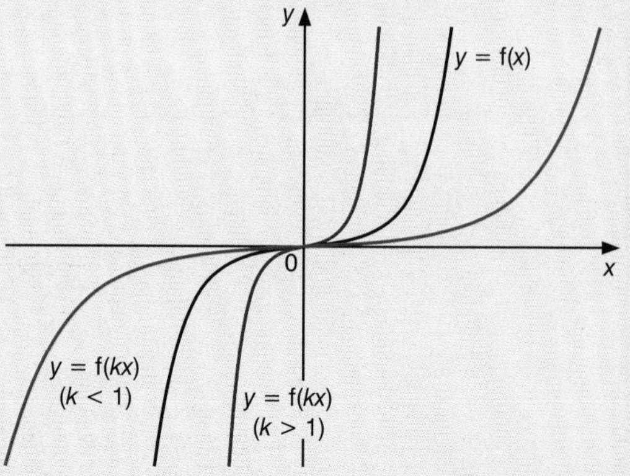

$y = f(x + a)$

Under this transformation the function is translated along the x-axis. If $a > 0$ then the graph of the function moves to the left (negative direction) and if $a < 0$ then the graph of the function moves to the right (positive direction).

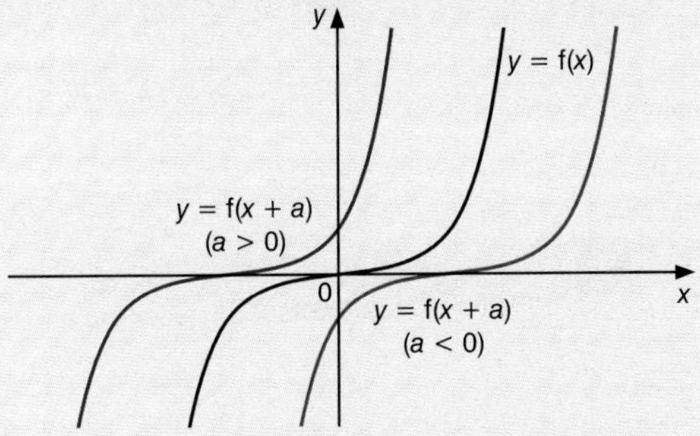

Check yourself

QUESTIONS

Sketch the following sets of graphs.

Q1 $y = x^2$ $y = x^2 - 3$ $y = \frac{1}{2}x^2$ $y = (2x)^2$

Q2 $y = x^2 + 2x$ $y = x^2 + 2x + 3$ $y = (x - 2)^2 + 2(x - 2)$ $y = (\frac{1}{2}x)^2 + 2(\frac{1}{2}x)$

Q3 $y = \sin x$ $y = \sin 4x$ $y = 4 \sin x$ $y = \sin (x + 90°)$

REMEMBER! Cover the answers if you want to.

ANSWERS

A1

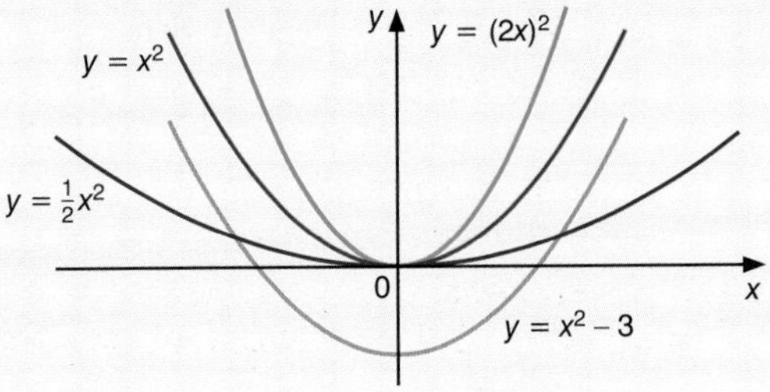

A2

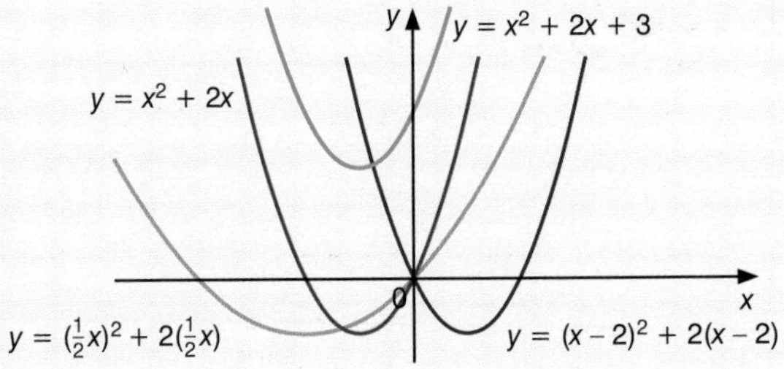

A3

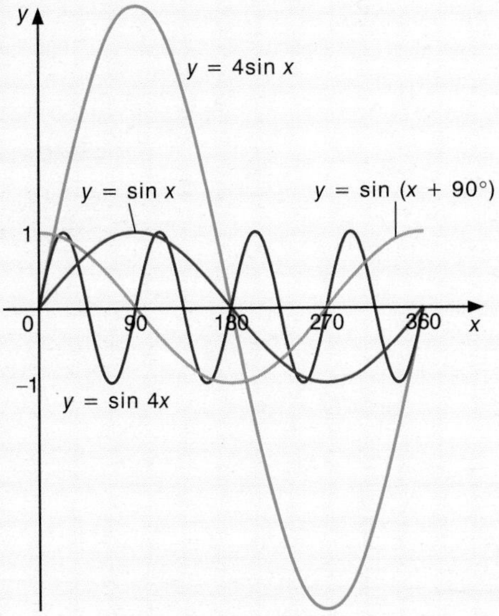

CHAPTER 51

EXAMINATION QUESTIONS 10

EXAM PRACTICE

Sample Student's Answers & Examiner's Comments

1 The graph shows how a car's speed, measured in metres per second, varies in the first 6 seconds after the car moves away from some traffic lights.

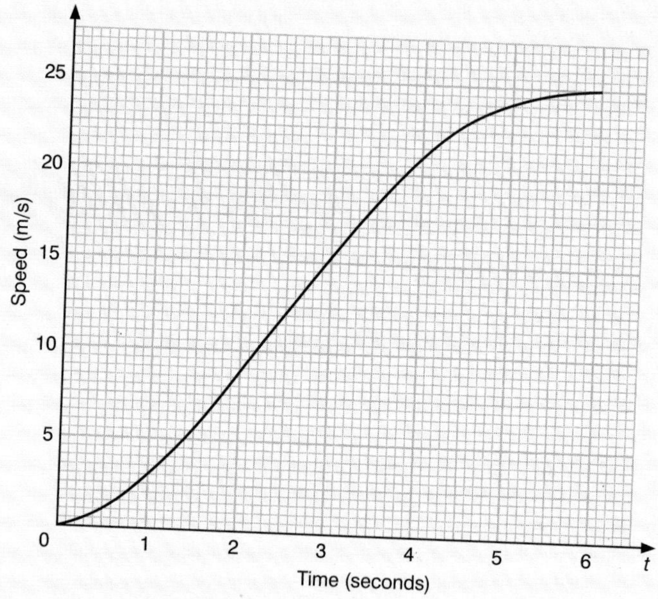

(a) Draw the tangent at the point on the curve where $t = 5$ seconds.

(b) Find the gradient of this tangent.

(c) What does this gradient represent?

(d) Making your method clear, estimate the area underneath the graph between $t = 0$ and $t = 6$. Hence estimate the distance travelled by the car in the first 6 seconds.

(MEG syllabus A, specimen paper 1998)

(a)

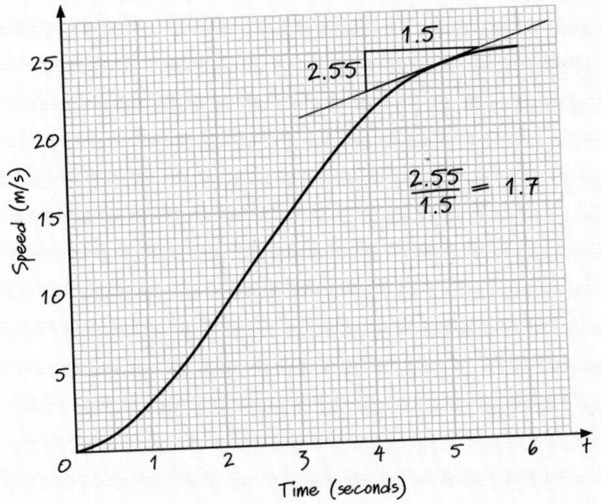

(b) Gradient = 1.7

(c) Gradient = m/s^2

(d) Area = $\frac{1}{2}h\{\text{ends} + 2 \times \text{middles}\}$

$= \frac{1}{2}\{f(0) + f(6) + 2[f(1) + f(2) + f(3) + f(4)]\}$

$= \frac{1}{2}\{0 + 25 + 2[3 + 9 + 15.5 + 20.5 + 24]\}$

$= \frac{1}{2}\{25 + 2 \times 72\}$

= 84.5 square units

Distance = 84.5 metres

EXAMINER'S COMMENTS

1 a)
The candidate has successfully drawn a tangent at the point t = 5 and used this tangent to obtain the value of the gradient.

b)
An answer of 1.6 to 1.8 was given full marks highlighting the need for considerable accuracy on this type of work. Most readings are expected to be accurate to the nearest half-square.

c)
She has given the units of acceleration but neglected to mention acceleration in her answer. The gradient of a speed/time (or velocity/time) graph is always acceleration.

d)
The candidate has used the trapezium rule to find the area under the graph.

She could have counted all of the squares under the curve as no mention was made of using the trapezium method although this would have been very laborious. An answer in the range 83 to 87 would gain full marks.

The area under the speed time graph represents the total distance travelled.

2 The percentage profit, p, on the sale of an item is given by the formula:

$$p = \frac{100(s - c)}{c}$$

where s is the selling price and c is the cost price.

Express c in terms of s and p.

(MEG syllabus A, specimen paper 1998)

$$p = \frac{100(s - c)}{c}$$

$$pc = 100s - 100c$$

$$pc + 100c = 100s$$

$$c(p + 100) = 100s$$

$$c = \frac{100s}{p + 100}$$

2 *The candidate has correctly applied the procedures for rearranging formula and gained full marks on this question.*

EXAMINER'S COMMENTS

3 Sue travels 70 miles at an average speed of x miles per hour. She then travels 50 miles at $(x + 10)$ miles per hour.

(a) Write down an expression in terms of x for the total time taken, in hours, for the whole journey.

(b) If the whole journey takes 3 hours, form an equation in x and show that it can be simplified to $3x^2 - 90x - 700 = 0$.

(c) Solve this equation to find the value of x.

(MEG syllabus B, specimen paper 1998)

$S = \frac{D}{T} \quad T = \frac{D}{S}$

For 70 miles: $T = \frac{70}{x}$

For 50 miles $T = \frac{50}{x + 10}$

(a) Time for whole journey $= \frac{70}{x} + \frac{50}{x + 10}$

(b) $\frac{70}{x} + \frac{50}{x + 10} = 3$

$\frac{70(x + 10) + 50x}{x(x + 10)} = 3$

$70(x + 10) + 50x = 3x(x + 10)$

$70x + 700 + 50x = 3x^2 + 30x$

$0 = 3x^2 + 30x - 70x - 700 - 50x$

$3x^2 - 90x - 700 = 0$

(c) $a = 3,\ b = {}^-90,\ c = {}^-700$

$x = \frac{-b \pm \sqrt{b^2 - 4ac}}{2a}$

$x = \frac{-{}^-90 \pm \sqrt{({}^-90)^2 - 4 \times 3 \times {}^-700}}{2 \times 3}$

$x = \frac{90 \pm \sqrt{16\,500}}{6}$

$x = \frac{90 + 128.4523258}{6}$

or $x = \frac{90 - 128.4523258}{6}$

$x = 36.4$ or $x = {}^-6.41$ (3 s.f.)

3 *The candidate has correctly identified that time = $\frac{\text{distance}}{\text{speed}}$.*

a) *He has appreciated that the total time is equal to the sum of the times for each part.*

b) *The candidate has equated the time to 3 hours and formed an equation in x.*

He has attempted to simplify this equation by multiplying by the denominators.

He has obtained the required quadratic formula.

c) *The candidate has reasonably rounded the answers to 3 s.f. but has failed to reject the negative solution which is inappropriate for this question.*

EXAMINER'S COMMENTS

4 The function $y = f(x)$ is defined for $0 < x < 2$.
The function is sketched below.

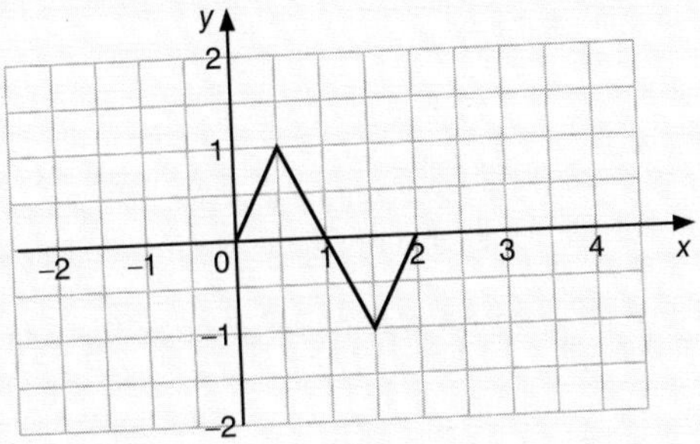

(a) Sketch $y = f(x + 1)$.

(b) Sketch $y = f(\frac{1}{2}x)$.

(SEG specimen paper 1998)

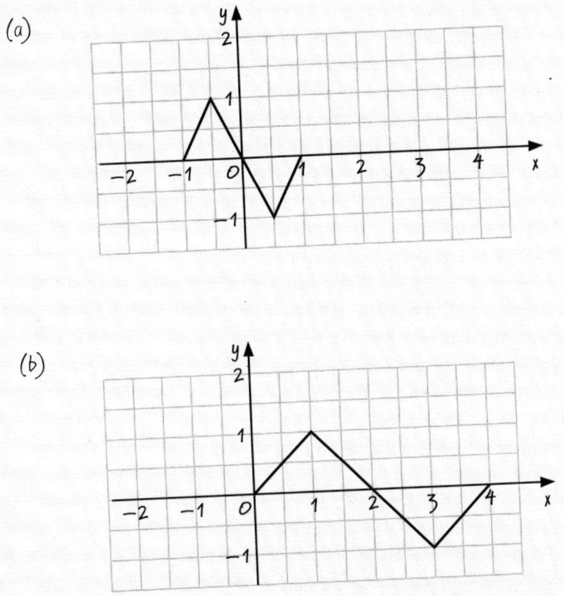

4 *The candidate has gained full marks on this work.*

EXAMINER'S COMMENTS

5 (a) Show that $x = 1 + \frac{11}{x-3}$ is a rearrangement of the equation $x^2 - 4x - 8 = 0$.

(b) Use the iterative formula:

$$x_{n+1} = 1 + \frac{11}{x_n - 3}$$

together with a starting value of $x_1 = {}^-2$ to obtain a root of the equation of $x^2 - 4x - 8 = 0$ accurate to 1 decimal place.

(c) Use the quadratic equation formula to solve the equation $x^2 - 4x - 8 = 0$.

(d) Comment on how your answer to part (c) compares with your answer to part (b).

(MEG syllabus A, specimen paper 1998)

(a) $x = 1 + \frac{11}{x-3}$

$x(x-3) = 1(x-3) + 11$

$x^2 - 3x = x - 3 + 11$

$x^2 - 3x - x + 3 - 11 = 0$

$x^2 - 4x - 8 = 0$ ✓

(b) $x_{n+1} = 1 + \frac{11}{x_n - 3}$

$x_1 = {}^-2$

$x_2 = {}^-1.2$

$x_3 = {}^-1.619\,047\,619$

$x_4 = {}^-1.381\,443\,299$

$x_5 = {}^-1.510\,588\,235$

$x_6 = {}^-1.438\,706\,312$

$x_7 = {}^-1.478\,199\,553$

$x_8 = {}^-1.456\,344\,312$

$x_9 = {}^-1.468\,390\,957$

(c) $a = 1,\ b = {}^-4,\ c = {}^-8$

$$x = \frac{-b \pm \sqrt{b^2 - 4ac}}{2a}$$

$$x = \frac{-{}^-4 \pm \sqrt{({}^-4)^2 - 4 \times 1 \times {}^-8}}{2 \times 1}$$

$$x = \frac{4 \pm \sqrt{48}}{2}$$

$$x = \frac{4 + 6.928\,203\,23}{2}$$

$$\text{or } x = \frac{4 - 6.928\,203\,23}{2}$$

$x = 5.46$ or $x = {}^-1.46$ (3 s.f.)

5 a)
The candidate has obtained the required quadratic formula.

b)
As x_7 and x_8 are both equal to ${}^-1.5$ (1 d.p.) then the candidate could have stopped here.
Unfortunately, no final answer has been provided and she should have concluded the question by stating that $x = {}^-1.5$ (1 d.p.).

c)
She has reasonably rounded the answers to 3 s.f. but has failed to comment on how the answer to part (c) compares with the answer to part (b).

The suggestion that the answer in part (c) was easier to calculate, or more accurate is acceptable as well as the fact that part (c) produces both solutions to the quadratic equation would be appropriate.

Questions to Answer

1 A formula used by scientists is $t = \frac{v + p}{v}$.

Change the subject of the formula to v.

(NEAB specimen paper 1998, part question)

2 (a) Rearrange the formula $v = u + 0.1t$ to make t the subject.

(b) Substitute this expression for t in the formula $s = ut + 0.05t^2$ to obtain a formula connecting u, v and s.

Show that this formula can be rearranged as $v^2 = u^2 + 0.2s$.

(MEG syllabus B, specimen paper 1998, part question)

3 The sum of the squares of two consecutive numbers is 685.
Find the larger number.

4 Solve the equation $5x^2 + 2x - 1 = 0$.

5 Write as a single fraction $\frac{1}{x + 3} + \frac{1}{x - 4}$.

6 Zarig took part in a 26-mile road race.

(a) He ran the first 15 miles at an average speed of x mph. He ran the last 11 miles at an average speed of $(x - 2)$ mph. Write down an expression, in terms of x, for the time he took to complete the 26-mile race.

(b) Zarig took 4 hours to complete the race. Using your answer to part (a), form an equation in terms of x.

(c) Simplify your equation and show that it can be written as $2x^2 - 17x + 15 = 0$.

(d) Solve this equation and obtain Zarig's average speed over the first 15 miles of the race.

(NEAB specimen paper 1998)

7 The graph of $y = f(x)$ where $f(x) = \frac{x}{x + 1}$ is sketched below.

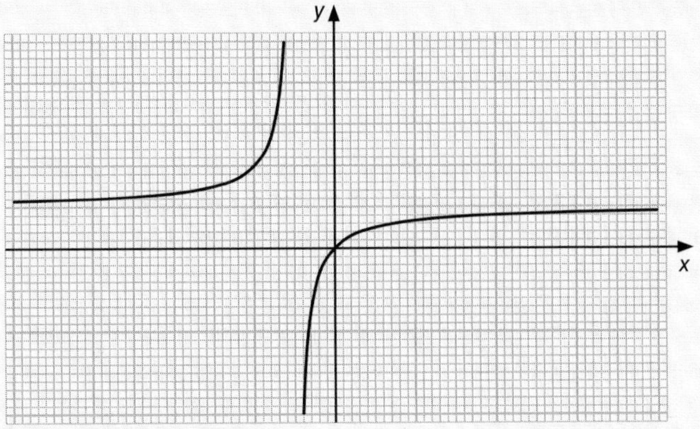

(a) Sketch $y = f(x - 1)$.

(b) Sketch $y = f(2x)$.

CHAPTER 52

GEOMETRIC TERMS

The following definitions are essential to the study of shape and space, and you need to understand them.

A point is the position where two lines meet.

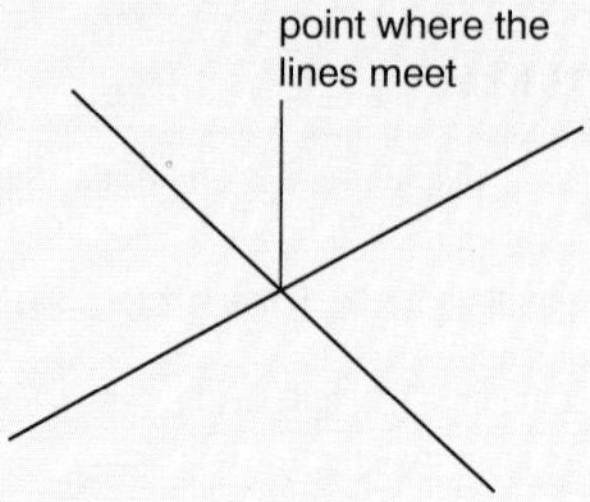

A line is where two planes meet.

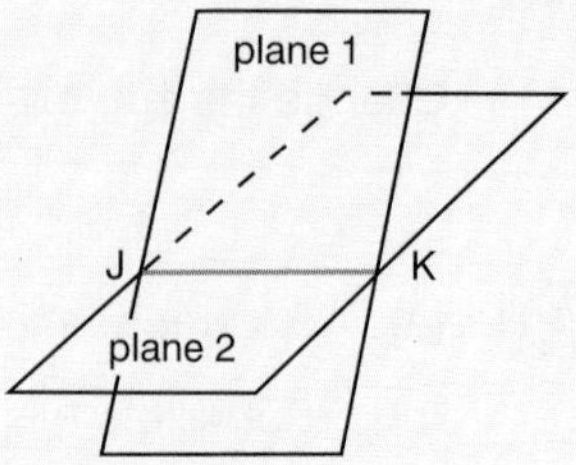

JK is the line where the two planes meet.

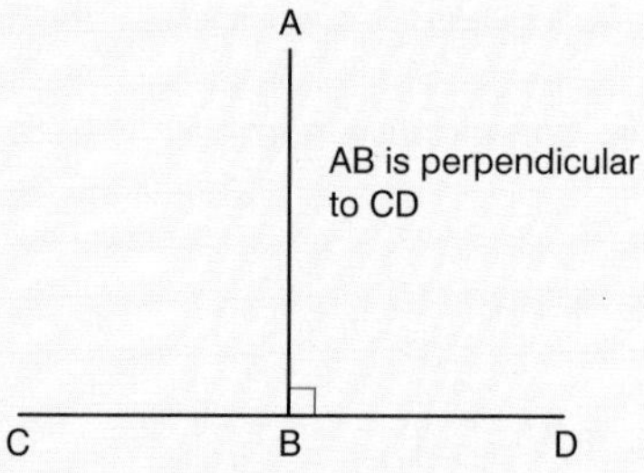

perpendicular lines

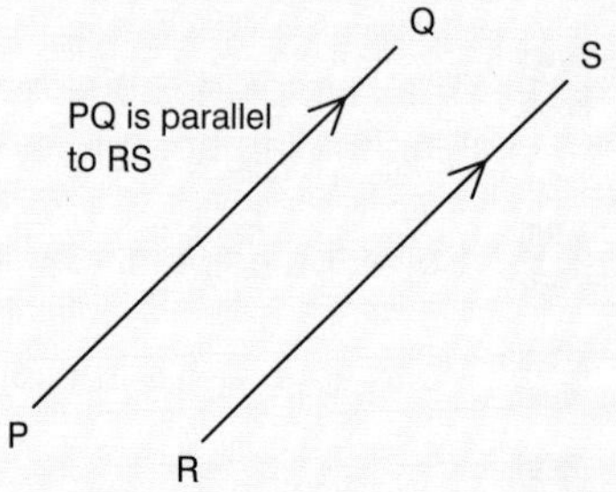

Two lines (or planes) are perpendicular if they meet at a right angle.

Two lines or planes are parallel if they are the same perpendicular distance apart everywhere.

ANGLES

A **right angle** is equal to 90°.

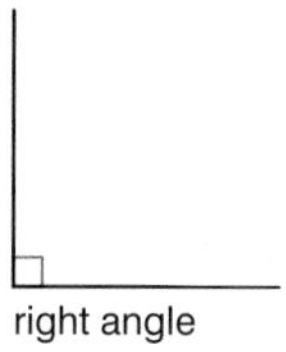

An **obtuse angle** is more than 90° but less than 180°.

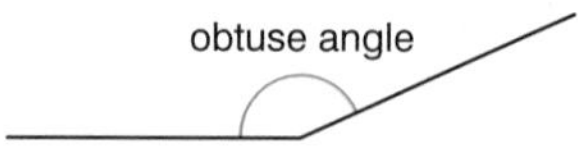

A **reflex angle** is more than 180° but less than 360°.

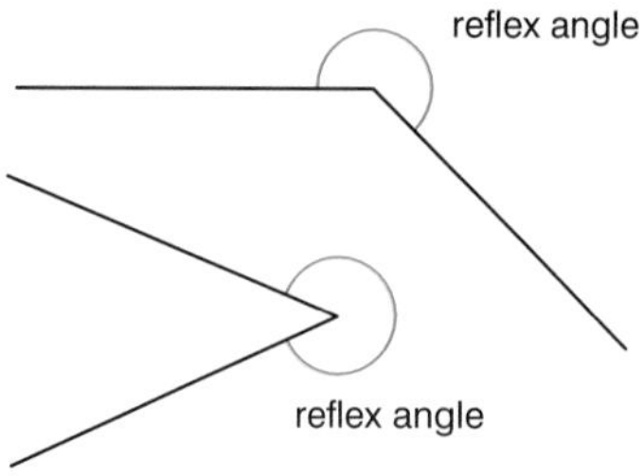

An **acute angle** is less than 90°.

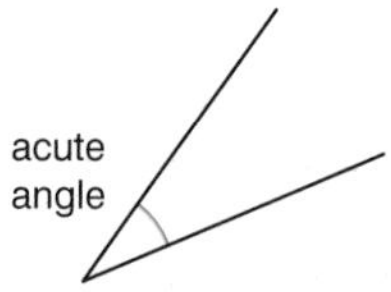

Complementary angles add up to 90°.

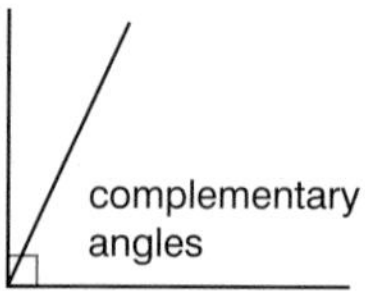

Supplementary angles add up to 180°.

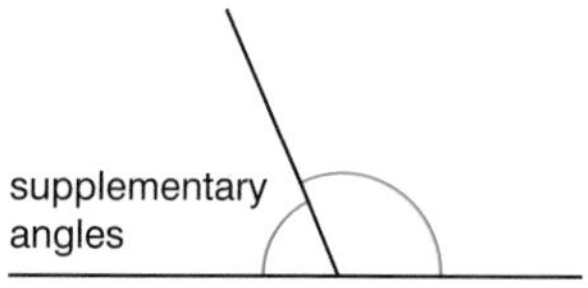

TRIANGLES

PROPERTIES OF TRIANGLES

The sum of the interior angles of a triangle is 180°.

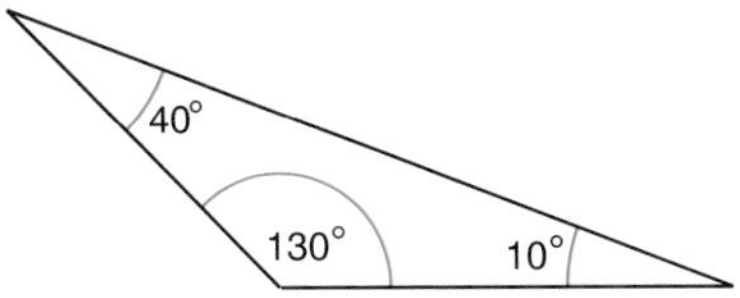

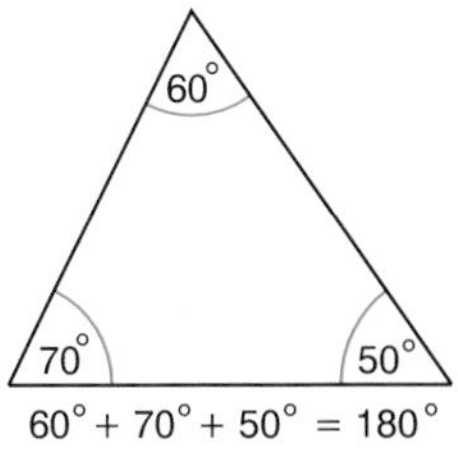

The smallest angle is always opposite the smallest side.

The largest angle is always opposite the largest side.

In general, the area of a triangle $= \frac{1}{2} \times$ base $\times$ perpendicular height.

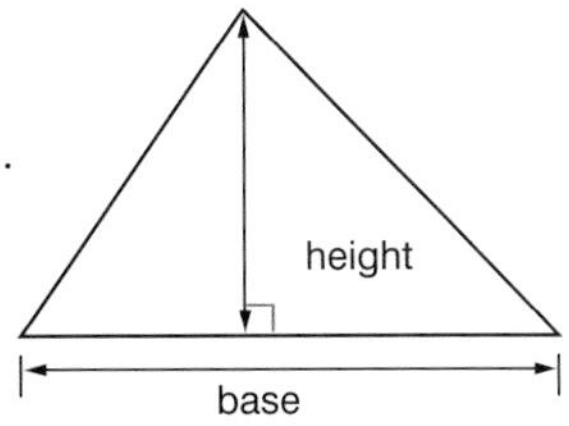

TYPES OF TRIANGLE

You need to know about the following triangles and their properties.

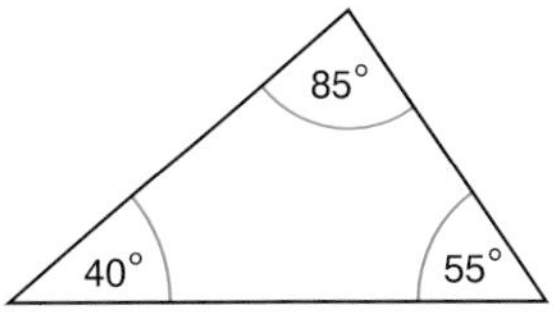

An **acute-angled triangle** has all of its angles less than 90°.

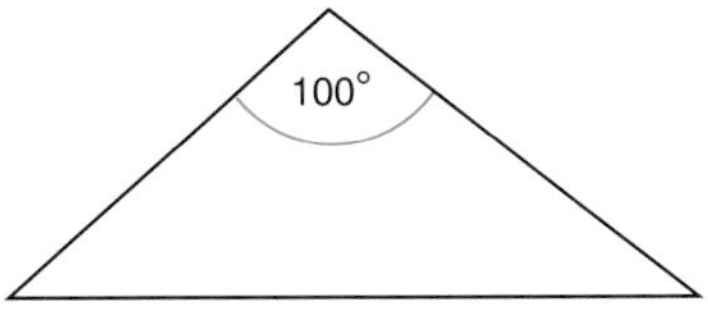

An **obtuse-angled triangle** has one of its angles greater than 90°.

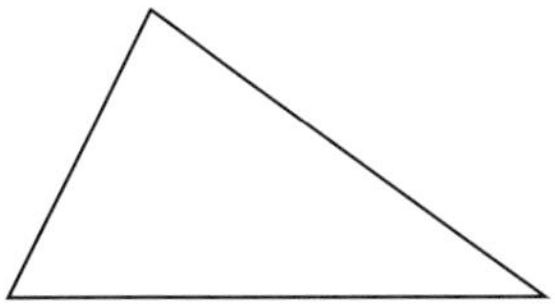

A **scalene triangle** has all sides and all angles different.

An **isosceles triangle** has two sides the same length and two angles the same size.

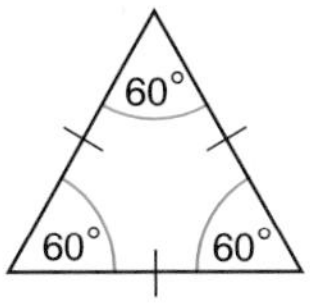

An **equilateral triangle** has all three sides the same length and all three angles the same size.

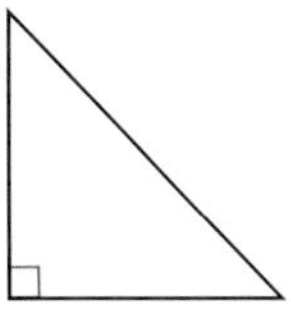

A **right-angled triangle** has one of its angles equal to 90°.

(a)

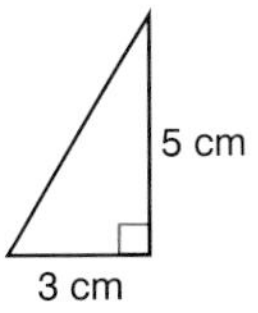

(b)

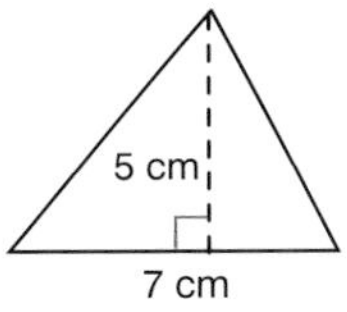

(c)

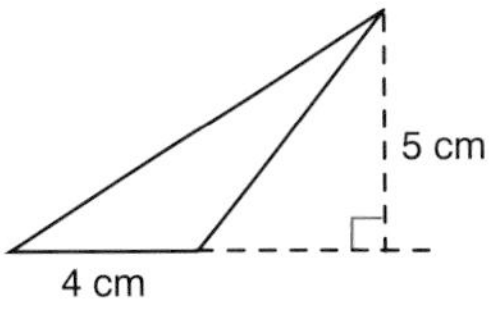

Worked example

Find the areas of the triangles.

(a) Area of a triangle $= \frac{1}{2} \times$ base $\times$ perpendicular height

$= \frac{1}{2} \times 3 \times 5$ Perpendicular height is 5 cm.

$= 7.5\,\text{cm}^2$ Remember to include the units of area.

(b) Area of a triangle $= \frac{1}{2} \times$ base $\times$ perpendicular height

$= \frac{1}{2} \times 7 \times 5$ Perpendicular height is 5 cm again.

$= 17.5\,\text{cm}^2$

(c) Area of a triangle $= \frac{1}{2} \times$ base $\times$ perpendicular height

$= \frac{1}{2} \times 4 \times 5$ Perpendicular height is still 5 cm.

$= 10\,\text{cm}^2$

QUADRILATERALS

PROPERTIES OF QUADRILATERALS

The sum of the interior angles of a quadrilateral is 360°.

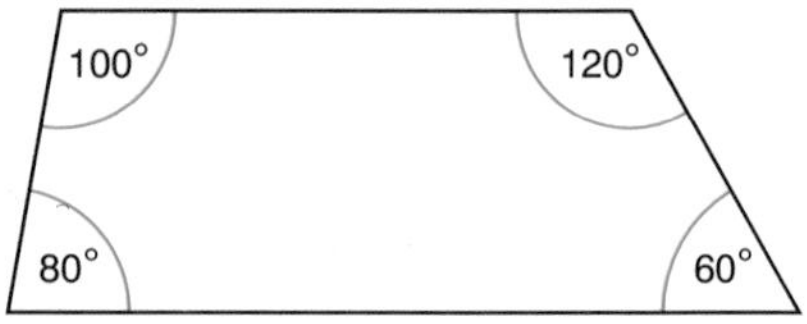

TYPES OF QUADRILATERAL

You need to know about the following quadrilaterals and their properties.

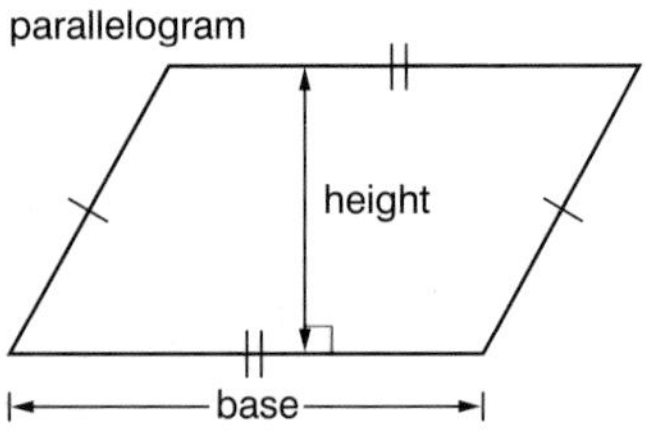

A **parallelogram** is a quadrilateral in which the two pairs of opposite sides are equal and parallel.

The area of a parallelogram = base × perpendicular height.

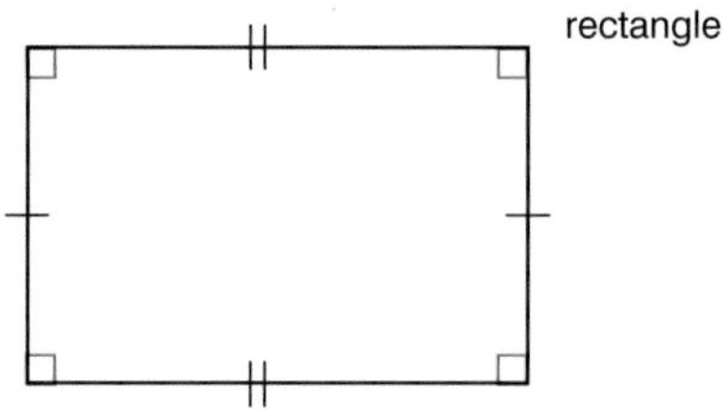

A **rectangle** is a parallelogram with four right angles.

The area of a rectangle = base × height.

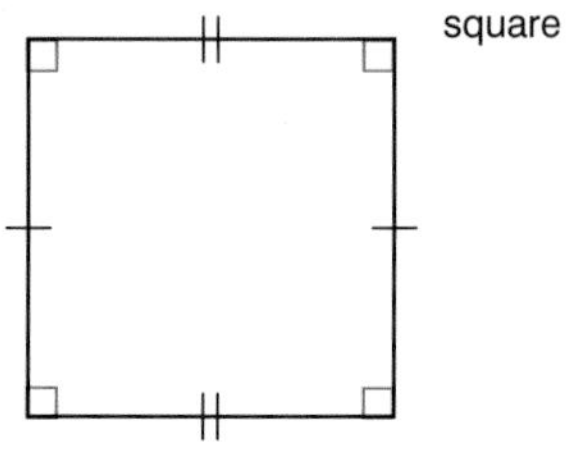

A **square** is a rectangle with four equal sides.

The area of a square = base × height (where base = height).

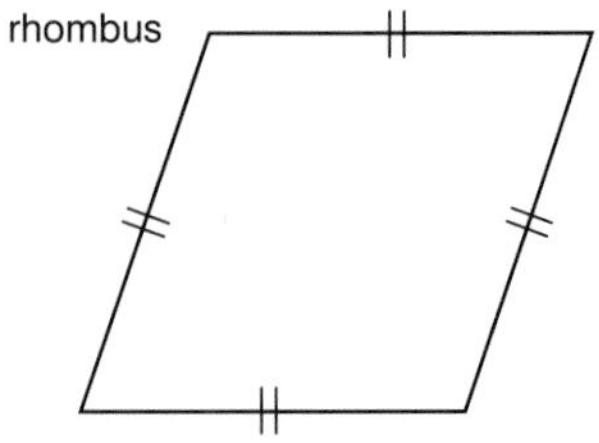

A **rhombus** is a parallelogram with four equal sides.

The area of a rhombus = base × perpendicular height.

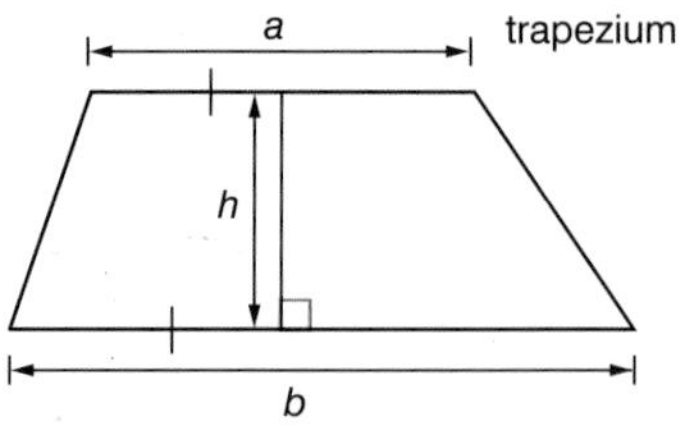

A **trapezium** is a quadrilateral with one pair of opposite sides parallel.

The area of a trapezium

$= \frac{1}{2} \times \left(\text{sum of parallel sides}\right) \times \text{height}.$

This is usually written as $\frac{1}{2}(a + b)h$

where a and b are the lengths of the parallel sides and h is the height.

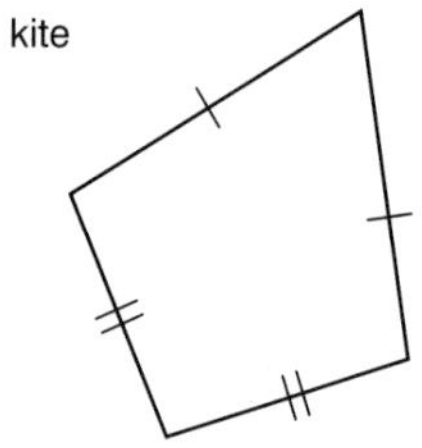

A **kite** is a quadrilateral with two pairs of adjacent sides equal.

Worked example

Find the areas of the quadrilaterals.

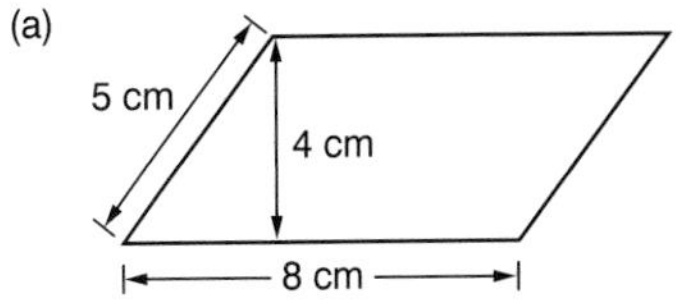

(a) Area of a parallelogram = base × perpendicular height

$= 8 \times 4$ Perpendicular height is 4 cm (not 5 cm).

$= 32\,\text{cm}^2$

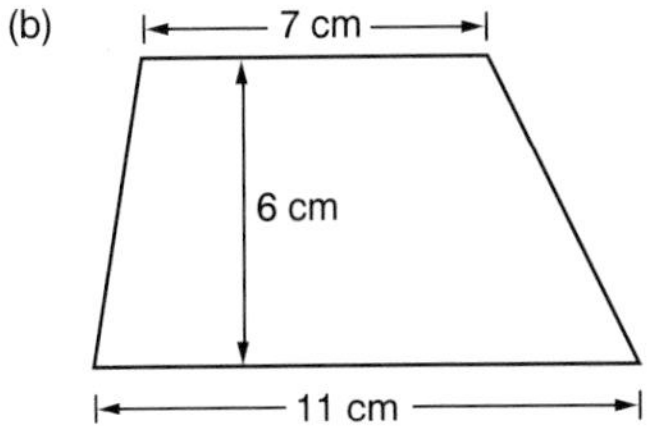

(b) Area of trapezium $= \frac{1}{2} \times$ (sum of parallel sides) × height

$= \frac{1}{2} \times (7 + 11) \times 6$ Perpendicular height is 6 cm.

$= \frac{1}{2} \times 18 \times 6$

$= 54\,\text{cm}^2$

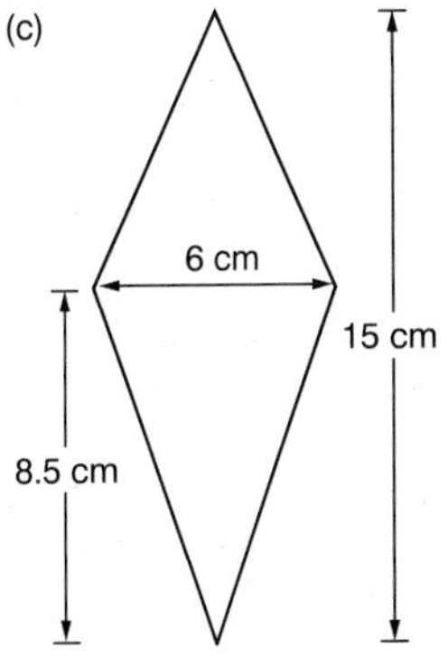

(c) To find the area of a kite it is helpful to split the shape into two triangles and use:

area of a triangle $= \frac{1}{2} \times$ base × perpendicular height

Area of kite $= \frac{1}{2} \times 6 \times 6.5 + \frac{1}{2} \times 6 \times 8.5$ Height of top triangle $= 15 - 8.5 = 6.5$ cm.

$= 19.5 + 25.5$

$= 45\,\text{cm}^2$

An easier way to find the area of a kite is to multiply the width by the height and divide by 2.

Area of kite $= \frac{1}{2} \times$ width × height

POLYGONS

Any shape enclosed by straight lines is called a **polygon**. Polygons are named according to their number of sides.

A **regular polygon** has all sides equal and all angles the same.

A **convex polygon** has no interior angle greater than 180°.

A **concave** (or **re-entrant**) **polygon** has at least one interior angle greater than 180°.

Number of sides	Name of polygon
3	triangle
4	quadrilateral
5	pentagon
6	hexagon
7	heptagon
8	octagon
9	nonagon
10	decagon

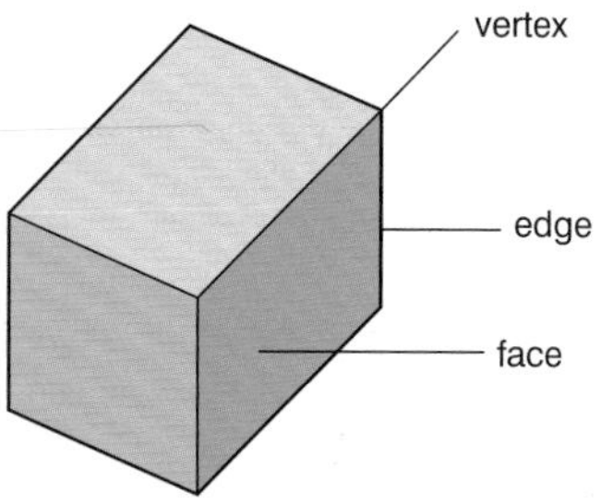

SOLIDS

A **solid** is a three-dimensional shape such as cube, cuboid, prism, cylinder, sphere, pyramid or cone.

You need to understand the following definitions relating to solids.

A **face** is the surface of a solid which is enclosed by edges.

An **edge** is a straight line where two faces meet.

A **vertex** is the point where three or more edges meet.

NOTE

Cubes and cuboids are also prisms as they have uniform cross-sections.

TYPES OF SOLIDS

You should be aware of the following solids and their properties.

A **cube** is a three-dimensional shape with six square faces.

A **cuboid** is a three-dimensional shape with six rectangular faces. Opposite faces are equal in size.

A **prism** is a three-dimensional shape with uniform cross-section. The prism is usually named after the shape of the cross-sectional area.

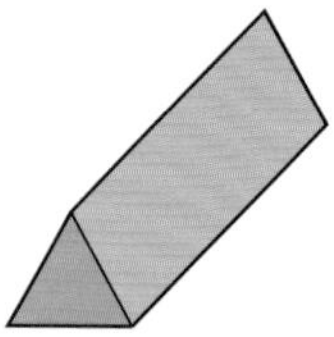
triangular prism

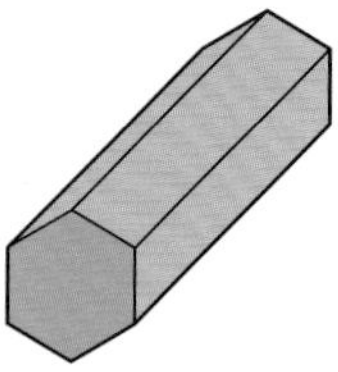
hexagonal prism

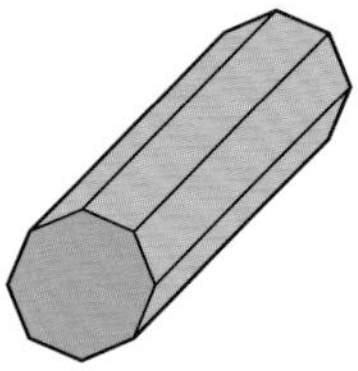
octagonal prism

NOTE

A hemisphere is one half of a sphere.

A **cylinder** is a prism with a uniform circular cross-section.

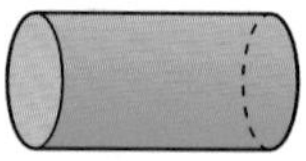

A **sphere** is a three-dimensional shape in which the surface is always the same distance from the centre.

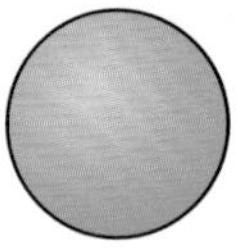
sphere

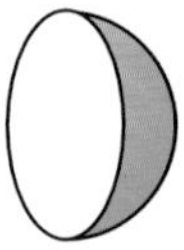
hemisphere

A **pyramid** is a three-dimensional shape with a polygon-shaped base and the remaining triangular faces meeting at a vertex. The pyramid is usually named after the shape of the polygon forming the base.

NOTE

There is more information about these solids in Chapter 67, Lengths, areas and volumes (intermediate level) and Chapter 69, Further areas and volumes (higher level).

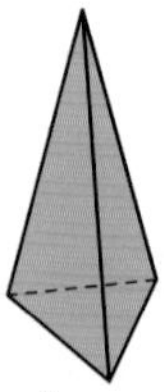
triangular pyramid

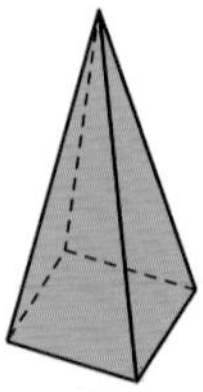
square-based pyramid

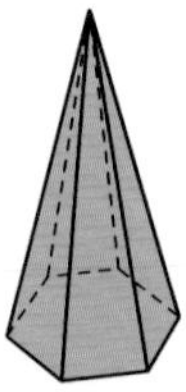
hexagon-based pyramid

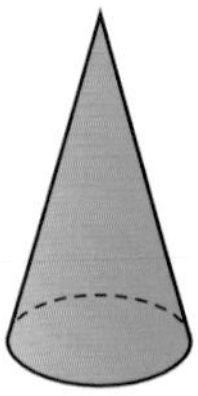
cone

A **cone** is a pyramid with a circular base.

NETS

A net is a pattern which can be cut out and folded to form a 3D shape. The following nets are the most common.

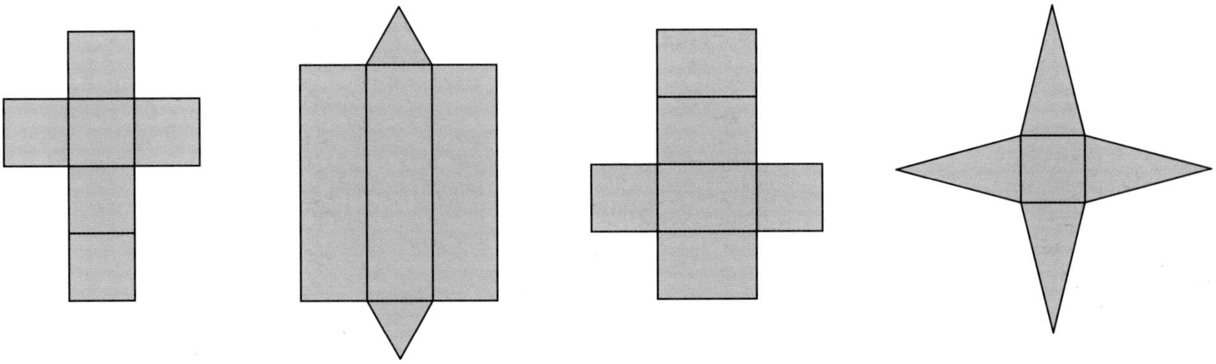

cube | triangular prism | cuboid | square-based pyramid

Check yourself

QUESTIONS

Q1 Find the value of the angle marked.

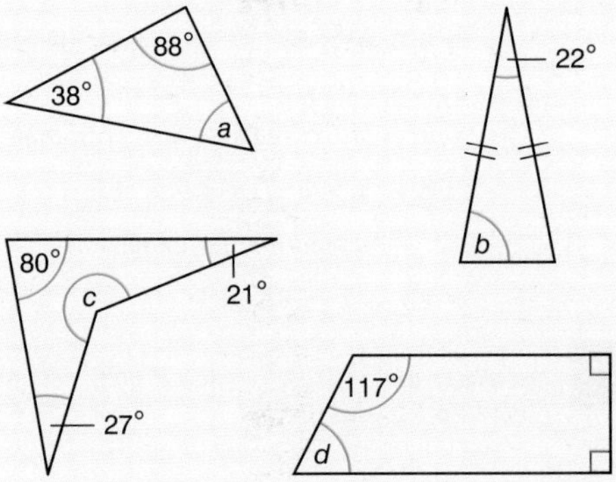

Q2 Write down the value of the smallest angle.

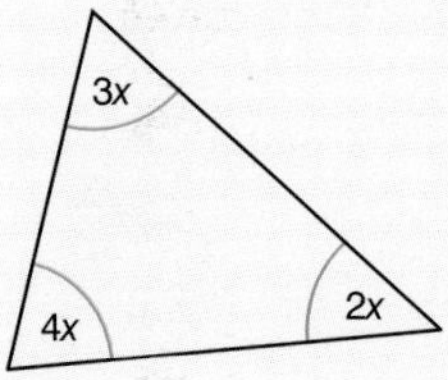

Q3 Find the areas of the following polygons.

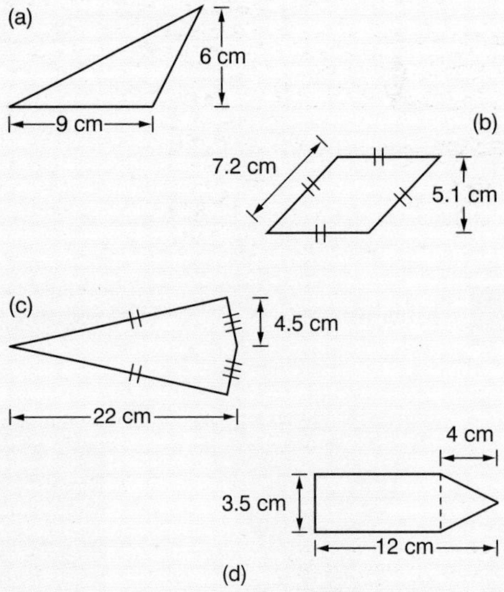

Q4 Draw nets of the following solids.

(a) tetrahedron (a pyramid in which all the faces are equilateral triangles)
(b) hexagonal prism

REMEMBER! Cover the answers if you want to.

ANSWERS

A1 $a = 54°$, $b = 79°$, $c = 232°$, $d = 63°$

A2 The smallest angle = 40°

A3
(a) Area of triangle = 27 cm^2
(b) Area of rhombus = 36.7 cm^2 (3 s.f.)
(c) Area of kite = 99 cm^2
(d) Area of polygon = 35 cm^2

A4 tetrahedron hexagonal prism

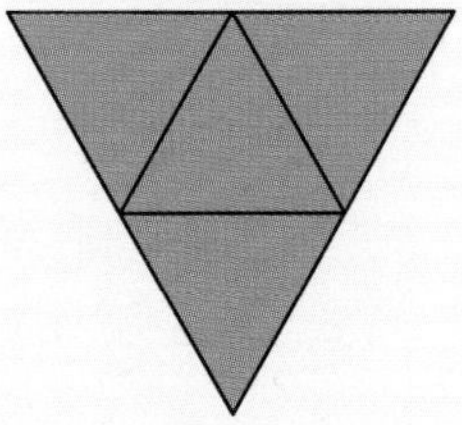
(a)

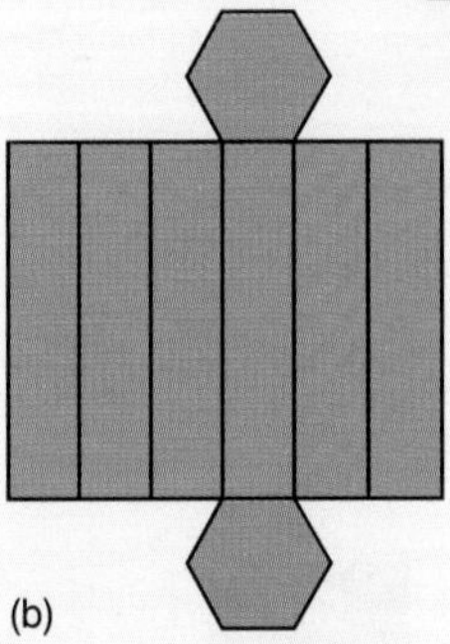
(b)

TUTORIAL

T1
$a = 54°$ *Angles of a triangle add up to 180°.*
$b = 79°$ *Angles of a triangle add up to 180° and the base angles of an isosceles triangles are equal.*
$c = 232°$ *Angles of a quadrilateral add up to 360° and the angle is clearly a reflex angle.*
$d = 63°$ *Angles of a quadrilateral add up to 360°.*

T2
The angles of a triangle add up to 180° so
$2x + 3x + 4x = 180°$
$9x = 180°$
$x = 20°$
The smallest angle $= 2x = 2 \times 20° = 40°$

T3
(a) *Area of triangle*
$= \frac{1}{2} \times$ *base* $\times$ *perpendicular height*
$= \frac{1}{2} \times 9 \times 6$
$= 27\ cm^2$

(b) *Area of rhombus = base × perpendicular height*
$= 7.2 \times 5.1$ *Perpendicular height is 5.1 cm (not 7.2 cm).*
$= 36.72 = 36.7\ cm^2$ (3 s.f.)
Rounding to an appropriate degree of accuracy.

(c) *Area of kite* $= \frac{1}{2}$ *width × height*
$= \frac{1}{2} \times 22 \times 9$ *Height* $= 2 \times 4.5 = 9$ *cm.*
$= 99\ cm^2$

(d) *Area of polygon*
= area of rectangle + area of triangle
$= 8 \times 3.5 + \frac{1}{2} \times 3.5 \times 4$
As length of rectangle $= 12 - 4 = 8$ *cm*
$= 28 + 7 = 35\ cm^2$

T4 *Use paper and scissors to cut out your nets and make them into the given solids.*

CHAPTER 53

USE OF MATHEMATICAL INSTRUMENTS

The following equipment will be useful to help you construct diagrams, produce scale drawings and find the locus of points.

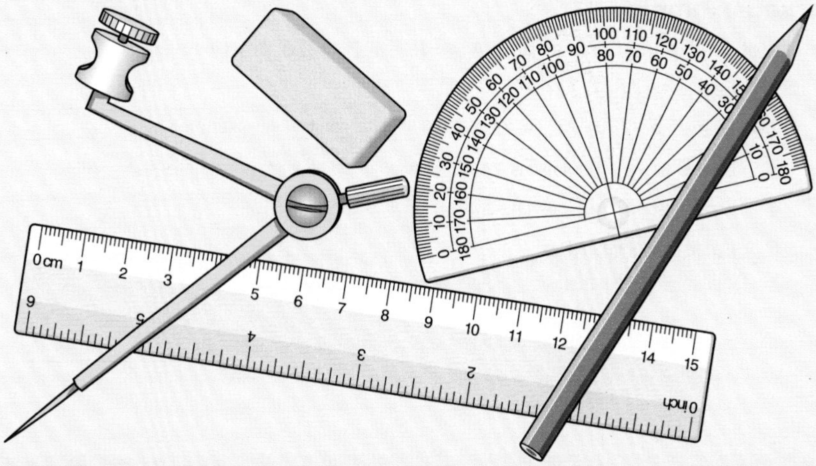

- **a sharp pencil and an eraser**
- **a ruler**
- **a pair of compasses**
- **a protractor**

You should always undertake construction work as accurately as you possibly can – the examiner expects that measurements of length will be accurate to the nearest millimetre and measurements of angles will be correct to the nearest degree.

CONSTRUCTING TRIANGLES

When constructing a triangle ABC, it is conventional to represent the side opposite angle A by the letter *a*, the side opposite angle B by the letter *b* and the side opposite angle C by the letter *c* etc. as shown here.

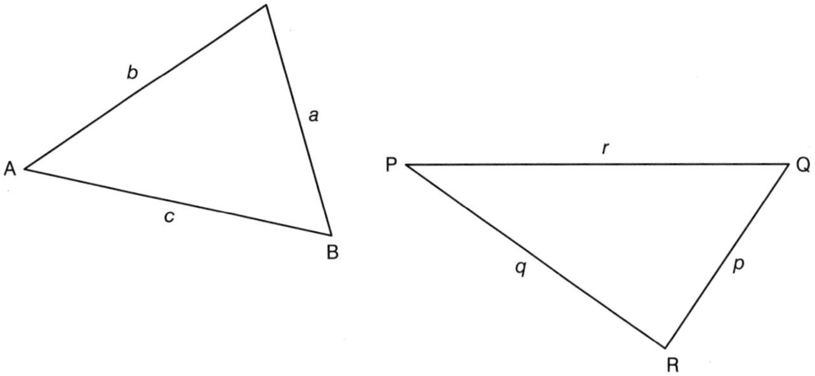

To construct a triangle you need to know:

- two angles and a side *or*
- two sides and the included angle *or*
- three sides.

HINT

Always leave your construction lines on your diagram.

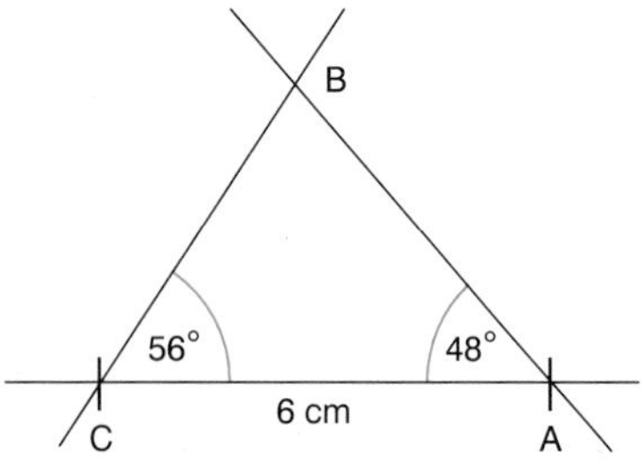

GIVEN TWO ANGLES AND A SIDE

Worked example

Construct triangle ABC where ∠A = 48°, ∠B = 76° and b = 6 cm.

1 Draw a rough sketch.
2 Draw the line AC = 6 cm.
3 Draw an angle of 56° at C.
4 Draw an angle of 48° at A.
5 Label the point of intersection as B.

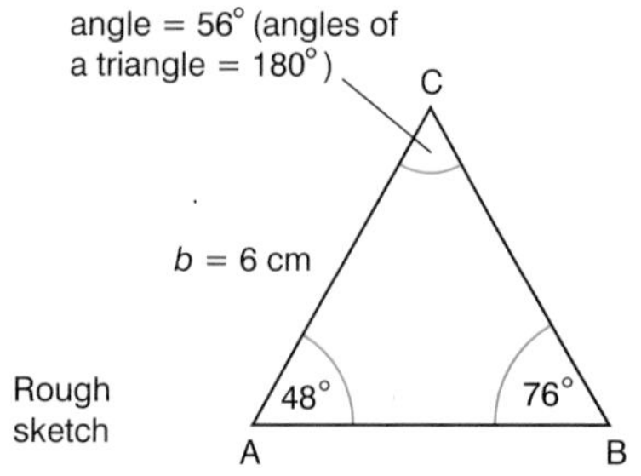

GIVEN TWO SIDES AND THE INCLUDED ANGLE

Worked example

Construct triangle PQR where p = 8 cm, q = 6.5 cm and ∠QRP = 53°.

1 Draw a rough sketch.
2 Draw the line QR = 8 cm.
3 Draw an angle of 53° at R.
4 Construct the point P which is 6.5 cm from R, using your compasses with centre R and radius 6.5 cm.
5 Join QP.

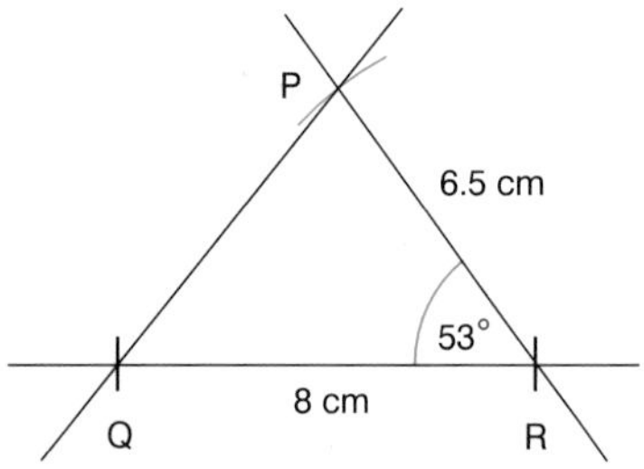

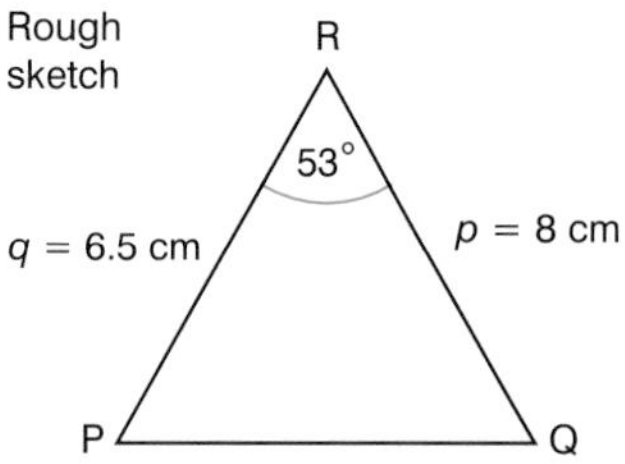

HINT

It is always best to start by drawing the longest side.

GIVEN THREE SIDES

Worked example

Construct triangle XYZ where x = 5.8 cm, y = 8.2 cm and z = 5.5 cm.

1 Draw a rough sketch.
2 Draw the line XZ = 8.2 cm (the longest side).
3 With centre Z and radius 5.8 cm, use your compasses to draw a circular arc giving points 5.8 cm from Z.
4 With centre X and radius 5.5 cm, use your compasses to draw a circular arc giving points 5.5 cm from X.
5 Label the point of intersection of the two arcs as Y.
6 Join ZY and XY.

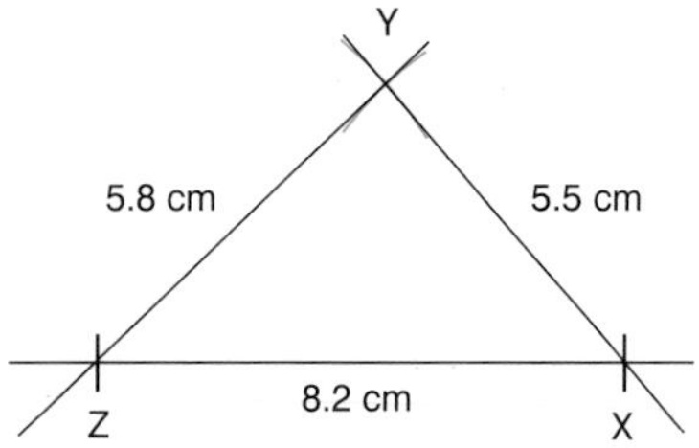

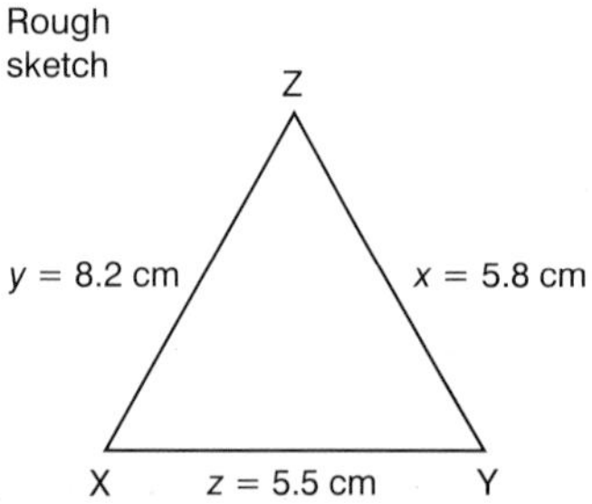

Worked example

Construct triangle ABC where $a = 7.6$ cm, $b = 5.6$ cm and $\angle ABC = 44°$.

Before constructing the triangle you need to remind yourself what you need to know:

- two angles and a side *or*
- two sides and the included angle *or*
- three sides.

In this particular case you do not have sufficient information to draw the triangle, since you are not given the included angle, but you can proceed as follows.

1 Draw a rough sketch.

2 Draw the line BC = 7.6 cm.

3 Draw an angle of 44° at B.

4 With compasses set to a radius of 5.6 cm, and centred on C, draw a circular arc giving points 5.6 cm from C.

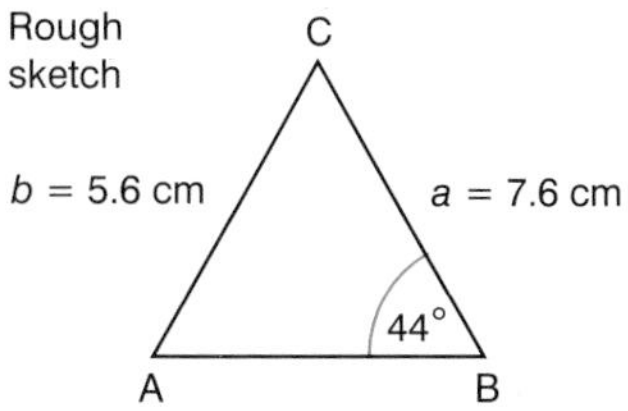

5 You will find that there are two points where this arc cuts the line from B. Either of these points will give a point A which satisfies the conditions for the triangle.

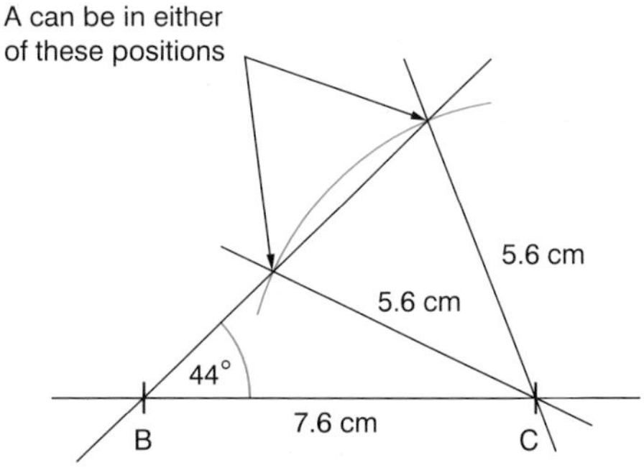

> **NOTE**
>
> Because there are two possible solutions to the triangle, this is sometimes called the ambiguous case.

GEOMETRICAL CONSTRUCTIONS

You should be able to carry out the following constructions using only a ruler and a pair of compasses:

- the perpendicular bisector of a line
- the perpendicular from a point on a straight line
- the perpendicular from a point to a straight line
- the angle bisector.

PERPENDICULAR BISECTOR OF A LINE

Worked example

Construct the perpendicular bisector of a line AB.

1 With the compasses set to a radius greater than half the length of AB, and centred on A, draw arcs above and below the line.

2 With the compasses still set to the same radius, and centred on B, draw arcs above and below the line, so that they cut the first arcs.

3 Join the points where these arcs cross (P and Q). This line is the perpendicular bisector of AB.

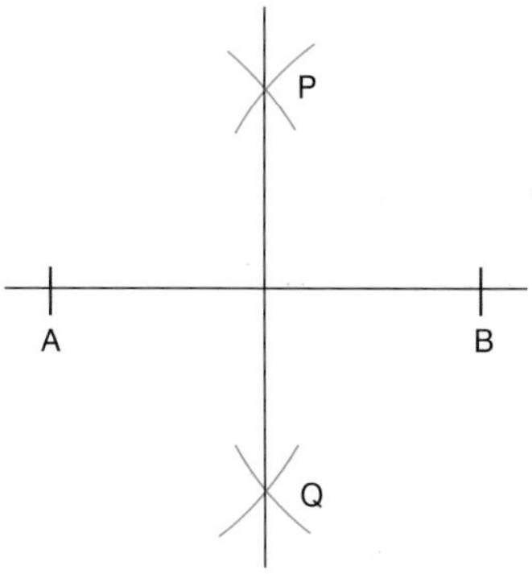

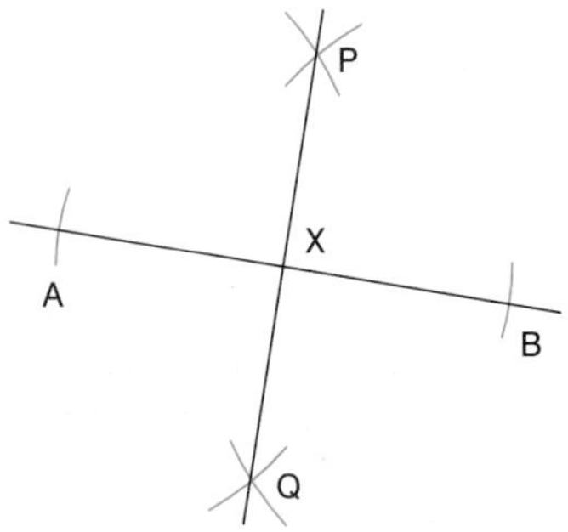

PERPENDICULAR FROM A POINT ON A STRAIGHT LINE

Worked example

Construct the perpendicular at the point X on a line.

1 With the compasses set to a radius of about 5 cm, and centred on X, draw arcs to cut the line at A and B.

2 Now construct the perpendicular bisector of the line segment AB, as above.

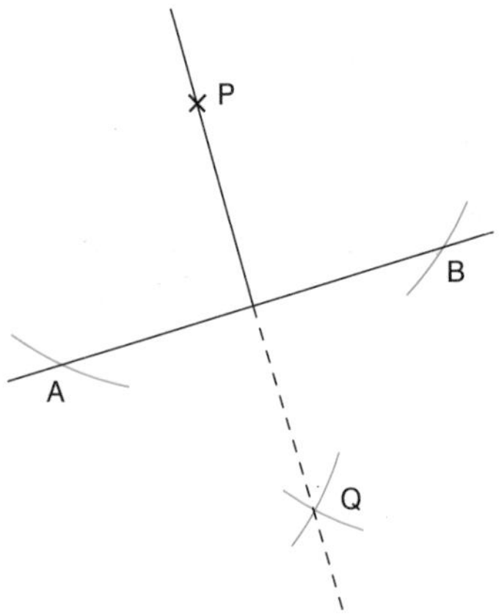

PERPENDICULAR FROM A POINT TO A STRAIGHT LINE

Worked example

Construct the perpendicular to the line from a point P above the line.

1 With the compasses set to a suitable radius, and centred on P, draw arcs to cut the line at A and B.

2 With the compasses set to a radius greater than half the length of AB, and centre

on A, draw an arc on the opposite side of the line from P.

3 With the compasses still set to the same radius, and centred on B, draw an arc to cut the arc drawn in step 2, at Q.

4 Join PQ.

THE ANGLE BISECTOR

Worked example

Construct the bisector of an angle ABC.

1 With the compasses set to a radius of about 5 cm, and centred on B, draw arcs to cut BA at L and BC at M.

2 With the compasses set to the same radius, and centred on L, draw an arc between BA and BC.

3 With the compasses still set to the same radius, and centred on M, draw an arc to cut the arc between BA and BC.

4 Label the point where the arcs cross as Q.

5 Join BQ. This is the bisector of the angle ABC.

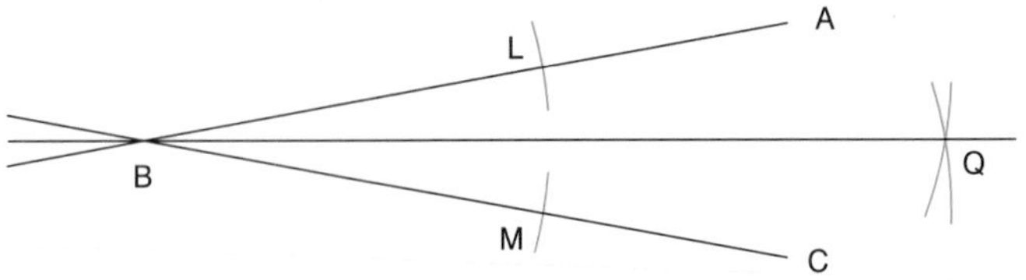

Angle bisector

Check yourself

QUESTIONS

Q1 Draw any triangle ABC, then construct the perpendicular bisectors of the two longest sides. Label the point of intersection of these perpendicular bisectors O. Draw the circle centre O and radius OA.

Q2 Construct triangle PQR where $p = 4.5$ cm, $q = 6$ cm and $r = 7.5$ cm. What do you notice about your triangle?

Q3 Draw the line PQ so that PQ = 8 cm. Construct the perpendicular bisector of PQ. Construct the points R and S so that they lie on the perpendicular bisector a distance of 3 cm from the line PQ.
Join PR, RQ, QS and SP.
What is the special name given to this quadrilateral?

REMEMBER! Cover the answers if you want to.

A1

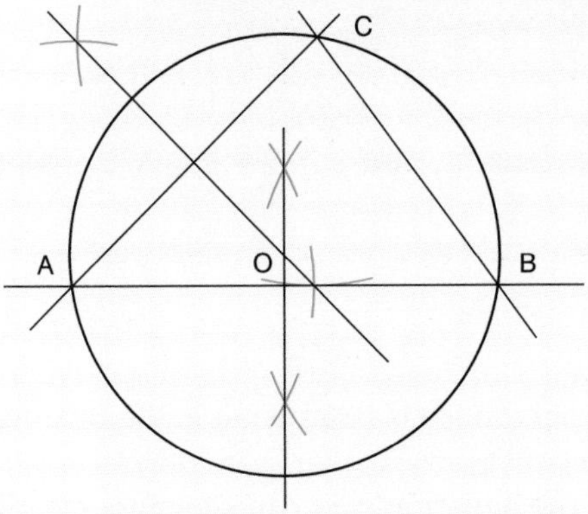

T1 *Accurate construction will result in the circle* ***circumscribing*** *(completely enclosing) triangle ABC and passing through the points A, B and C of the triangle.*

A2

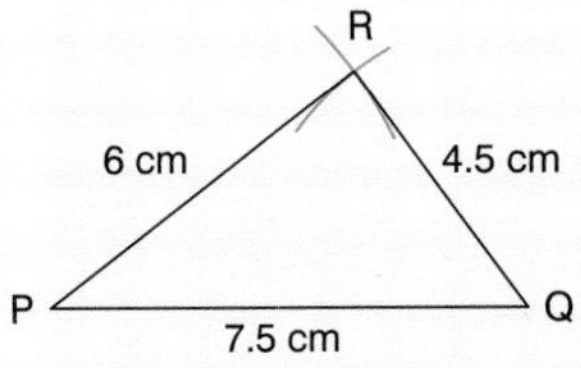

The triangle PQR is right-angled.

T2 *You should follow the instructions given under the heading 'Constructing a triangle given three sides'.*

A3

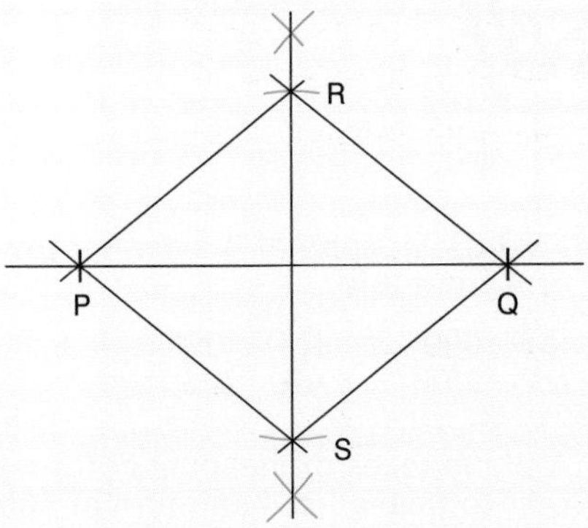

The quadrilateral is called a rhombus.

T3 *Accurate construction will result in a rhombus PRQS.*

CHAPTER 54

THREE-FIGURE BEARINGS

Bearings are a useful way of describing directions.

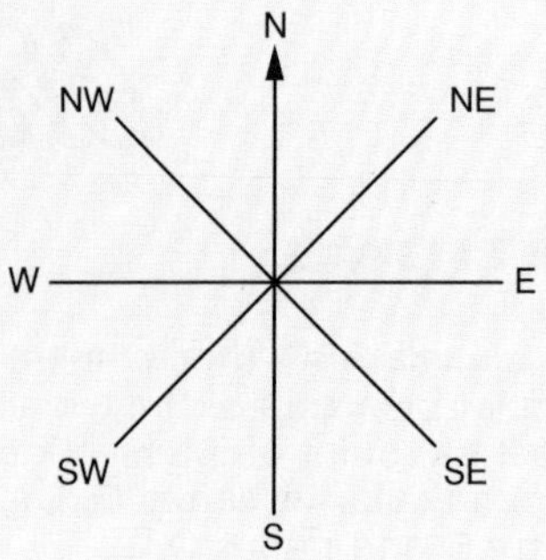

Bearings can be given in terms of the points on a compass or else as directions measured from north in a clockwise direction. Bearings are usually given as three-figure numbers, so a bearing of 50° would usually be written as 050°.

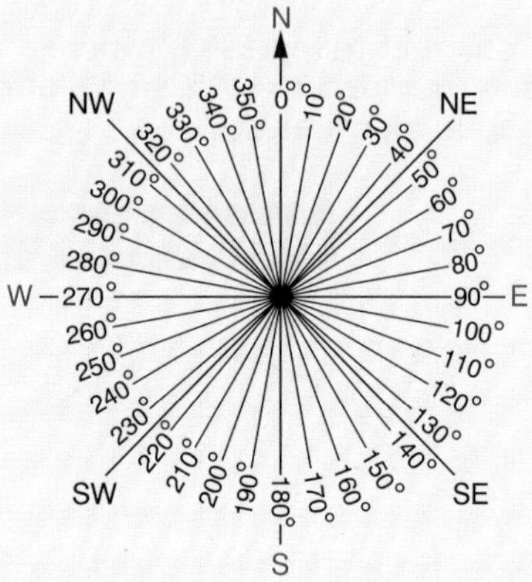

Worked example

The bearing of a ship from a lighthouse is 035°. What is the bearing of the lighthouse from the ship?

By drawing a sketch of the situation, you can see more clearly that the required bearing is 215°.

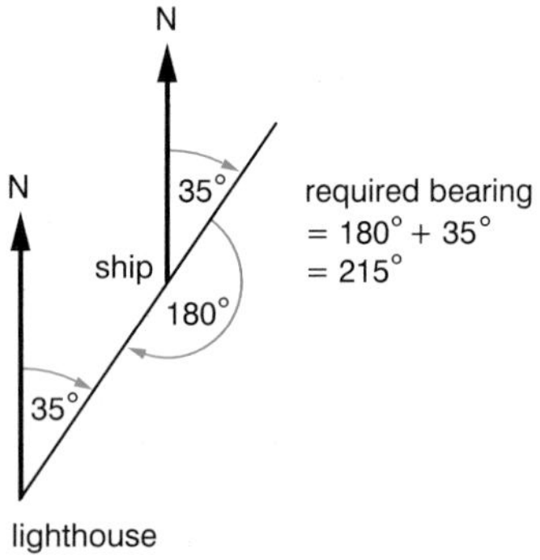

Check yourself

QUESTIONS

Q1 What bearing is the same as:

(a) east (b) south-west
(c) north-west?

Q2 What compass point is the same as:

(a) 180° (b) 045° (c) 135°?

Q3 Find the bearing of B from A if the bearing of A from B is:

(a) 090° (b) 120°
(c) 355° (d) 203°.

Q4 A plane takes off heading north-west and is told to take a left-hand turn at 5000 feet. What bearing is the plane now headed on?

ANSWERS

A1 (a) 090° (b) 225° (c) 315°

A2 (a) south (b) north-east
(c) south-east

A3 (a) 270° (b) 300°
(c) 175° (d) 023°

A4 225°

NOTE

As the two north lines are parallel you can use the fact that interior angles between parallel lines add up to 180°.

TUTORIAL

T1 *See the diagram given at the start of this chapter.*

T2 *See the diagram given at the start of this chapter.*

T3 (*a*)

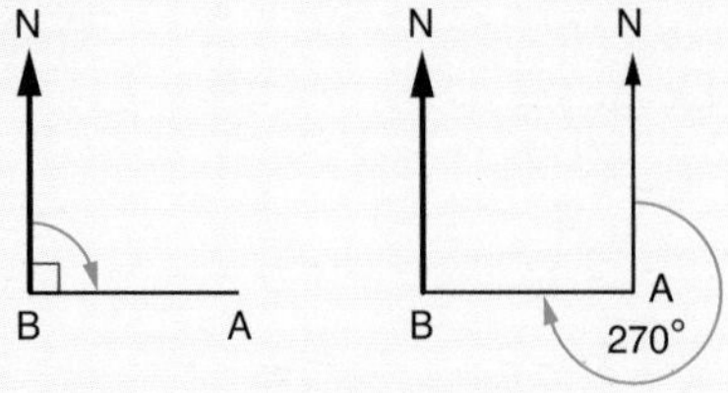

(*b*)

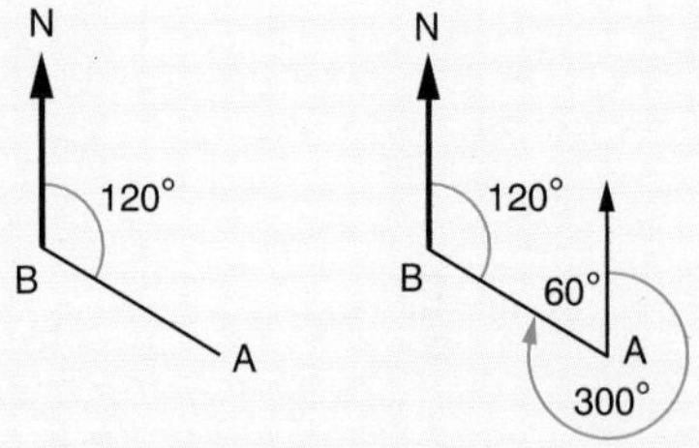

(*c*)

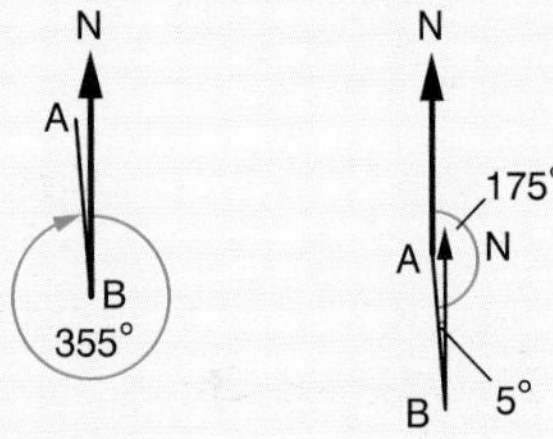

(*d*)

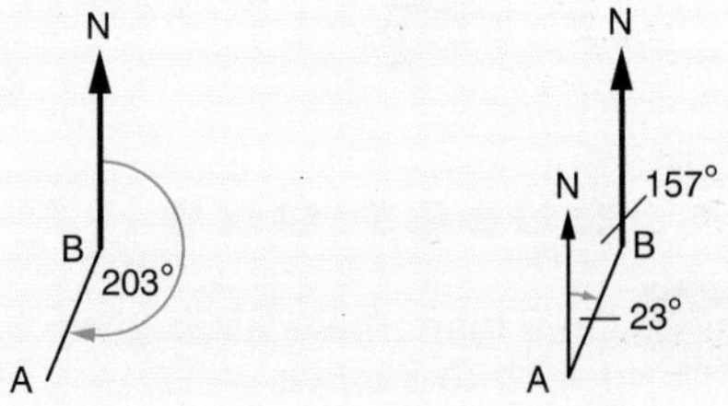

T4

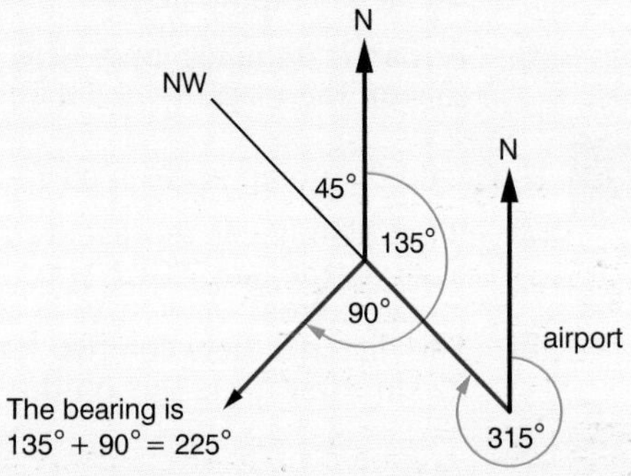

A *south-west direction is equivalent to a bearing of 225°.*

SHAPE, SPACE AND MEASURES

CHAPTER 55

SCALE DRAWINGS

Scale drawings are drawings or diagrams that are scaled according to some rule. The scale for the drawing or diagram must be clearly stated.

Worked example

A boat sails due east from a point A for 12 km to a position B, then for 6 km on a bearing of 160° to a point C. Using a scale of 1 cm to represent 2 km, show this on a scale drawing and use this to find the bearing and distance from A to C.

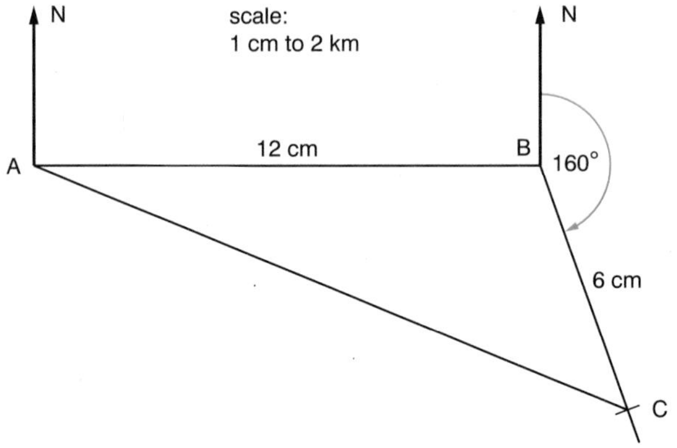

The bearing from A to C is measured as 112° and the distance from A to C is 7.6 cm which represents 15.2 km (remembering to convert back to kilometres using the given scale).

Check yourself

QUESTIONS

Q1 On a map, the scale is given as 1 inch to 5 miles.

(a) What distance is represented by $4\frac{1}{2}$ inches on the map?

(b) A road is 36 miles long. How long is this on the map?

Q2 The scale on a map is given as 1 : 500 000.

(a) What distance is represented by 5 cm on the map? Give your answer in kilometres.

(b) The distance between two farms is 22 km. How far is this on the map?

Q3 Two explorers set off from the same point one morning. One explorer travels at 4.5 mph on a bearing of 036° and the other explorer at 5.5 mph on a bearing of 063°. Using a scale of 0.5 cm to 1 mile, calculate how far they are apart after 2 hours.

ANSWERS

A1 (a) $22\frac{1}{2}$ miles (b) 7.2 inches

A2 (a) 25 km (b) 4.4 cm

A3 The distance between them is 5.1 miles.

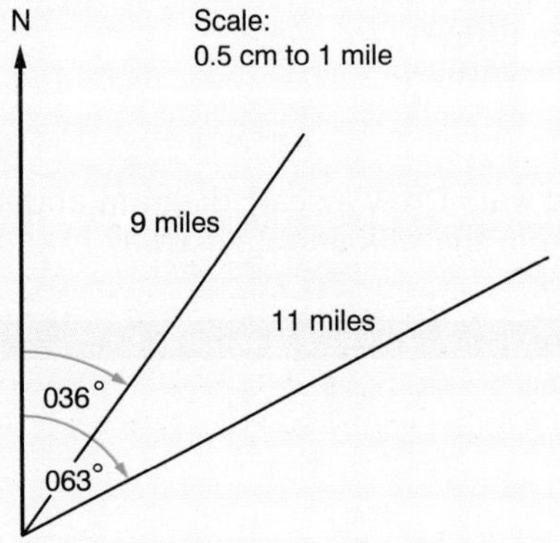

TUTORIAL

T1 (*a*) *If 1 inch represents 5 miles*

then $4\frac{1}{2}$ *inches represents* $4\frac{1}{2} \times 5$ *miles*

$= 22\frac{1}{2}$ *miles*

(*b*) *As 5 miles represents 1 inch*

then 1 mile represents $\frac{1}{5}$ *inch* *Dividing both sides by 5.*

36 miles represents $36 \times \frac{1}{5}$ *inches*

$= 7.2$ *inches*

T2 (*a*) *Here 1 cm represents 500 000 cm so*
5 cm represents $5 \times 500\,000$ *cm*

$= 2\,500\,000$ *cm*

$= 25\,000$ *m* *As 100 cm = 1 m.*

$= 25$ *km* *As 1000 m = 1 km.*

(*b*) *As 500 000 cm represents 1 cm*

then 5000 m represents 1 cm *As 100 cm = 1 m.*

5 km represents 1 cm *As 1000 m = 1 km.*

1 km represents $\frac{1}{5}$ *cm* *Dividing both sides by 5.*

22 km represents $22 \times \frac{1}{5}$ *cm*

$= 4.4$ *cm*

T3 *After two hours the explorers travel 9 miles on a bearing of 036° and 11 miles on a bearing of 063°. The distance between them is 2.55 cm which converts to 5.1 miles, using the given scale.*

CHAPTER 56

LOCUS OF POINTS

A locus is the path followed by points that satisfy some given rule. The following examples illustrate the most common loci.

Worked example

Find the locus of a point moving so that it is a fixed distance from a point O.

The locus of a point moving so that it is a fixed distance from a point O is a circle with centre O.

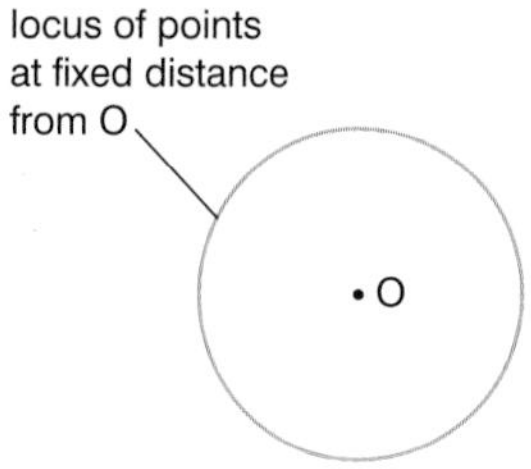

> **NOTE**
>
> 'Equidistant from' means 'the same distance from'.

Worked example

Find the locus of a point moving so that it is always equidistant from two fixed points A and B.

The locus of a point moving so that it is equidistant from two fixed points A and B is the perpendicular bisector of the line AB.

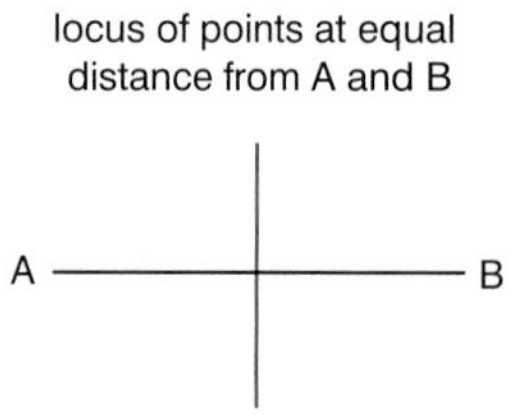

Worked example

Find the locus of a point moving so that it is a fixed distance from the line PQ.

The locus of a point moving so that it is a fixed distance from the line PQ is a pair of lines parallel to PQ, along with the semicircles at the points P and Q.

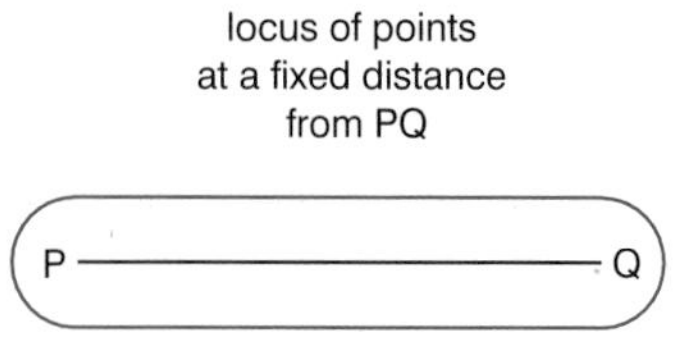

Worked example

Find the locus of a point moving so that it is a fixed distance from two lines AB and CD.

The locus of a point moving so that it is a fixed distance from two lines AB and CD is the pair of bisectors of the angles between the two lines (drawn from the point where the lines cross).

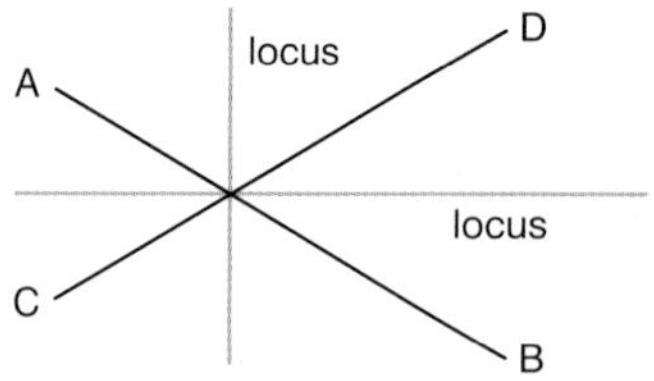

Worked example

A tree is situated 4 m away from a wall. Draw a scale diagram and identify the points which are 3 metres away from the wall and $2\frac{1}{2}$ m away from the tree.

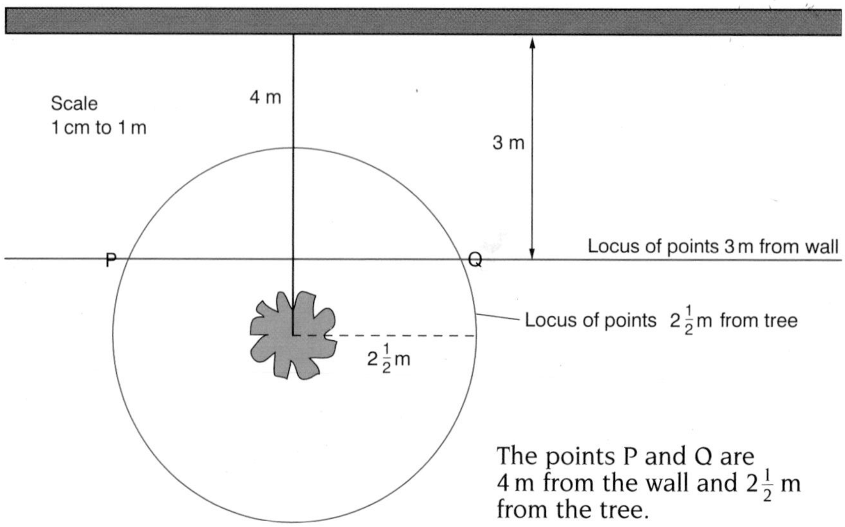

The points P and Q are 4 m from the wall and $2\frac{1}{2}$ m from the tree.

Check yourself

QUESTIONS

Q1 Find the locus of points which are 3 cm from the line AB, which is of length 5 cm.

Q2 Find the locus of points which are equidistant from two points P and Q which are 3 cm apart.

Q3 XY is a fixed line of length 8 cm and P is a variable point. The distance of P from X is 6.2 cm and the distance of P from Y is 3.7 cm. Show all of the possible positions of P on a diagram.

Q4 Find the locus of points less than 2 cm from a square of side $4\frac{1}{2}$ cm.

Q5 XYZ is an equilateral triangle of sides 5 cm. Show all of the points which are less than 2 cm from the edges of the triangle.

REMEMBER! Cover the answers if you want to.

ANSWERS

A1

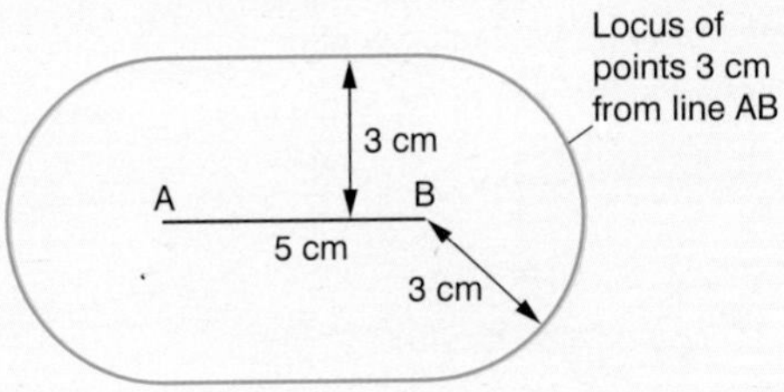

A2

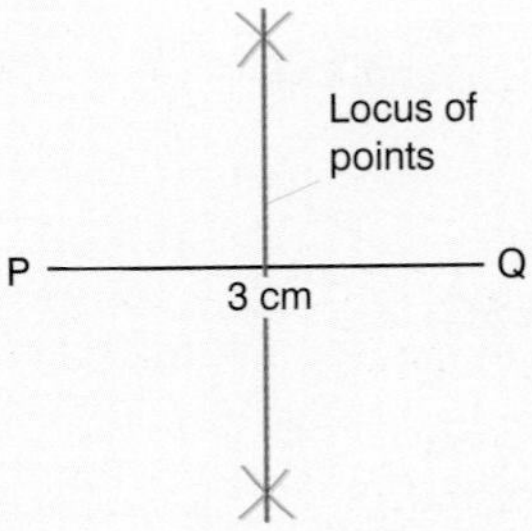

A3

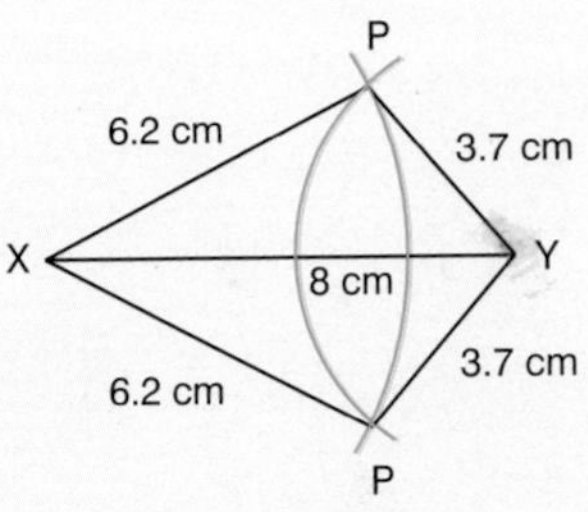

P is found at the intersection of the two arcs (two possible positions).

TUTORIAL

Sufficient detail is included on the diagrams to help you complete these accurately.

A4

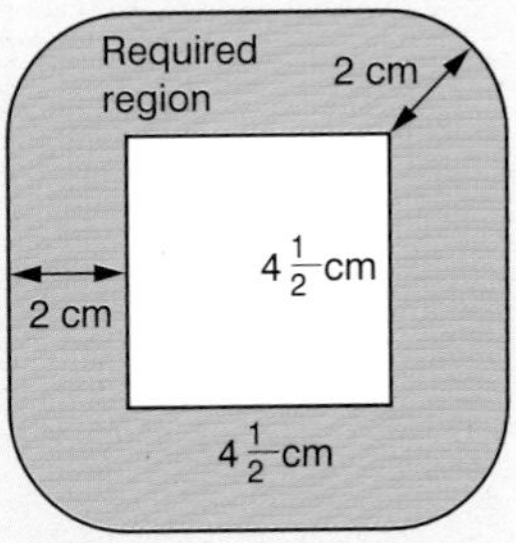

A5

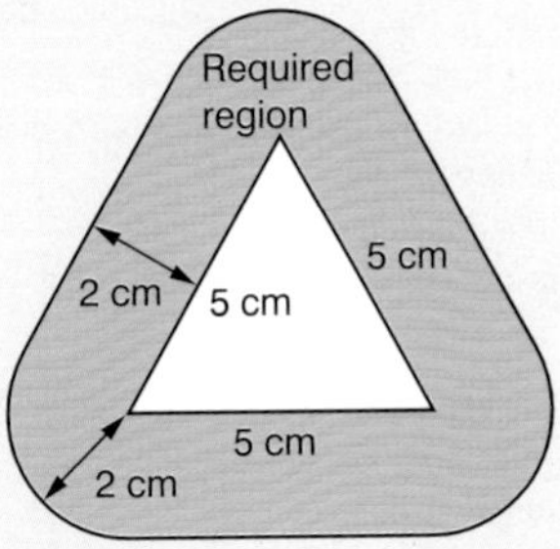

CHAPTER 57

EXAMINATION QUESTIONS 11

EXAM PRACTICE

Sample Student's Answers & Examiner's Comments

1 The diagram shows a square-based pyramid.

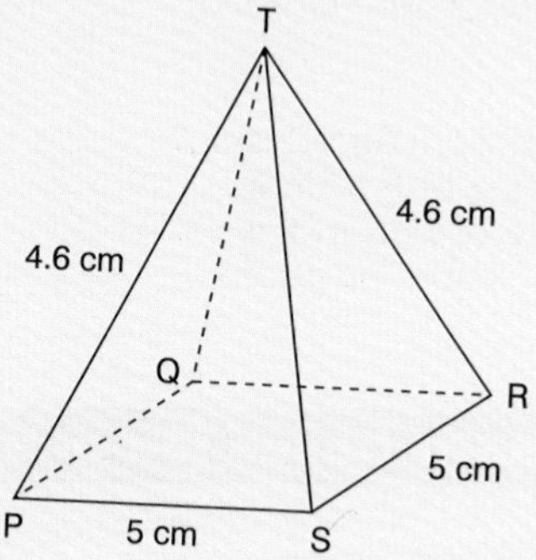

not to scale

(a) Draw a sketch of the net for the pyramid PQRST.
(b) Make an accurate drawing of the triangle PQT.
(c) Use your drawing to find the size of the angle PQT.

(a)

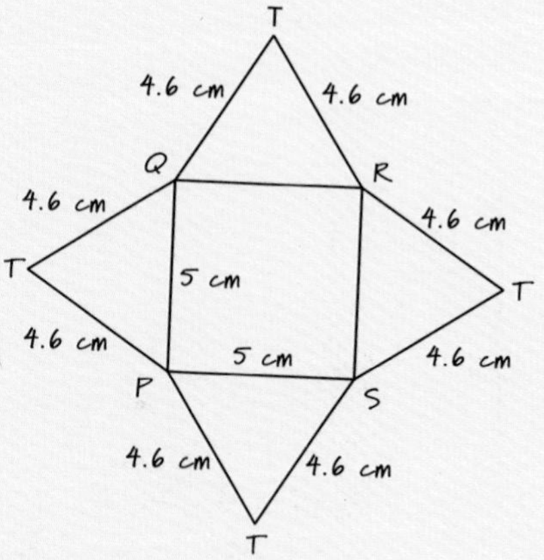

(b)

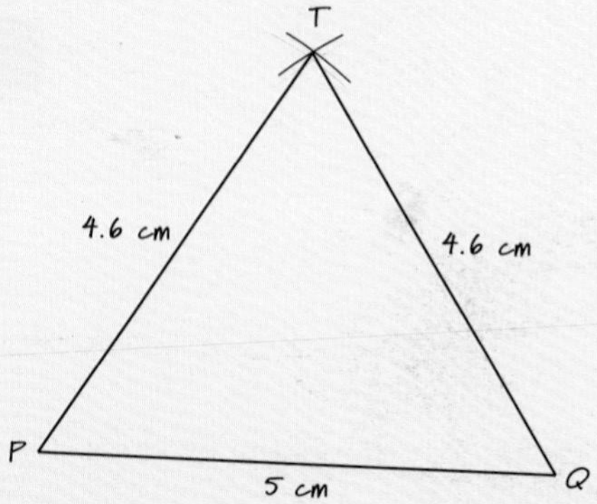

(c) ∠PQT = 57°

EXAMINER'S COMMENTS

1 **a)**
The candidate has correctly drawn the net of the pyramid and labelled the sides appropriately.

b)
The candidate has drawn the triangle PQT with lengths drawn to the required accuracy of ±1 mm, and angles drawn to the required accuracy of ±1°. The candidate has left the construction lines on the drawing as these will contribute to the marks awarded.

The candidate has correctly found the size of the angle PQT from the drawing. It is important that this angle is found from the drawing as required by the question and not found by using trigonometry – even though this answer might well be more accurate.

2 Two radio transmitters A and B are 60 kilometres apart and B is on a bearing of 080° from A.

(a) Calculate the bearing of A from B.

Each radio transmitter can broadcast to a distance of 45 kilometres.

(b) Using a scale of 1 cm to 10 km construct a scale drawing and show the area within which each transmitter can broadcast.

A car is situated so that it is on a bearing of 150° from A and a bearing of 225° from B.

(c) Which radio broadcasts can the car receive? Give reasons for your answer.

(a) Bearing = 180° + 80° = 260°

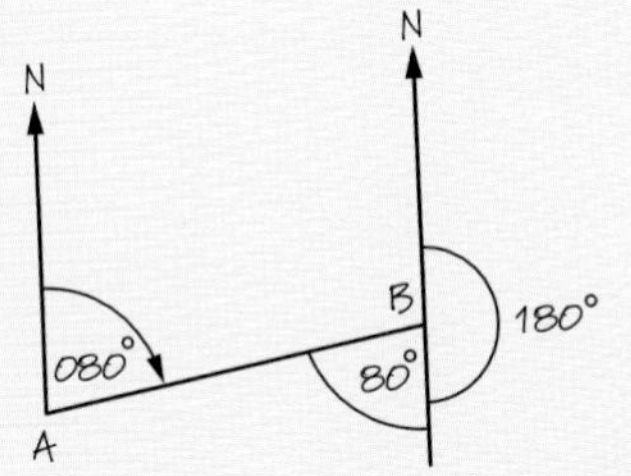

The bearing of A from B is 260°.

(b)

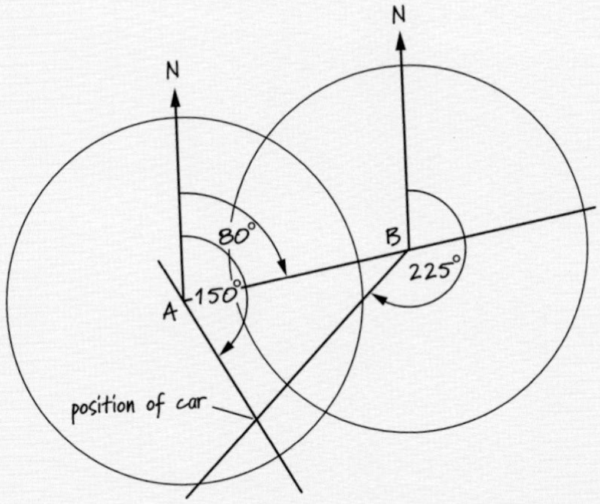

(c) The car can receive broadcasts from transmitter A only.
This can be seen from the diagram which shows the position of the car within the transmission area for transmitter A but outside the transmission area for transmitter B.

EXAMINER'S COMMENTS

2 **a)** *The candidate has drawn a sketch to ascertain the bearing of A from B.*

b) *She has constructed the drawing using the given scale. The circles represent the areas to which each transmitter can broadcast.*

c) *The candidate has drawn and marked the position of the car by constructing a bearing of 150° from A and a bearing of 225° from B.*

She gives the correct answer and gives reasons for her answer by reference to the diagram.

Questions to Answer

1 A playground is in the shape of a rhombus with diagonals of length 500 m and 800 m. Using a suitable scale, construct a scale drawing of the playground and use it to find the perimeter.

2 Draw accurately the net of the following prism.

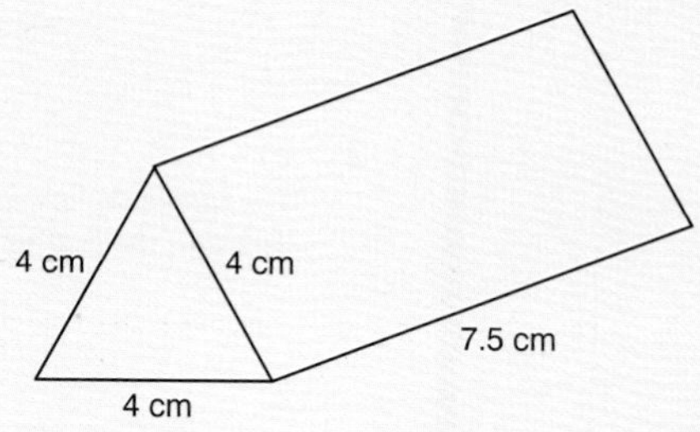

3 Three villages P, Q and R are shown on the following map.

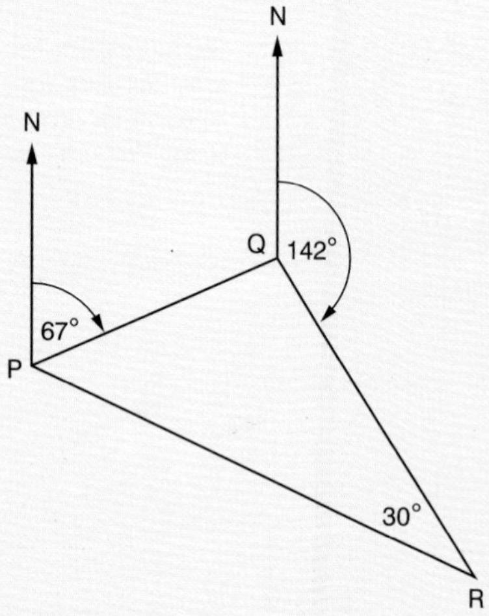

Use the information given to find:

(a) the bearing of Q from P
(b) the bearing of P from Q
(c) the bearing of R from P.

4 XYZ is an equilateral triangle of side 4.5 cm. Show all of the points which are less than 2.5 cm from the edges of the triangle.

5 On any triangle ABC, construct angle bisectors on the two smallest angles. Label the point of intersection of these perpendicular bisectors O. Draw the circle centre O which touches the sides AB, BC, AC and is **inscribed** in (i.e. lies completely inside) the triangle ABC.

CHAPTER 58

SYMMETRY

There are two types of symmetry, called **line symmetry** and **rotational symmetry**. The following work gives examples of both of them.

LINE SYMMETRY

When a shape can be folded so that one half fits exactly over the other half, the shape is **symmetrical** and the fold line is called a **line of symmetry**.

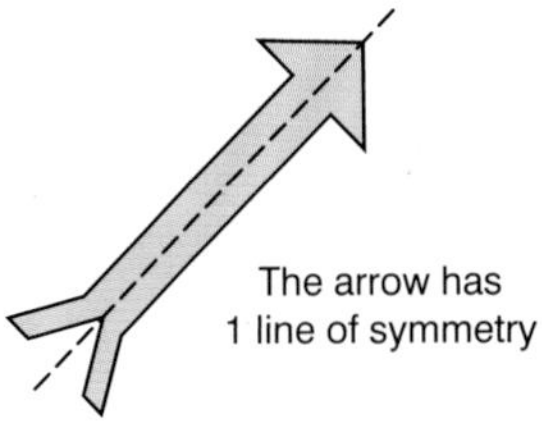
The arrow has
1 line of symmetry

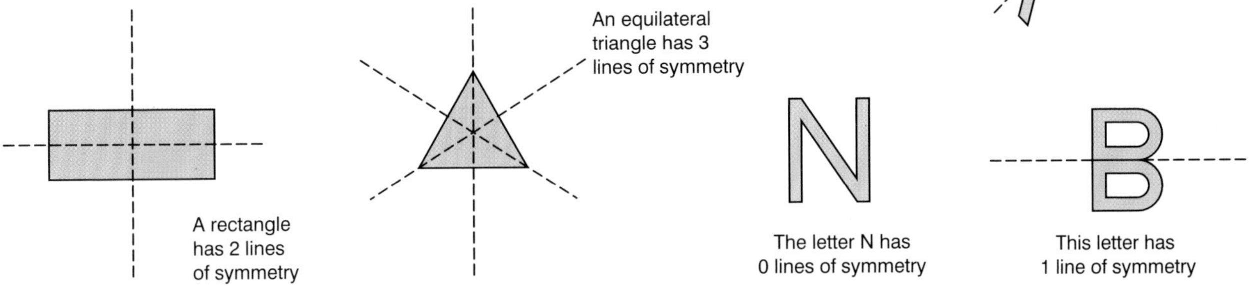
An equilateral triangle has 3 lines of symmetry
A rectangle has 2 lines of symmetry
The letter N has 0 lines of symmetry
This letter has 1 line of symmetry

Tracing paper is useful when you need to identify lines of symmetry or to draw reflections. You can use tracing paper in the mathematics examination.

ROTATIONAL SYMMETRY

When a shape can be rotated about its centre to a new position, so that it fits exactly over its original position, then the shape has **rotational symmetry**.

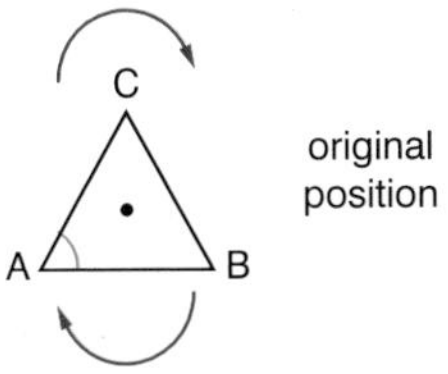
C
A
B
original position

The number of different positions tells you the **order** of rotational symmetry. An equilateral triangle has rotational symmetry of order 3.

Once again, tracing paper is useful when you need to identify rotational symmetry. You can use tracing paper in the mathematics examination.

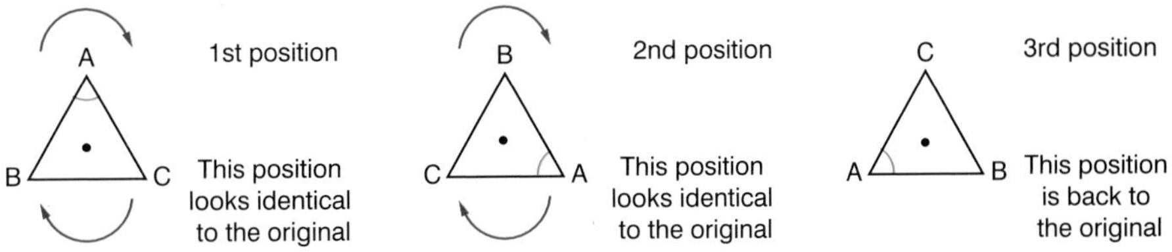
A
B
C
1st position
This position looks identical to the original
B
C
A
2nd position
This position looks identical to the original
C
A
B
3rd position
This position is back to the original

PLANES OF SYMMETRY

A plane of symmetry divides a solid into two equal halves.

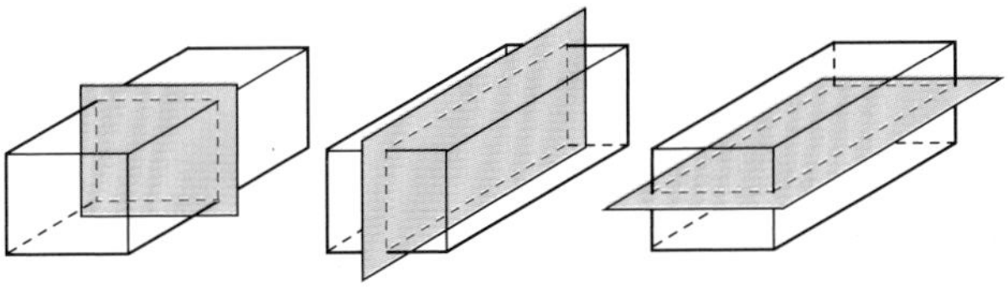

A cuboid has 3 planes of symmetry

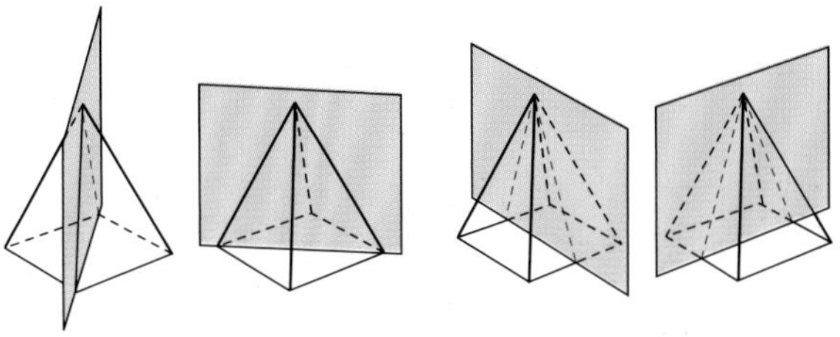

A square-based pyramid has 4 planes of symmetry

Check yourself

QUESTIONS

Q1 For each of the following shapes write down:
(i) the number of lines of symmetry
(ii) the order of rotational symmetry.

(a) square
(b) rectangle
(c) parallelogram
(d) scalene trapezium
(e) isosceles trapezium
(f) equilateral triangle
(g) isosceles triangle
(h) scalene right-angled triangle

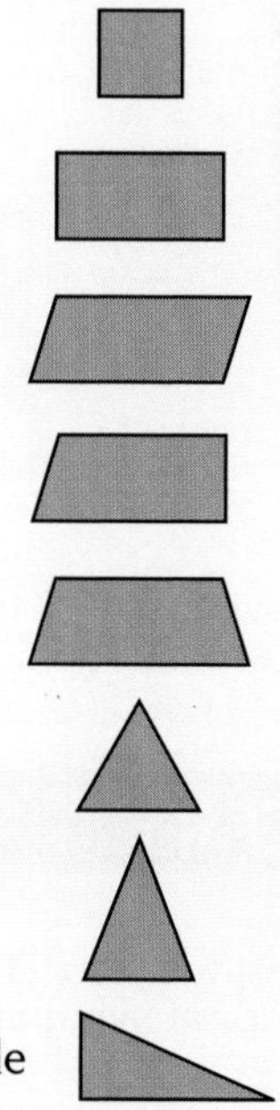

Q2 How many lines of symmetry does a 50p piece have?

Q3 Sketch a quadrilateral with one line of symmetry.

Q4 If a regular polygon has eight lines of symmetry, how many sides does it have?

Q5 How many planes of symmetry do the following solids have?
(a) triangular prism
(b) hexagonal prism

REMEMBER! Cover the answers if you want to.

ANSWERS

A1

		(i)	(ii)
(a)	square	4	4
(b)	rectangle	2	2
(c)	parallelogram	0	2
(d)	scalene trapezium	1	1
(e)	isosceles trapezium	0	1
(f)	equilateral triangle	3	3
(g)	isosceles triangle	1	1
(h)	scalene right-angled triangle	0	1

TUTORIAL

T1

		(i)	*(ii)*
(a)	*square*	4	4
(b)	*rectangle*	2	2
(c)	*parallelogram*	0	2
(d)	*scalene trapezium*	0	1
(e)	*isosceles trapezium*	1	1
(f)	*equilateral triangle*	3	3
(g)	*isosceles triangle*	1	1
(h)	*scalene right-angled triangle*	0	1

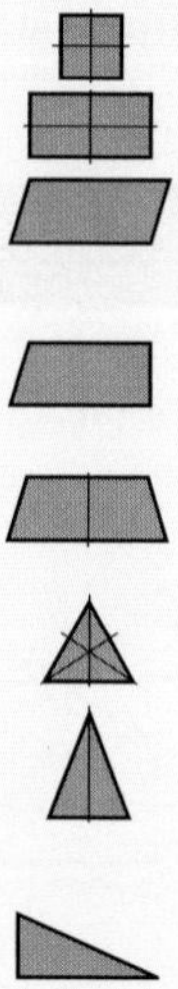

ANSWERS

TUTORIAL

A2 A 50p piece has seven lines of symmetry.

T2

A3

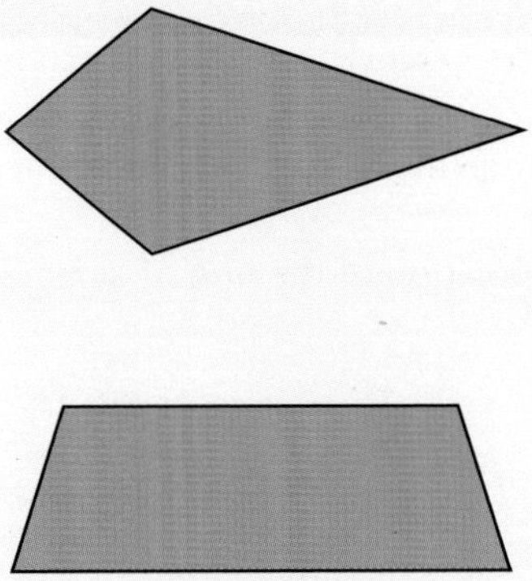

T3

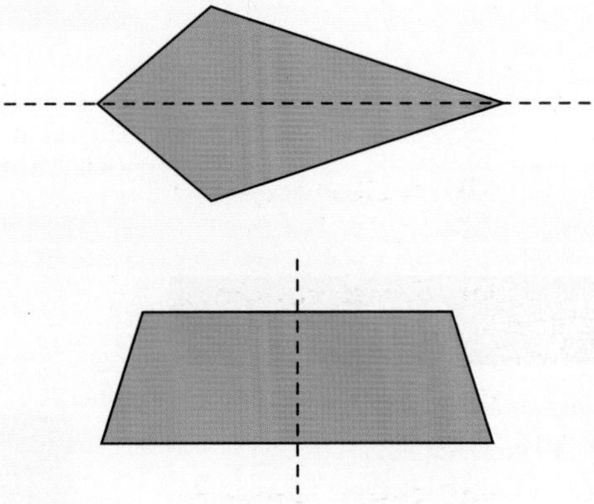

A4 8

T4

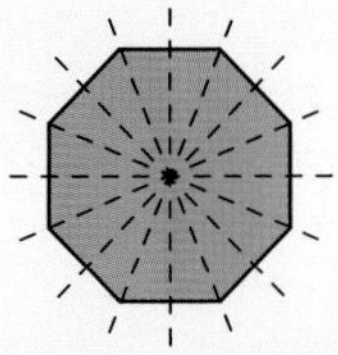

In general an n-sided regular polygon will have n lines of symmetry.

A5

(a) A triangular prism has four planes of symmetry (provided the triangle is regular).

(b) A hexagonal prism has seven planes of symmetry (provided the hexagon is regular).

T5

(a)

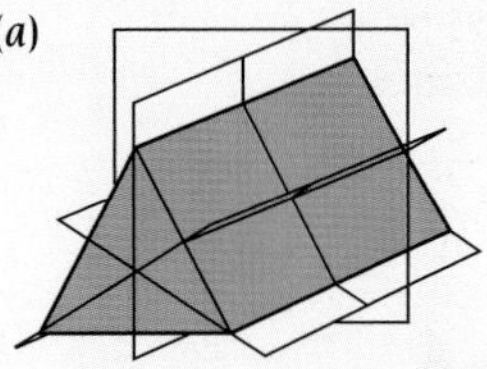

(b)

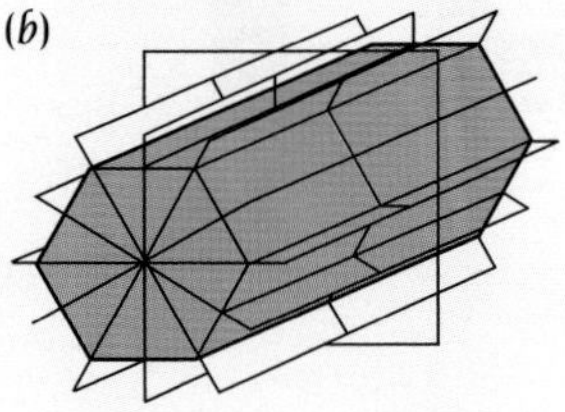

SHAPE, SPACE AND MEASURES

CHAPTER 59

TRANSFORMATIONS

If a point or a collection of points is moved from one position to another then it is said to have undergone a transformation.

The object is the point or collection of points before the transformation and the image is the point or collection of points after the transformation.

A variety of transformations are discussed here including:

- **reflection**
- **rotation**
- **enlargement**
- **translation.**

REFLECTION

A reflection is a transformation in which any two corresponding points on the object and image are the same distance away from a fixed line (called the **line of symmetry** or **mirror line**).

A reflection is defined by giving the position of the line of symmetry.

HINT

Take care with non-vertical or non-horizontal lines – tracing paper is useful in these situations.

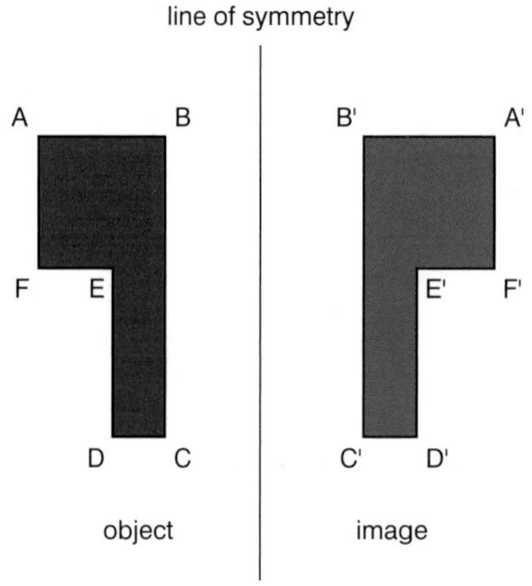

The image of ABCDE is labelled A'B'C'D'E' and corresponding points in the image and the object are equidistant from the line of symmetry.

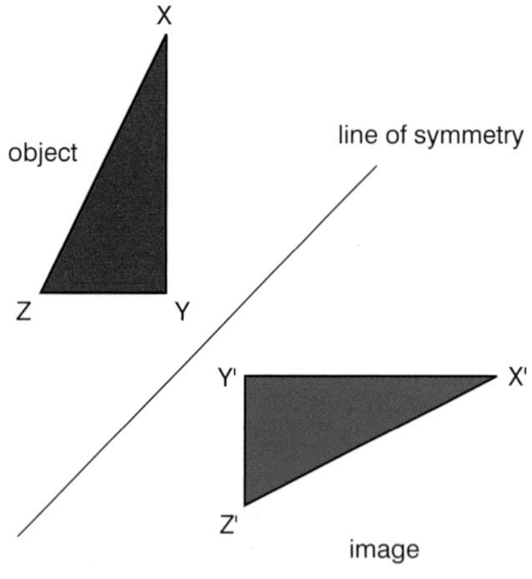

HINT

Subsequent transformations on ABC... are usually written as A'B'C'..., A"B"C"... or else $A_1B_1C_1$..., $A_2B_2C_2$... etc.

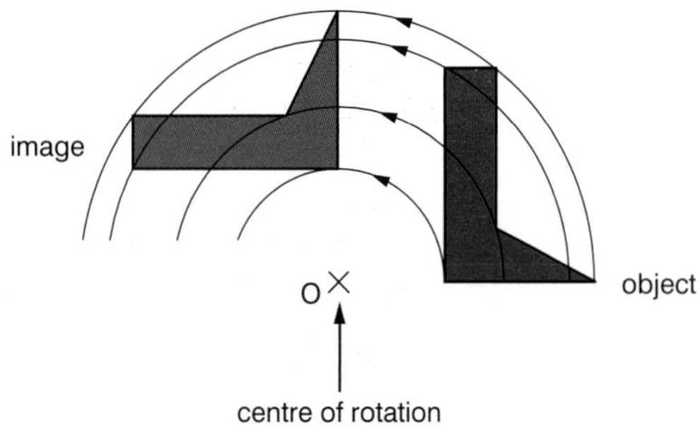

ROTATION

A rotation is a transformation in which lines from any two corresponding points on the object and image make the same angle at a fixed point (called the **centre of rotation**).

A rotation is defined by giving the position of the centre of rotation, along with the angle and direction of the rotation.

The diagram shows a rotation through 90° in an anticlockwise direction (called a rotation of $^{+}90°$) about the centre of rotation O.

To find the centre of rotation you should join corresponding points on the object and image with straight lines and draw the **perpendicular bisectors** of these lines. The centre of rotation lies on the intersection of these bisectors.

To find the angle of rotation you should join corresponding points on the object and image to the centre of rotation. The angle between these lines is the angle of rotation.

HINT

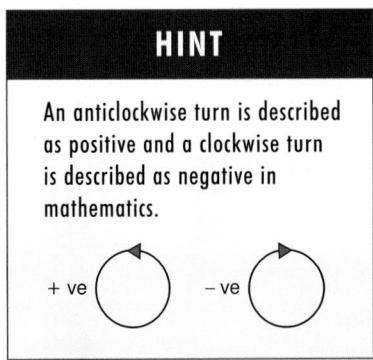

An anticlockwise turn is described as positive and a clockwise turn is described as negative in mathematics.

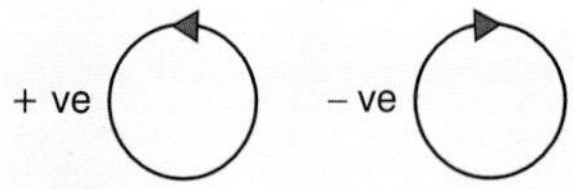

ENLARGEMENT

An enlargement is a transformation in which the distance between a point on the image and a fixed point (called the **centre of enlargement**) is a factor of the distance between the corresponding point on the object and the fixed point.

An enlargement is defined by giving the position of the centre of enlargement along with the factor (called the **scale factor**).

This diagram on the right shows an enlargement, scale factor 3, based on the centre of enlargement O.

OA = 1.1 cm OA' = 3 × 1.1 = 3.3 cm

OB = 1.7 cm OB' = 3 × 1.7 = 5.1 cm

OC = 1.35 cm OC' = 3 × 1.35 = 4.15 cm

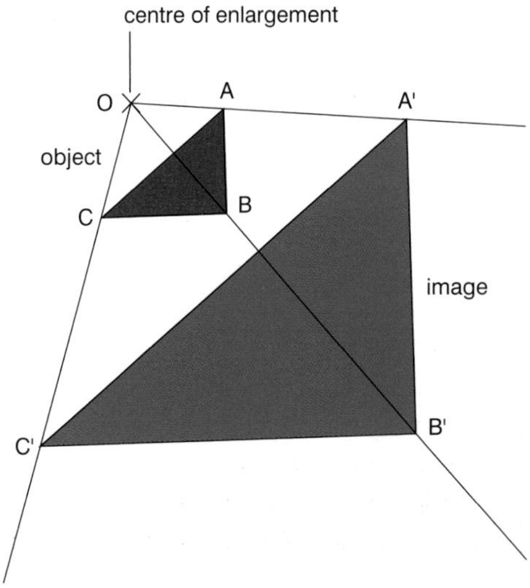

To find the centre of enlargement you join corresponding points – on the object and image – with straight lines. The centre of enlargement lies at the intersection of these straight lines.

The scale factor (SF) of an enlargement can be found as follows:

$$\text{SF} = \frac{\text{distance from the centre of a point on the image}}{\text{distance from the centre of the corresponding point on the object}}$$

$$\text{or SF} = \frac{\text{distance between two points on the image}}{\text{distance between two corresponding points on the object}}$$

A scale factor greater than 1 will enlarge the object.

A scale factor less than 1 will have the effect of reducing the object.

Worked example

The points A(3, 8), B(7, 8), C(7, $^{-}4$) and D(3, $^{-}2$) are joined to form a trapezium which is enlarged, scale factor $\frac{1}{2}$, with ($^{-}5$, $^{-}6$) as the centre of enlargement.
Draw ABCD on a graph and hence find A'B'C'D'.

The solution is shown on the right.

OA = 16.12 $OA^1 = \frac{1}{2} \times 16.12 = 8.06$

OB = 18.44 $OB^1 = \frac{1}{2} \times 18.44 = 9.22$

OC = 12.2 $OC^1 = \frac{1}{2} \times 12.2 = 6.1$

OD = 8.94 $OD^1 = \frac{1}{2} \times 8.94 = 4.57$

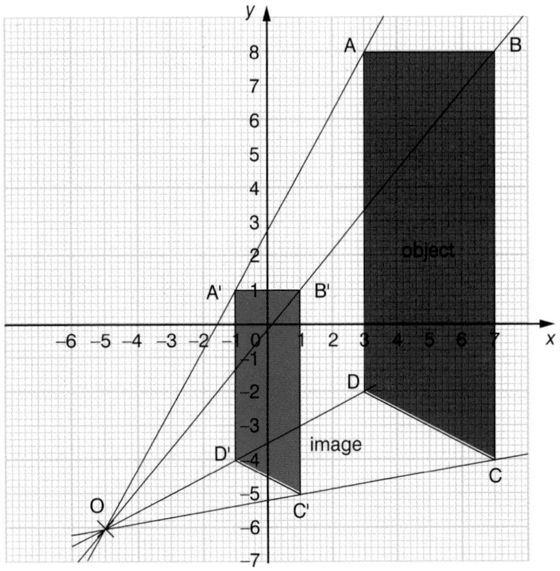

TRANSLATION

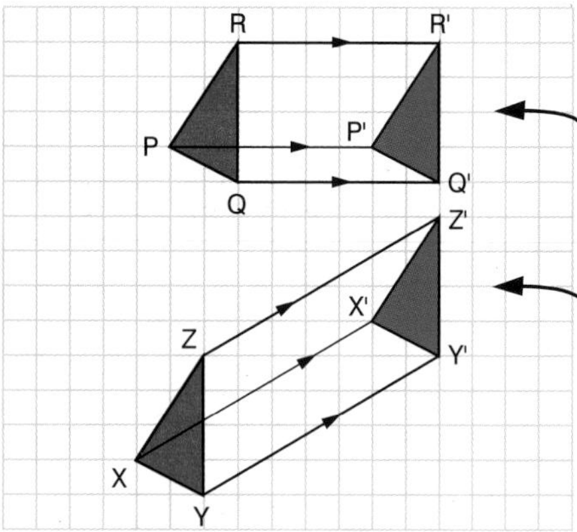

A translation is a transformation in which the distance and direction between any two corresponding points on the object and image are the same.

A translation is defined by giving the distance and direction of the translation.

This translation can be described as a movement of five units to the right.

This translation can be described as a movement of seven units to the right and four units upwards.

You can use **vector notation** to write a movement of seven units to the right and four units up as $\begin{pmatrix} 7 \\ 4 \end{pmatrix}$.

In general, you can write:

$$\begin{pmatrix} a \\ b \end{pmatrix} \Rightarrow \begin{pmatrix} \text{number of units to the right (in the positive } x\text{-direction)} \\ \text{number of units upwards (in the positive } y\text{-direction)} \end{pmatrix}$$

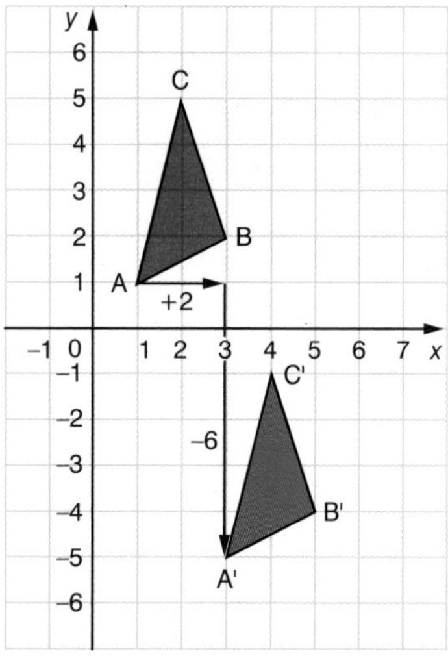

Worked example

The triangle ABC with coordinates (1, 1), (3, 2) and (2, 5) undergoes a translation of $\begin{pmatrix} 2 \\ ^-6 \end{pmatrix}$ to A'B'C'. Sketch the triangles ABC and A'B'C' and write down the translation which will return A'B'C' to ABC.

The vector $\begin{pmatrix} 2 \\ ^-6 \end{pmatrix}$ describes a movement of 2 units to the right and $^-6$ units upwards (i.e. 6 units downwards).

The translation which will return A'B'C' to ABC is $\begin{pmatrix} ^-2 \\ 6 \end{pmatrix}$.

COMBINATIONS OF TRANSFORMATIONS

Combinations of the same type of transformation or combinations of different types of transformation can usually be described as a single transformation.

Worked example

If R is a reflection in the y-axis and T is a rotation about the origin of $^+90°$, show (on separate diagrams) the image of the triangle XYZ with vertices X(2, 1), Y(2, 5) and Z(4, 2), under the combined transformations:

(a) R followed by T (b) T followed by R.

Which single transformation will return each of these combined transformations back to their original position?

(a) The single transformation that will return A"B"C" to ABC is a reflection in the line $y = {}^-x$.

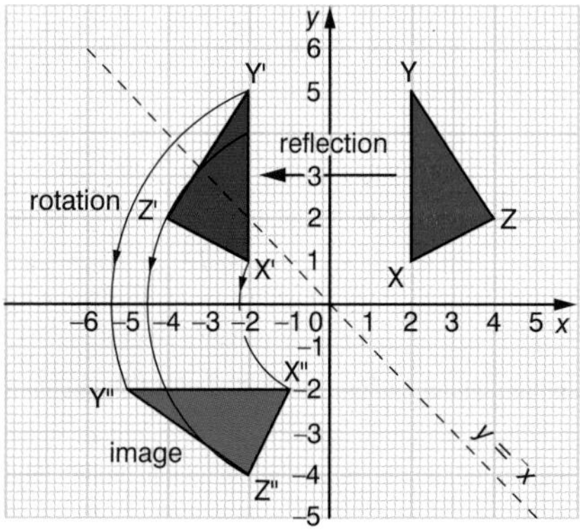

(b) The single transformation that will return A"B"C" to ABC is a reflection in the line $y = x$.

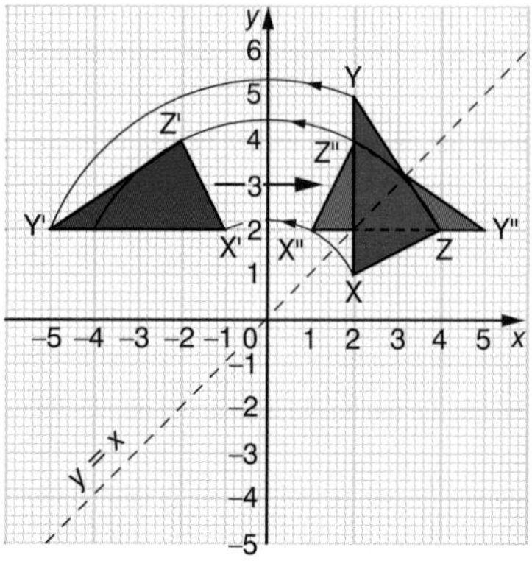

Check yourself

QUESTIONS

Q1 Find the images of the point (3, 4) reflected in the following lines.

(a) $x = 0$ (b) $y = 0$ (c) $x = 2$
(d) $y = 2$ (e) $y = x$

Q2 Draw the images of the following objects after:

(i) a rotation of $^{+}90°$ about the given centre of rotation O
(ii) a rotation of $^{-}90°$ about the given centre of rotation O.

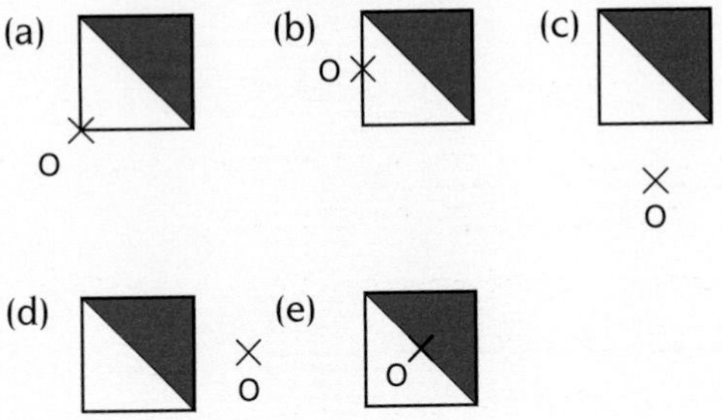

Q3 Find the centre of rotation and the angle of rotation for the following transformation.

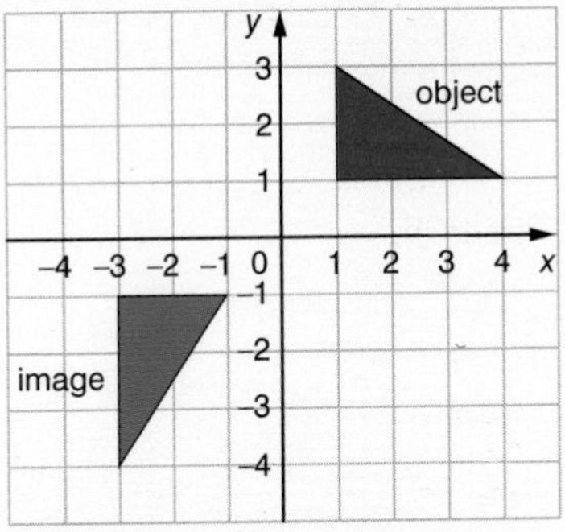

Q4 Draw the images of the following objects after an enlargement, scale factor $\frac{3}{2}$, with centre of enlargement O.

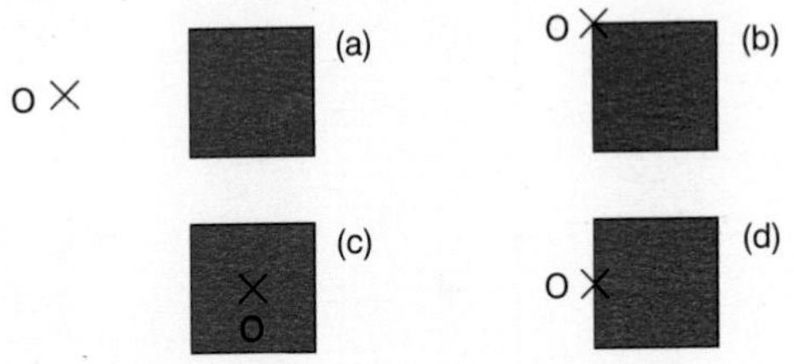

Q5 Draw the images of ΔABC (triangle ABC) after the following translations.

(a) $\begin{pmatrix}3\\2\end{pmatrix}$ (b) $\begin{pmatrix}^{-}4\\0\end{pmatrix}$ (c) $\begin{pmatrix}^{-}2\\^{-}5\end{pmatrix}$

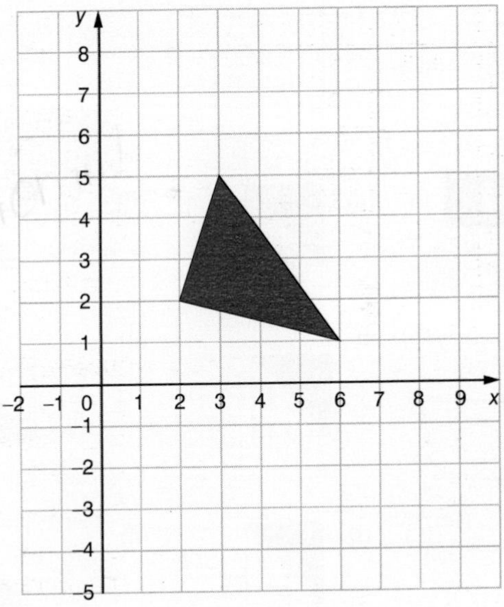

Write down the translation which will return the image to ABC in each case.

Q6 The triangle ABC with coordinates A(3, 1), B(4, 1) and C(4, 4) is reflected in the line $y = x$ then the image undergoes a reflection in the line $x = 0$. The resulting triangle is then rotated through $^{-}90°$ about ($^{-}1$, 4) to A'B'C'. What single transformation would map ABC onto A'B'C'?

ANSWERS

A1 (a) $(^{-}3, 4)$ (b) $(3, ^{-}4)$ (c) $(1, 4)$
(d) $(3, 0)$ (e) $(4, 3)$

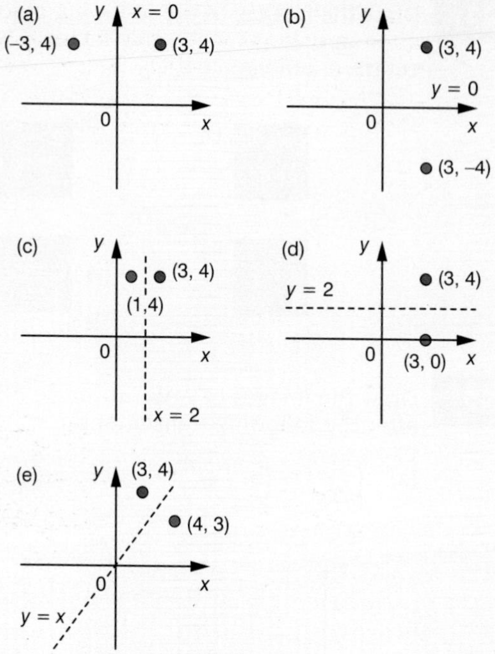

A2

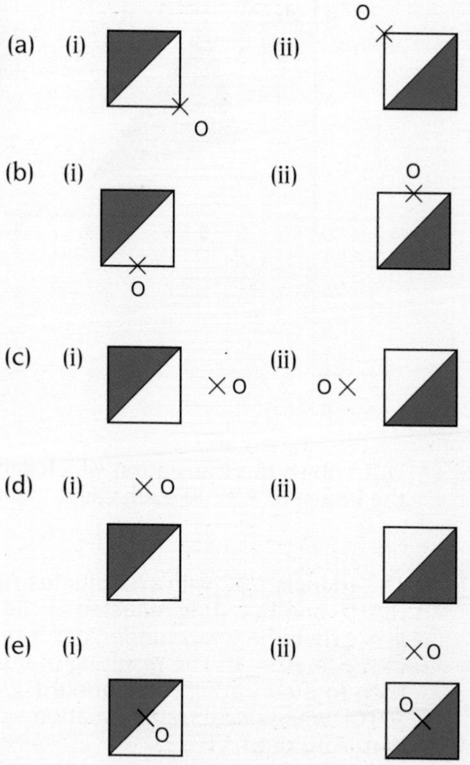

TUTORIAL

T1 *A reflection is such that the object and image are the same distance away from the given line.*

T2 *A solution is that any two corresponding points on the object and image make the same angle ($^{+}90°$ in part (i) and $^{-}90°$ in part (ii)) at the centre of rotation.*

ANSWERS

A3

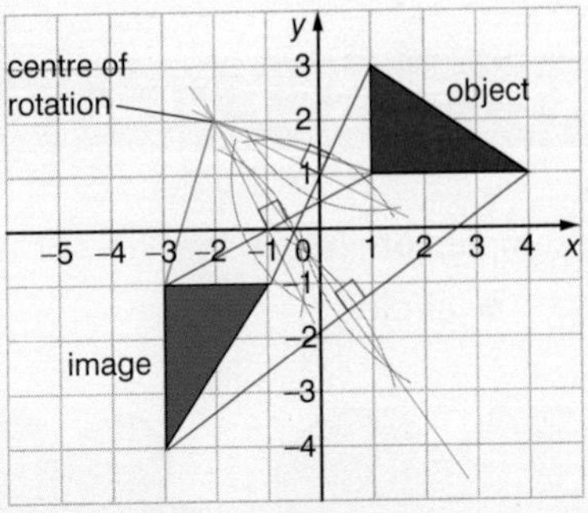

Centre of rotation is (⁻2, 2).
Angle of rotation is 90°.

A4

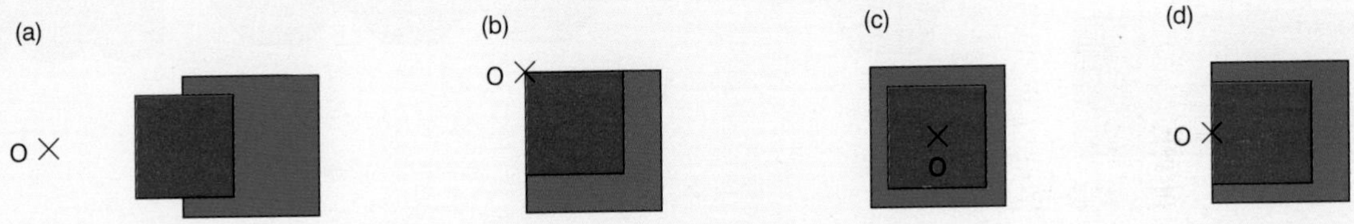

A5

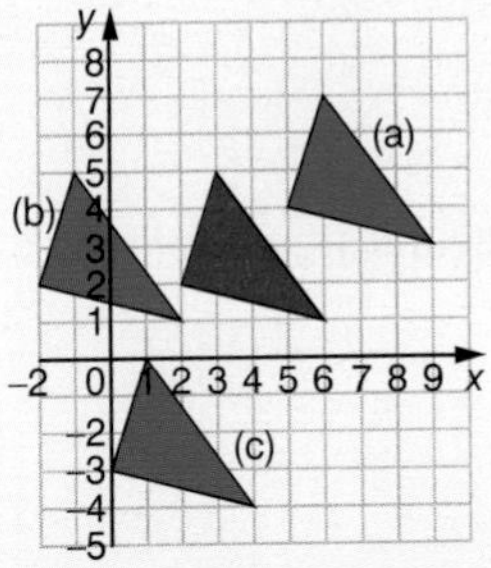

The translations which will return the image to ABC are given by:

(a) $\begin{pmatrix} ^-3 \\ ^-2 \end{pmatrix}$ (b) $\begin{pmatrix} 4 \\ 0 \end{pmatrix}$ (c) $\begin{pmatrix} 2 \\ 5 \end{pmatrix}$

A6 The single transformation is a translation of $\begin{pmatrix} ^-5 \\ 3 \end{pmatrix}$.

TUTORIAL

T3 *To find the centre of rotation you should join corresponding points on the object and image with straight lines and draw the perpendicular bisectors of these lines. The centre of rotation lies on the intersection of these perpendicular bisectors. To find the angle of rotation you should join corresponding points on the object and image to the centre of rotation. The angle between these lines is the angle of rotation.*

T5 *After a translation of* $\begin{pmatrix} a \\ b \end{pmatrix}$ *the translation that will return the image to the object is* $\begin{pmatrix} ^-a \\ ^-b \end{pmatrix}$.

T6

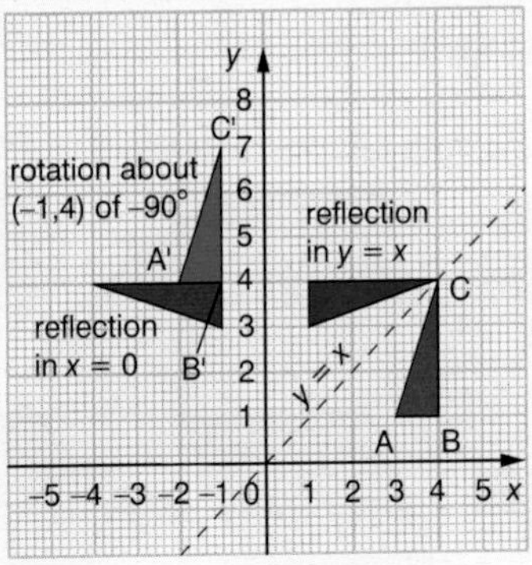

CHAPTER 60

TESSELLATIONS

If congruent shapes fit together exactly to cover an area completely, then the shapes are said to tessellate.

SHAPES THAT TESSELLATE

These are examples of shapes that tessellate.

- isosceles triangle
- rectangle
- hexagon

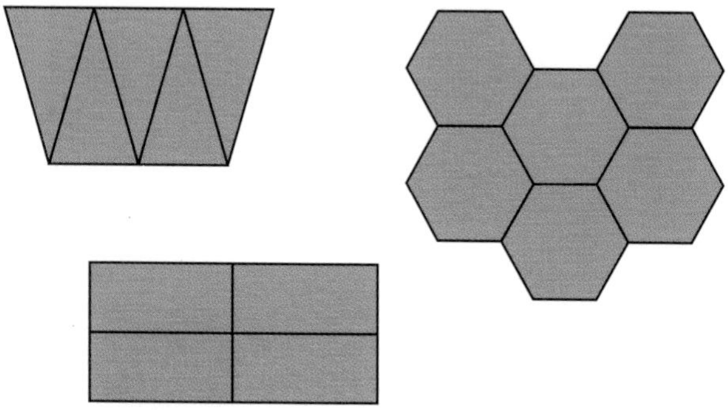

These are examples of shapes that do not tessellate.

- octagons
- circles

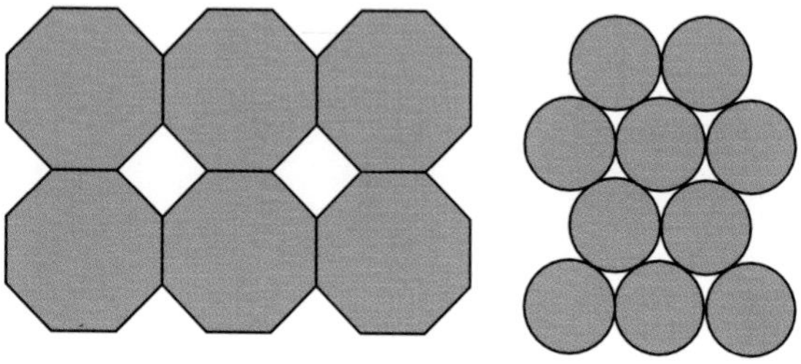

Check yourself

QUESTIONS

Q1 Which of the following shapes tessellate?

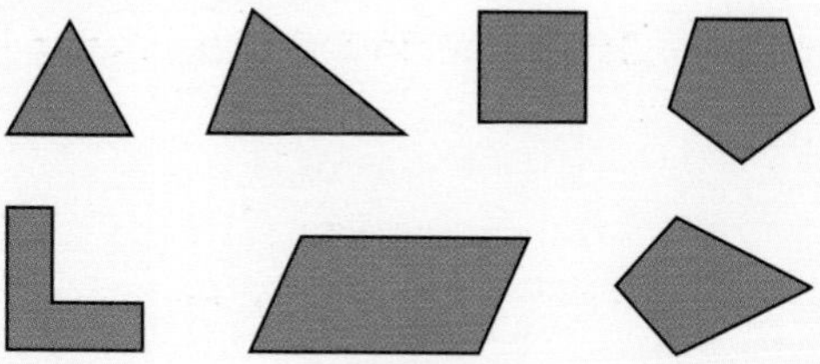

Q2 Explain why regular pentagons do not tessellate.

REMEMBER! Cover the answers if you want to.

ANSWERS

A1 The following shapes tessellate.

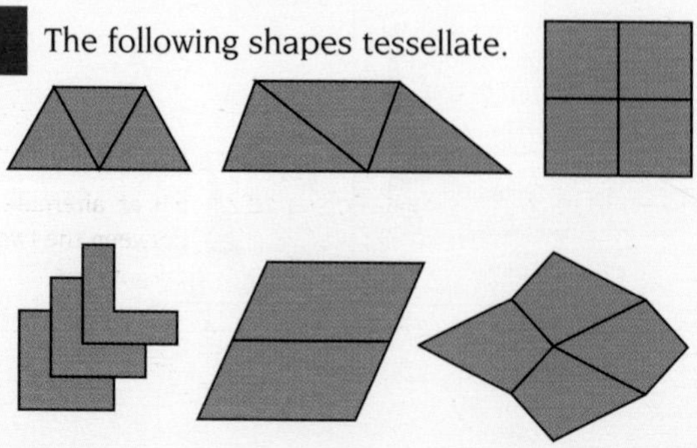

A2 Regular pentagons do not tessellate as the angles at their vertices cannot be placed together to form a total of 360°.

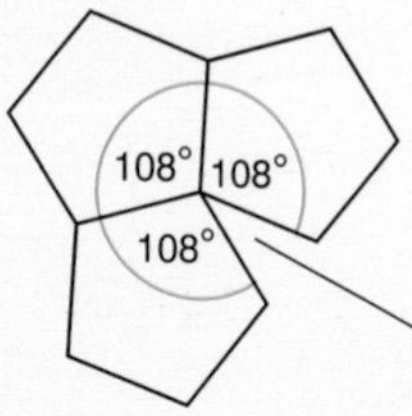
108°
108°
108°
Gap here means that regular pentagons cannot tessellate

CHAPTER 61

ANGLE PROPERTIES

You need to know these angle properties, and use them when explaining work with lines and angles.

PROPERTIES OF ANGLES

Angles on a straight line add up to 180°.

$90° + 90° = 180°$

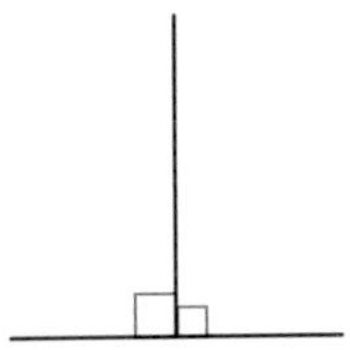

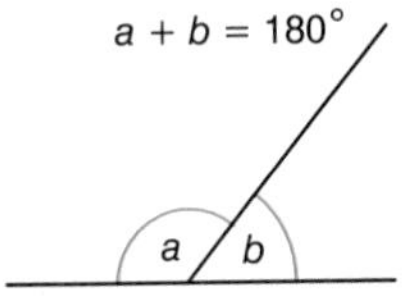

Angles at a point add up to 360°.

$90° + 90° + 90° + 90° = 360°$

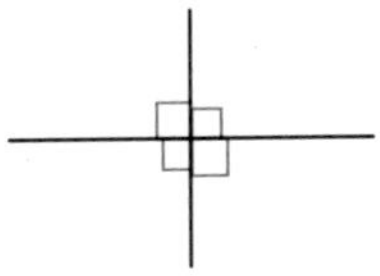

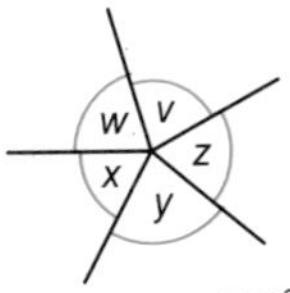

$v + w + x + y + z = 360°$

When two straight lines intersect the (vertically) opposite angles are equal.

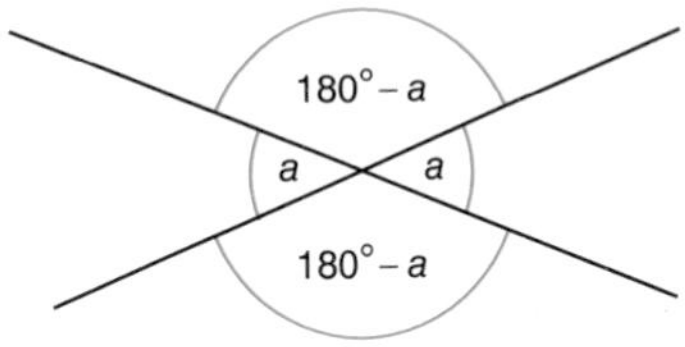

ANGLES BETWEEN PARALLEL LINES

A **transversal** is a line which cuts two or more parallel lines.

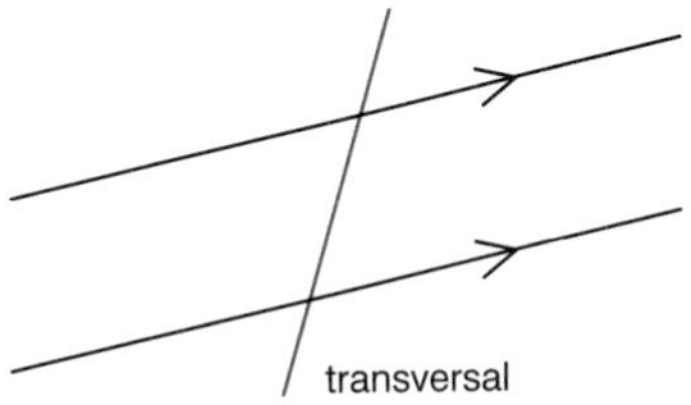

Corresponding angles are equal.

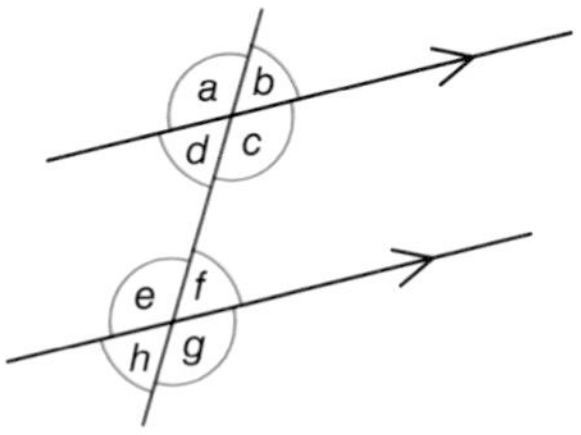

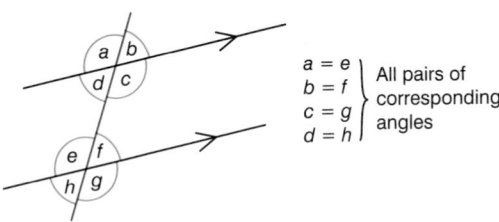

$a = e$, $b = f$, $c = g$, $d = h$ } All pairs of corresponding angles

Alternate angles (or Z angles) are equal.

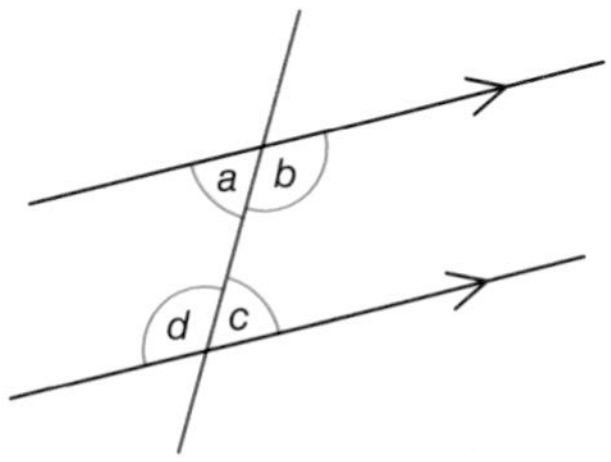

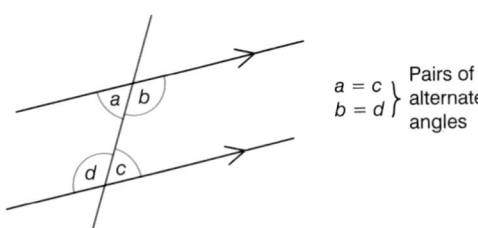

$a = c$, $b = d$ } Pairs of alternate angles

Interior angles add up to 180° (i.e. interior angles are **supplementary**).

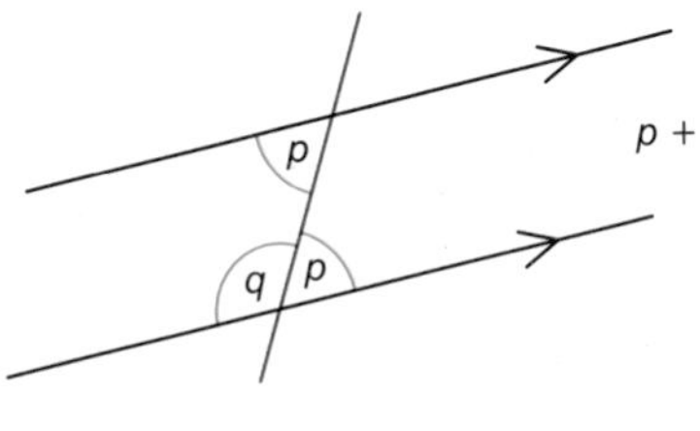

$p + q = 180°$

Worked example

Find the size of angles *a*, *b* and *c* in this diagram. Give reasons for your answers.

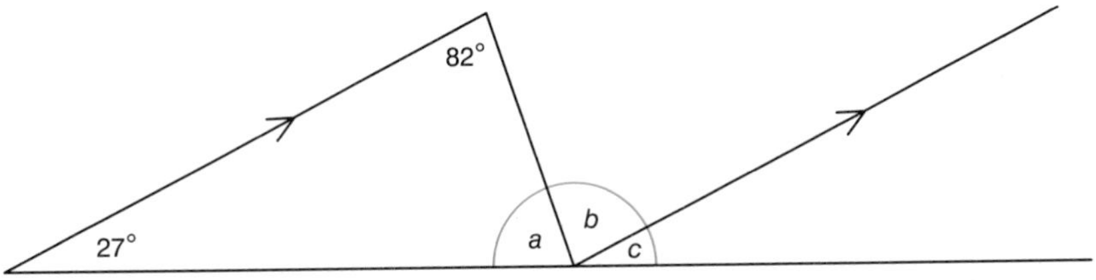

Angle $a = 71°$	The angles of the triangle add up to 180°.
Angle $b = 82°$	*b* is an alternate angle between the two given parallel lines.
Angle $c = 27°$	The angles on a straight line add up to 180°.

Check yourself

QUESTIONS

Give reasons in all of your answers.

Q1 Write down the value of the missing angles.

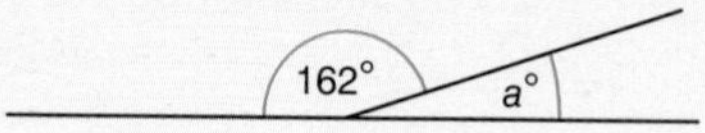

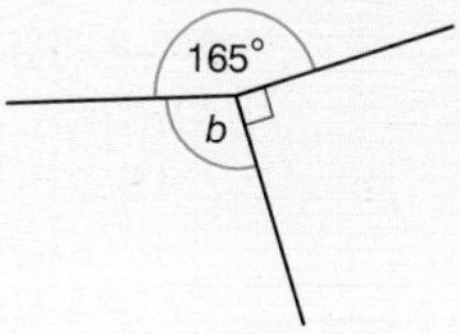

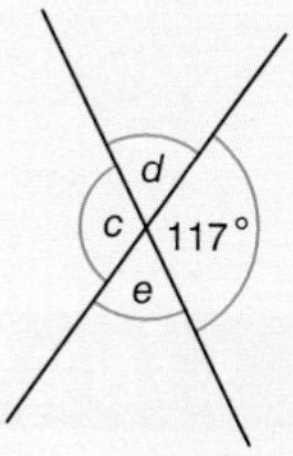

Q2 Write down the value of each of the missing angles.

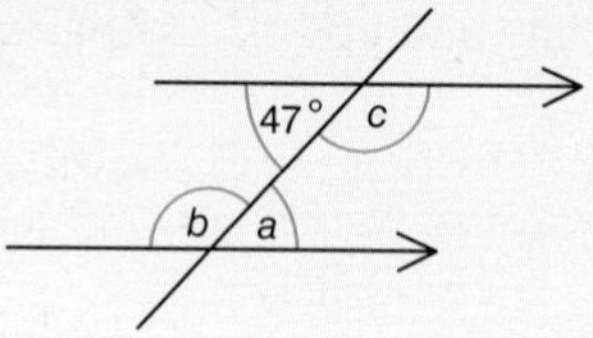

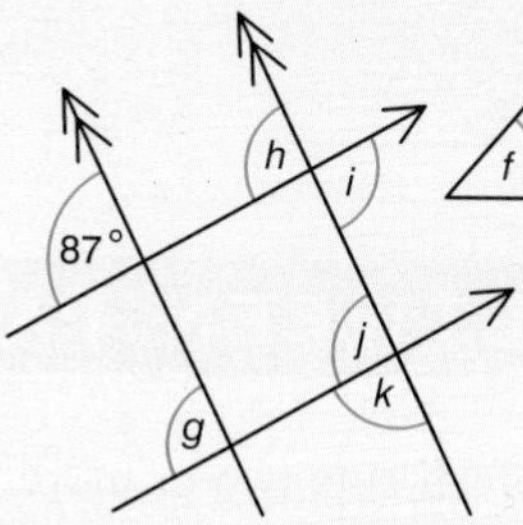

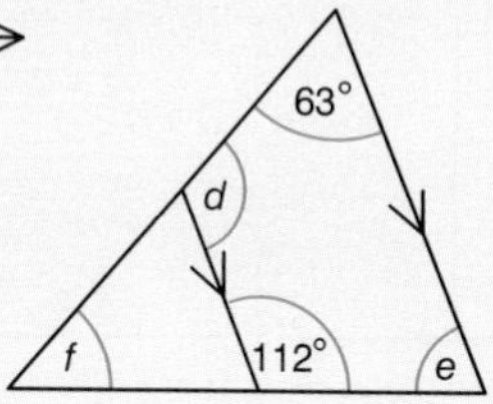

REMEMBER! Cover the answers if you want to.

ANSWERS

A1

$a = 18°$ $b = 105°$ $c = 117°$
$d = 63°$ $e = 63°$

A2

$a = 47°$
$b = 133°$
$c = 133°$
$d = 117°$
$e = 68°$
$f = 49°$
$g = 87°$
$h = 87°$
$i = 87°$
$j = 87°$
$k = 93°$

TUTORIAL

T1

$a = 18°$	*Angles on a straight line add up to 180°.*
$b = 105°$	*Angles at a point add up to 360°.*
$c = 117°$	*Vertically opposite angle.*
$d = 63°$	*Angles on a straight line add up to 180°.*
$e = 63°$	*Vertically opposite angle or angles on a straight line add up to 180°.*

T2

$a = 47°$	*Alternate angle between parallel lines with given angle of 47°.*
$b = 133°$	*Interior angles between parallel lines add up to 180°. (180° – 47° = 133°)*
$c = 133°$	*Angles on a straight line add up to 180°, or alternate angle with angle b, or interior angles between parallel lines with angle a.*
$d = 117°$	*Interior angles between parallel lines add up to 180°. (180° – 63° = 117°)*
$e = 68°$	*Interior angles between parallel lines add up to 180°. (180° – 112° = 68°)*
$f = 49°$	*Angles in a triangle add up to 180°. (180° – 63° – 68° = 49°)*
$g = 87°$	*Corresponding angles between parallel lines.*
$h = 87°$	*Corresponding angles between parallel lines (the other pair).*
$i = 87°$	*Vertically opposite angle to angle h.*
$j = 87°$	*Alternate angle between parallel lines with angle i.*
$k = 93°$	*Angles on a straight line add up to 180°.*

CHAPTER 62

ANGLES OF POLYGONS

You have already seen that the interior angles of a triangle add up to 180° and the interior angles of a quadrilateral add up to 360°.

The exterior angle plus the adjacent interior angle add up to 180° (angles on a straight line) and the sum of all the exterior angles of a polygon is 360°.

ANGLE SUM OF A POLYGON

The sum of the angles of a polygon can be found by dividing the polygon into a series of triangles where each triangle has an angle sum of 180°.

> **HINT**
>
> The exterior angle of a polygon is found by continuing the line of the polygon externally.

A four-sided polygon can be split into two triangles.

Angle sum = 2 × 180° = 360°

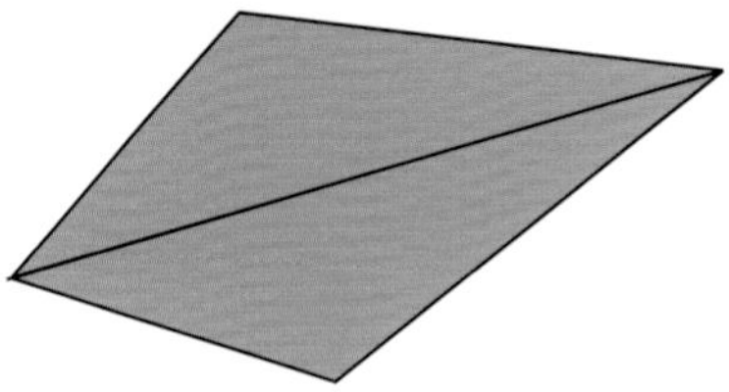

A six-sided polygon can be split into four triangles.

Angle sum = 4 × 180° = 720°

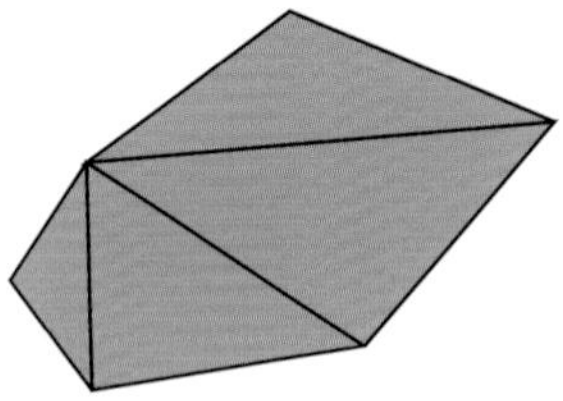

A seven-sided polygon can be split into five triangles.

Angle sum = 5 × 180° = 900°

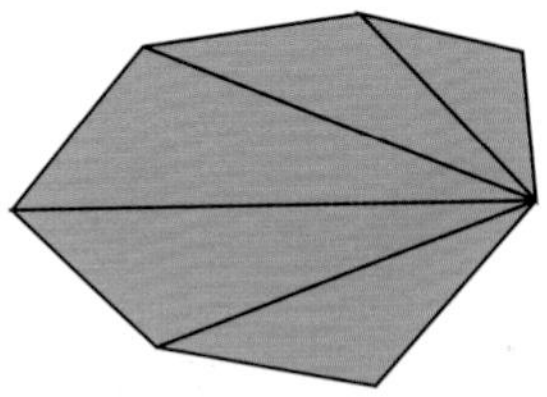

From the above, you can see that an n-sided polygon can be split into $(n - 2)$ triangles.

Angle sum of an n-sided polygon = $(n - 2) \times 180°$

> **HINT**
>
> The angle sum of an n-sided polygon is sometimes presented as $(2n - 4)$ right angles.
> This formula gives exactly the same results as $(n - 2) \times 180°$.

EXTERIOR ANGLES

Another useful result involves the exterior angles. Since the sum of the exterior angles is 360°, each exterior angle of a regular polygon is $360° \div n$.

This fact can be used to find the interior angle of a regular polygon.

Interior angle = 180° – exterior angle

The angle sum of an n-sided regular can be found in the same way as the angle sum of a non-regular polygon.

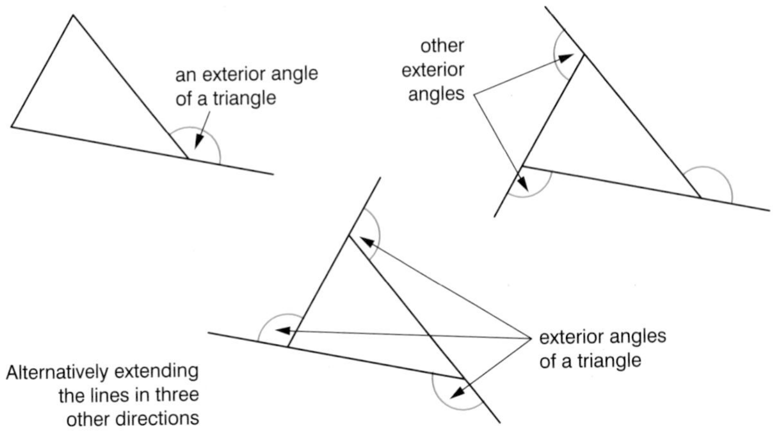

Check yourself

QUESTIONS

Q1 Find the angle sum of a five-sided polygon.

Q2 Find the angle sum of a nonagon.

Q3 Calculate the interior angle of a regular decagon.

Q4 Calculate the exterior angle of a regular decagon.

REMEMBER! Cover the answers if you want to.

ANSWERS

A1 Angle sum = 540°

A2 Angle sum = 1260°

A3 Each interior angle = 144°

A4

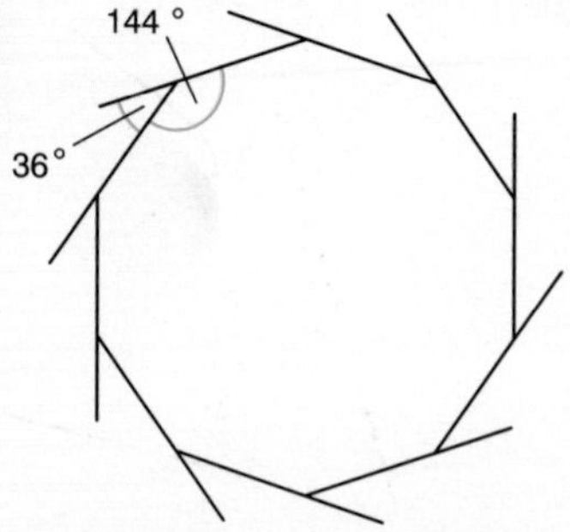

Each exterior angle = 36°

TUTORIAL

T1 *A five-sided polygon can be split into three triangles.*

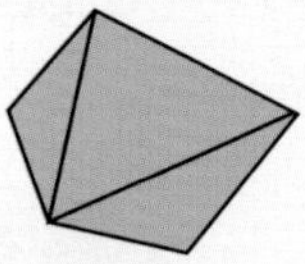

Angle sum = 3 × 180° = 540°.

T2 *A nonagon is a nine-sided shape.*
A nine-sided polygon can be split into seven triangles.

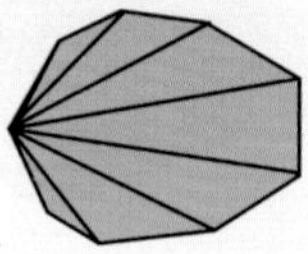

Angle sum = 7 × 180°
= 1260°

Alternatively using the formula:
Angle sum of an n-sided polygon = (n – 2) × 180°
Angle sum of a nine-sided polygon = (9 – 2) × 180°
= 7 × 180°
= 1260°

T3 *A decagon is a ten-sided shape.*
Using the formula:
Angle sum of an n-sided polygon = (n – 2) × 180°
Angle sum of a ten-sided polygon = (10 – 2) × 180°
= 8 × 180°
= 1440°

As the polygon is a regular polygon then each interior angle is equal so 1440° ÷ 10 = 144°.

T4 *Drawing a diagram helps to see the situation.*

The previous question concludes that the interior angle of a decagon is 144° so the exterior angle = 36° (as the sum of the interior and exterior angle is 180°).

Alternatively, you could use the fact that the sum of the exterior angles of a polygon is 360°.
As the polygon is a regular polygon then all the exterior angles are equal so 360° ÷ 10 = 36°.

CHAPTER 63

CONGRUENCE AND SIMILARITY

You need to understand the terms 'congruent' and 'similar' for the examination.

CONGRUENT TRIANGLES

Two triangles are **congruent** if one of the triangles can be fitted exactly over the other triangle so that all corresponding angles and corresponding sides are equal.

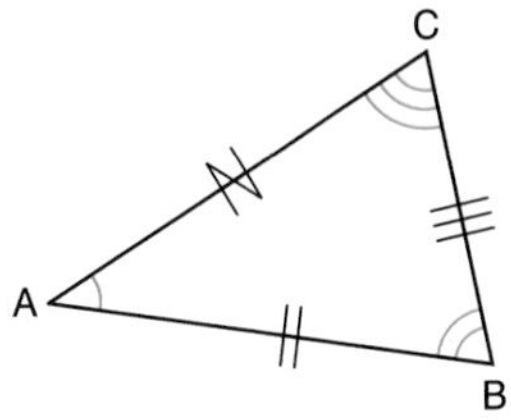

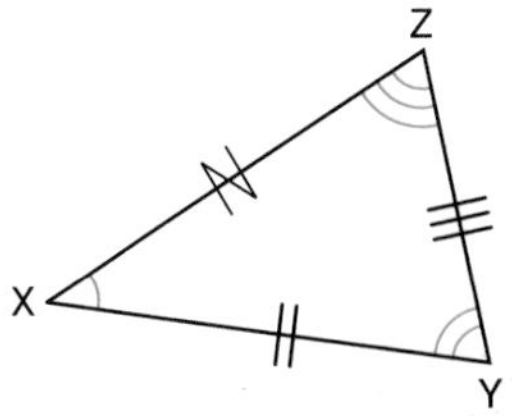

$AB = XY \quad \angle A = \angle X$

$BC = YZ \quad \angle B = \angle Y$

$CA = ZX \quad \angle C = \angle Z$

If the two triangles ABC and XYZ are congruent then you can write $\Delta ABC \equiv \Delta XYZ$.

To prove that two triangles are congruent, it is not necessary to prove that all of the above conditions are true. The following minimum conditions are sufficient to show that two triangles are congruent.

Two triangles are congruent if:

- two angles and a side of one triangle are equal to two angles and the corresponding side of the other (AAS)
- two sides and the included angle of one triangle are equal to two sides and the included angle of the other (SAS)
- the three sides of one triangle are equal to the three sides of the other (SSS).

If the triangles are right-angled then the two triangles are congruent if:

- the hypotenuses and one other side are equal on both triangles (RHS).

NOTE

The symbol $\equiv$ is a short way of writing 'is congruent to'.

SIMILAR TRIANGLES

Two triangles are **similar** if one of the triangles is an enlargement of the other triangle so that all corresponding angles are equal and corresponding sides are in the same ratio.

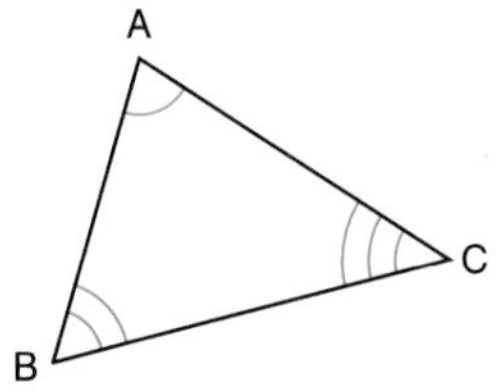

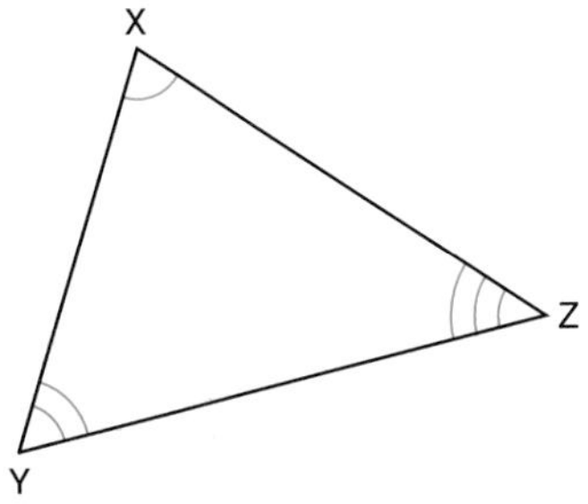

$\angle A = \angle X$

$\angle B = \angle Y$

$\angle C = \angle Z$

If two triangles are similar then the ratios of the corresponding sides are equal.

$$\frac{AB}{XY} = \frac{BC}{YZ} = \frac{CA}{ZX} \quad \text{or} \quad \frac{XY}{AB} = \frac{YZ}{BC} = \frac{ZX}{CA}$$

To prove that two triangles are similar, it is not necessary to prove that all of the above conditions are true. The following minimum conditions are sufficient to show that two triangles are similar.

Two triangles are similar if:

- two angles of one triangle are equal to two angles of the other
- two pairs of sides are in the same ratio and the included angles are equal
- three pairs of sides are in the same ratio.

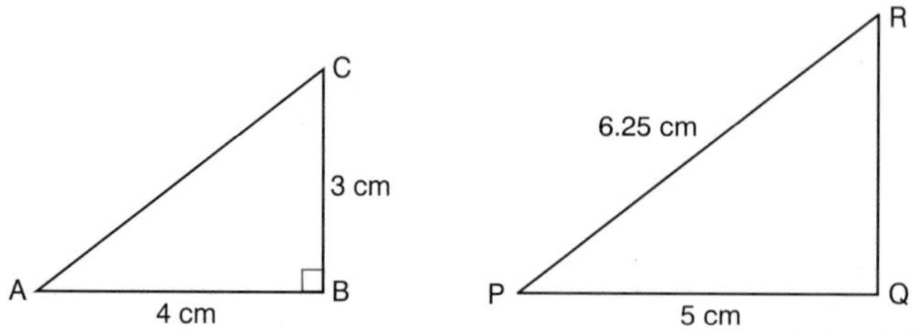

Worked example

Triangles ABC and PQR are similar.

Calculate:

(a) $\angle PQR$ (b) QR (c) AC.

What can you say about the ratio of the lengths of corresponding sides of the two triangles?

What can you say about the ratio of the areas of the two triangles?

(a) $\angle PQR = \angle ABC = 90°$ As the triangles are similar then the corresponding angles are equal.

(b) In order to find QR, you need to know the ratio of corresponding sides.

$$AB : PQ = 4 : 5 \text{ or } \frac{AB}{PQ} = \frac{4}{5}$$

so $\frac{BC}{QR} = \frac{AB}{PQ} = \frac{4}{5}$

or $\frac{BC}{QR} = \frac{4}{5}$

$\frac{3}{QR} = \frac{4}{5}$ As BC = 3 cm.

$\frac{QR}{3} = \frac{5}{4}$ Turning both sides upside-down.

$QR = \frac{5}{4} \times 3$ Multiplying both sides by 3.

$QR = 3.75\text{ cm}$

NOTE

You could have used Pythagoras' theorem to find this last length, but it is probably better to use similar triangles, as this is what the question is about.

(c) In order to find AC, you need to use the ratio of corresponding sides.

$\frac{AB}{PQ} = \frac{4}{5}$

so $\frac{AC}{PR} = \frac{4}{5}$ As $\frac{AC}{PR} = \frac{AB}{PQ}$

$\frac{AC}{6.25} = \frac{4}{5}$ As PR = 6.25 cm.

$AC = \frac{4}{5} \times 6.25$ Multiplying both sides by 6.25.

$AC = 5\text{ cm}$

Using the fact that the area of a triangle

$= \frac{1}{2} \times \text{base} \times \text{perpendicular height}$

area of triangle ABC $= \frac{1}{2} \times 4 \times 3 = 6\text{ cm}^2$

area of triangle PQR $= \frac{1}{2} \times 5 \times 3.75 = 9.375\text{ cm}^2$

ratio of lengths = 4 : 5

ratio of areas = 6 : 9.375

= 6000 : 9375 Multiplying both sides of the ratio by 1000 to get whole numbers.

= 16 : 25 Cancelling both sides.

You should notice that the ratio of the areas is equal to the ratio of the lengths squared.

Check yourself

QUESTIONS

Q1 Write down a pair of triangles which are congruent. Explain why they are congruent.

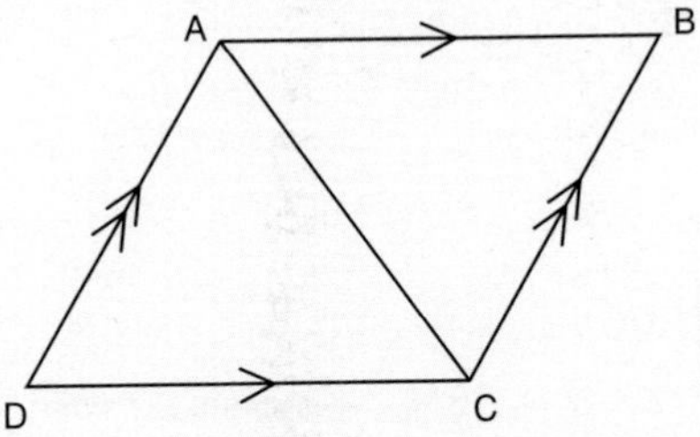

Q2

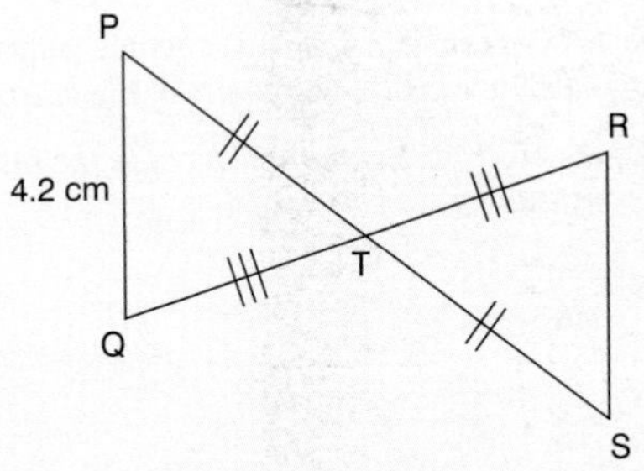

Show that $\Delta PQT \equiv \Delta SRT$ and hence find the length RS.

Q3 Find the length AC in the following diagram.

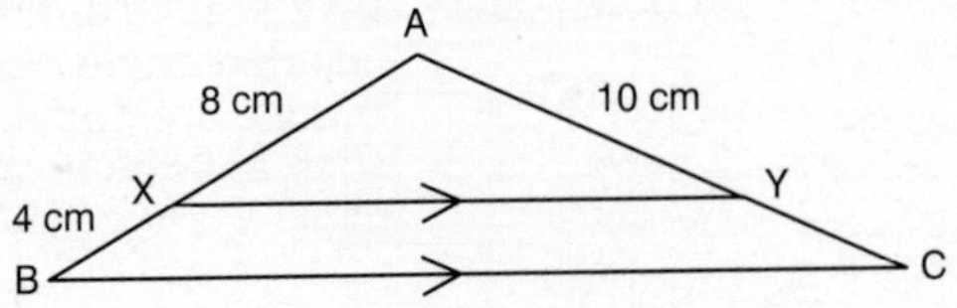

Q4 Find the length MN in the following diagram.

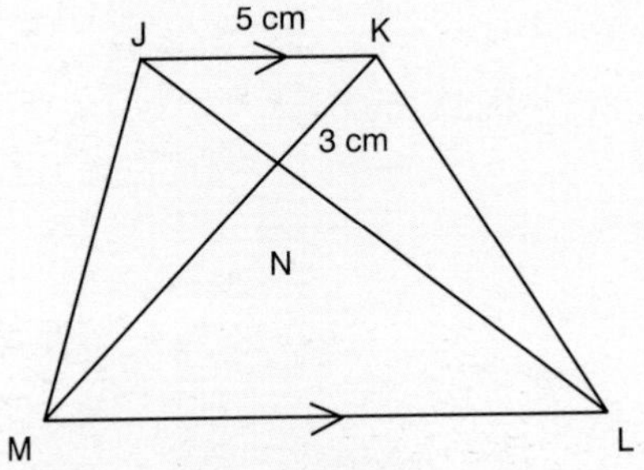

REMEMBER! Cover the answers if you want to.

ANSWERS

A1

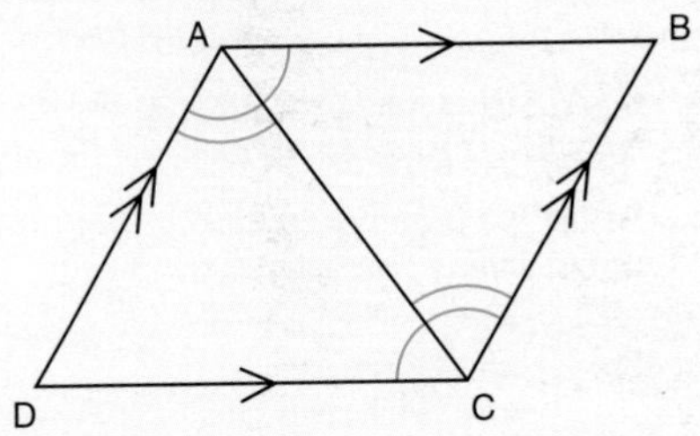

$\Delta ABC \equiv \Delta ADC$

Two angles and a side of one triangle are equal to two angles and the corresponding side of the other (AAS).

TUTORIAL

T1

$\angle BAC = \angle DCA$	*Alternate angles between parallel lines AB and DC.*
$\angle DAC = \angle ACB$	*Alternate angles between parallel lines AD and BC .*
$AC = AC$	*Common line.*
so $\Delta ABC \equiv \Delta ADC$	*As two angles and a side of one triangle are equal to two angles and the corresponding side of the other (AAS).*

ANSWERS

A2

∠PTQ = ∠RTS	Vertically opposite angles.
TQ = TR	Given.
TP = TS	Given.
so ΔPQT ≡ ΔSRT	As two sides and the included angle of one triangle are equal to two sides and the included angle of the other (SAS).

RS = 4.2 cm

A3

AC = 15 cm

A4

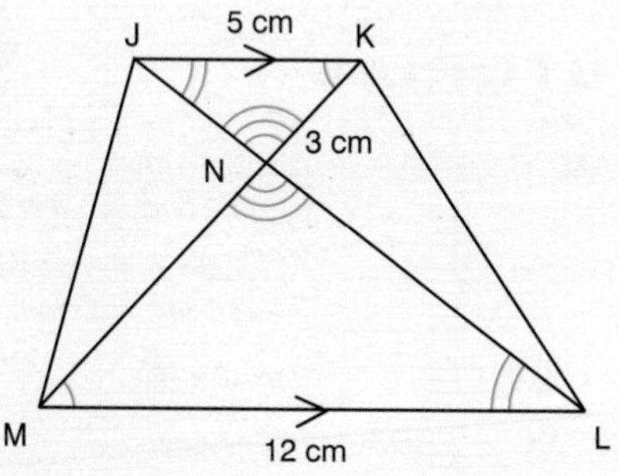

MN = 7.2 cm

TUTORIAL

T2

ΔPQT ≡ ΔSRT

As ΔPQT ≡ ΔSRT *then* RS = PQ = 4.2 *cm*.

T3

ΔAXY *is similar to* ΔABC.

∠XAY = ∠BAC	*They are the same angle.*
∠AXY = ∠ABC	*Corresponding angles between parallel lines.*

so ΔAXY *is similar to* ΔABC *As two angles of one triangle are equal to two angles of the other.*

As the two triangles are similar then the ratios of the corresponding sides are equal.

So $\frac{AX}{AB} = \frac{AY}{AC}$

$\frac{8}{12} = \frac{10}{AC}$

$\frac{12}{8} = \frac{AC}{10}$ *Turning both sides upside-down.*

$AC = \frac{12}{8} \times 10$ *Multiplying both sides by* 10.

AC = 15 *cm*

T4

Drawing a diagram helps to see the situation clearly.

∠JNK = ∠LNM	*Vertically opposite angles.*
∠KJL = ∠JLM	*Alternate angles between parallel lines.*

So ΔLMN *is similar to* ΔJKN *as two angles of one triangle are equal to two angles of the other.*

As the two triangles are similar then the ratios of the corresponding sides are equal.

So $\frac{JK}{ML} = \frac{NK}{MN}$

$\frac{5}{12} = \frac{3}{MN}$

$\frac{12}{5} = \frac{MN}{3}$ *Turning both sides upside-down.*

$MN = \frac{12}{5} \times 3$ *Multiplying both sides by* 3.

MN = 7.2 *cm*

EXAMINATION QUESTIONS 12

EXAM PRACTICE

Sample Student's Answers & Examiner's Comments

1 (a) Reflect triangle A in the x-axis. Label the reflection B.

(b) Reflect triangle B in the line $y = x$. Label the reflection C.

(c) Describe fully the single transformation which maps triangle A onto triangle C.

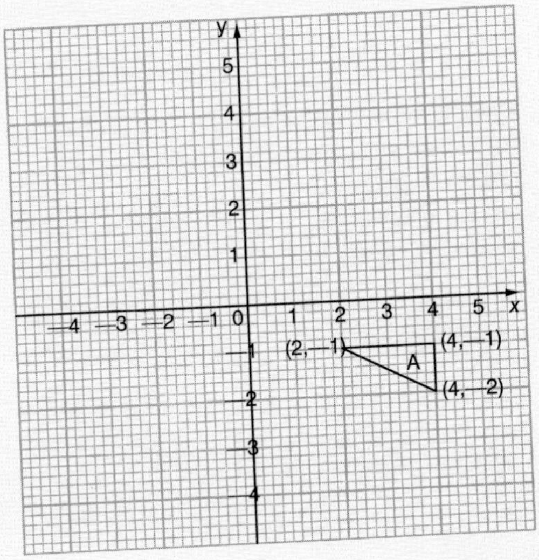

(London, specimen paper 1998)

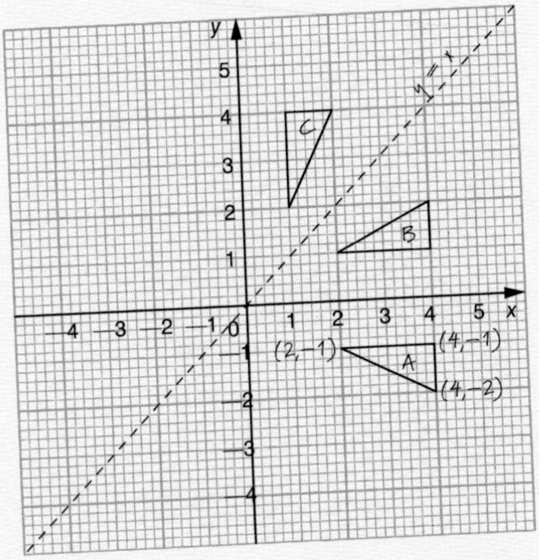

The single transformation is a +90° rotation about (0, 0).

EXAMINER'S COMMENTS

1 *The candidate has followed the given instructions to find the various transformations and noticed that the single transformation is a +90° rotation about (0, 0).*

Alternatively he might have found the centre by joining corresponding points on the object and image with straight lines and drawing the perpendicular bisectors of these lines. The centre of rotation lies on the intersection of these bisectors.

The candidate has remembered that a rotation needs to be described in terms of the position of the centre of rotation, the angle of the rotation and the direction of the rotation.

2 In the diagram BA is parallel to DE.
AEC and BDC are straight lines.
AC = 4.5 cm, DE = 0.8 m, CD = 2.4 m, BC = 6 m.
Angle ABC = 47°, angle BCA = 32°.

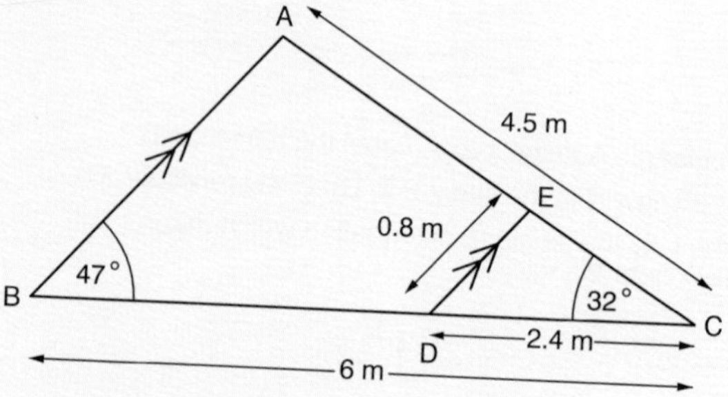

(a) (i) Calculate the size of angle DEC.
(ii) Explain how you obtained your answer.
(b) Calculate the length of EC.
(c) Calculate the length of AB.

(London, specimen paper 1998)

(a) $\angle BAC = 101°$

$\angle DEC = 101°$

(b) ΔABC is similar to ΔEDC

(c) $\frac{BC}{DC} = \frac{AC}{EC}$

$\frac{6}{2.4} = \frac{4.5}{EC}$

$\frac{2.4}{6} = \frac{EC}{4.5}$

$EC = \frac{2.4}{6} \times 4.5$

$EC = 1.8$

$\frac{BC}{DC} = \frac{AB}{ED}$

$\frac{6}{2.4} = \frac{AB}{0.8}$

$AB = \frac{6}{2.4} \times 0.8$

$AB = 2$

EXAMINER'S COMMENTS

2 *The candidate should appreciate that much of the wording to this question is merely 'setting the scene'. The information is more clearly seen on the diagram.*

a)
The candidate has not given any reason or shown any working so marks will be lost, especially as the question asks the candidate to explain how they obtained the answer.

In this case:
$\angle BAC = 101°$ *as the angles of triangle ABC must add up to 180°.*
$\angle DEC = 101°$ *as it is a corresponding angle to* $\angle BAC$ *(BA and DE are parallel).*

b)
Again the candidate gives no reason but:
$\angle BCA = \angle DCE$
they are the same angle
$\angle BAC = \angle DEC$
corresponding angles (already proved)

so ΔABC *is similar to* ΔEDC *as two angles of one triangle equal two angles of the other.*

As the two triangles are similar then the ratios of the corresponding sides are equal.

Units should always be included in the question as these will be tested in the examination.

An answer of EC = 1.8 m provides the complete solution.

c)
Again, the ratios of the corresponding sides are equal.

Units should again be included so that AB = 2 m.

Questions to Answer

1 A B C D E F G H I J K

Write down all of these letters that have:

(a) a horizontal line of symmetry
(b) a vertical line of symmetry
(c) both horizontal and vertical lines of symmetry
(d) no line symmetry
(e) rotational symmetry of order 2.

2 The triangles in the following diagram are all congruent.

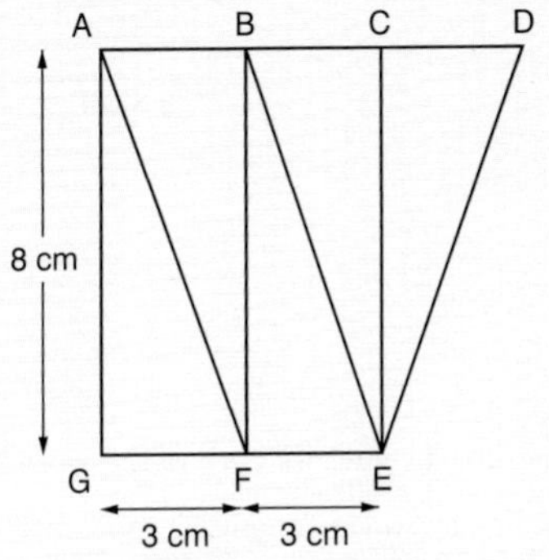

Describe fully the transformation which maps:

(a) ΔAGF onto ΔBFE (b) ΔBCE onto ΔDCE
(c) ΔAGF onto ΔFBA (d) ΔAFB onto ΔEBF.

3 Use squared paper to draw the image of the square with vertices at (3, 1), (3, 5), (${}^{-}1$, 5), (${}^{-}1$, 1) under the following rotations.

(a) Angle ${}^{-}90°$ about centre of rotation (0, 0)
(b) Angle ${}^{+}60°$ about centre of rotation (${}^{-}1$, 1)

4 A rectangle measures 14 cm by 8 cm. A similar rectangle has sides 4 cm and *x* cm. What are the possible values of *x*?

5 Two sides AB and DC of a regular pentagon ABCDE when produced meet at a point P. Calculate ∠BPC.

6 A hexagon has interior angles of 98°, 103°, 129° and 140°. The sizes of the remaining angles are in the ratio 2 : 3. How big are the remaining angles?

7 How many planes of symmetry does the following solid have?

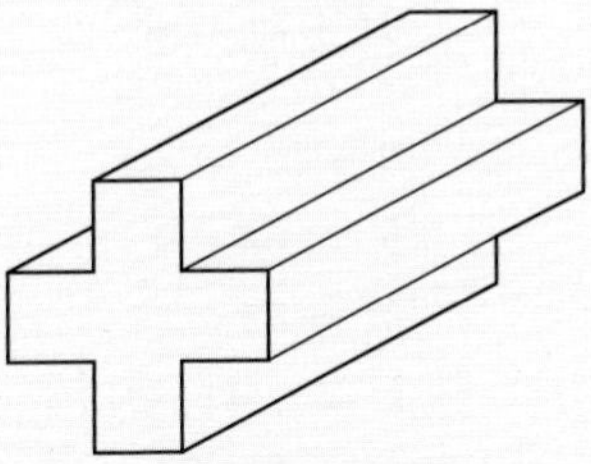

CHAPTER 65

PYTHAGORAS' THEOREM – IN TWO DIMENSIONS

For any right-angled triangle, the square of the length of the hypotenuse is equal to the sum of the squares of the lengths of the other two sides.

$$a^2 + b^2 = c^2$$

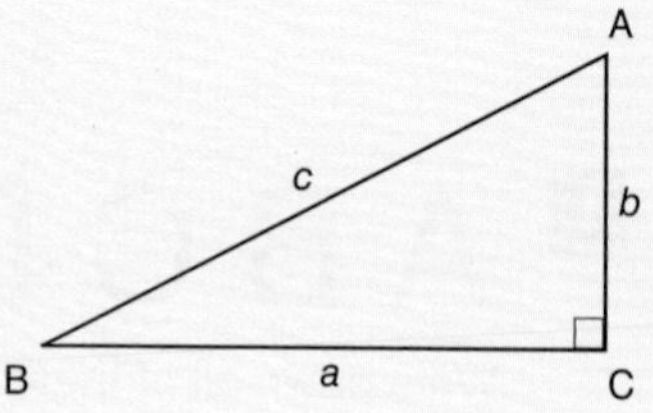

NOTE

For a right-angled triangle, the side opposite the right angle is called the hypotenuse and this is always the longest side.

Worked example

Find the length of the hypotenuse in this triangle.

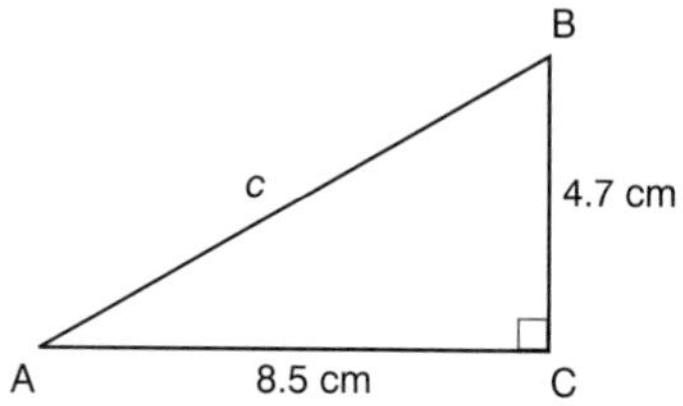

Using Pythagoras' theorem:

$a^2 + b^2 = c^2$

$c^2 = a^2 + b^2$

$c^2 = 4.7^2 + 8.5^2$ Substituting values for each of the given lengths.

$c^2 = 22.09 + 72.25$ Squaring individual lengths.

$c^2 = 94.34$

$c = 9.712\,878$ Taking square roots of both sides to find c.

$c = 9.71$ cm (3 s.f.) Rounding to an appropriate degree of accuracy.

Worked example

Calculate the height of this isosceles triangle.

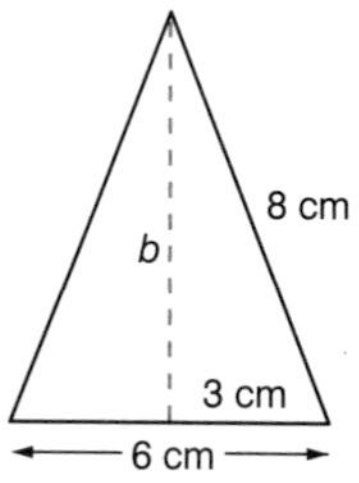

The isosceles triangle can be split into two right-angled triangles to find the height.

Using Pythagoras' theorem:

$a^2 + b^2 = c^2$

$3^2 + b^2 = 8^2$ Substituting values for each of the given lengths – in this case the length of the hypotenuse is known.

$9 + b^2 = 64$ Squaring individual lengths.

$b^2 = 64 - 9$ Making the height the subject.

$b^2 = 55$

$b = 7.416\,198\,5$ Taking square roots of both sides to find the height.

$b = 7.42$ cm (3 s.f.) Rounding to an appropriate degree of accuracy.

Check yourself

QUESTIONS

Q1 Calculate the length of a diagonal of a square of side 4.5 cm.

Q2 Find the area of an equilateral triangle with side 6 cm.

Q3 A plane flies 24 km on a bearing of 020°, then a further 16.7 km on a bearing of 110°. How far is the plane away from its starting point?

REMEMBER! Cover the answers if you want to.

ANSWERS

A1

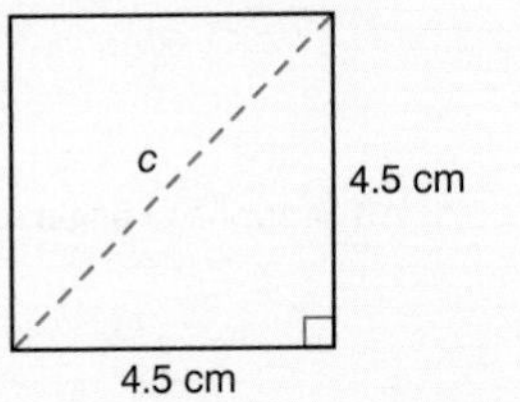

The length of a diagonal = 6.36 cm (3 s.f.)

A2

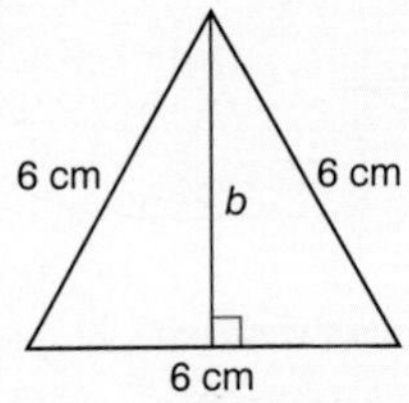

Area of triangle = 15.6 cm^2 (3 s.f.)

A3

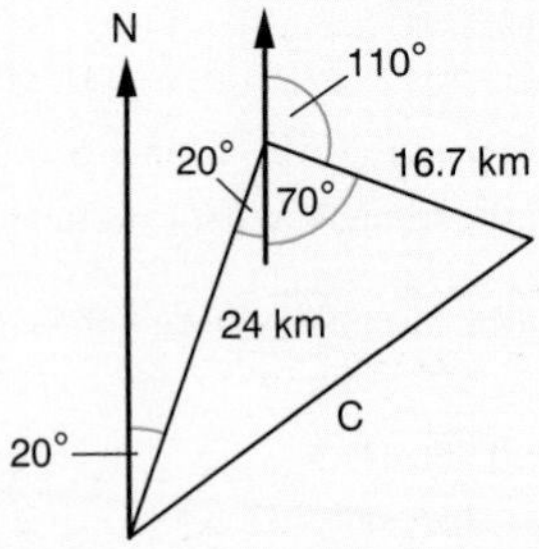

Distance = 29.2 km (3 s.f.)

TUTORIAL

T1 *Drawing a diagram helps to see the situation.*

Using Pythagoras' theorem:

$a^2 + b^2 = c^2$

$4.5^2 + 4.5^2 = c^2$ *Substituting values for a and b.*

$c^2 = 40.5$ *Squaring 4.5 and adding.*

$c = 6.363\,961$ *Taking square roots to find c.*

The length of a diagonal = 6.36 cm (3 s.f.)

T2 *The triangle can be split into two right-angled triangles to find the height. Drawing a diagram helps to see the situation.*

Using Pythagoras' theorem:

$a^2 + b^2 = c^2$

$3^2 + b^2 = 6^2$ *Length of the hypotenuse is 6 cm.*

$9 + b^2 = 36$ *Squaring individual lengths.*

$b^2 = 36 - 9$ *Making the height the subject.*

$b^2 = 27$

$b = 5.196\,152\,4$ *Taking square roots to find b.*

Area of triangle $= \frac{1}{2} \times$ *base* $\times$ *perpendicular height*

$= \frac{1}{2} \times 6 \times 5.196\,152\,4$

$= 15.588\,457$

$= 15.6\ cm^2$ (3 s.f.)

T3 *Drawing a diagram helps to see the situation.*

Since the diagram includes a right-angled triangle you can use Pythagoras' theorem:

$a^2 + b^2 = c^2$

$16.7^2 + 24^2 = c^2$ *Substituting values for a and b.*

$c^2 = 854.89$ *Squaring and adding.*

$c = 29.238\,502$ *Taking square roots to find c.*

Distance = 29.2 km (3 s.f.)

CHAPTER 66

SINE, COSINE AND TANGENT – RIGHT-ANGLED TRIANGLES

The sides of a right-angled triangle are given special names.

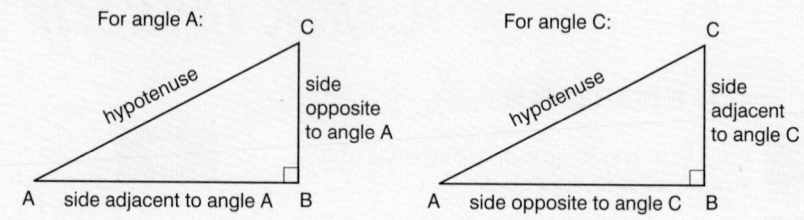

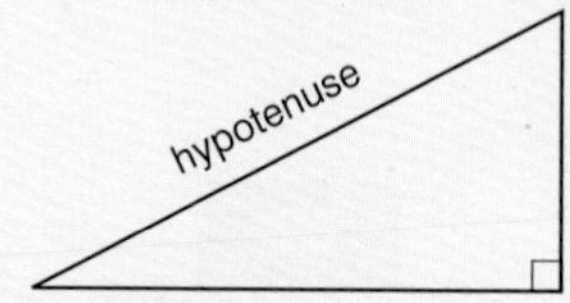

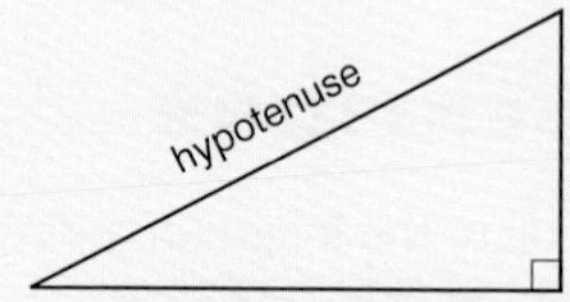

SINE OF AN ANGLE

The sine of an angle (usually abbreviated as sin) $= \dfrac{\text{length of opposite side}}{\text{length of hypotenuse}}$.

So $\quad \sin A = \dfrac{\text{length of opposite side}}{\text{length of hypotenuse}} = \dfrac{\text{BC}}{\text{AC}}$

and $\quad \sin C = \dfrac{\text{length of opposite side}}{\text{length of hypotenuse}} = \dfrac{\text{AB}}{\text{AC}}$ As AB is opposite angle C.

Worked example

Use the sine ratio to find p and q in these right-angled triangles.

(a)

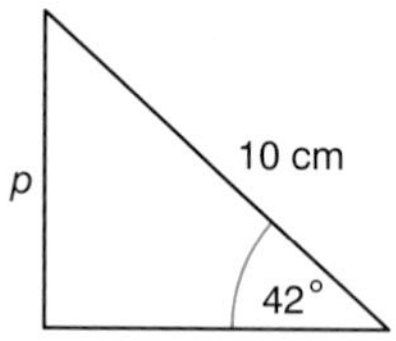

(b)

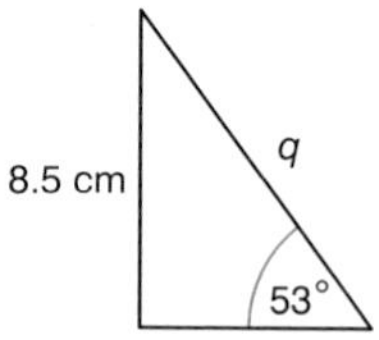

(a) Using the formula $\sin A = \dfrac{\text{length of opposite side}}{\text{length of hypotenuse}}$:

$\sin 42° = \dfrac{p}{10}$

$p = 10 \times \sin 42°$ Rearranging the formula to make p the subject.

$p = 10 \times 0.669\,130\,6$ As $\sin 42° = 0.669\,130\,6$

$p = 6.691\,306$

$p = 6.69$ cm (3 s.f.) Including the units and rounding to an appropriate degree of accuracy.

(b) Using the formula $\sin A = \dfrac{\text{length of opposite side}}{\text{length of hypotenuse}}$:

$\sin 53° = \dfrac{8.5}{q}$

$q \times \sin 53° = 8.5$ Rearranging the formula to get q on the top line.

$q = \dfrac{8.5}{\sin 53°}$ Rearranging the formula to make q the subject.

$q = \dfrac{8.5}{0.798\,635\,5}$ As $\sin 53° = 0.798\,635\,5$

$q = 10.643\,153$

$q = 10.6$ cm (3 s.f.) Including the units and rounding to an appropriate degree of accuracy.

Worked example

Use the sine ratio to find the angle θ.

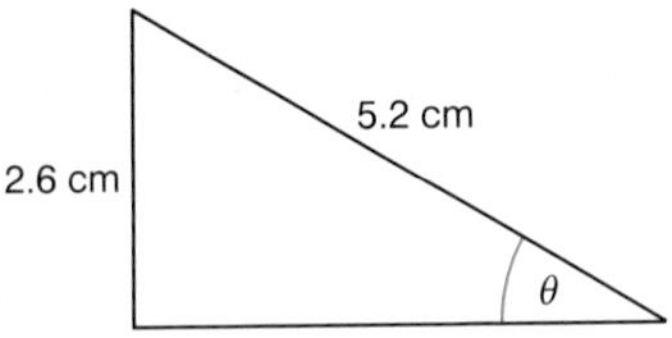

> **HINT**
>
> Angles are frequently represented by Greek letters of the alphabet such as $\alpha, \beta, \gamma, \delta, \theta, \phi$, etc.

Using the formula $\sin A = \dfrac{\text{length of opposite side}}{\text{length of hypotenuse}}$:

$\sin\theta = \dfrac{2.6}{5.2}$

$\sin\theta = 0.5$ — As $\dfrac{2.6}{5.2} = 0.5$

$\theta = \sin^{-1} 0.5$ — Use $\sin^{-1} 0.5$ (or arcsin 0.5) to show that you are working backwards to find the angle, given the sine of the angle (finding the inverse).

$\theta = 30°$ — Use the inverse button ($\sin^{-1}$ or arcsin) on your calculator but make sure your calculator is in DEG (degree) mode

COSINE OF AN ANGLE

The cosine of an angle (usually abbreviated as cos) $= \dfrac{\text{length of adjacent side}}{\text{length of hypotenuse}}$.

So $\quad \cos A = \dfrac{\text{length of adjacent side}}{\text{length of hypotenuse}} = \dfrac{AB}{AC}$

and $\quad \cos C = \dfrac{\text{length of adjacent side}}{\text{length of hypotenuse}} = \dfrac{BC}{AC}$

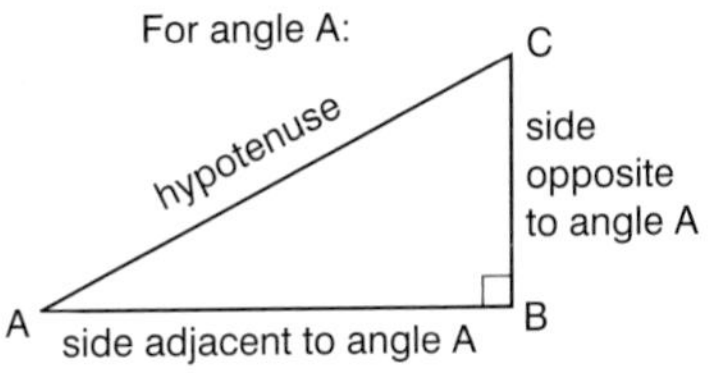

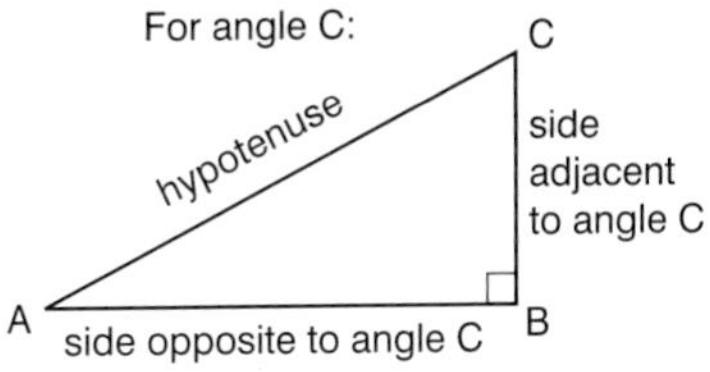

Worked example

Use the cosine ratio to find the lengths of the sides indicated on the following right-angled triangles.

(a)

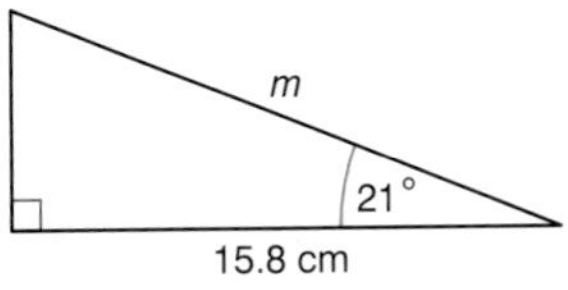

(b)

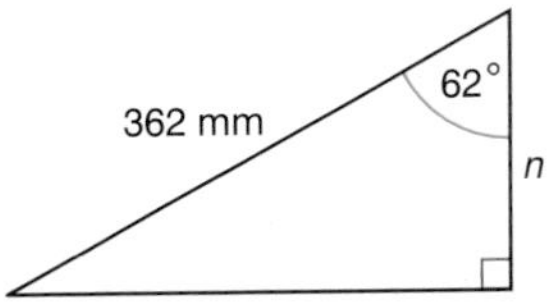

(a) Using the formula $\cos A = \dfrac{\text{length of adjacent side}}{\text{length of hypotenuse}}$:

$\cos 21° = \dfrac{15.8}{m}$

$m \times \cos 21° = 15.8$ — Rearranging the formula to get m on the top line.

$m = \dfrac{15.8}{\cos 21°}$ — Rearranging the formula to make m the subject.

$m = \dfrac{15.8}{0.933\,580\,4}$ — As $\cos 21° = 0.933\,580\,4$

$m = 16.924\,091$

$m = 16.9\text{ cm}$ (3 s.f.) — Including the units and rounding to an appropriate degree of accuracy.

(b) Using the formula $\cos A = \dfrac{\text{length of adjacent side}}{\text{length of hypotenuse}}$:

$\cos 62° = \dfrac{n}{362}$ — As the side labelled n is adjacent to the angle 62°.

$n = 362 \times \cos 62°$ — Rearranging the formula to make n the subject.

$n = 362 \times 0.469\,471\,6$ — As $\cos 62° = 0.469\,471\,6$

$n = 169.948\,71$

$n = 170$ mm (3 s.f.) — Including the units and rounding to an appropriate degree of accuracy.

NOTE

The value of n might also be found by using $\sin 28° = \dfrac{n}{362}$ where 28° is the missing angle.

Worked example

Use the cosine ratio to find the angles α and β.

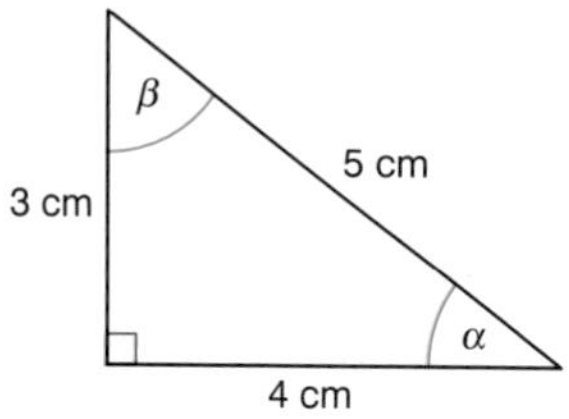

Using the formula $\cos A = \dfrac{\text{length of adjacent side}}{\text{length of hypotenuse}}$:

$\cos \alpha = \dfrac{4}{5}$

$\cos \alpha = 0.8$

$\alpha = \cos^{-1} 0.8$ — Use $\cos^{-1} 0.8$ (or arccos 0.8) to show that you are working backwards to find the angle, given the cosine of the angle (finding the inverse).

$\alpha = 36.869\,898°$ — Use the inverse button ($\cos^{-1}$ or arccos) on your calculator.

$\alpha = 36.9°$ (3 s.f.)

Similarly $\cos \beta = \dfrac{3}{5}$

$\cos \beta = 0.6$

$\beta = \cos^{-1} 0.6$ — Use $\cos^{-1} 0.6$ (or arccos 0.6) to show that you are working backwards to find the angle, given the cosine of the angle.

$\beta = 53.130\,102°$ — Use the inverse button ($\cos^{-1}$ or arccos) on your calculator.

$\beta = 53.1°$(3 s.f.)

HINT

Once the angle α had been found then it is possible to find β by using the fact that $\alpha + \beta + 90° = 180°$ (the angle sum of a triangle = 180°)

TANGENT OF AN ANGLE

The tangent of an angle (usually abbreviated as tan) $= \dfrac{\text{length of opposite side}}{\text{length of adjacent side}}$.

So that $\tan A = \dfrac{\text{length of opposite side}}{\text{length of adjacent side}} = \dfrac{BC}{AB}$

and $\tan C = \dfrac{\text{length of opposite side}}{\text{length of adjacent side}} = \dfrac{AB}{BC}$

For angle A:

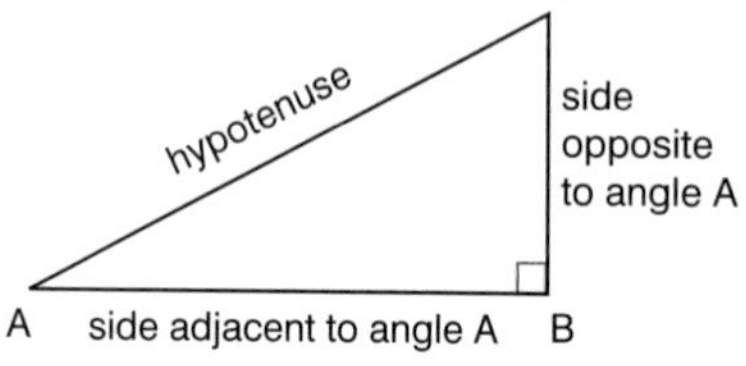

For angle C:

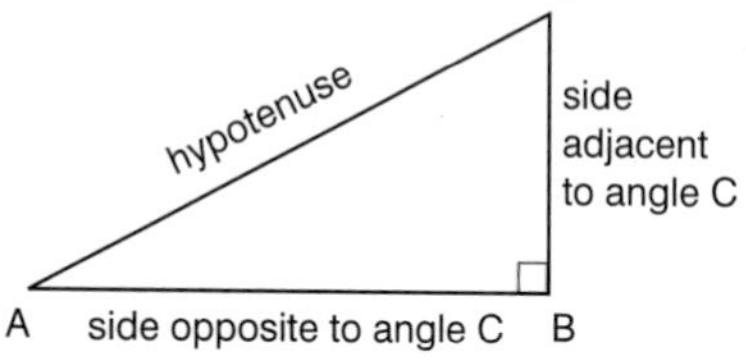

Worked example

Given that AD = 30 cm, BC = 10 cm and ∠DBC = 61° find:

(a) CD (b) ∠DAC.

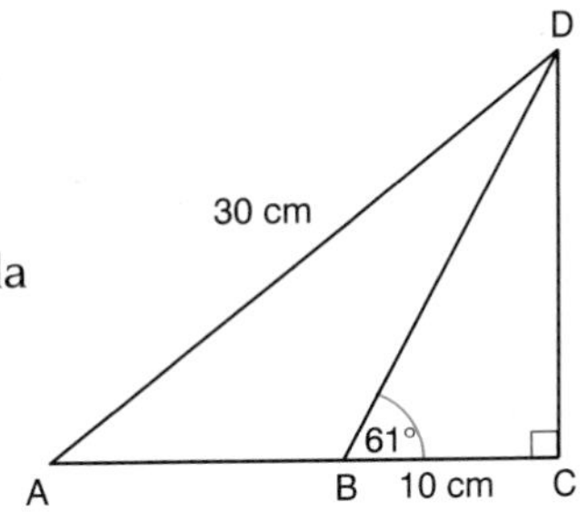

(a) For the right-angled triangle DBC, using the formula

$$\tan DBC = \frac{\text{length of opposite side}}{\text{length of adjacent side}}:$$

$$\tan 61° = \frac{CD}{10}$$

CD = 10 × tan 61° — Rearranging the formula to make CD the subject.

CD = 10 × 1.804 047 8 — As tan 61° = 1.804 047 8

CD = 18.040 478

CD = 18.0 cm (3 s.f.) — Including the units and rounding to an appropriate degree of accuracy.

HINT

Learn how to use the memory buttons on your calculator. You can use more exact values, and it makes you less likely to make a keying error.

(b) For the right-angled triangle DAC, using the formula

$$\sin A = \frac{\text{length of opposite side}}{\text{length of hypotenuse}}:$$

$$\sin DAC = \frac{18.040\,478}{30}$$ Using the original value of CD and not the rounded value.

sin DAC = 0.601 349 3

∠DAC = $\sin^{-1}$ 0.601 349 3

∠DAC = 36.966 593° — Use the inverse button ($\sin^{-1}$ or arcsin) on your calculator.

∠DAC = 37.0° (3 s.f.) — Rounding to an appropriate degree of accuracy.

Worked example

A rectangle measures 10 cm by 5 cm. What angle does the diagonal make with the longer sides?

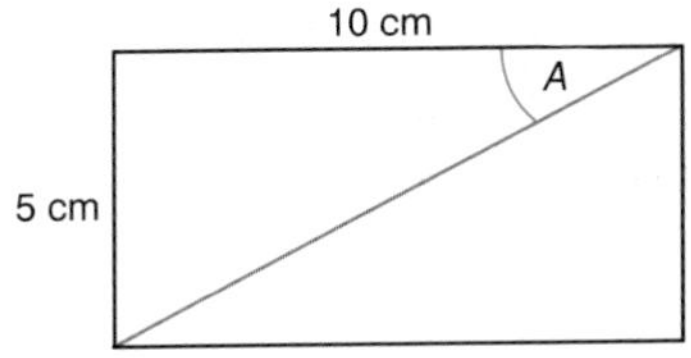

Start by drawing a sketch of the rectangle. Let the required angle be *A*.

Using the formula $\tan A = \frac{\text{length of opposite side}}{\text{length of adjacent side}}$:

$$\tan A = \frac{5}{10}$$

tan *A* = 0.5

A = $\tan^{-1}$ 0.5 — Use $\tan^{-1}$ 0.5 (or arctan 0.5) to show that you are working backwards to find the angle, given the tangent of the angle (finding the inverse).

A = 26.565 051°

The angle which the diagonal makes with the longest side is 26.6° (3 s.f.).

ANGLES OF ELEVATION AND DEPRESSION

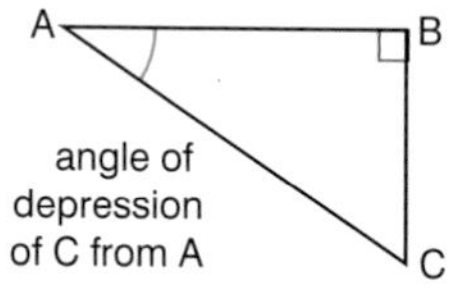

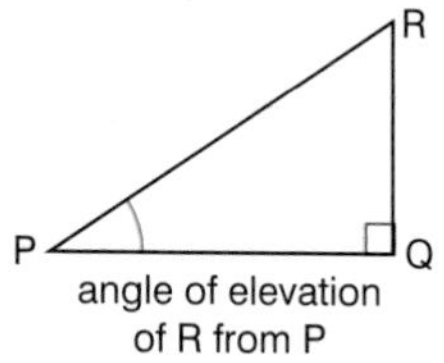

The angle of elevation is the angle up from the horizontal.

The angle of depression is the angle down from the horizontal.

Worked example

An object is situated 25 metres from the foot of a mast of height 32 metres. What is the angle of depression of the object from the top of the mast?

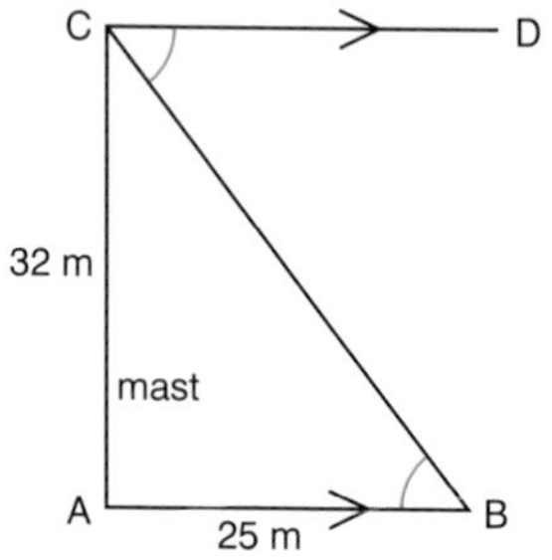

Start by drawing a diagram of the situation.

The angle of depression of the object from the top of the mast is $\angle DCB$.

$\angle DCB = \angle ABC$ — As these are alternate angles between parallel lines AB and CD.

$\tan ABC = \frac{32}{25}$ — As the tangent of an angle $= \frac{\text{length of opposite side}}{\text{length of adjacent side}}$.

$\tan ABC = 1.28$

$\angle ABC = \tan^{-1} 1.28$ — Use the inverse button ($\tan^{-1}$ or arctan) on your calculator.

$\angle ABC = 52.001\,268°$

The angle of depression is 52.0° (3 s.f.).

Check yourself

QUESTIONS

Q1 Find the lengths of the sides in the following right-angled triangles.

(a)

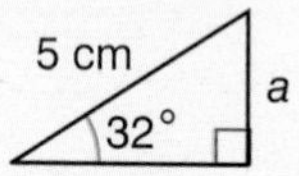

(b)

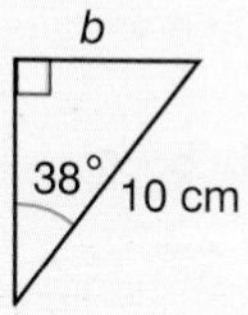

(c)

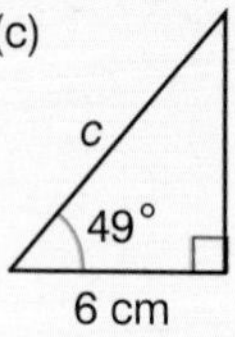

(d)

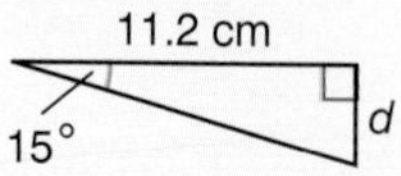

Q2 A ladder 6 m long reaches $4\frac{1}{2}$ metres up a vertical wall. Find the angle between the ladder and the horizontal.

Q3 Find the angles in the following triangles.

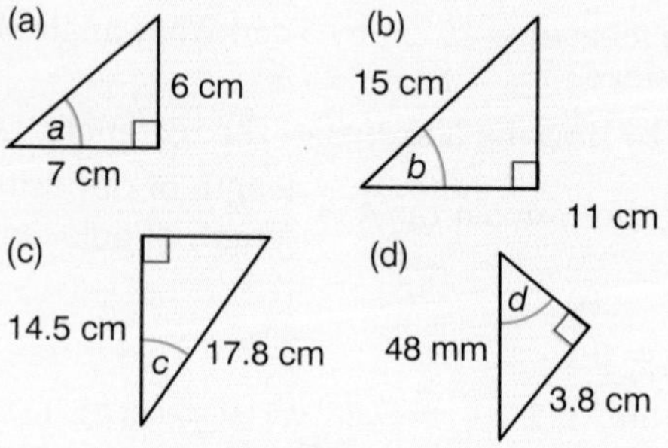

Q4 A tree of height 30 feet casts a shadow which is 36 feet long. What is the angle of elevation of the sun?

REMEMBER! Cover the answers if you want to.

ANSWERS

A1
(a) $a = 2.65$ cm (3 s.f.)
(b) $b = 6.16$ cm (3 s.f.)
(c) $c = 9.15$ cm (3 s.f.)
(d) $d = 3.00$ cm (3 s.f.)

A2 48.6° (3 s.f.)

A3
(a) $a = 40.6°$ (3 s.f.)
(b) $b = 42.8°$ (3 s.f.)
(c) $c = 35.5°$ (3 s.f.)
(d) $d = 52.3°$ (3 s.f.)

A4 39.8° (3 s.f.)

TUTORIAL

T1
(a) $\sin 32° = \frac{a}{5}$ $a = 2.65$ cm (3 s.f.)

(b) $\sin 38° = \frac{b}{10}$ $b = 6.16$ cm (3 s.f.)

(c) $\cos 49° = \frac{6}{c}$ $c = \frac{6}{\cos 49°} = 9.15$ cm (3 s.f.)

(d) $\tan 15° = \frac{d}{11.2}$ $d = 3.00$ cm (3 s.f.)

T2

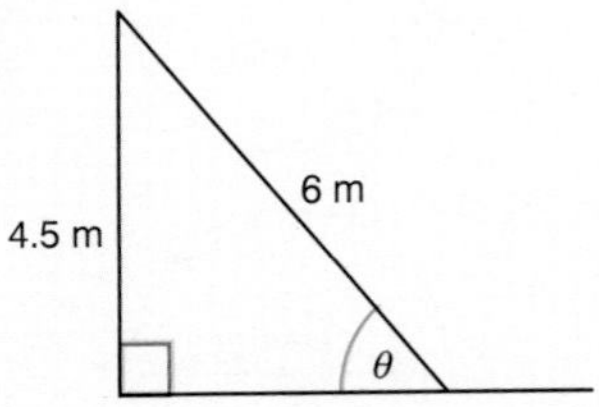

Labelling the required angle θ:

$\sin\theta = \frac{4.5}{6}$

$\theta = 48.590\,378° = 48.6°$ (3 s.f.)

T3
(a) $\tan a = \frac{6}{7}$ $a = \tan^{-1}\frac{6}{7}$

$a = \tan^{-1} 0.857\,142\,8... = 40.6°$ (3 s.f.)

(b) $\cos b = \frac{11}{15}$ $b = \cos^{-1}\frac{11}{15}$

$b = \cos^{-1} 0.733\,333\,3... = 42.8°$ (3 s.f.)

(c) $\cos c = \frac{14.5}{17.8}$ $c = \cos^{-1}\frac{14.5}{17.8}$

$c = \cos^{-1} 0.814\,606\,7\ ... = 35.5°$ (3 s.f.)

(d) $\sin d = \frac{3.8}{4.8}$ *Converting 48 mm to 4.8 cm.*

$d = \sin^{-1}\frac{3.8}{4.8} = \sin^{-1} 0.791\,666\,6...$

$d = 52.3°$ (3 s.f.)

T4

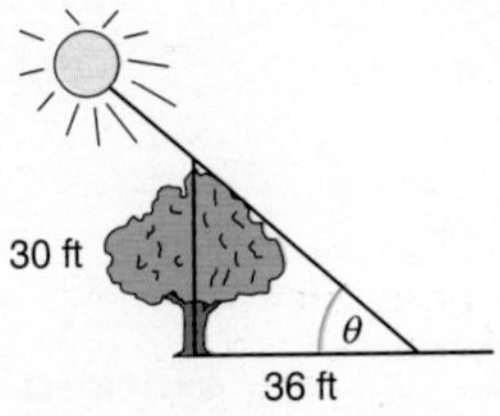

Let the required angle be θ.

$\tan\theta = \frac{30}{36}$

$\theta = 39.805\,571° = 39.8°$ (3 s.f.)

CHAPTER 67

LENGTHS, AREAS AND VOLUMES

Information on areas of triangles and quadrilaterals is given in Chapter 52, Geometric terms.

The following information will be useful to you for the examination paper at the Intermediate level. The Area of trapezium and Volume of prism will be given to you on the examination paper at that level.

Circumference of circle
$= \pi \times$ diameter
$= 2 \times \pi \times$ radius

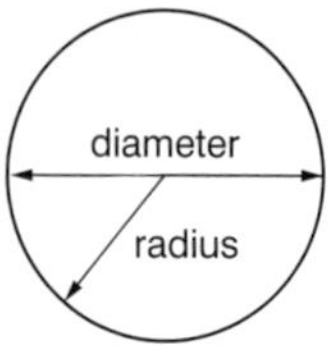

Area of circle
$= \pi \times (\text{radius})^2$

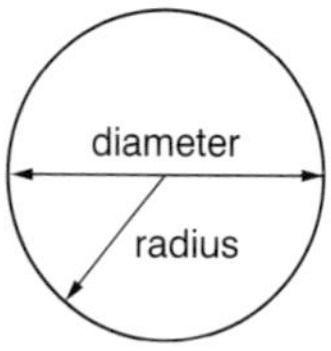

Area of parallelogram
$=$ base $\times$ height

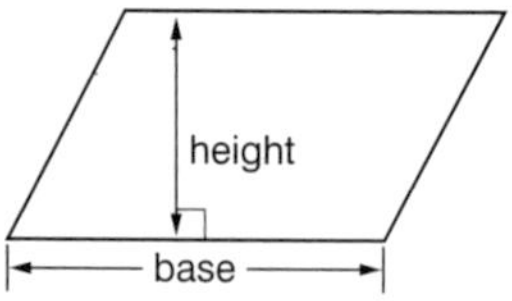

Area of trapezium $= \frac{1}{2}(a + b)h$

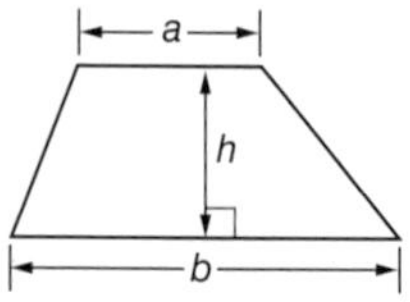

Volume of cuboid
$=$ length $\times$ width $\times$ height

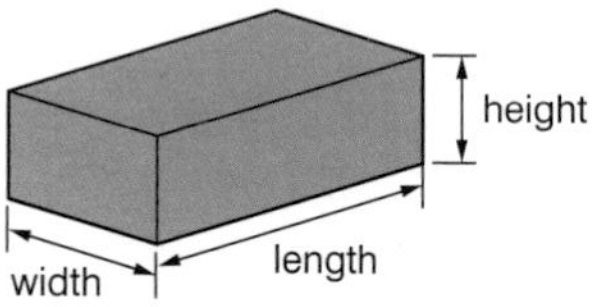

Volume of a cylinder $= \pi r^2 h$

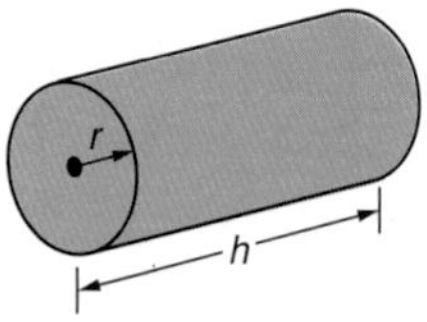

Volume of prism
$=$ area of cross section $\times$ length

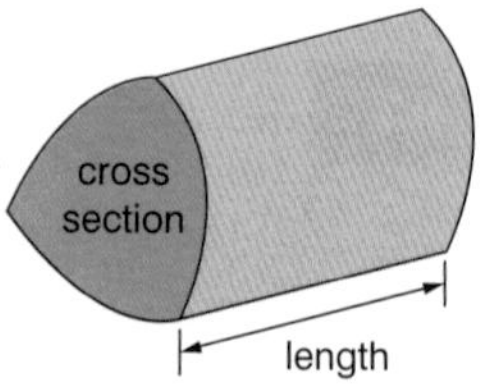

Worked example

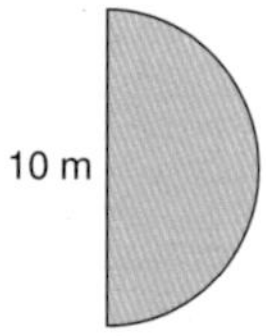

Find the area and perimeter of a semicircle of diameter 10 metres.

Area of a semicircle $= \frac{1}{2} \times$ area of a circle

$= \frac{1}{2} \times \pi r^2$ Area of a circle $= \pi r^2$

$= \frac{1}{2} \times \pi \times 5^2$ Radius $= \frac{1}{2} \times$ diameter

$= \frac{1}{2} \times 10\,\text{m} = 5\,\text{m}$

$= 39.269\,908 = 39.3\,\text{m}^2$ (3 s.f.)

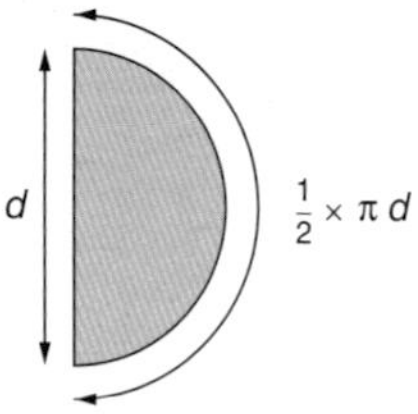

Perimeter of a semicircle $= \frac{1}{2} \times$ circumference of circle $+$ diameter

Perimeter $= \frac{1}{2} \times 2\pi r + d$ Circumference of a circle $= 2\pi r$ and diameter $= d$.

$= \frac{1}{2} \times 2 \times \pi \times 5 + 10 = 15.707\,963 + 10 = 25.707\,963$

$= 25.7\,\text{m}$ (3 s.f.)

Worked example

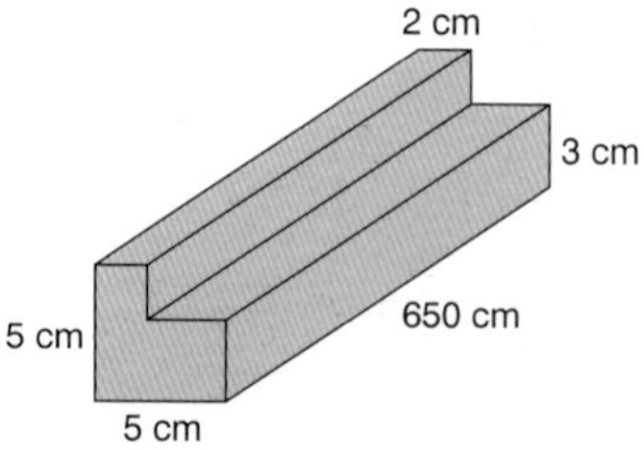

Find the volume of the prism with dimensions as shown.

Volume of a prism $=$ area of cross-section $\times$ length.

Area of cross-section $= 5 \times 3 + 2 \times 2 = 15 + 4 = 19\,\text{cm}^2$

Volume of a prism $=$ area of cross-section $\times$ length

$= 19 \times 65$ Writing all lengths in the same units with $10\,\text{mm} = 1\,\text{cm}$.

$= 1235\,\text{cm}^3 = 1240\,\text{cm}^3$ (3 s.f.)

UNITS FOR LENGTH, AREA AND VOLUME

Length, area and volume can all be identified by looking at the formulae or units, as area is two-dimensional (the product of two lengths) and volume is three-dimensional (the product of three lengths).

By ignoring constants (including π) you should be able to identify length, area and volume as shown in the table.

Lengths	Areas	Volumes
d	lb	lbh
$2\pi r$	$\frac{1}{2}lb$	$\frac{4}{3}\pi r^3$
	$\frac{1}{2}(a+b)h$	
	πr^2	$\pi r^2 h$
	$2\pi rh$	
	πrl	$\frac{1}{3}\pi r^2 h$
	πab	

NOTE

The sum of two lengths is still length and the sum of two areas is still area etc.

Check yourself

QUESTIONS

Q1 Calculate the areas of the shaded parts of the following shapes.

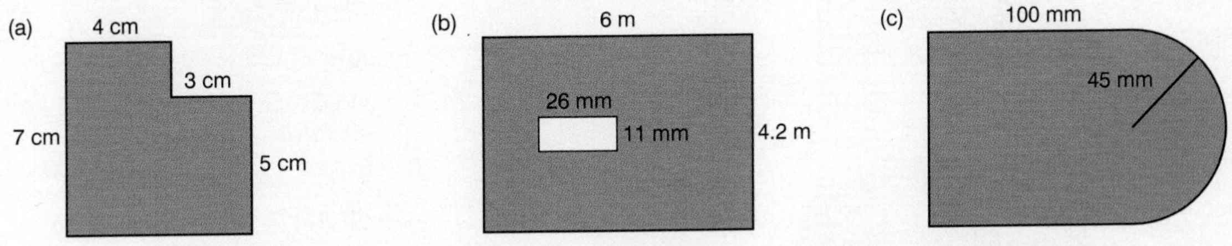

Q2 The cross-section of a block of wood is a trapezium with measurements as shown on the diagram. Find the area of the cross-section and the volume of the block.

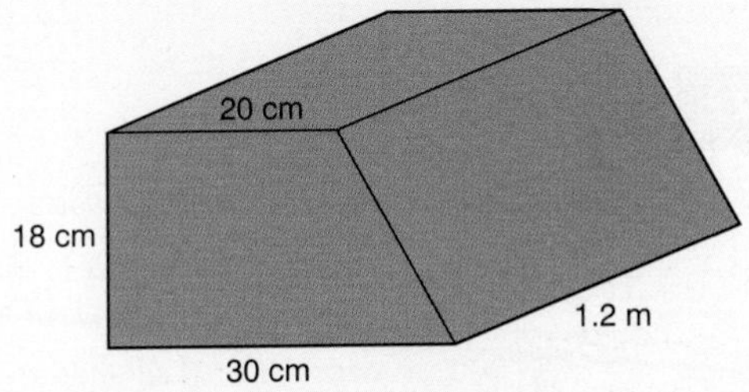

Q3 The diagram shows the plan of a front room. Calculate the perimeter and the area of the floor

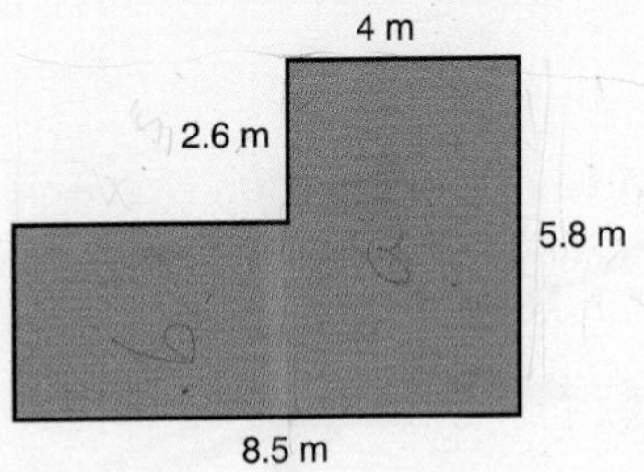

QUESTIONS

Q4 PQRS is a rhombus of side 10 cm. The length of the longer diagonal is 16 cm. Find the area of the rhombus.

Q5 A circular disc is cut from a square of side 7 cm to leave a minimum amount of waste. What is the area of the waste and what is this as a percentage of the original area?

Q6 A circular pond has a surface area of 400 m^2. Calculate the diameter of the pond to an appropriate degree of accuracy.

Q7 Identify the following as length, area or volume.

(a) $2\pi r(r + h)$ (b) $\frac{\theta}{360} \times 2\pi r$

(c) $\frac{\pi}{4}d^2h$ (d) $\sqrt{r^2 + h^2}$

REMEMBER! Cover the answers if you want to.

ANSWERS

A1 (a)

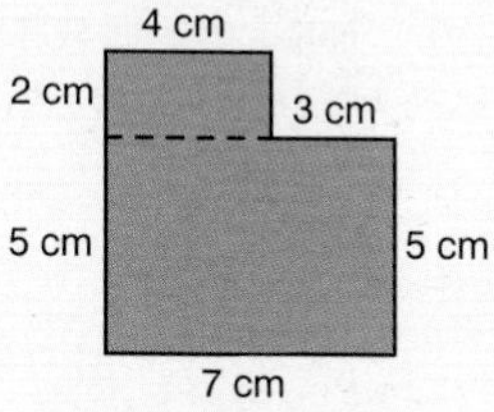

Area = 43 cm^2

(b)

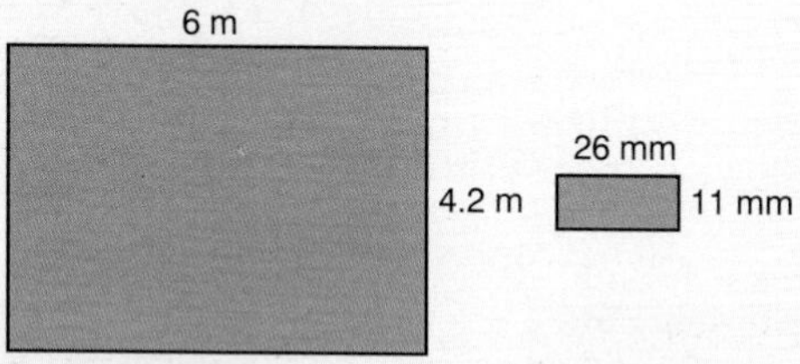

Area = 25 200 cm^2 or 25.2 m^2 (3 s.f.)

(c)

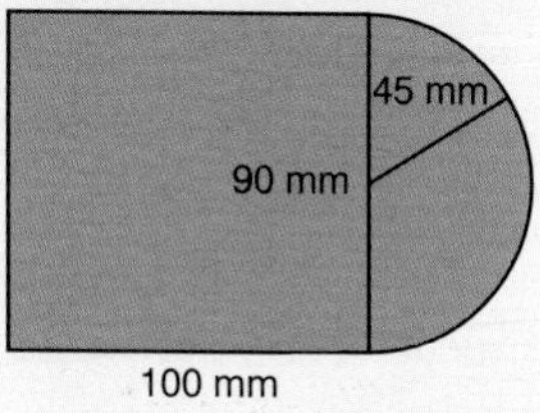

Area = 12 200 mm^2 or 122 cm^2 (3 s.f.)

A2 Area of cross-section = 450 cm^2

Volume of the block = 54 000 cm^3 or 0.054 m^3

TUTORIAL

T1 (a) *Drawing a diagram helps to see the situation clearly.*

Area = $4 \times 2 + 5 \times 7 = 8 + 35 = 43$ *cm*^{2}

(b) *Drawing a diagram helps to see the situation clearly.*

Area = $600 \times 420 - 2.6 \times 1.1$ *Writing all lengths in the same units.*

= 252 000 − 2.86

= 251 997.14

= 25 200 *cm*^{2} (3 *s.f.*) *or* 25.2 *m*^{2} (3 *s.f.*) *if all units are converted into metres.*

(c) *Drawing a diagram helps to see the situation clearly.*

Area = $100 \times 90 + \frac{1}{2}(\pi \times 45^2)$

Height of rectangle = 45 + 45 *from radius of semicircle.*

Area of a semicircle = $\frac{1}{2} \times \pi r^2$

So area = 9000 + 3180.8626

= 12 180.863

= 12 200 *mm*^{2} (3 *s.f.*)

or 122 *cm*^{2} (3 *s.f.*) *if all units are converted into centimetres.*

T2 *Area of cross-section* = $\frac{1}{2}(20 + 30) \times 18 = 450$ *cm*^{2}.

Volume of the block = area of cross-section × length

= 450 × 120 *Writing all lengths in the same units with* 1.2 *m* = 120 *cm.*

= 54 000 *cm*^{3} *or* 0.054 *m*^{3} *if all units are converted into metres*

ANSWERS

A3

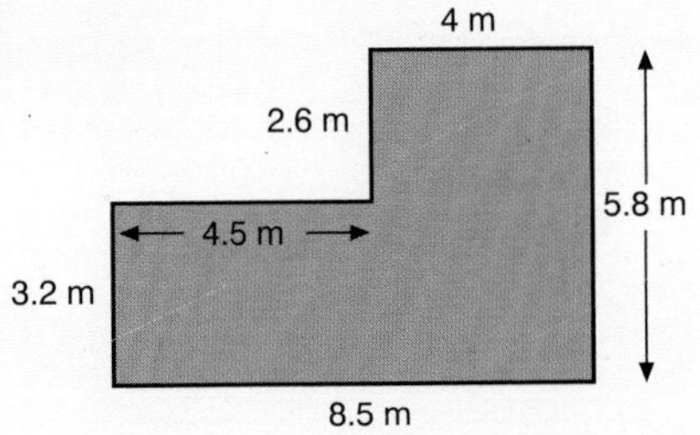

Perimeter = 28.6 m, area = 37.6 m^2

A4

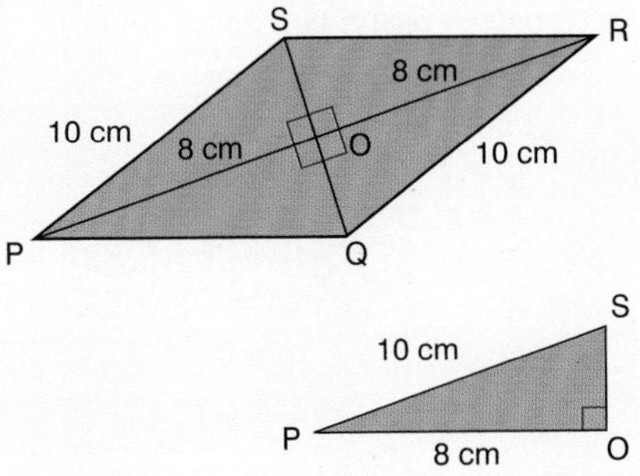

Area of rhombus PQRS = 96 cm^2

A5

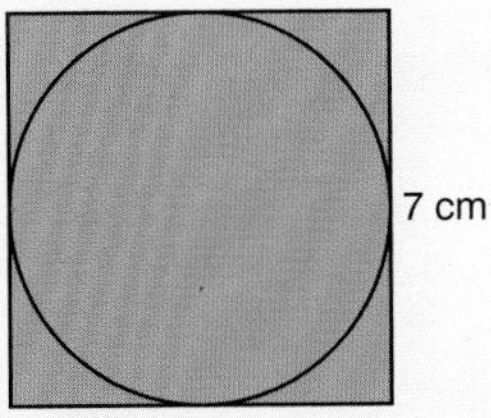

Area of waste = 10.5 cm^2

Percentage waste = 21.5% (3 s.f.)

TUTORIAL

T3

Drawing a diagram helps to see the situation clearly.

Perimeter = 5.8 + 8.5 + 3.2 + 4.5 + 2.6 + 4 *working in a clockwise direction from the top right-hand corner*

= 28.6 *m*

Alternatively you might appreciate that opposite sides of the plan have equal lengths so perimeter = 5.8 + 8.5 + 5.8 + 8.5 + 28.6 *m.*

Area of floor = 4 × 2.6 + 8.5 × 3.2
= 10.4 + 27.2
= 37.6 *m*^{2}

T4

Drawing a diagram helps to see the situation clearly. Then using the fact that the diagonals of a rhombus bisect each other at right angles, the triangle POS can be extracted from the diagram.

Using Pythagoras' theorem OS = 6 *cm*:

Area of triangle POS

$= \frac{1}{2} \times$ *base* × *perpendicular height*

$= \frac{1}{2} \times 8 \times 6 = 24$ *cm*^{2}

Area of rhombus PQRS = 4 × *area of triangle POS as area of rhombus is made up from four similar triangles.*

= 4 × 24 = 96 *cm*^{2}

T5

Drawing a diagram helps to see the situation clearly.
Area of waste = area of square – area of circle

$= 7 \times 7 - \pi \times 3.5^2$

= 10.515 49 *cm*^{2}

= 10.5 *cm*^{2} (3 *s.f.*)

Percentage waste = $\frac{10.51549}{49} \times 100\%$

As original area = 49 *cm*^{2}

= 21.460 184%

= 21.5% (3 *s.f.*)

ANSWERS

A6 Diameter = 22.6 m (3 s.f.)

A7
(a) area
(b) length
(c) volume
(d) length

TUTORIAL

T6 *Area of pond* = πr^2

$\pi r^2 = 400$

$r^2 = \frac{400}{\pi}$

$r^2 = 127.323\,95$

$r = \sqrt{127.323\,95}$

$r = 11.283\,792$

$d = 2 \times 11.283\,792$ *As diameter = 2 × radius*

$d = 22.567\,583$

$d = 22.6\,m$ (3 s.f.)

An answer correct to 3 s.f. or 2 s.f. would seem most appropriate in view of the original data.

T7

(a) $2\pi r(r + h)$ — *$r + h$ = length and 2π is a constant, $r(r + h)$ gives units of length × length = area*

(b) $\frac{\theta}{360} \times 2\pi r$ — *$\frac{\theta}{360} \times 2\pi$ is a constant giving units of length.*

(c) $\frac{\pi}{4}d^2h$ — *$\frac{\pi}{4}$ is a constant and d^2h gives units of length × length × length = volume*

(d) $\sqrt{(r^2 + h^2)}$ — *$r^2 + h^2$ gives units of area + area = area.*

The square root of area = length.

EXAM PRACTICE

Sample Student's Answers & Examiner's Comments

1 (a) Calculate the distance AB.

(b) Calculate the angle x.

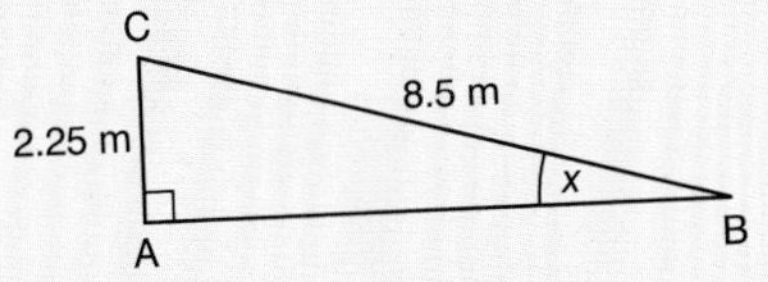

(NEAB specimen paper 1998)

(a) $8.5^2 = 2.25^2 + AB^2$

$72.25 = 5.0625 + AB^2$

$AB^2 = 72.25 - 5.0625$

$= 67.1875$

$= 8.1967982$

(b) $\sin x = \frac{2.25}{8.5}$

$x = 15$

2 A ship sails from a point P a distance of 20 km on a bearing of 048°. Calculate how far the ship is:

(a) to the north of P (b) to the east of P.

(a) $\cos 48° = \frac{PS}{20}$

$PS = 20 \times \cos 48°$

$PS = 13.382612$

(b) $\sin 48° = \frac{SR}{20}$

$SR = 20 \times \sin 48°$

$SR = 14.862897$

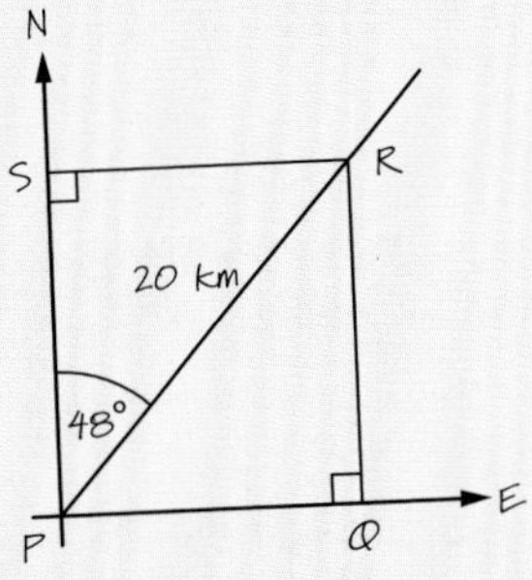

EXAMINER'S COMMENTS

1 **a)**
The candidate correctly makes use of Pythagoras' theorem although this is not mentioned in the answer.

Here she is correctly finding the value of AB but the method is not very clear. A better presentation of the work would be:

$AB^2 = 72.25 - 5.0625$

$AB^2 = 67.1875$

$AB = 8.196\,798\,2$

and the final answer should be given to 3 s.f. or 2 s.f. along with the units.

b)
The candidate correctly identifies the use of sine to find the angle. She is using the calculator to find the angle but is not showing sufficient working to gain credit if any mistakes are made.

$\sin x = 0.264\,705\,9$ *or*
$x = \sin^{-1} 0.264\,705\,9$

would allow some credit if an error occurred in the final answer (although full marks are always given for correct solutions). An answer of 15°, 15.3° or equivalent would be acceptable.

2 **a)**
The candidate has provided a diagram to support the work and allow a better understanding of the question.

He has worked out the distance north but does not make it clear in the answer that PS is the distance north. An answer of 13.4 km (3 s.f.) would indicate that he was aware of the accuracy of the answer.

b)
The candidate has appreciated that PQ, the distance east, equals SR.

Again, he has worked out the distance east but has not made it clear in the answer that SR is the distance east. An answer of 14.9 km (3 s.f.) would indicate an awareness of the accuracy of the answer.

EXAMINER'S COMMENTS

3 From the diagram:

(a) find BC (b) find ∠CAB.

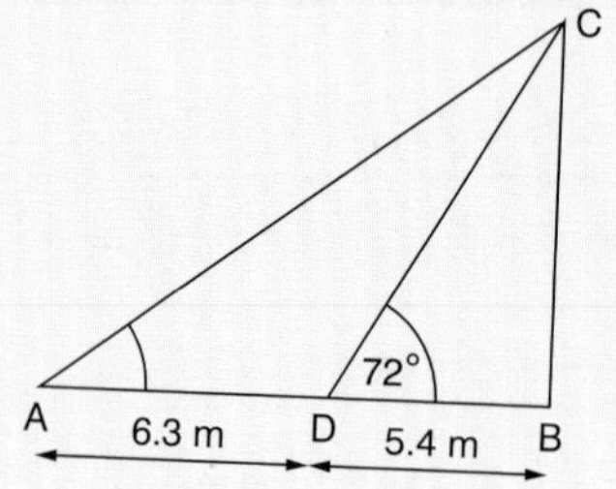

(a) $\tan 72° = \frac{BC}{5.4}$

$BC = 5.4 \times \tan 72°$

$BC = 16.619491$

$BC = 16.6\text{ m}$ (3 s.f.)

(b) $\tan CAB = \frac{BC}{AB}$

$\tan CAB = \frac{16.6}{11.7}$

$\tan CAB = 1.4188034$

$\angle CAB = 54.823081°$

$\angle CAB = 54.8°$ (3 s.f.)

3 a)
The candidate has correctly identified the use of tangent to find the length BC.

The final answer for BC is appropriately rounded to 3 s.f.

b)
The candidate has correctly identified the use of tangent to find the angle ∠CAB.

She has used the approximated value for BC rather than the original value of BC = 16.619 491.

Such an error, which is called premature approximation, will be penalised in the examination and should be carefully avoided.

The correct solution is given by:

$\tan CAB = \frac{16.619491}{11.7}$

$\tan CAB = 1.4204693$

$\angle CAB = 54.854735°$

$\angle CAB = 54.9°$ (3 s.f.)

4 In the diagram ABCD is a parallelogram. Calculate the area of the parallelogram.

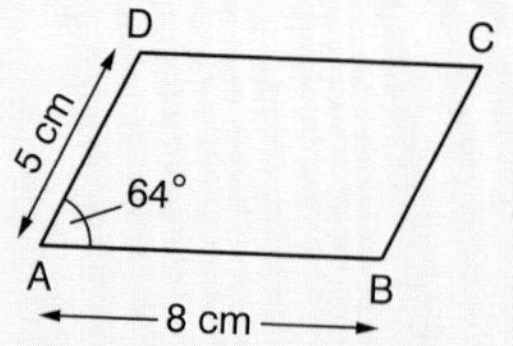

(MEG syllabus A, specimen paper 1998)

Area of a parallelogram = base × perpendicular height

$\sin 64° = \frac{\text{height}}{5}$

height = $5 \times \sin 64°$

height = 4.4939702

Area of parallelogram

= base × perpendicular height

= 8×4.4939702

= 35.95176185

= 36.0 cm^2 (3 s.f.)

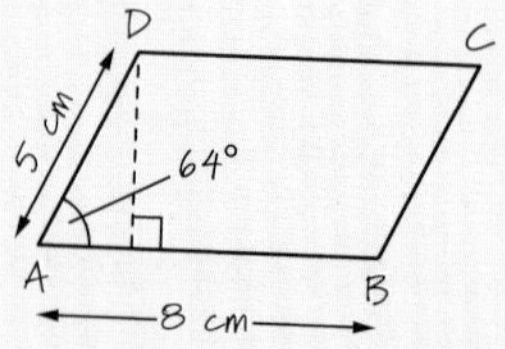

4 *The candidate has correctly identified the use of sine to find the perpendicular height.*

The value for the perpendicular height is substituted in the formula which is given on the examination paper.

The answer is rounded to 3 s.f. or else 2 s.f. to refect the accuracy of the information given in the question.

Questions to Answer

1 Given that T is the midpoint of QR, PQ = 15 cm and PR = 20 cm, find the length PT.

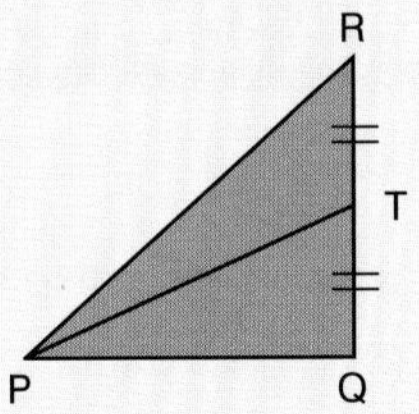

2 ABCD is a kite with AB = AD = 15 cm, AO = 12 cm and CO = 18 cm. Find the area of the kite and the angle BAD.

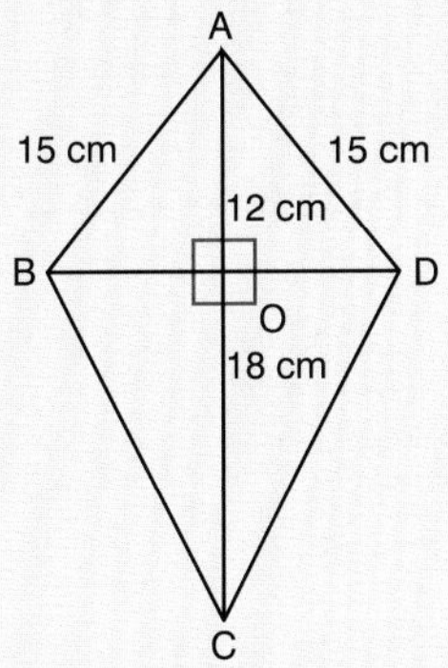

3 A vertical cliff is 485 metres high. The angle of depression of a boat at sea is 20°. What is the distance of the boat from the foot of the cliff?

4 A pole is held vertical by ropes which are 2.8 metres long and which are secured to the ground at a distance 1.9 metres from the pole. If the ropes are secured 0.7 metres from the top of the pole then calculate the height of the pole.

5 A washer has an outside diameter of 12 mm and an inside diameter of 6 mm. Calculate the cross-sectional area of the washer.

6 A packet of tea measures 8 cm by 4 cm by 12 cm. Calculate the volume. There are 125 grams of tea in the packet. What is the volume of a packet containing 100 grams?

7 The diameter of a £1 coin is 22 mm. The coin is 3 mm thick. Work out the volume of the coin.

(In the style of MEG syllabus B, specimen paper 1998)

8 Find the volume of the prism of which the length is 2 metres and the cross-section is an equilateral triangle of side 6 centimetres.

9 Explain why the expression $\frac{4}{3}\pi r^2$ cannot represent the volume of a sphere.

SHAPE, SPACE AND MEASURES

CHAPTER 69

FURTHER AREAS AND VOLUMES

NOTE

The formulae for the surface area and volume of a sphere and a cone will be given to you on the examination paper at the higher level.

Volume of sphere $= \frac{4}{3}\pi r^3$

Surface area of sphere $= 4\pi r^2$

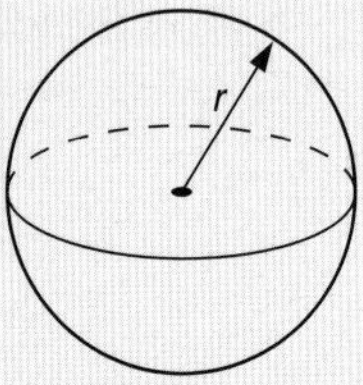

Volume of cone $= \frac{1}{3}\pi r^2 h$

Curved surface area of cone $= \pi r l$

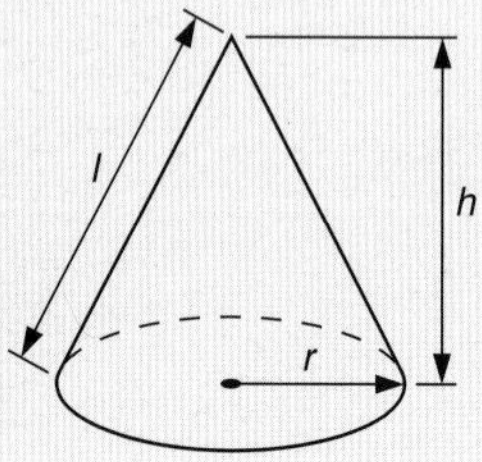

Information on areas of triangles and quadrilaterals is given in Chapter 52, Geometric terms and Chapter 67, Lengths, areas and volumes.

AREA AND VOLUME

Worked example

Find the volume and surface area of a hemisphere of diameter 10 metres.

Volume of a hemisphere $= \frac{1}{2} \times$ volume of a sphere

Volume $= \frac{1}{2} \times \frac{4}{3}\pi r^3$ — Volume of a sphere $= \frac{4}{3}\pi r^3$

$= \frac{1}{2} \times \frac{4}{3} \times \pi \times 5^3$ — Radius $= \frac{1}{2} \times$ diameter $= \frac{1}{2} \times 10\,\text{m} = 5\,\text{m}$

$= 261.799\,39 = 262\,\text{m}^3$ (3 s.f.)

Surface area of a hemisphere

$= \frac{1}{2} \times$ surface area of a sphere + area of a circle

$= \frac{1}{2} \times 4\pi r^2 + \pi r^2$

Surface area of a sphere $= 4\pi r^2$ and area of a circle $= \pi r^2$

$= \frac{1}{2} \times 4 \times \pi \times 5^2 + \pi \times 5^2$

$= 235.619\,45\,\text{m}^2$

$= 236\,\text{m}^2$ (3 s.f.)

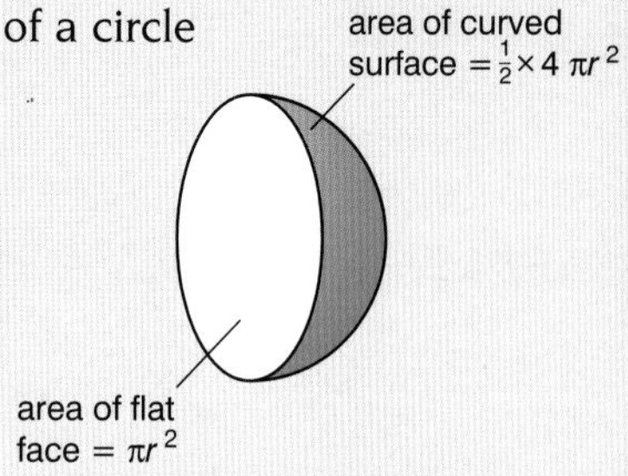

Worked example

The dimensions of a cone are given in the diagram.

Calculate:

(a) the area of the curved surface of the cone

(b) the volume of the cone.

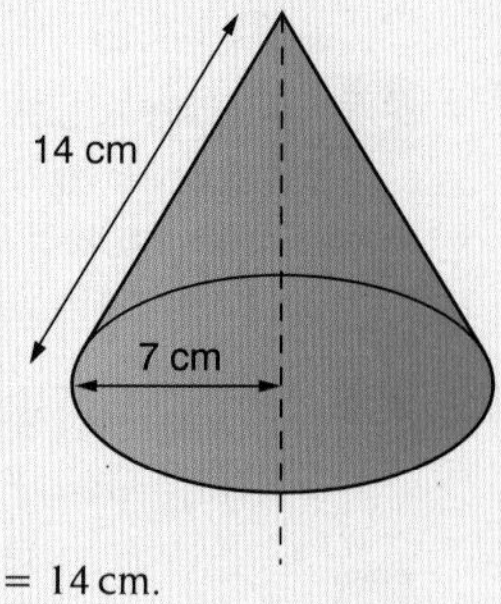

(a) Curved surface area of cone

$= \pi r l$

$= \pi \times 7 \times 14$ — As radius = 7 cm and slant height = 14 cm.

$= 307.876\,08$

$= 308\,\text{cm}^2$ (3 s.f.)

(b) Volume of cone $= \frac{1}{3}\pi r^2 h$ — Where h is the perpendicular height.

Using Pythagoras' theorem:

$l^2 = r^2 + h^2$

$14^2 = 7^2 + h^2$

$196 = 49 + h^2$

$h^2 = 196 - 49 = 147$

$h = 12.124\,356$ — Taking square roots on both sides.

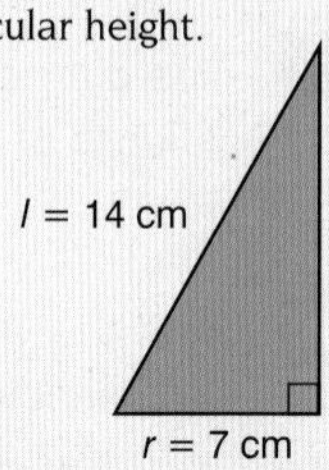

Volume of cone $= \frac{1}{3}\pi r^2 h$

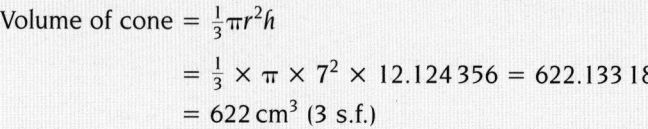

$= \frac{1}{3} \times \pi \times 7^2 \times 12.124\,356 = 622.133\,18$

$= 622\,\text{cm}^3$ (3 s.f.)

SCALE FACTORS OF LENGTH, AREA AND VOLUME

Two solids are said to be **similar** if the ratios of their corresponding linear dimensions are equal.

In general:

- the corresponding areas of similar solids are proportional to the squares of their linear dimensions
- the corresponding volumes of similar solids are proportional to the cubes of their linear dimensions.

So if an object is enlarged by a scale factor of s:

lengths are multiplied by s

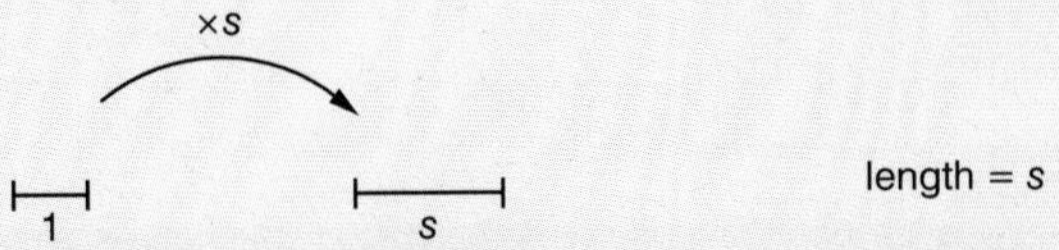

areas are multiplied by s^2

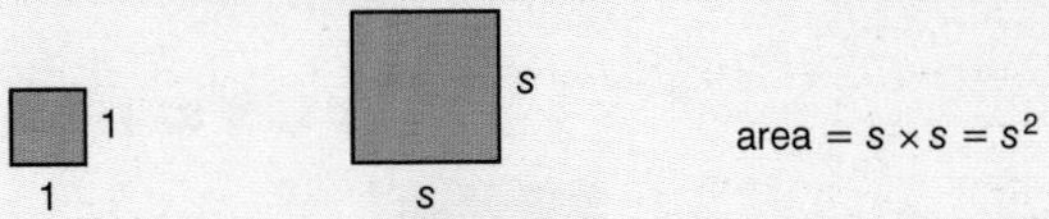

volumes are multiplied by s^3.

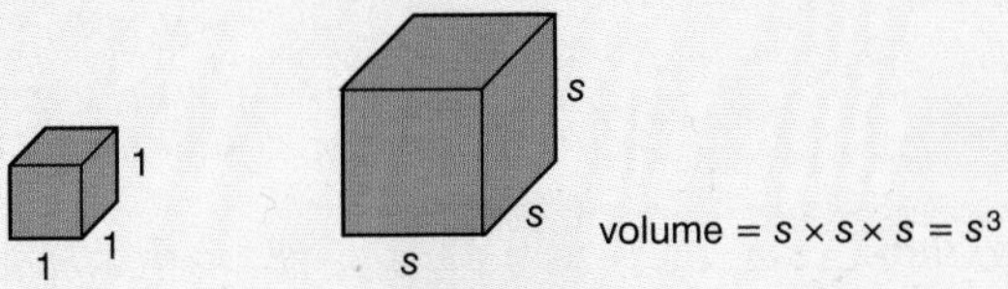

Worked example

The ratio of the surface areas of two similar cylinders is 16 : 25. Calculate the ratio of their volumes and work out the volume of the larger cylinder given that the smaller cylinder has a volume of 540 mm^2.

If the ratio of the lengths is $x : y$

then the ratio of the areas is $x^2 : y^2$

and the ratio of the volumes is $x^3 : y^3$

Here $x^2 : y^2 = 16 : 25$

so $x : y = 4 : 5$

and $x^3 : y^3 = 4^3 : 5^3$

$= 64 : 125$

So the ratio of the volumes is 64 : 125.

$64 : 125 = 1 : \frac{125}{64}$ Rewriting as an equivalent ratio.

$= 1 \times 540 : \frac{125}{64} \times 540$ Rewriting as an equivalent ratio with 540 on the left-hand side.

$= 540 : 1054.6875$

The volume of larger cylinder is 1050 mm^3 (3 s.f.).

Check yourself

QUESTIONS

Q1 The corresponding lengths of two similar solids are in the ratio 3 : 5. What is the ratio of:

(a) their surface areas
(b) their volumes?

Q2 The curved surface area of a cone is 1165 mm^3. What is the curved surface area of a similar cone with height three times the height of the original cone?

Q3 The volume of a cylinder of height 17.5 cm is 432 cm^3. What is the volume of a similar cylinder of height 5 cm?

REMEMBER! Cover the answers if you want to.

ANSWERS

A1 (a) Ratio of their surface areas = 9 : 25
(b) Ratio of their volumes = 27 : 125

A2 Curved surface area = 10 485 mm^3

A3 Volume = 10.1 cm^3 (3 s.f.)

TUTORIAL

T1 *If the ratio of corresponding lengths = 3 : 5 then ratio of corresponding areas = $3^2 : 5^2$ ratio of corresponding volumes = $3^3 : 5^3$.*

(a) *Ratio of their surface areas* = $3^2 : 5^2 = 9 : 25$
(b) *Ratio of their volumes* = $3^3 : 5^3 = 27 : 125$

T2 *If the height of the similar cone is three times that of the original then the area of the similar cone is 3^2 (= 9) times that of the original.*

Curved surface area = $9 \times 1165\ mm^3 = 10\,485\ mm^3$.

T3 *Ratio of lengths*

= 17.5 : 5 *Using the respective heights.*
= 7 : 2 *Rewriting as an equivalent ratio.*

Ratio of volumes

$= 7^3 : 2^3$

= 343 : 8

$= 1 : \frac{8}{343}$ *Making an equivalent ratio by dividing both sides by 243.*

$= 432 : 432 \times \frac{8}{343}$ *Making an equivalent ratio by dividing both sides by 432.*

= 432 : 10.075 802

The volume of a similar cylinder of height 5 cm is 10.1 cm^3 (3 s.f.).

CHAPTER 70

THREE-DIMENSIONAL TRIGONOMETRY

For the Higher paper examination you will be required to work with trigonometry in three dimensions. For this work, it is useful to identify right-angled triangles and show them diagrammatically to answer the questions.

Worked example

The pyramid OABCD has a square base of length 15 cm and a vertical height of 26 cm. Calculate:

(a) the length OA (b) the angle OAC.

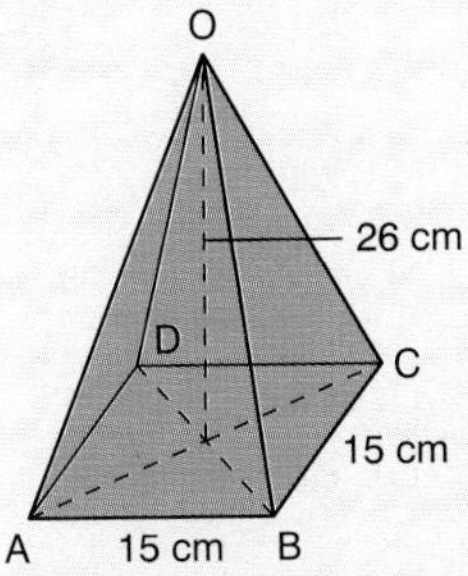

(a) Consider the base ABCD (where X is the centre of the square base).

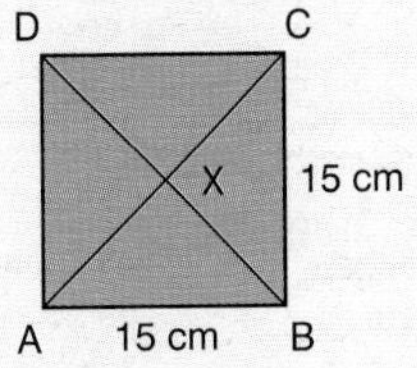

$AC^2 = 15^2 + 15^2$ Using Pythagoras' theorem on the right-angled triangle ABC.

$AC^2 = 225 + 225$

$AC^2 = 450$

$AC = 21.213\,203$ Taking square roots on both sides.

$AX = \frac{1}{2} \times AC$

$AX = 10.606\,602$

Now consider triangle OAX.

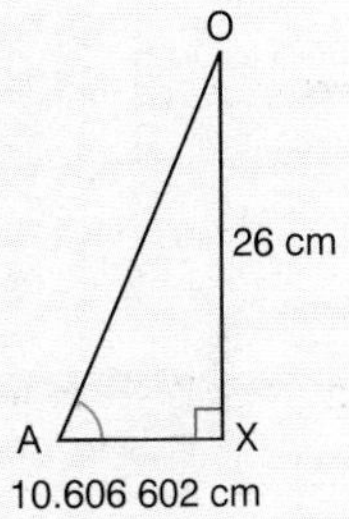

$OA^2 = AX^2 + OX^2$ Using Pythagoras' theorem on the right-angled triangle AXO where OX is the vertical height.

$OA^2 = 10.606\,602^2 + 26^2$

$OA^2 = 788.5$

$OA = 28.080\,242$ Taking square roots on both sides.

$OA = 28.1$ cm (3 s.f.) Rounding to an appropriate degree of accuracy.

(b) The angle OAC is the same as the angle OAX in the previous diagram.

$\tan OAC = \frac{OX}{AX}$ As $\tan\theta = \frac{\text{length of opposite side}}{\text{length of adjacent side}}$

$\tan OAC = \frac{26}{10.606\,602}$

$\tan OAC = 2.451\,303\,4$

$\angle OAC = 67.807\,182$ Using $\tan^{-1}$ or arctan to find the angle.

$\angle OAC = 67.8°$ (3 s.f.) Rounding to an appropriate degree of accuracy.

Check yourself

QUESTIONS

Q1 Find the following lengths in the cuboid shown in the diagram.

(a) AF (b) AC (c) AG

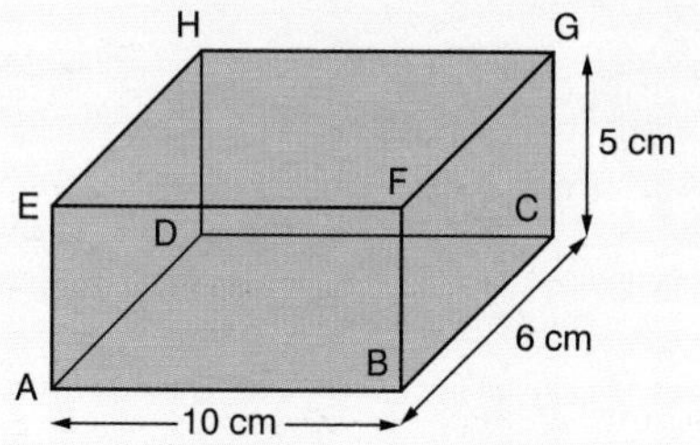

Q2 Find the area of the plane ADGF in this diagram.

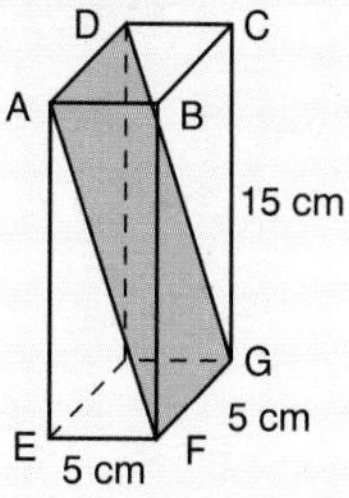

Q3 Find the angle which the line EC in the diagram makes with the base of the cube.

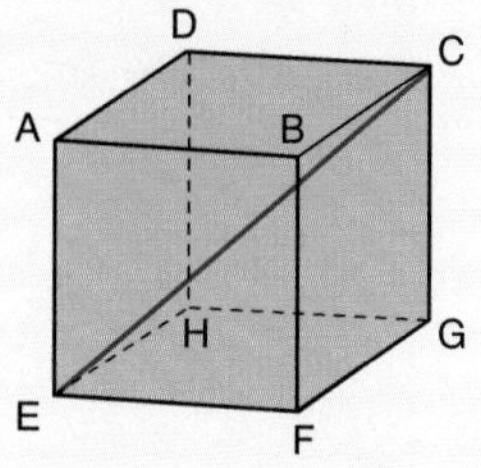

Q4 An electric pylon XY is 30 m high. From a point S, due south of the pylon, the angle of elevation is 26° and from a point W, due west of the pylon, the angle of elevation is 32°. Find the distance WS.

REMEMBER! Cover the answers if you want to.

ANSWERS

A1 (a)

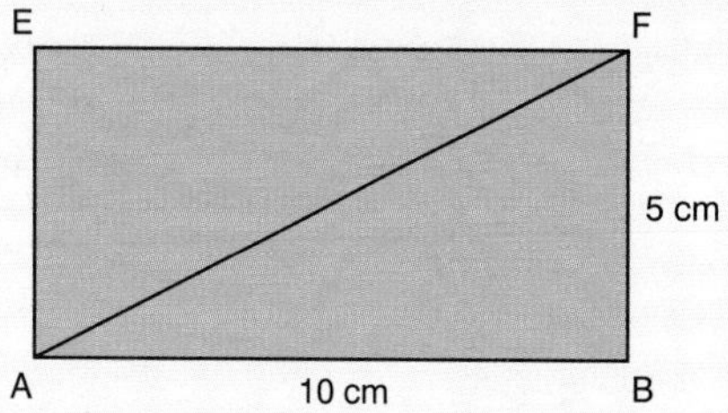

AF = 11.2 cm (3 s.f.)

TUTORIAL

T1 *(a)* *Drawing a diagram helps to see the situation clearly.*

$AF^2 = AB^2 + BF^2$	*Using Pythagoras' theorem on the right-angled triangle ABF.*
$AF^2 = 10^2 + 5^2$	*As BF = CG.*
$AF^2 = 125$	
AF = 11.180 340	*Taking square roots on both sides.*
AF = 11.2 *cm* (3 *s.f.*)	*Rounding to an appropriate degree of accuracy.*

ANSWERS

A1 (b)

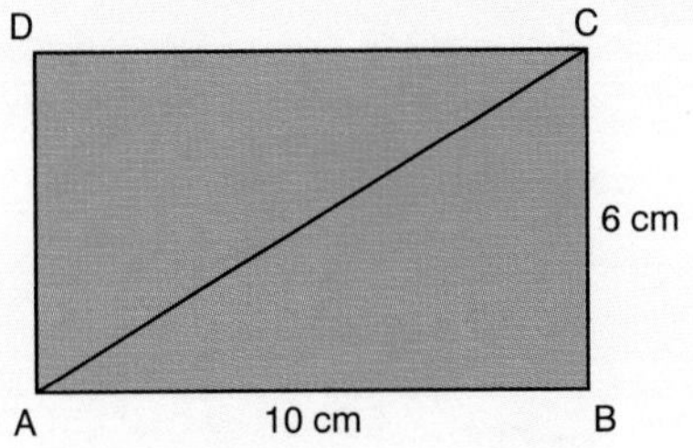

AC = 11.7 cm (3 s.f.)

(c)

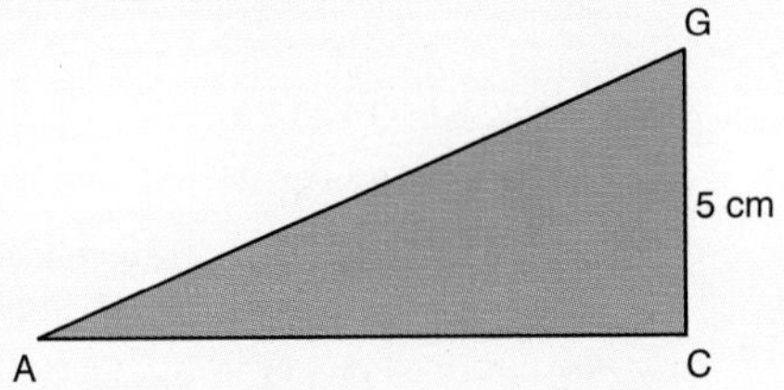

AG = 12.7 cm (3 s.f.)

A2

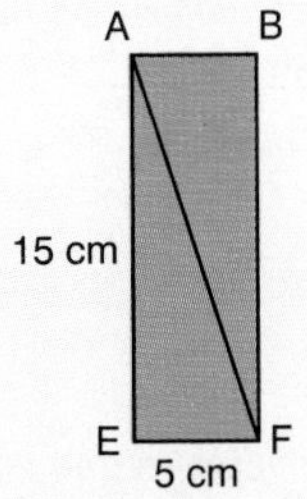

Area of ADGF = 79.1 cm^2

TUTORIAL

T1 (*b*) *Drawing a diagram helps to see the situation clearly.*

$AC^2 = AB^2 + BC^2$ — *Using Pythagoras' theorem on the right-angled triangle ABC.*

$AC^2 = 10^2 + 6^2$

$AC^2 = 136$

AC = 11.661 904 — *Taking square roots on both sides.*

AC = 11.7 *cm* (3 *s.f.*) — *Rounding to an appropriate degree of accuracy.*

(*c*) *Drawing a diagram helps to see the situation clearly.*

$AG^2 = AC^2 + CG^2$ — *Using Pythagoras' theorem on the right-angled triangle ACG.*

$AG^2 = 136 + 5^2$ — *Note that AC^2 = 136 from (b).*

$AG^2 = 161$

AG = 12.688 578 — *Taking square roots on both sides.*

AG = 12.7 *cm* (3 *s.f.*) — *Rounding to an appropriate degree of accuracy.*

T2 *Drawing a diagram helps to see the situation clearly.*

$AF^2 = AE^2 + EF^2$ — *Using Pythagoras' theorem on the right-angled triangle AEF.*

$AF^2 = 15^2 + 5^2$

$AF^2 = 250$

AF = 15.811 388 — *Taking square roots on both sides.*

Area of ADGF = AF × FG — *As the plane is a rectangle.*

= 15.811 388 × 5

= 79.056 942

= 79.1 cm^2 — *Rounding to an appropriate degree of accuracy.*

ANSWERS

A3

HINT

Assume the cube to have sides of length one unit (since any other cube will be similar and therefore give the same value for the required angle).

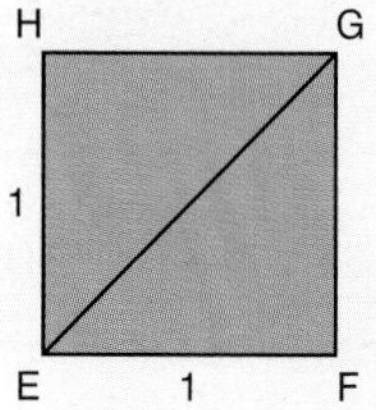

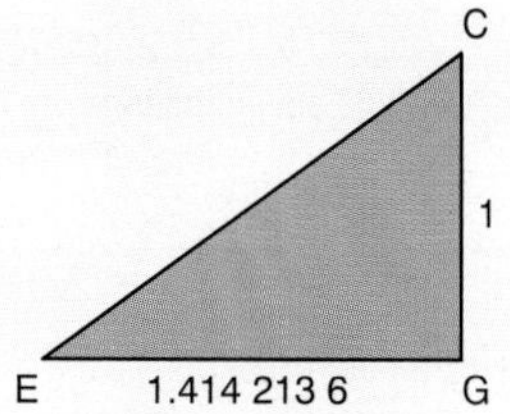

Angle = 35.3° (3 s.f.)

A4

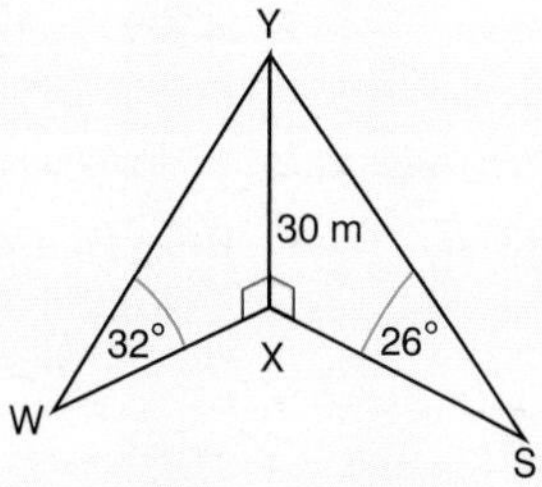

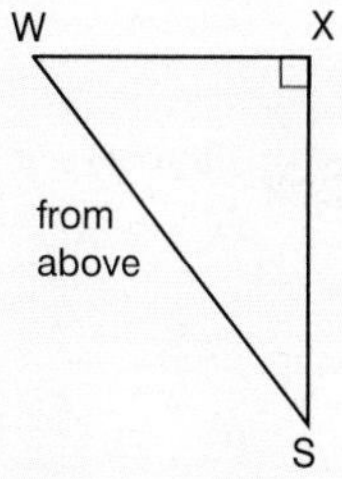

WS = 78.0 m (3 s.f.)

TUTORIAL

T3

Drawing a diagram helps to see the situation clearly.

$EG^2 = EF^2 + FG^2$ *Using Pythagoras' theorem on the right-angled triangle EFG.*

$EG^2 = 1^2 + 1^2$

$EG^2 = 2$

$EG = 1.4142136$ *Taking square roots on both sides.*

$\tan CEG = \dfrac{CG}{EG}$

As $\tan\theta = \dfrac{\text{length of opposite side}}{\text{length of adjacent side}}$.

$\tan CEG = \dfrac{1}{1.4142136}$

$\tan CEG = 0.7071068$

$\angle CEG = 35.264390$ *Using* $\tan^{-1}$ *or arctan to find the angle.*

$\angle CEG = 35.3°$ (3 s.f.) *Rounding to an appropriate degree of accuracy.*

The angle which the line EC makes with the base of the cube = 35.3° (3 s.f.).

T4

To find WS, you need to find WX and XS as follows.

$\tan 32° = \dfrac{YX}{WX}$ *Using tangent on the right-angled triangle WXY.*

$\tan 32° = \dfrac{30}{WX}$ *As the height YX = 30.*

$WX = \dfrac{30}{\tan 32°}$ *Rearranging the equation.*

$WX = 48.010036$

$\tan 26° = \dfrac{YX}{XS}$ *Using tangent on the right-angled triangle XYS.*

$\tan 26° = \dfrac{30}{XS}$ *As the height YX = 30.*

$XS = \dfrac{30}{\tan 26°}$ *Rearranging the equation.*

$XS = 61.509115$

$WS^2 = WX^2 + XS^2$ *Using Pythagoras' theorem on the right-angled triangle WXS.*

$WS^2 = 48.010036^2 + 61.509115^2$

$WS^2 = 6088.3348$

$WS = 78.027782$ *Taking square roots on both sides.*

$WS = 78.0$ *m* (3 *s.f.*) *Rounding to an appropriate degree of accuracy.*

CHAPTER 71

SINE, COSINE AND TANGENT – FOR ANGLES OF ANY SIZE

The sine, cosine and tangent can be found for any angle by using a calculator or else utilising the properties of sine, cosine and tangent curves as shown in the following graphs.

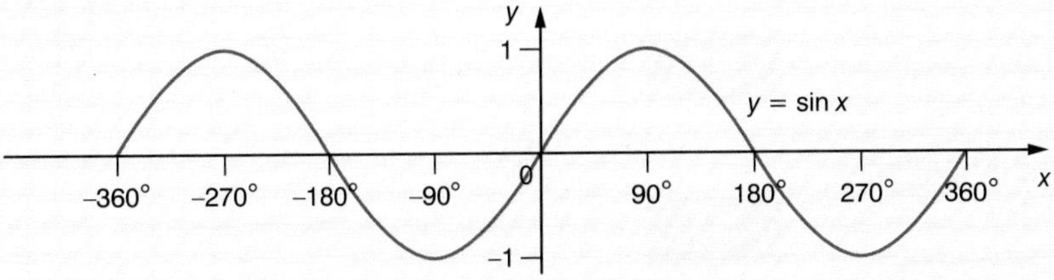

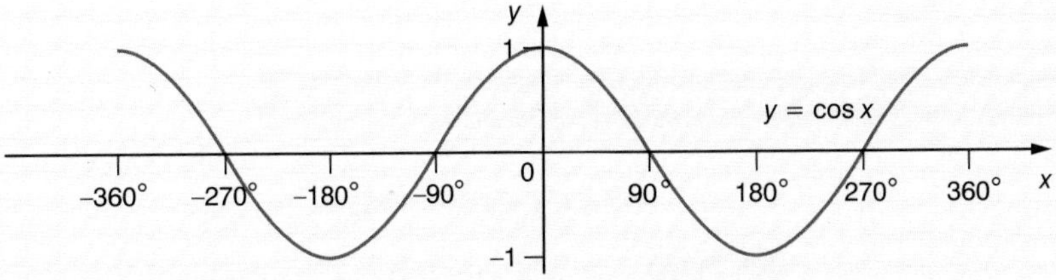

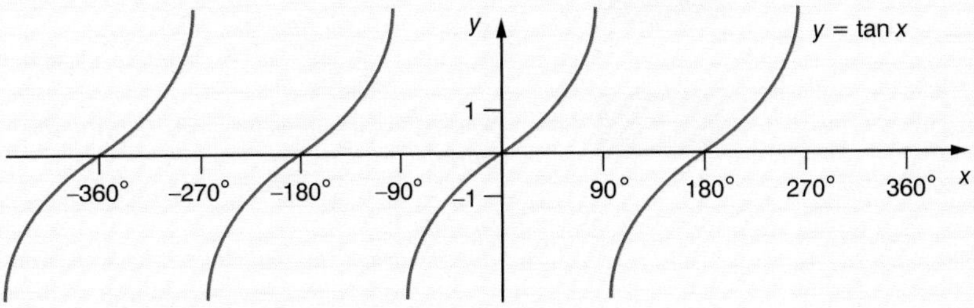

You can use your calculator to find the sine, cosine and tangent of any angle although the reverse process is not so easy as the calculator only gives answers in a specified range.

Worked example

Find the values of θ in the range $^-360° \leqslant \theta \leqslant 360°$ such that $\tan\theta = 1$.

Using the calculator:

$\tan\theta = 1$

$\theta = \tan^{-1} 1$

$\theta = 45°$

However, this is only one of many solutions, as can be seen from the following graph of $\tan\theta$.

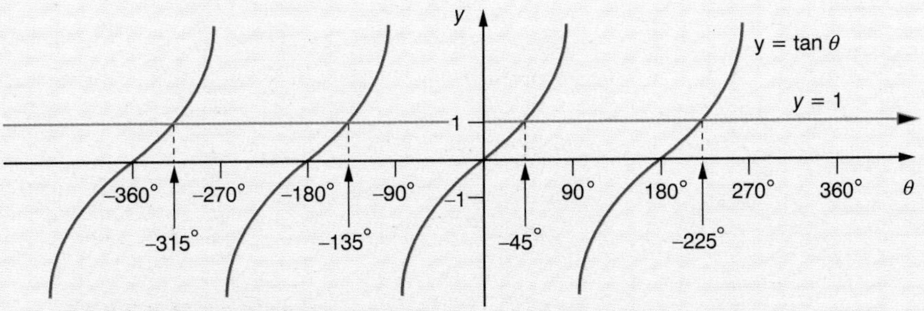

From the graph you can see that the solutions in the range $^-360° \leqslant \theta \leqslant 360°$ are $^-315°$, $^-135°$, 45° and 225°.

HINT

You can check these on your calculator by evaluating tan $^-315°$, tan $^-135°$ and tan 225°.

Check yourself

QUESTIONS

Q1 Find (a) sin 120° (b) cos 240° (c) tan 308°.

Q2 Find all the values of x between 0° and 360° for which:

(a) $\sin x = 0.6$ (b) $\sin x = {}^{-}0.6$
(c) $\cos x = 0.2$ (d) $\tan x = {}^{-}2.5$.

Q3 Solve these equations for ${}^{-}360° \leqslant x \leqslant 360°$.

(a) $\tan x = 2$ (b) $\sin x = 0.525$

Q4 Solve these equations for $0° \leqslant x \leqslant 360°$.

(a) $\sin x \geqslant 0.5$ (b) $\cos x < 0.5$

REMEMBER! Cover the answers if you want to.

ANSWERS

A1
(a) 0.866 (3 s.f.)
(b) ${}^{-}0.5$
(c) ${}^{-}1.28$ (3 s.f.)

A2
(a) $x = 36.9°$ and $143.1°$ (1 d.p.)

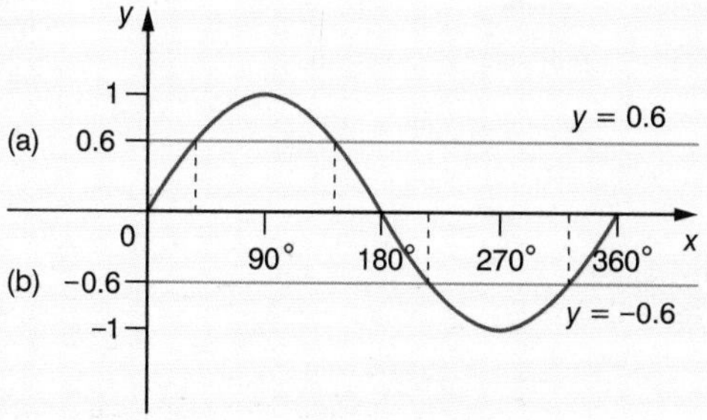

(b) $x = 216.9°$ and $323.1°$ (1 d.p.)

(c) $x = 78.5°$ and $281.5°$ (1 d.p.)

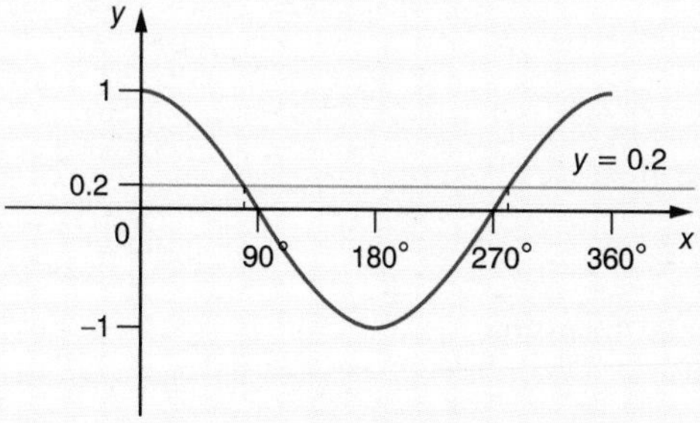

(d) $x = 111.8°$ and $291.8°$ (1 d.p.)

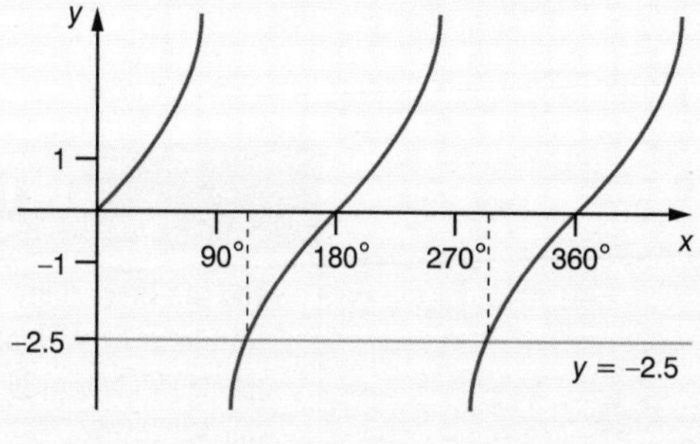

TUTORIAL

T1
(a) $\sin 120° = 0.866\,025\,4$ (from the calculator)
(b) $\cos 240° = {}^{-}0.5$ (from the calculator)
(c) $\tan 308° = {}^{-}1.279\,941\,6$ (from the calculator)

T2
(a) Drawing a diagram helps to see the situation clearly.
$x = \sin^{-1} 0.6$, $x = 36.869\,898$ (from the calculator)
From the graph the solutions in the given range are $x = 36.9°$ and $143.1°$ (1 d.p.).

(b) $x = \sin^{-1}\ {}^{-}0.6$
$x = {}^{-}36.869\,898°$ (from the calculator)
From the graph the solutions in the given range are $216.9°$ and $323.1°$ (1 d.p.).

(c) Drawing a diagram helps to see the situation clearly.
$x = \cos^{-1} 0.2$
$x = 78.463\,041°$ (from the calculator)
From the graph, the solutions in the given range are $78.5°$ and $281.5°$ (1 d.p.).

(d) Drawing a diagram helps to see the situation clearly.
$x = \tan^{-1}\ {}^{-}2.5$
$x = {}^{-}68.198\,591°$ (from the calculator)
From the graph, the solutions in the given range are $111.8°$ and $291.8°$ (1 d.p.).

ANSWERS

A3 (a) $x = {}^-296.6^\circ, {}^-116.6^\circ, 63.4^\circ$ and 243.4° (1 d.p.)

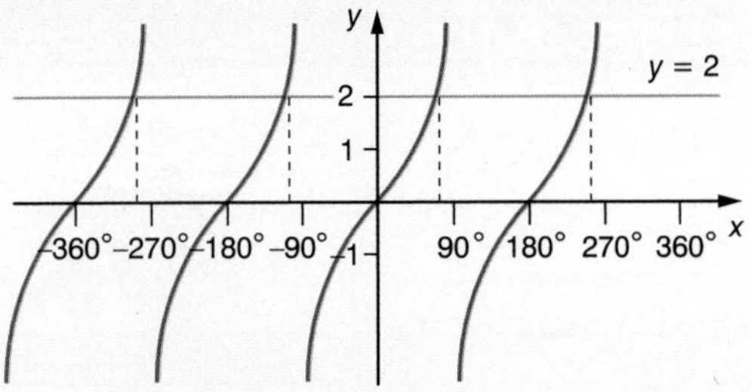

(b) $x = {}^-328.3^\circ, {}^-211.7^\circ, 31.7^\circ$ and 148.3° (1 d.p.)

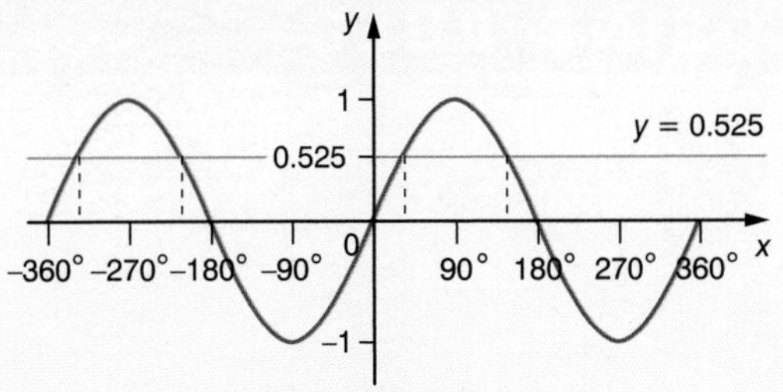

A4 (a) $30^\circ \leqslant x \leqslant 150^\circ$

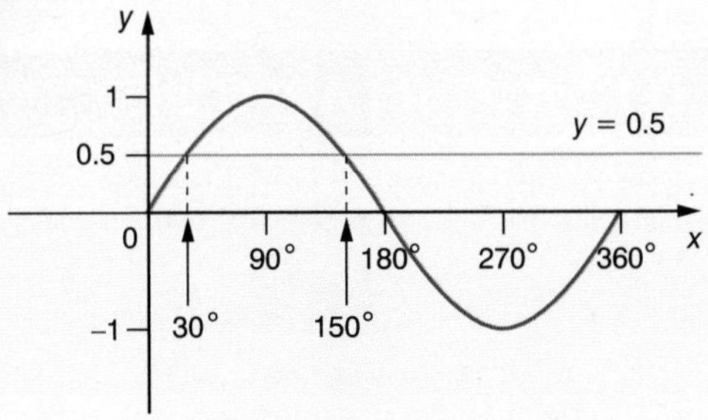

(b) $60^\circ < x < 300^\circ$

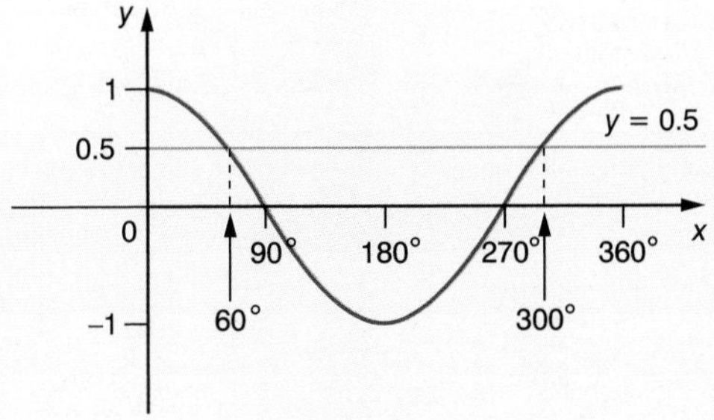

TUTORIAL

T3 (*a*) *Drawing a diagram helps to see the situation clearly.*
Using the calculator:
$\tan x = 2$
$x = \tan^{-1} 2$
$x = 63.434\,949^\circ$
From the graph, the solutions in the given range are ${}^-296.6^\circ, {}^-116.6^\circ, 63.4^\circ$ *and* 243.4° *(1 d.p.).*

(*b*) *Drawing a diagram helps to see the situation clearly.*
Using the calculator:
$\sin x = 0.525$
$x = \sin^{-1} 0.525$
$x = 31.668\,243^\circ$
From the graph, the solutions in the given range are ${}^-328.3^\circ, {}^-211.7^\circ, 31.7^\circ$ *and* 148.3° *(1 d.p.).*

T4 (*a*) *Drawing a diagram helps to see the situation clearly.*
$\sin x \geqslant 0.5$
From the graph, x lies in the range $30^\circ \leqslant x \leqslant 150^\circ$.

(*b*) *Drawing a diagram helps to see the situation clearly.*
$\cos x < 0.5$
From the graph, x lies in the range $60^\circ < x < 300^\circ$.

CHAPTER 72

SINE AND COSINE RULES

Right-angled triangles can be solved using sine, cosine and tangent as introduced in Chapter 66, Sine, cosine and tangent – right-angled triangles. The sine and cosine rules are used to solve triangles which are not right-angled.

SINE RULE

The sine rule states that:

$$\frac{a}{\sin A} = \frac{b}{\sin B} = \frac{c}{\sin C}$$

Sometimes it is useful to use the alternative form:

$$\frac{\sin A}{a} = \frac{\sin B}{b} = \frac{\sin C}{c}$$

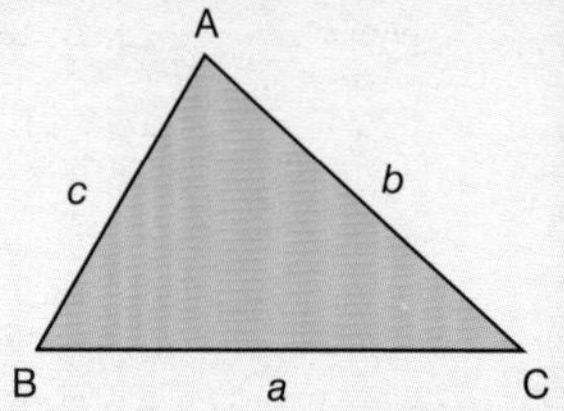

COSINE RULE

The cosine rule states that:

$$a^2 = b^2 + c^2 - 2bc\cos A$$

$$\text{or } \cos A = \frac{b^2 + c^2 - a^2}{2bc}$$

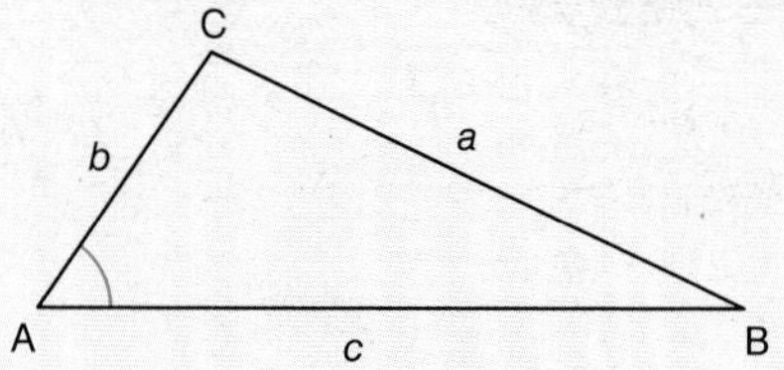

NOTE

The sine rule should be used for situations involving two sides and two angles.

The cosine rule should be used for situations involving three sides and one angle.

AREA OF A TRIANGLE

NOTE

The angle C is the 'included' angle between sides a and b.

The sine and cosine rules can be extended to find the area of a triangle so that:

area of a triangle = $\frac{1}{2}ab\sin C$

Worked example

Find the angle A in the triangle ABC where AB is 11 cm, BC is 17.5 and angle C is 36°.

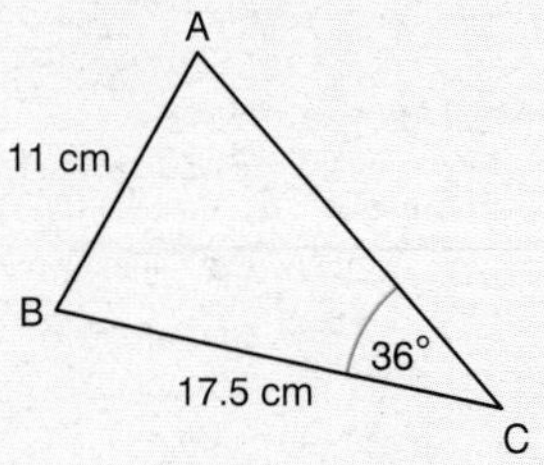

Since the information involves two sides and two angles then it is appropriate to use the sine rule.

$$\frac{a}{\sin A} = \frac{b}{\sin B}$$

$$\frac{\sin A}{a} = \frac{\sin B}{b}$$ Reciprocating both sides.

$$\frac{\sin A}{17.5} = \frac{\sin 36°}{11}$$

$$\sin A = 17.5 \times \frac{\sin 36°}{11}$$ Multiplying both sides by 17.5.

$$\sin A = 0.935\,112\,9$$

$$A = 69.246\,393°$$ Using the inverse button ($\sin^{-1}$ or arcsin) on your calculator.

Unfortunately the value of angle $A = 69.246\,392°$ is not unique since there is another possible value for angle A which satisfies $\sin A = 0.935\,112\,9$.

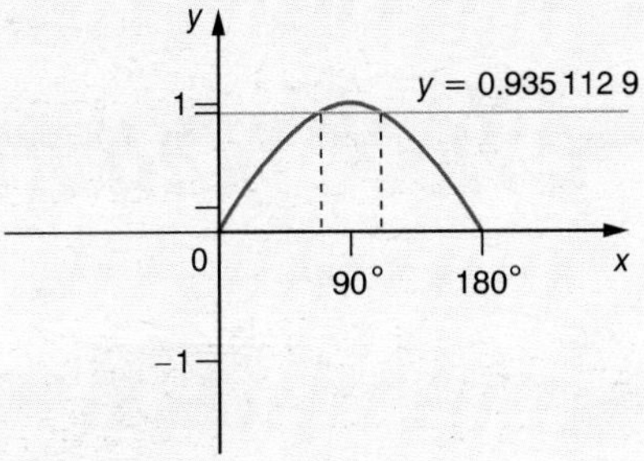

From the graph you can see that another possible value of A is $110.753\,61°$.

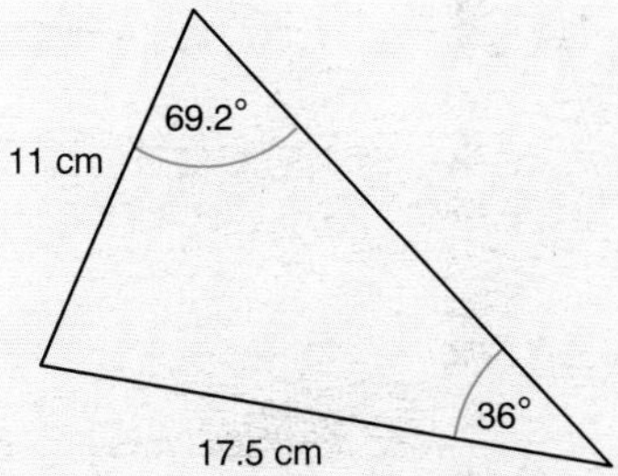

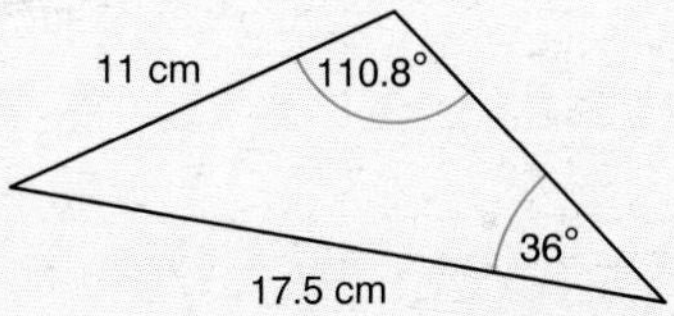

Angle $A = 69.2°$ or $110.8°$ (1 d.p.)

NOTE

Whenever you use the sine rule you need to be aware of the possibility that two solutions exist. This problem does not arise when using the cosine rule and can be avoided when using the sine rule by finding the smallest angles first (if possible).

Worked example

Find the angle P and the area of the triangle.

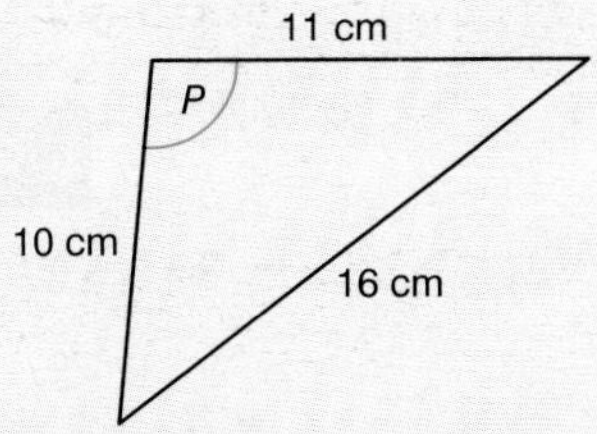

In this case, since the information involves three sides and one angle, it is appropriate to use the cosine rule.

$$\cos A = \frac{b^2 + c^2 - a^2}{2bc}$$

$$\cos P = \frac{10^2 + 11^2 - 16^2}{2 \times 10 \times 11}$$ Substituting the given lengths.

$$\cos P = \frac{100 + 121 - 256}{220}$$

$$\cos P = \frac{^{-}35}{220}$$

$\cos P = {}^{-}0.159\,090\,9$ The negative value shows that the angle P is obtuse.

$P = 99.154\,133°$ Using the inverse button ($\cos^{-1}$ or arccos) on the calculator.

$P = 99.2°$ (1 d.p.)

To find the area of the triangle use:

Area of a triangle $= \frac{1}{2}ab\sin C$

$= \frac{1}{2} \times 10 \times 11 \times \sin 99.154\,133°$ Remembering to use the most accurate value of P.

$= 54.299\,517$

$= 54.3\text{ cm}^2$ (3 s.f.) Rounding to an appropriate degree of accuracy.

Check yourself

QUESTIONS

Q1 Calculate the missing angles and sides in the following triangles.

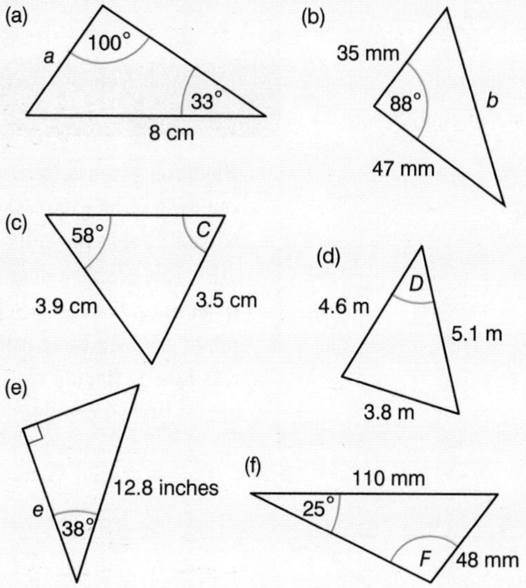

Q2 Calculate the areas of the given shapes.

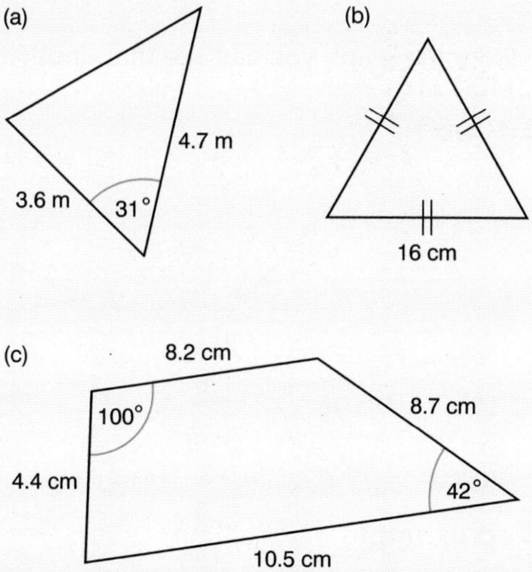

Q3 A ship leaves port and sails on a bearing of 035°. After 11 km the ship changes direction and sets sail on a bearing of 067°. The ship sails a distance of 16.5 km on this bearing. What is the distance and bearing of the ship from the port?

REMEMBER! Cover the answers if you want to.

ANSWERS

A1

(a) $a = 4.42$ cm (3 s.f.)
(b) $b = 57.6$ mm (3 s.f.)
(c) C = 70.9° (1 d.p.)

TUTORIAL

T1

(*a*) *Using the sine rule:*

$$\frac{a}{\sin 33°} = \frac{8}{\sin 100°}$$ *Substituting the given values.*

$a = 4.4243278$

(*b*) *Using the cosine rule:*
$b^2 = 47^2 + 35^2 - (2 \times 47 \times 35 \times \cos 88°)$
Substituting the given lengths and taking care with the $-2bc\cos A$ *term.*
$b = 57.612331$

(*c*) *Using the sine rule and both sides.*

$$\frac{a}{\sin A} = \frac{b}{\sin B} \text{ or } \frac{\sin A}{a} = \frac{\sin B}{b}$$

$$\frac{\sin C}{3.9} = \frac{\sin 58°}{3.5}$$ *Substituting the given values.*

$$\sin C = 3.9 \times \frac{\sin 58°}{3.5}$$ *Multiplying both sides by 3.9.*

C = 70.903321°

ANSWERS

A1

(d) $D = 45.8°$ (1 d.p.)
(e) $e = 10.1$ inches (3 s.f.)
(f) $F = 75.6°$ or $104.4°$ (1 d.p.)

A2

(a) Area $= 4.36\,\text{m}^2$ (3 s.f.)
(b) Area $= 111\,\text{cm}^2$ (3 s.f.)
(c)

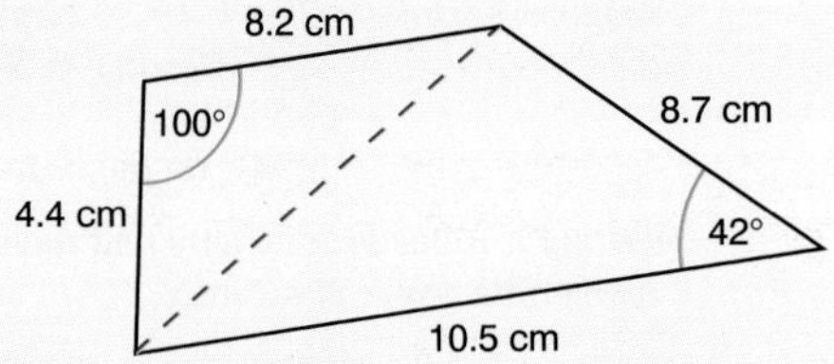

Area $= 48.3\,\text{cm}^2$

TUTORIAL

T1

(d) *Using the cosine rule:*

$$\cos D = \frac{4.6^2 + 5.1^2 - 3.8^2}{2 \times 4.6 \times 5.1}$$ *Substituting the given lengths.*

$D = 45.767\,605°$

(e) *Using* $\cos\theta = \frac{\text{length of adjacent side}}{\text{length of hypotenuse}}$:

As the triangle is right-angled.

$\cos 38° = \frac{e}{12.8}$ *Substituting the given lengths.*

$e = 12.8 \times \cos 38°$ *Multiplying both sides by 12.8.*

$e = 10.086\,538$

(f) *Using the sine rule and reciprocating both sides:*

$\frac{\sin F}{110} = \frac{\sin 25°}{48}$ *Substituting the given values.*

$\sin F = 110 \times \frac{\sin 25°}{48}$ *Multiplying both sides by 110.*

$\sin F = 0.968\,500\,1$

$F = 75.580\,895°$ *or* $104.4191°$
As sin F has two possible solutions here.

T2

(a) *Area of a triangle* $= \frac{1}{2}ab\sin C$

$= \frac{1}{2} \times 3.6 \times 4.7 \times \sin 31° = 4.357\,222\,1$

(b) *Area of a triangle* $= \frac{1}{2}ab\sin C$

$= \frac{1}{2} \times 16 \times 16 \times \sin 60°$
As the angles of an equilateral triangle are all 60°.
$= 110.851\,25$

(c) *Drawing a diagram helps to see the situation clearly.*

Split the quadrilateral into two triangles.

Area of each triangle $= \frac{1}{2}ab\sin C$

Area of quadrilateral
$= \frac{1}{2} \times 8.2 \times 4.4 \times \sin 100°$
$+ \frac{1}{2} \times 10.5 \times 8.7 \times \sin 42°$
$= 48.328\,472$

As the area of the quadrilateral equals the sum of the areas of the two triangles.

ANSWERS

A3

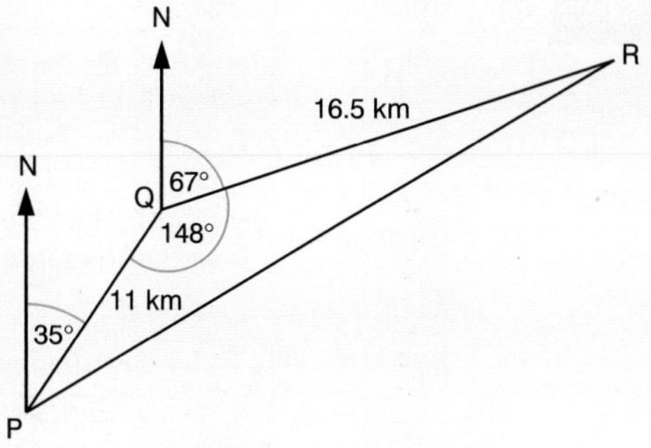

Distance = 26.5 km (3 s.f.)
Bearing = 054° (to the nearest whole degree)

TUTORIAL

T3

By drawing a diagram of the situation you can see that PR *represents the required distance and that the bearing can be found from the angle* QPR.

The bearing of P *from* Q *is* 215° *and the angle* PQR = 148°, *as* 215° – 67° = 148°.

Using the cosine rule on the triangle PQR:

$a^2 = b^2 + c^2 - 2bc\cos A$

$PR^2 = 11^2 + 16.5^2 - (2 \times 11 \times 16.5 \times \cos 148°)$
Substituting the given lengths and taking care with the $-2bc\cos A$ *term.*

PR = 26.478 132

So the distance = 26.5 *km* (3 *s.f.*).

To find angle QPR, *using the sine rule in the form*

$\frac{\sin A}{a} = \frac{\sin B}{b}$:

$\frac{\sin QPR}{QR} = \frac{\sin 148°}{PR}$

$\frac{\sin QPR}{16.5} = \frac{\sin 148°}{26.478132}$ *Substituting the given lengths and* ∠PQR.

$\sin QPR = 16.5 \times \frac{\sin 148°}{26.478\,132}$ *Multiplying both sides by* 16.5.

sin QPR = 0.330 222 23

∠QPR = 19.282 265°

Required bearing = 35° + 19.282 265°
Bearing = 35° + *angle* QPR
= 54.282 265°
= 054° (*to the nearest whole degree*)

Writing it in the bearing form and rounding to an appropriate degree of accuracy.

EXAMINATION QUESTIONS 14

EXAM PRACTICE

Sample Student's Answers & Examiner's Comments

1 The diagram represents a metal cone of height 90 cm made in two parts, labelled T and S. The top part, T, has a height of 45 cm.

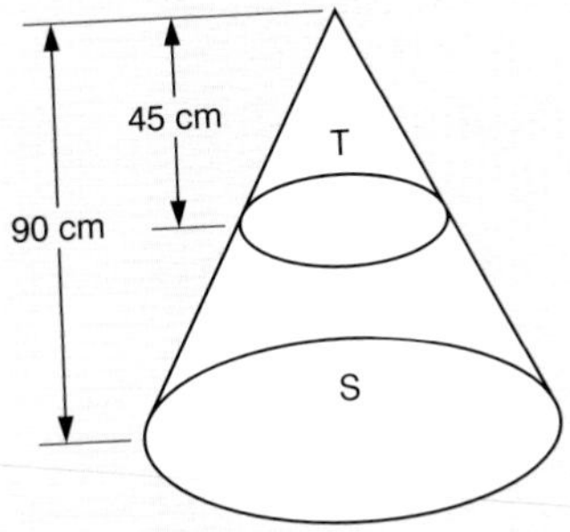

The volume of S and T together is $10\,500\pi$ cm^3.

(a) Calculate the volume of S.

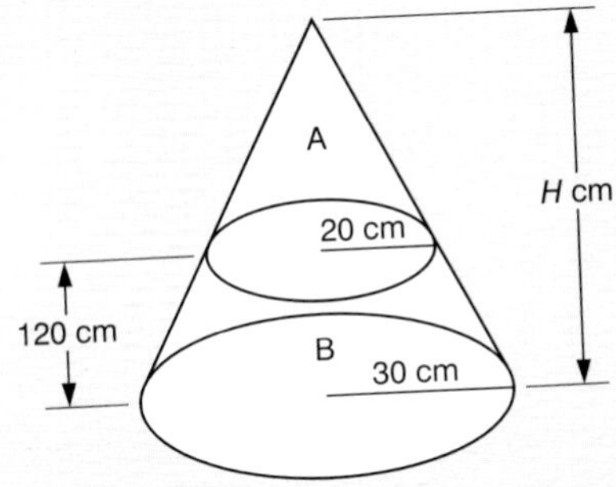

A second cone is also made up of two parts A and B. The radius of the top part is 20 cm and the radius of the base of B is 30 cm. The height of the bottom part B is 120 cm.

(b) Calculate the height, H cm, of the cone.

(London specimen paper 1998)

EXAMINER'S COMMENTS

1 *The candidate has pursued the information given to find the base radius r but has soon realised that this enquiry is not going anywhere. He has abandoned this line of investigation, appreciating that the task is about ratios of similar shapes.*

(a) Volume of cone $= \frac{1}{3}\pi r^2 h = 10\,500\pi$

$\frac{1}{3} \times \pi \times r^2 \times 90 = 10\,500\pi$

$\pi \times r^2 \times 90 = 3 \times 10\,500\pi$

$r^2 \times 90 = 3 \times 10\,500$

$r^2 = 3 \times \frac{10\,500}{90}$

$r^2 = 350$

$r = 18.708\,287$

a)
The candidate has correctly identified corresponding heights and cancelled this down to form a more manageable ratio. The ratio of corresponding volumes is correctly calculated.

(a) Ratio of corresponding heights of top cone to cone

$T = 45 : 90 = 1 : 2$

Ratio of corresponding volumes of top cone to whole cone

$T = 1^3 : 2^3$

$= 1 : 8$

The volume of S is correctly identified and he has given the answer in terms of π as given in the question. An answer of 28 863.383 giving 28 900 cm³ (3 s.f.) is also acceptable.

Volume of $T = \frac{1}{8} \times$ volume of whole cone

Volume of $S = \frac{7}{8} \times$ volume of whole cone

$= \frac{7}{8} \times 10\,500\pi$

$= 9187.5\ \pi\ \text{cm}^3$

b)
The candidate has correctly identified corresponding ratios and cancelled down.

(b) Ratio of corresponding lengths of top cone to whole cone

$A = 20 : 30$

$= 2 : 3$

Ratio of corresponding heights of top cone to whole cone

$A = H - 120 : H$

So $H - 120 : H = 2 : 3$

He has equated the two ratios and rewritten them as a fraction which is manipulated to form an equation which can be solved to find H.

or $\frac{H - 120}{H} = \frac{2}{3}$

$3(H - 120) = 2H$

$3H - 360 = 2H$

$H = 360\text{ cm}$

EXAMINER'S COMMENTS

2 The 'folly' tower in the grounds of Poldark Castle is in the shape of a tetrahedron VPQR standing on a prism FGHPQR.

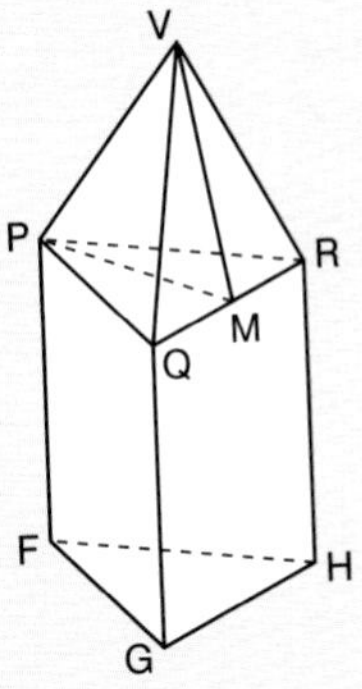

The cross-section PQR is an equilateral triangle of side 9.0 m.

VP = VQ = VR = 20.5 m

PF = QG = RH = 28.0 m

M is the midpoint of QR.

(a) (i) Use triangle PQR to find the length of PM.
(ii) Use triangle VQR to find the length of VM.

(b) Find the size of angle VPM.

(c) Find the height of V above the base FGH.
Give your answer to an appropriate degree of accuracy.

(MEG syllabus B, specimen paper 1998)

(a) (i)

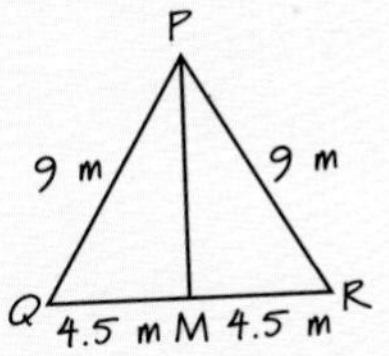

$PQ^2 = QM^2 + PM^2$

$9^2 = 4.5^2 + PM^2$

$PM^2 = 81 - 20.25$

$PM^2 = 60.75$

$PM = 7.794\,228\,6$ m

(ii)

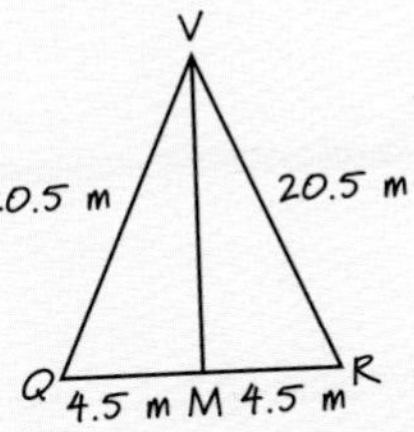

$VQ^2 = QM^2 + VM^2$

$20.5^2 = 4.5^2 + VM^2$

$420.25 = 20.25 + VM^2$

$VM^2 = 420.25 - 20.25$

$VM^2 = 400$

$VM = 20$ m

2 a)
The use of individual diagrams helps to break down the task and allows the candidate to see more clearly what she is doing.

The candidate correctly identifies the information to find PM although it would be better to round this answer to an appropriate degree of accuracy such as 7.79 m (3 s.f.) or 7.8 m (2 s.f.).

The candidate has correctly identified the information to find VM.

EXAMINER'S COMMENTS

2 b)

The candidate has avoided the mistake of assuming that the triangle is right-angled and has used the cosine rule formula provided in the examination.

(b)

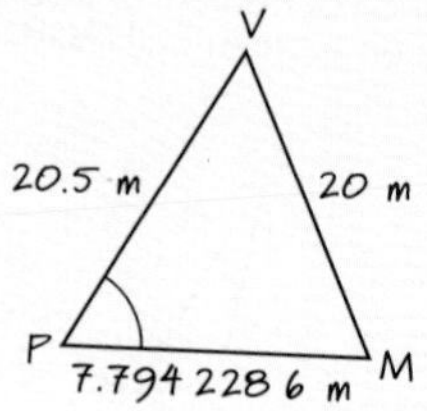

$$\cos A = \frac{b^2 + c^2 - a^2}{2bc}$$

$$\cos P = \frac{20.5^2 + 7.794\,228\,6^2 - 20^2}{2 \times 20.5 \times 7.794\,228\,6}$$

$$\cos P = \frac{80.999\,999}{319.563\,37}$$

$$\cos P = 0.253\,470\,8$$

$$P = 75.317\,005°$$

$$\angle VPM = 75.3° \text{ (3 s.f.)}$$

The candidate has completed the question by considering the accuracy given in the question, and has rounded off to 3 s.f.

The use of the diagram allows her to see the problem more clearly and to identify the necessary information required to solve the problem.

She remembers to use the original value for P rather than the rounded value which might be penalised for premature approximation.

(c)

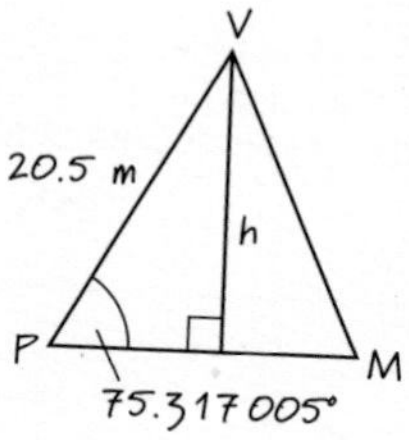

$$\sin P = \frac{h}{20.5}$$

$$\sin 75.317\,005° = \frac{h}{20.5}$$

$$h = 20.5 \times \sin 75.317\,005°$$

$$h = 19.830\,532$$

$$h = 19.8\text{m (3 s.f.)}$$

She has rounded off the answer appropriately but has forgotten to add the height of the prism which adds a further 28 m to the overall height. She should have reread the question to check that the answer had been completed properly. An answer of 47.8 m or 48 m would gain full marks.

Questions to Answer

1 A metal cylinder of radius 15 cm and height 22 cm is melted down and made into spheres of radius 2 cm. How many such spheres can be made?

2 A spinning top which consists of a cone of base radius 5 cm, height 9 cm and a hemisphere of radius 5 cm is illustrated here.

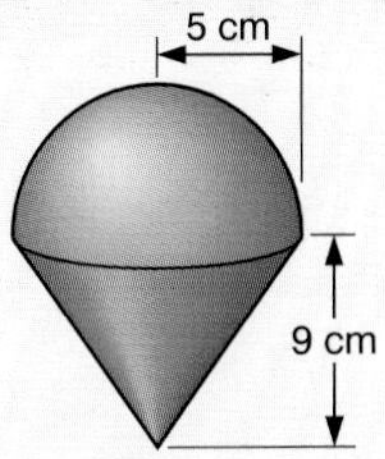

(a) Calculate the volume of the spinning top.

(b) Calculate the total surface area of the spinning top.

(SEG specimen paper 1998)

3 A symmetrically-shaped timer is made from hollow hemispheres, cylinders and cones joined together as shown in the diagram. It contains sand just sufficient to fill the top cone and cylinder sections. Calculate the volume of sand.

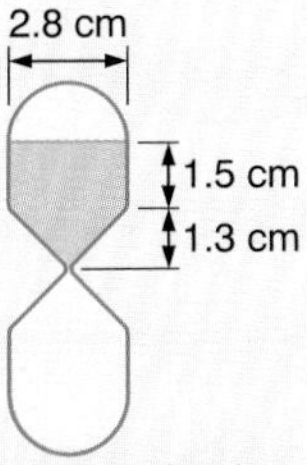

When all the sand has run through, it collects as shown in the diagram. Calculate the height, h cm, of sand in the cylindrical part of the timer.

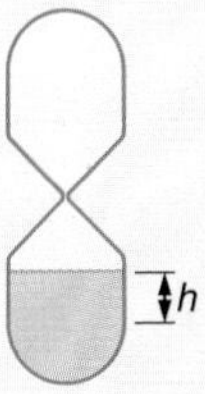

(MEG syllabus A, specimen paper 1998)

4 A square-based pyramid is shown. The horizontal base has side length 3 centimetres and the vertex P is 7 centimetres vertically above the centre of the base.

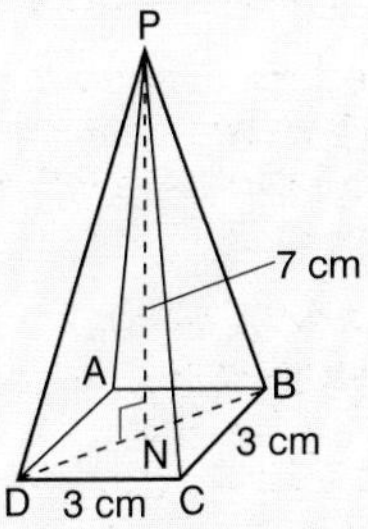

(a) Calculate the angle edge CP makes with the base.

(b) Calculate the angle face APD makes with the base.

(c) Calculate the area of one of the triangular faces of the pyramid.

(SEG modular syllabus, specimen paper 1998)

5 Sketch the graphs of $y = \sin 2\theta$ and $y = \tan\theta$ to find solutions of $\sin 2\theta = \tan\theta$ in the range $0 \leqslant \theta \leqslant 180°$.

6 A helicopter leaves a heliport H and its measuring instruments show that it flies 3.2 km on a bearing of 128° to a checkpoint C. It then flies 4.7 km on a bearing of 066° to its base B.

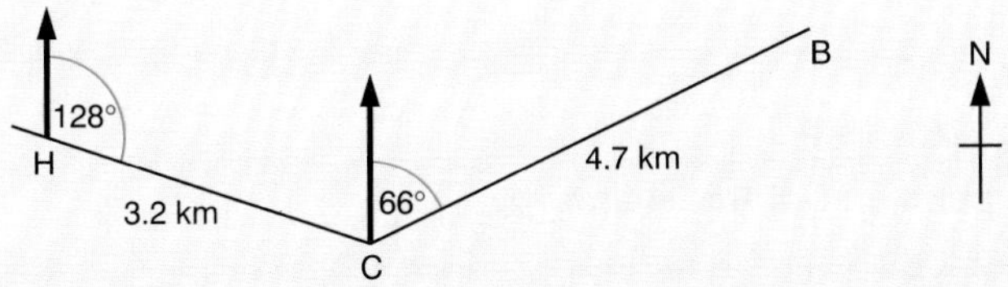

(a) Show that angle HCB is 118°.

(b) Calculate the direct distance from the heliport H to the base B.

(NEAB specimen paper 1998)

SHAPE, SPACE AND MEASURES

CHAPTER 74

ARC, SECTOR AND SEGMENT

It will be useful if you know the definitions in this section.

You need to know the definitions of parts of the circle. They are given in the diagrams below.

PARTS OF A CIRCLE

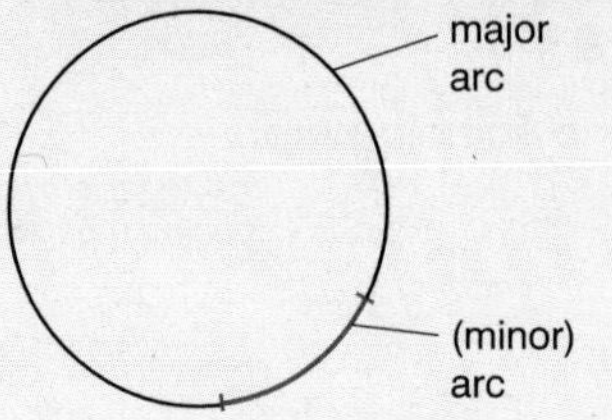

Arcs are part of the circumference of the circle.

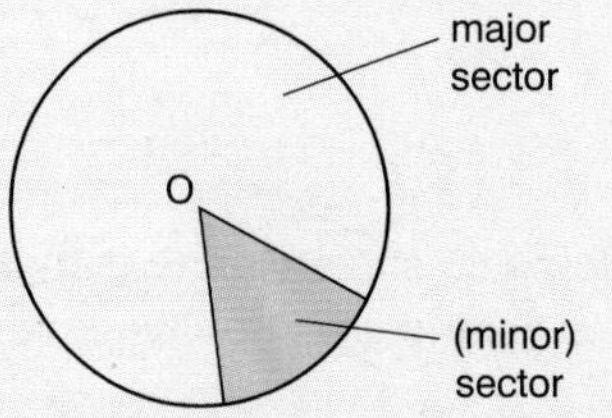

Sectors are parts of circles formed when two radii are drawn in.

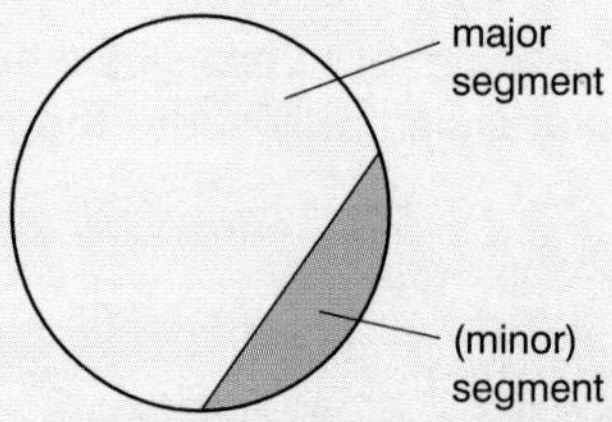

Segments are formed when chords divide the circle into different parts.

ARC

An arc is a part of the circumference of a circle.

$$\text{Arc length} = \frac{\text{angle subtended at centre}}{360} \times 2\pi r$$

$$\text{Arc length} = \frac{\theta}{360} \times 2\pi r$$

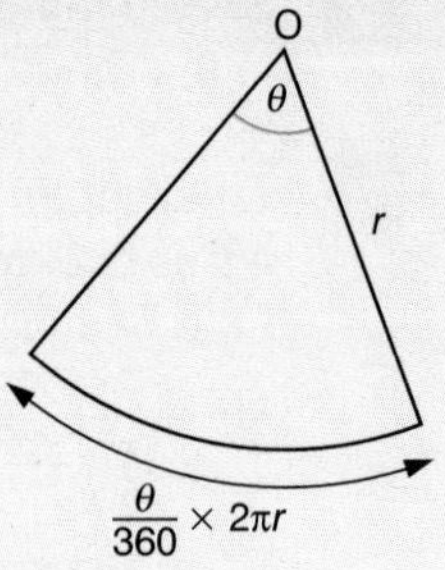

Worked example

Find the perimeter of this shape.

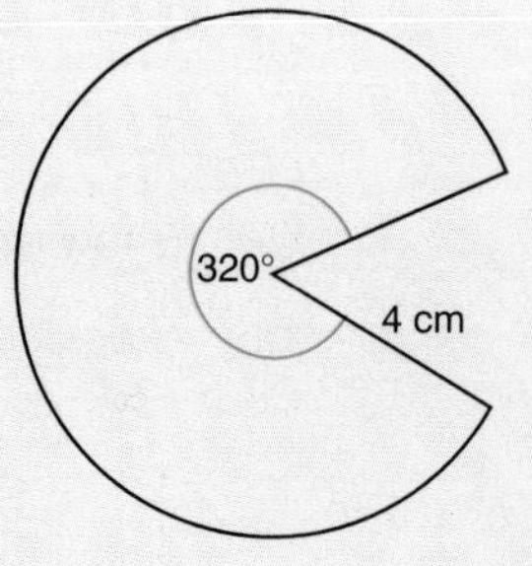

Perimeter = arc + radius + radius

$= \frac{320}{360} \times 2\pi r + r + r$

$= \frac{320}{360} \times 2 \times \pi \times 4 + 4 + 4$

$= 30.340\,214$

$= 30.3\,\text{cm (3 s.f.)}$

SECTOR

A sector is the area enclosed between an arc and two radii.

Sector area $= \frac{\text{angle subtended at centre}}{360} \times \pi r^2$

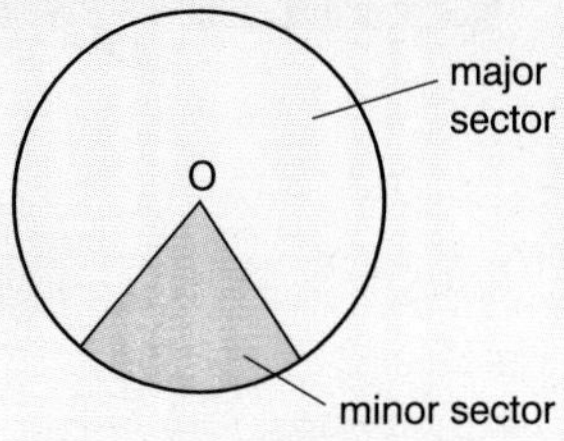

Sector area $= \frac{\theta}{360} \times \pi r^2$

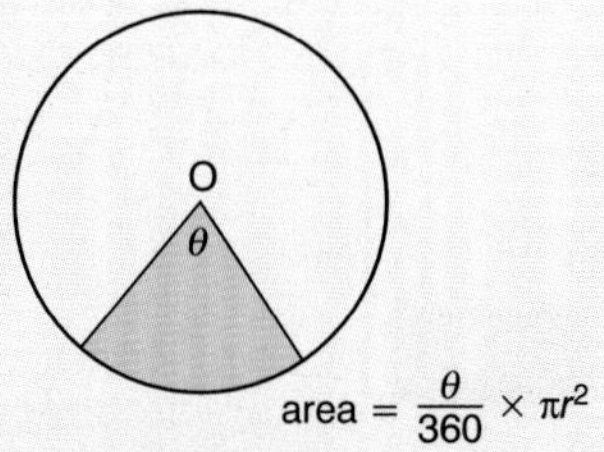

SEGMENT

A segment is the area enclosed between an arc and a chord.

The following example illustrates how to find the area of a segment.

Worked example

Find the area of the segment shaded in this circle of radius 5 cm.

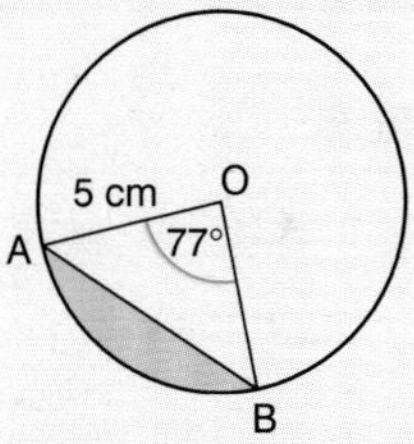

Area of segment = area of sector AOB – area of triangle AOB

Area of sector AOB $= \frac{77}{360} \times \pi r^2$

$= \frac{77}{360} \times \pi \times 5^2$

$= 16.798\,794$

Area of triangle AOB $= \frac{1}{2}ab\sin\theta$

$= \frac{1}{2} \times 5 \times 5 \times \sin 77°$ Where a and b are equal to the radius of the circle.

$= 12.179\,626$

Area of segment = area of sector AOB – area of triangle AOB

$= 16.798\,794 - 12.179\,626$

$= 4.619\,168$

$= 4.62\text{ cm}^2$ (3 s.f.)

Check yourself

QUESTIONS

Q1 Find the arc length and sector area for each of the following.

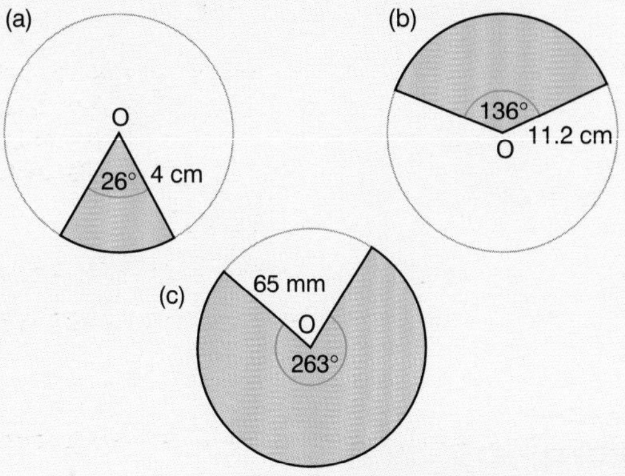

Q2 Calculate the segment area in this diagram.

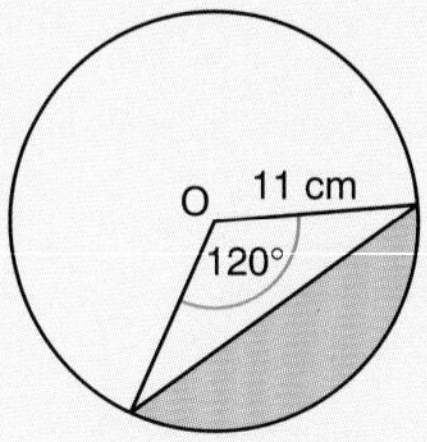

Q3 The arc length between two points A and B on the circumference of a circle is 8.6 cm. What is the angle subtended at the centre of the circle if the radius of the circle is 12 cm?

Q4 The sector of a circle is folded to make a cone of slant height 16 cm and base radius 12 cm. Calculate the arc length and the angle of the original sector.

REMEMBER! Cover the answers if you want to.

ANSWERS

A1

	Arc length	Sector area
(a)	1.82 cm	3.63 cm^2
(b)	26.6 cm	149 cm^2
(c)	298 mm	9700 mm^2

All to 3 s.f.

TUTORIAL

T1

(a) *Arc length* $= \frac{26}{360} \times 2 \times \pi \times 4 = 1.815\,142\,4$

Sector area $= \frac{26}{360} \times \pi \times 4^2 = 3.630\,284\,8$

(b) *Arc length* $= \frac{136}{360} \times 2 \times \pi \times 11.2$

$= 26.584\,855$

Sector area $= \frac{136}{360} \times \pi \times 11.2^2$

$= 148.875\,19$

(c) *Arc length* $= \frac{263}{360} \times 2 \times \pi \times 65$

$= 298.364\,04$

Sector area $= \frac{263}{360} \times \pi \times 65^2$

$= 9696.8312$

ANSWERS

A2 Area $= 74.3\,\text{cm}^2$

A3

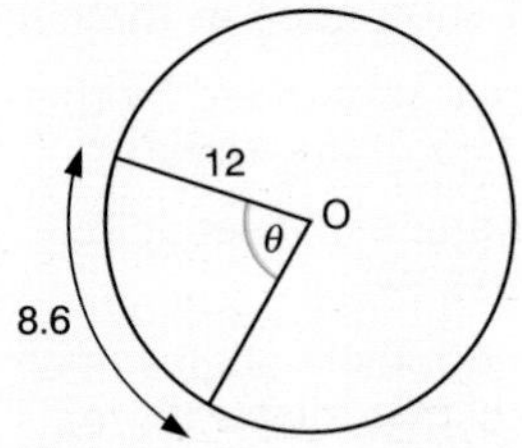

Angle $= 41.1°$ (1 d.p.)

A4

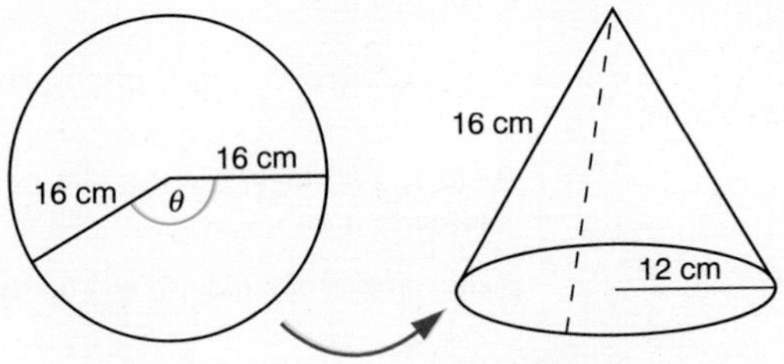

Arc length $= 24\pi\,\text{cm}^2$ or $75.4\,\text{cm}^2$ (3 s.f.), angle $= 270°$

> **HINT**
>
> Writing the circumference as 24π allows you to use this 'exact' value in the second part of the work.

TUTORIAL

T2 *Area of segment = area of sector – area of triangle*

Area of sector $= \frac{120}{360} \times \pi r^2 = \frac{120}{360} \times \pi \times 11^2$

$= 126.7109$

Area of triangle $= \frac{1}{2}ab\sin\theta$

$= \frac{1}{2} \times 11 \times 11 \times \sin 120°$ *Where a and b are equal to the radius of the circle.*

$= 52.394\,537$

Area of segment = area of sector AOB *– area of* ΔAOB

$= 126.7109 - 52.394\,537$

$= 74.316\,363$

T3 *Drawing a diagram helps to see the situation clearly.*

Arc length $= \frac{\theta}{360} \times 2\pi r$

$8.6 = \frac{\theta}{360} \times 2 \times \pi \times 12$

$\theta = 41.061\,975°$

Angle subtended by arc $= 41.1°$ (3 *s.f.*)

T4 *Drawing a diagram helps to see the situation clearly. From the diagram you can see that:*

(*i*) *the radius of the circle is the same as the slant height of the cone (given as 16 cm)*

(*ii*) *the circumference of the base of the cone is the same as the arc cut out from the circle.*

Circumference of the base of the cone

$= 2\pi r$

$= 2 \times \pi \times 12$

$= 24\pi$ *or* $75.398\,224$

Arc length $= \frac{\theta}{360} \times 2\pi r$

$24\pi = \frac{\theta}{360} \times 2 \times \pi \times 16$ *As arc length = circumference of base of cone.*

$\theta = 270°$ *Cancelling* π *on both sides and working out.*

SHAPE, SPACE AND MEASURES

CHAPTER 75

ANGLE AND TANGENT PROPERTIES OF CIRCLES

This diagram shows the main parts of a circle and the names given to them.

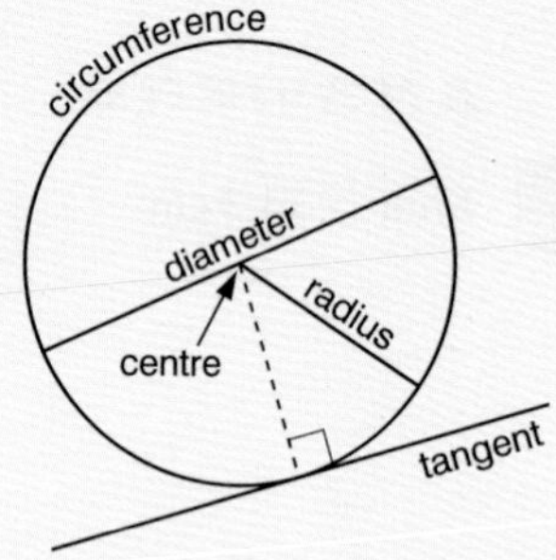

ANGLES IN CIRCLES

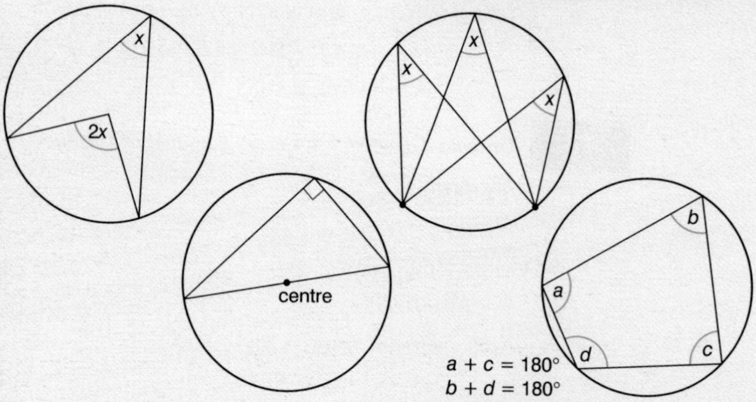

You need to know the following angle properties.

- The angle subtended by an arc (or chord) at the centre is twice that subtended at the circumference.
- Angles subtended by the same arc (or chord) are equal.
- The angle in a semicircle is always 90°.
- The opposite angles of a cyclic quadrilateral are **supplementary** (they add up to 180°).

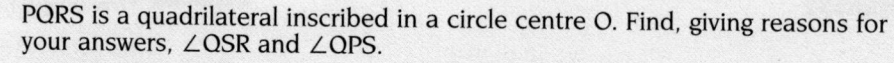

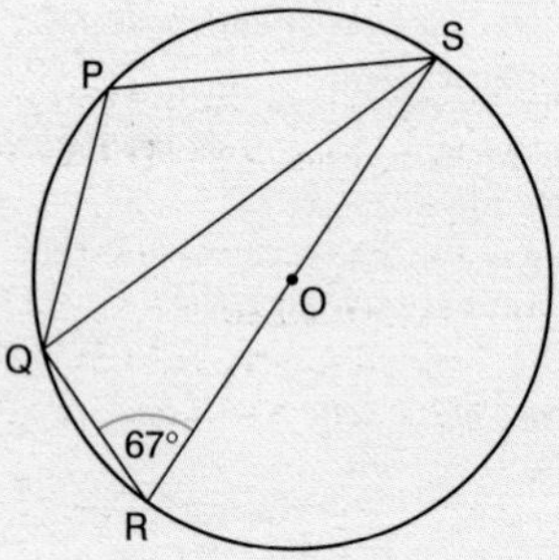

Worked example

PQRS is a quadrilateral inscribed in a circle centre O. Find, giving reasons for your answers, ∠QSR and ∠QPS.

∠SQR = 90°	As RS is a diameter and the angle in a semicircle is 90°.
∠QSR = 180° – (67° + 90°)	As the angles of a triangle add up to 180°.
= 23°	
∠QPS = 180° – 67°	As the opposite angles of a cyclic quadrilateral PQRS add up to 180°.
= 113°	

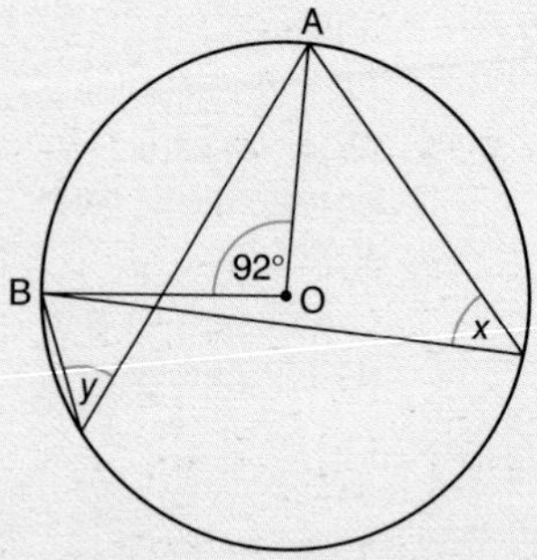

Worked example

The angle subtended by the arc AB at the centre of the circle is 92°. Find the values of x and y.

$x = 46°$	As the angle subtended by the arc AB at the centre is twice that subtended at the circumference.
$y = 46°$	As the angle subtended by the arc AB at the centre is twice that subtended at the circumference.
or else ...	As the angles subtended by the arc AB at the circumference are equal and $x = 46°$, so $y = 46°$.

Check yourself

QUESTIONS

Q1 For the following circles, where O marks the centre of the circle, find the missing angles.

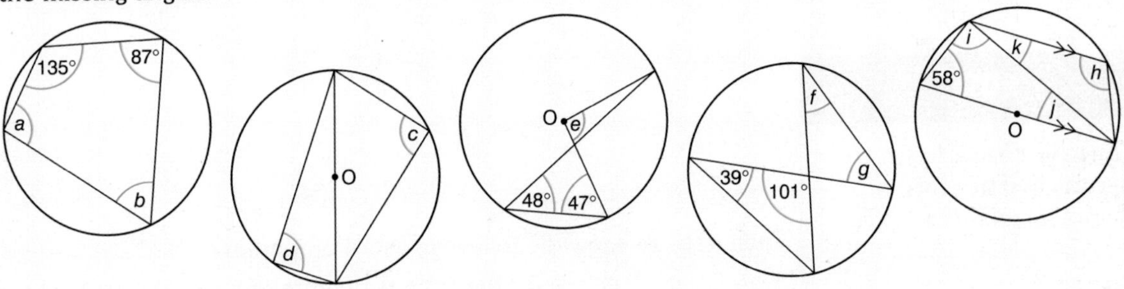

Q2 In the triangle ABC, AB is a diameter of length 11 cm and AC measures 3.5 cm.

Find (a) BC (b) ∠BAC (c) ∠ABC.

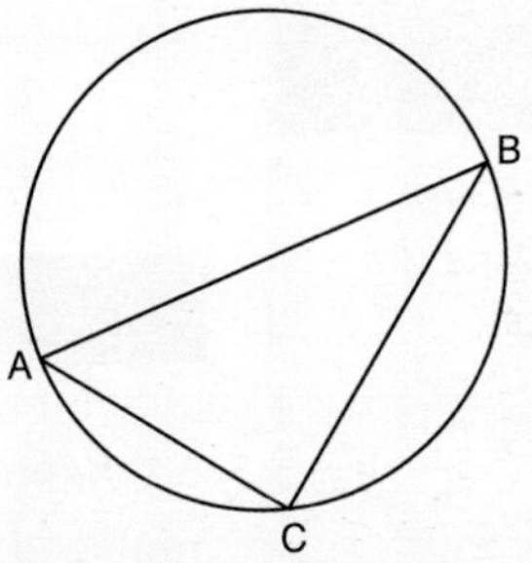

REMEMBER! Cover the answers if you want to.

ANSWERS AND TUTORIAL

A1

$a = 93°$ As opposite angles of a cyclic quadrilateral add up to 180°.
$b = 45°$ As opposite angles of a cyclic quadrilateral add up to 180°.
$c = 90°$ As the angle in a semicircle is 90°.
$d = 90°$ As the angle in a semicircle is 90°.
$e = 96°$ As the angle subtended by an arc at the centre is twice that subtended at the circumference.
$f = 39°$ As the angles subtended by the same arc at the circumference are equal.
$g = 40°$ As vertically opposite angles are equal and the angles of a triangle add up to 180°.
$h = 122°$ As opposite angles of a cyclic quadrilateral add up to 180°.
$i = 90°$ As the angle in a semicircle is 90°.
$j = 32°$ As the angles of a triangle add up to 180°.
$k = 32°$ As j and k are alternate angles between two parallel lines.

A2

(a) BC = 10.4 cm (3 s.f.)
(b) ∠BAC = 71.4° (3 s.f.)
(c) ∠ABC = 18.6° (3 s.f.)

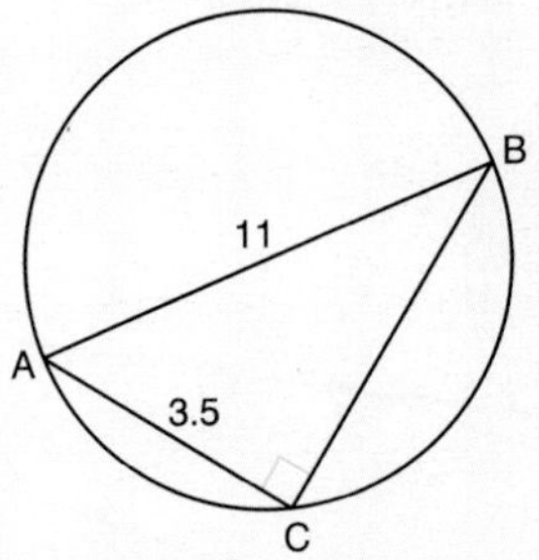

T2

(a) *Drawing a diagram helps to see the situation clearly.*

∠ACB = 90° *As the angle in a semicircle is 90°.*

$AB^2 = AC^2 + BC^2$
Applying Pythagoras' theorem to triangle ABC.

$11^2 = 3.5^2 + BC^2$ *so* BC = 10.428 327 *cm*

(b) *Using* $\cos\theta = \frac{\text{length of adjacent side}}{\text{length of hypotenuse}}$:

cos BAC $= \frac{3.5}{11}$ *so* ∠BAC = 71.446 995°

(c) ∠ABC = 18.553 005°
As the angles of a triangle add up to 180°.

> **NOTE**
>
> A diameter is a chord which passes through the centre of the circle.

> **NOTE**
>
> Conversely, a perpendicular bisector of a chord passes through the centre of the circle.

> **NOTE**
>
> Conversely, chords which are equidistant from the centre of a circle are equal in length.

CHORD PROPERTIES

A chord is a straight line joining two points on the circumference of a circle.

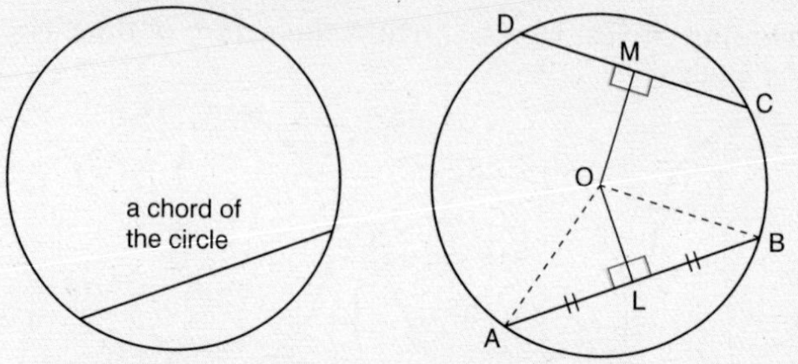

If AB = CD then OL = OM and conversely if OL = OM then AB = CD

You need to know the following chord properties.

- A perpendicular from the centre of a circle to a chord bisects the chord.
- Chords which are equal in length are equidistant from the centre of the circle.

TANGENT PROPERTIES

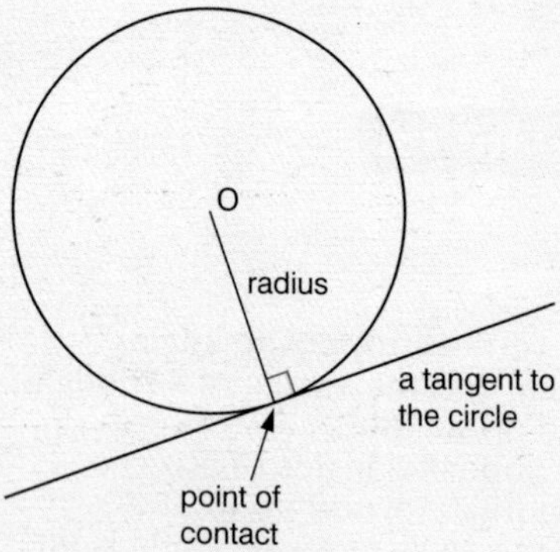

A **tangent** is a straight line which touches a circle at one point only.

You need to know the following tangent properties.

- A tangent to a circle is perpendicular to the radius at the point of contact.
- Tangents to a circle from an external point are equal in length.

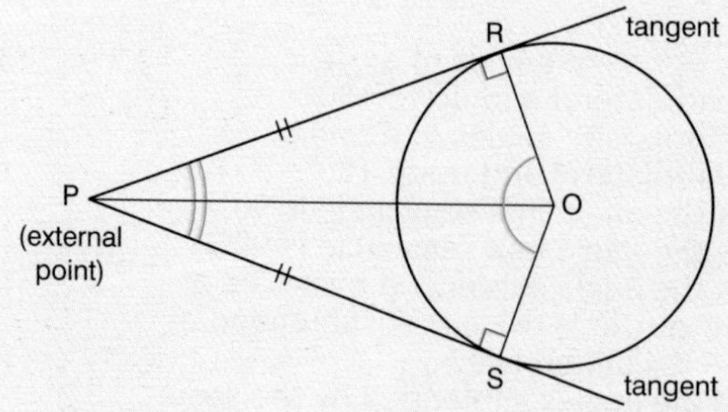

R and S are points of contact
PR = PS
∠RPO = ∠SPO
∠ROP = ∠SOP
RO = SO as they are both radii of the circle

- The angle between a tangent and a chord equals the angle subtended by the chord in the alternate segment.

  This property is known as the **alternate segment theorem**.

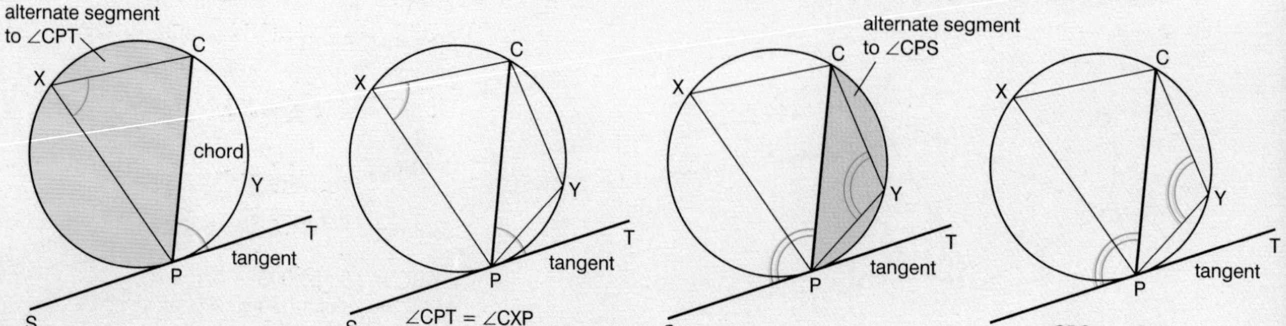

∠CPT = ∠CXP (alternate segment theorem)

∠CPS = ∠CYP (alternate segment theorem)

Worked example

Two chords PQ and RS are parallel to each other on opposite sides of a circle of radius 12 cm. If PQ = 18 cm and RS = 11 cm, find the distance between the chords.

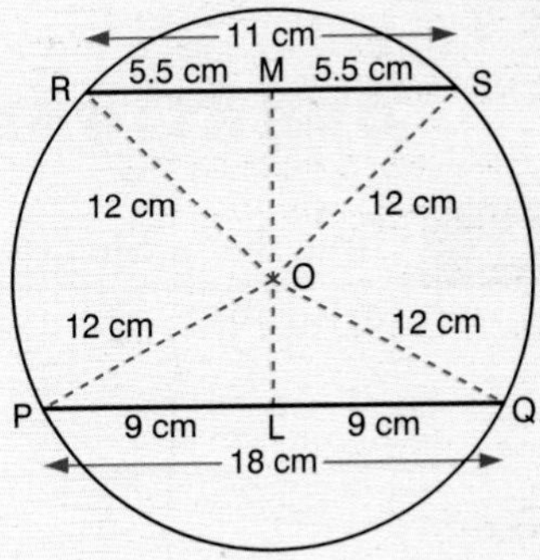

Show this information on a diagram and use the chord properties to show the respective lengths.

For the right-angled triangle OPL:

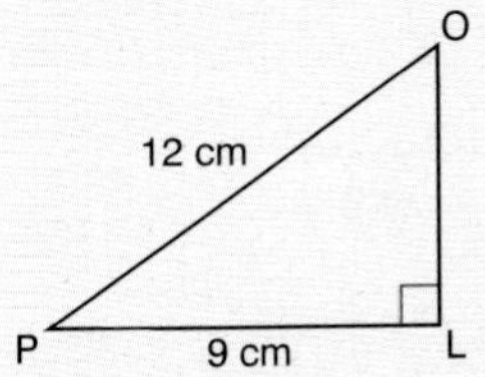

$12^2 = 9^2 + OL^2$ Applying Pythagoras' theorem to the right-angled triangle.

$144 = 81 + OL^2$

$OL^2 = 63$

$OL = 7.937\,254$

For the right-angled triangle OMS:

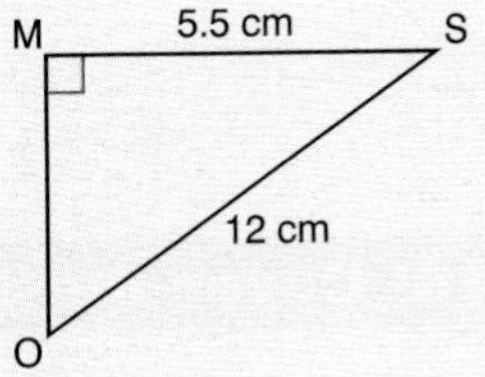

$12^2 = 5.5^2 + OM^2$ Applying Pythagoras' theorem to the right-angled triangle.

$144 = 30.25 + OM^2$

$OM^2 = 113.75$

$OM = 10.665\,365$

Distance between the two chords $= OL + OM$

$= 7.937\,254 + 10.665\,365$

$= 18.602\,619$

$= 18.6$ cm (3 s.f.)

Worked example

The line ST is a tangent to the circle at P. Find the value of ∠XPC.

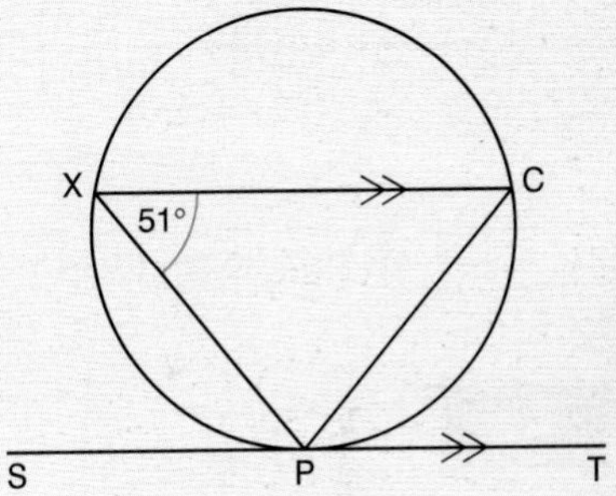

Using the alternate segment theorem:

$\angle CPT = \angle CXP = 51°$ Alternate segment theorem.

$\angle XCP = \angle CPT = 51°$ Alternate angles between parallel lines XC and PT.

$\angle XPC = 180° - (51° + 51°)$ Angles of triangle XCP add up to 180°.

$\angle XCP = 78°$

Check yourself

QUESTIONS

Q1 A chord AB is drawn on a circle of radius 6 cm. If the chord is 4.5 cm from the centre of the circle, calculate the length of the chord.

Q2 Given that PR and PS are tangents to a circle and the points of contact are R and S respectively, find:

(a) ∠POS (b) ∠OPR (c) ∠OPS.

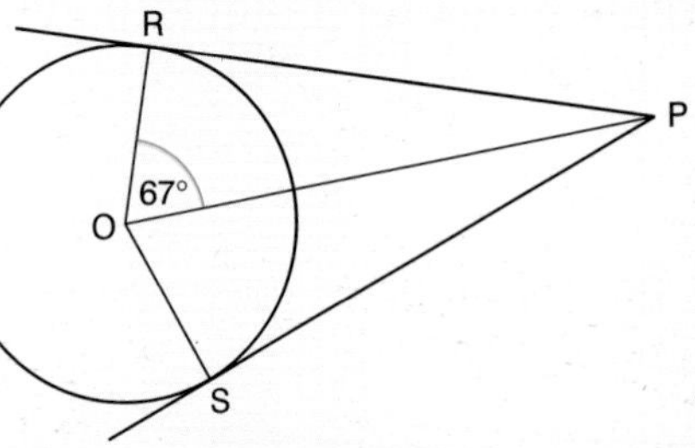

QUESTIONS

Q3 Given that angle CPT = 61° find:

(a) ∠PXC (b) ∠PYC.

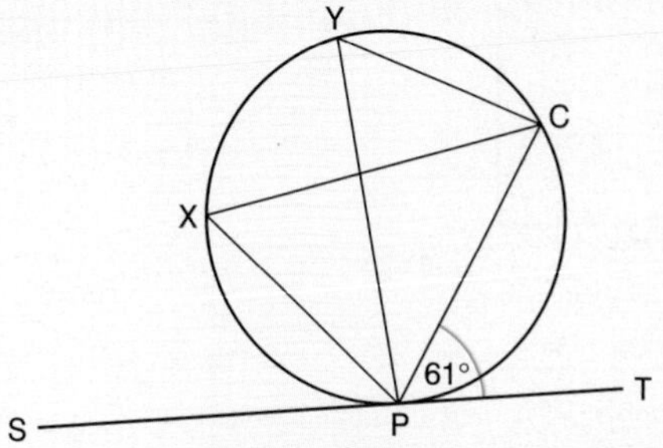

Q4 Given that angle CDP = 38° find:

(a) ∠COP (b) ∠CPT (c) ∠OCP.

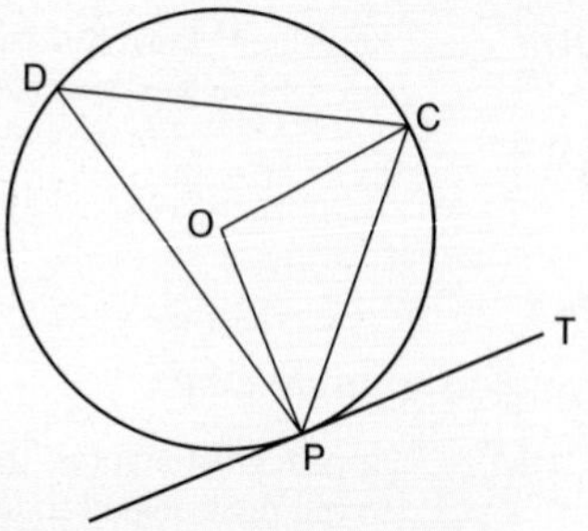

REMEMBER! Cover the answers if you want to.

ANSWERS

A1

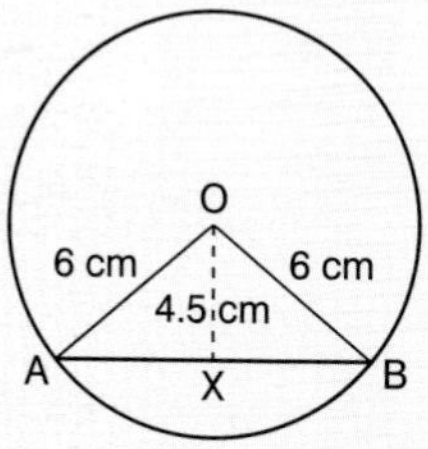

AB = 7.94 cm (3 s.f.)

A2

(a) ∠POS = 67°
(b) ∠OPR = 23°
(c) ∠OPS = 23°

A3

(a) ∠PXC = 61°
(b) ∠PYC = 61°

A4

(a) ∠COP = 76°
(b) ∠CPT = 38°
(c) ∠OCP = 52°

TUTORIAL

T1 *Drawing a diagram helps to see the situation clearly. Let O be the centre of the circle, OX be the perpendicular bisector of the chord.*

For the right-angled triangle OAX:

$6^2 = 4.5^2 + AX^2$ *Applying Pythagoras' theorem to the right-angled triangle.*

AX = 3.968 627

AB = 2 × AX = 2 × 3.968 627 = 7.937 254

AB = 7.94 *cm* (3 *s.f.*)

T2

(*a*) ∠POS = 67° ∠POS = ∠POR *as triangle* ORP *and triangle* OSP *are congruent triangles* (RHS).

(*b*) ∠OPR = 23° ∠ORP = 90° *and angles of the triangle* RPO = 180°.

(*c*) ∠OPS = 23° ∠OPS = ∠OPR *as triangle* ORP *and triangle* OSP *are congruent triangles* (RHS).

T3

(*a*) ∠PXC = 61° ∠PXC = ∠CPT *by the alternate segment theorem.*

(*b*) ∠PYC = 61° ∠PYC = ∠PXC *as these are angles subtended by the same arc* PC.

T4

(*a*) ∠COP = 76° *The angle subtended by the arc* PC *at the centre is twice that subtended at the circumference.*

(*b*) ∠CPT = 38° ∠CPT = ∠CDP *by the alternate segment theorem.*

(*c*) ∠OCP = 52° ∠OCP = ∠OPC *as they are base angles of the isosceles triangle* COP *where* ∠COP = 76°.

CHAPTER 76

ENLARGEMENT – NEGATIVE SCALE FACTORS

An enlargement with a negative scale factor means that the enlargement is situated on the opposite side of the centre of enlargement.

Enlargements with positive scale factors have already been discussed in Chapter 60, Transformations.

Worked example

The triangle PQR with vertices P(2, 2), Q(4, 2) and R(2, 6) is enlarged with a scale factor of $^{-}2$ about the origin. Draw PQR and hence P'Q'R'.

The points P, Q and R are drawn and the enlargement, scale factor $^{-}2$, is produced on the opposite side of O to give P'Q'R' as shown.

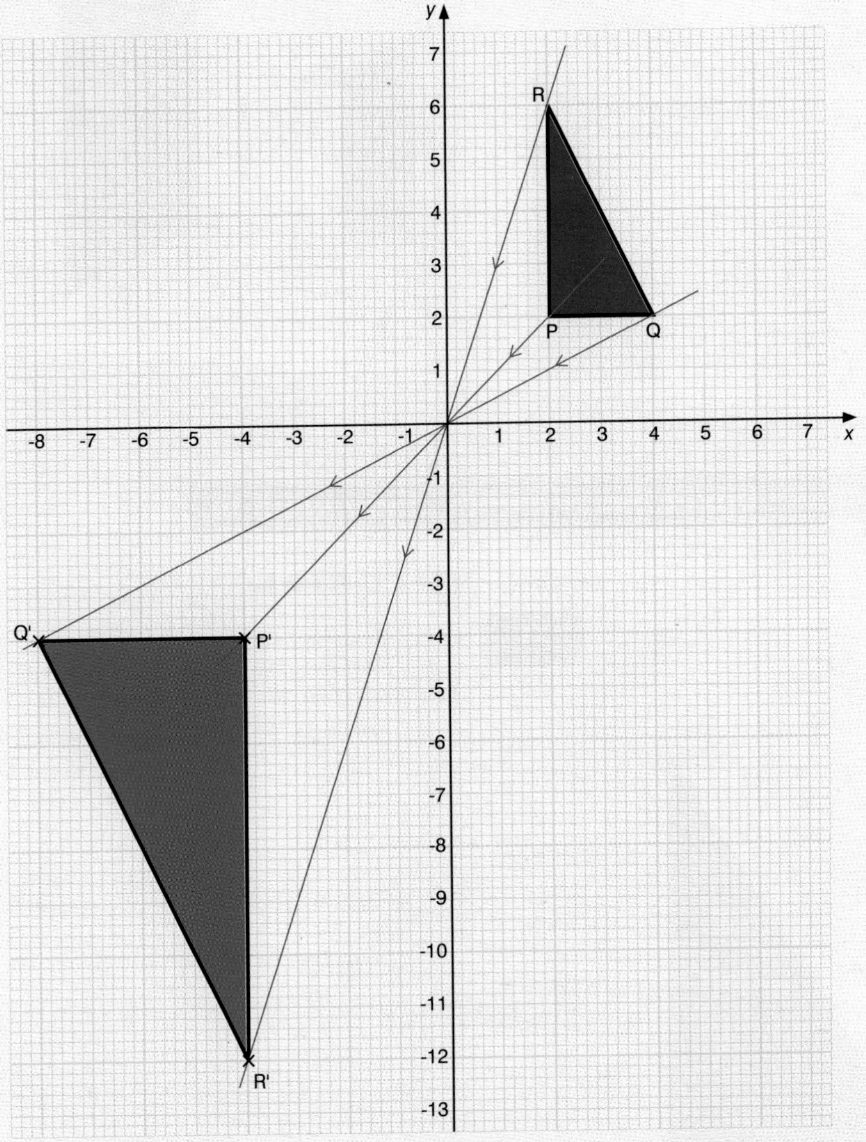

Check yourself

QUESTIONS

Q1 Draw the image of the following after an enlargement, scale factor $^{-}2$ about the given centre of enlargement O.

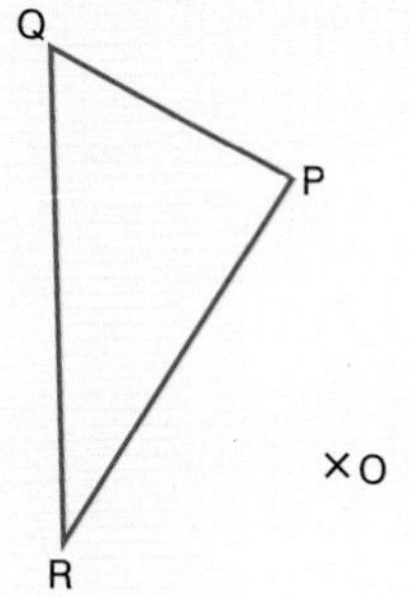

Q2 On graph paper draw the triangle R with vertices (1, 1), (3, 1) and (1, 2).

Enlarge the triangle R with scale factor 3 and centre (0, 0) to form S.

Enlarge the triangle S with scale factor $^{-}\frac{1}{3}$ and centre (0, 0) to form T.

What single transformation maps R onto T?

Q3 Find the centre of enlargement and the scale factor of the following.

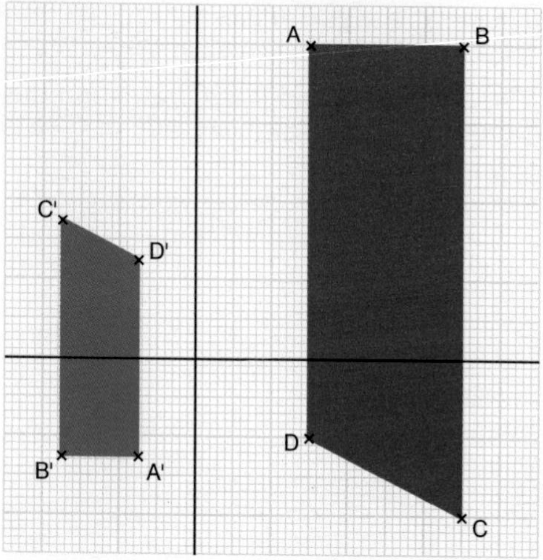

REMEMBER! Cover the answers if you want to.

ANSWERS

A1

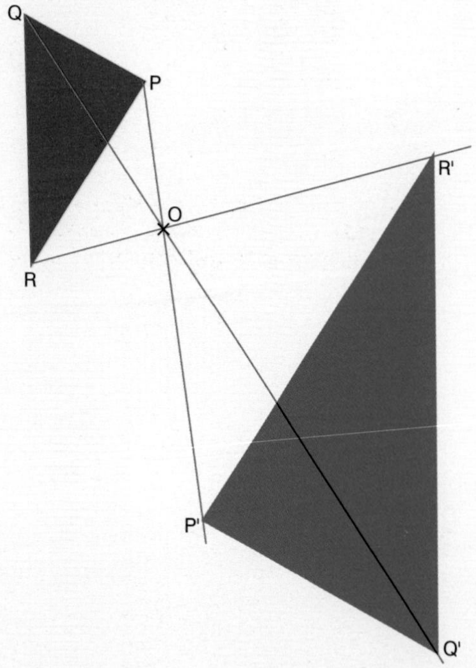

TUTORIAL

T1 *An enlargement of scale factor $^{-}2$ will double the length of all of the sides but the enlargement will be on the opposite side of the centre of enlargement.*

ANSWERS

A2 The single transformation is a rotation of 180° about the origin.

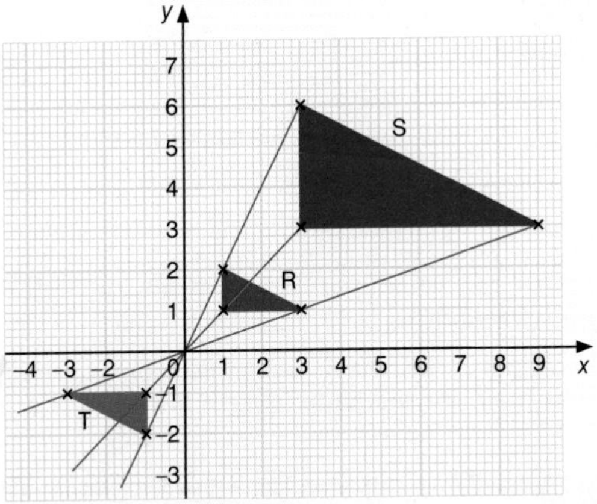

A3 Centre (0, 1), enlargement $^{-}\frac{1}{2}$

TUTORIAL

T2 *The enlargement of scale factor $^{-}\frac{1}{3}$ has the effect of producing an enlargement where the new lengths are $\frac{1}{3}$ of the original lengths. The negative enlargement means that the enlargement occurs on the opposite side of the centre of enlargement.*

T3 *The centre of enlargement is (0, 1).*

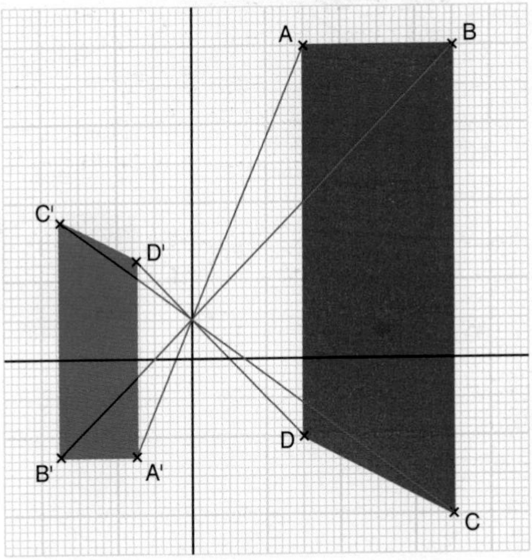

Corresponding lengths on A'B'C'D' are half of those on ABCD and the position of the enlargement on the opposite side of the centre suggest an enlargement scale factor $^{-}\frac{1}{2}$.

SHAPE, SPACE AND MEASURES

CHAPTER 77

VECTORS AND VECTOR PROPERTIES

A vector is a quantity which has magnitude (length) and direction (as indicated by the arrow).

Displacement, velocity, acceleration, force and momentum are all examples of vectors.

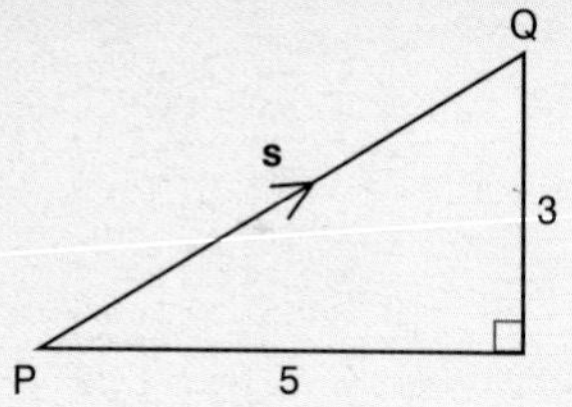

The vector on the left can be represented in a number of ways:

- **PQ** or you can write $\overrightarrow{PQ}$
- **s** or you can write $\underline{s}$
- as a column vector $\begin{pmatrix}5\\3\end{pmatrix}$ as seen in the work on translations in Chapter 59, Transformations.

Two vectors are said to be equal if they have the same magnitude and direction, which means that they are the same length and they are parallel.

COMPONENTS OF A VECTOR

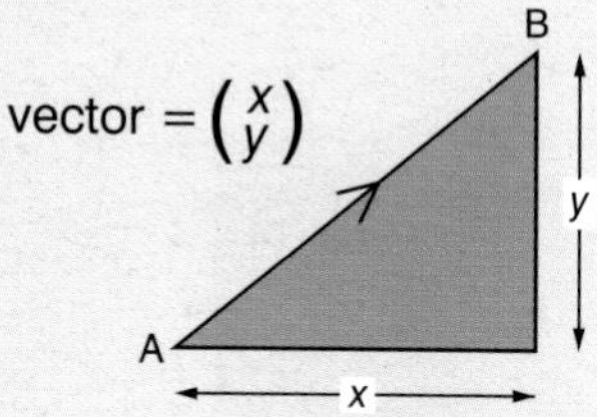

The components of a vector are usually described in terms of:

- the number of units moved in the x-direction
- the number of units moved in the y-direction.

These units are best expressed as a column vector, $\begin{pmatrix}\text{change in } x\text{-value}\\ \text{change in } y\text{-value}\end{pmatrix}$.

ADDITION AND SUBTRACTION OF VECTORS

Vectors can be added or subtracted by placing them end to end, so that the arrows point in the same direction or lead on from one to the next.

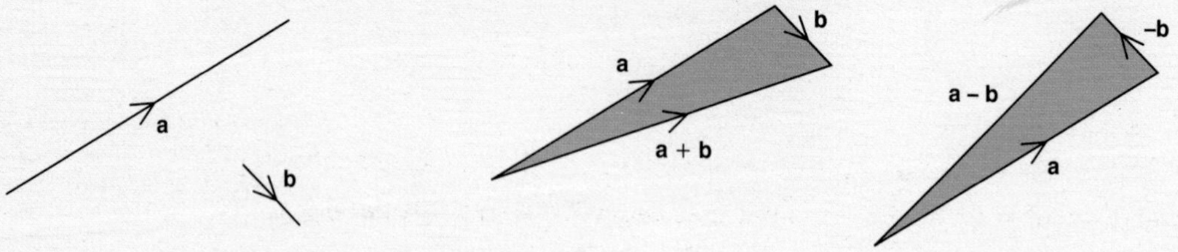

a – **b** is the same as **a** + (–**b**) where –**b** is the same as **b** but in the opposite direction

An easier way to add vectors is to write them as column vectors so that if

a $= \begin{pmatrix}5\\3\end{pmatrix}$ and **b** $= \begin{pmatrix}1\\-1\end{pmatrix}$ then:

a + **b** $= \begin{pmatrix}5\\3\end{pmatrix} + \begin{pmatrix}1\\-1\end{pmatrix} = \begin{pmatrix}6\\2\end{pmatrix}$

a – **b** $= \begin{pmatrix}5\\3\end{pmatrix} - \begin{pmatrix}1\\-1\end{pmatrix} = \begin{pmatrix}4\\4\end{pmatrix}$

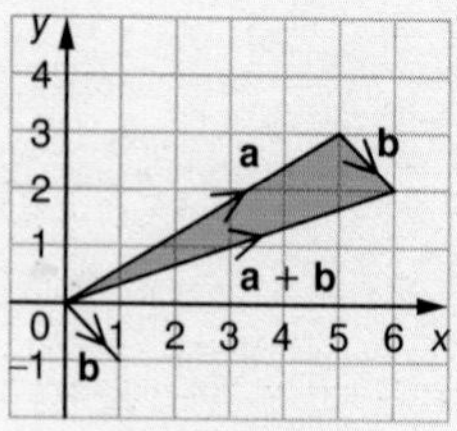

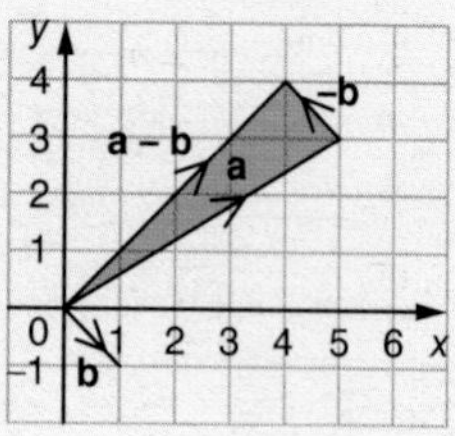

MAGNITUDE OF A VECTOR

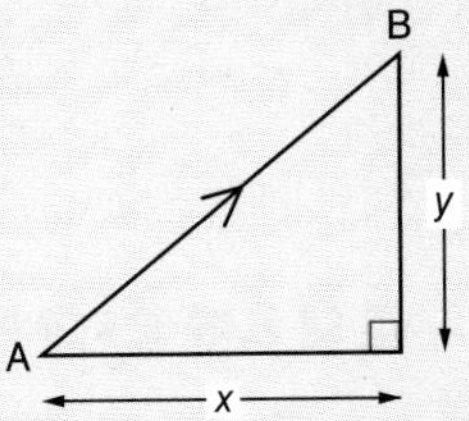

The **magnitude** (length) of a vector can be found by using Pythagoras' theorem.

Suppose that $\overrightarrow{AB} = \begin{pmatrix} x \\ y \end{pmatrix}$.

Then the length of the vector $\overrightarrow{AB} = \sqrt{x^2 + y^2}$ and you can write $|\overrightarrow{AB}| = \sqrt{x^2 + y^2}$ where the two horizontal lines stand for 'magnitude' or length.

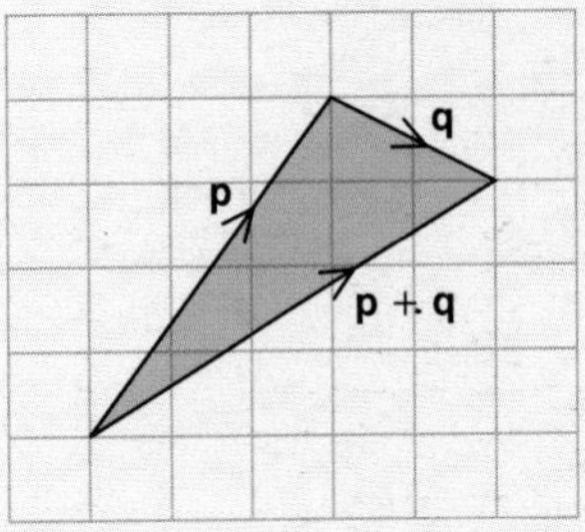

Worked example

Given that $\mathbf{p} = \begin{pmatrix} 3 \\ 4 \end{pmatrix}$ and $\mathbf{q} = \begin{pmatrix} 2 \\ -1 \end{pmatrix}$, find $\mathbf{p} + \mathbf{q}$ and $|\mathbf{p} + \mathbf{q}|$.

$$\mathbf{p} + \mathbf{q} = \begin{pmatrix} 3 \\ 4 \end{pmatrix} + \begin{pmatrix} 2 \\ -1 \end{pmatrix} = \begin{pmatrix} 5 \\ 3 \end{pmatrix}$$

$$|\mathbf{p} + \mathbf{q}| = \sqrt{5^2 + 3^2} = \sqrt{34} \text{ units}$$

MULTIPLICATION OF A VECTOR

Vectors cannot be multiplied by other vectors but they can be multiplied by a constant (sometimes called **scalar multiplication**).

Worked example

Given that $\mathbf{p} = \begin{pmatrix} 3 \\ 4 \end{pmatrix}$ and $\mathbf{q} = \begin{pmatrix} 2 \\ -1 \end{pmatrix}$ find:

(a) $2\mathbf{p}$ (b) $2\mathbf{p} - 3\mathbf{q}$.

(a) $2\mathbf{p} = 2 \times \begin{pmatrix} 3 \\ 4 \end{pmatrix} = \begin{pmatrix} 6 \\ 8 \end{pmatrix}$

(b) $2\mathbf{p} - 3\mathbf{q} = 2 \times \begin{pmatrix} 3 \\ 4 \end{pmatrix} - 3 \times \begin{pmatrix} 2 \\ -1 \end{pmatrix} = \begin{pmatrix} 6 \\ 8 \end{pmatrix} - \begin{pmatrix} 6 \\ -3 \end{pmatrix} = \begin{pmatrix} 0 \\ 11 \end{pmatrix}$

VECTORS IN GEOMETRY

Vectors are often used to prove geometrical theorems.

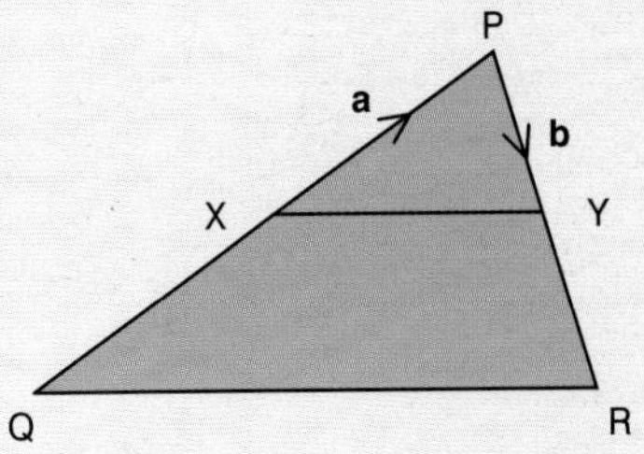

Worked example

In the triangle PQR, X and Y are the midpoints of PQ and PR respectively.

Given that $\overrightarrow{XP} = \mathbf{a}$ and $\overrightarrow{PY} = \mathbf{b}$ show that QR = 2XY and QR is parallel to XY.

From the diagram:

$\overrightarrow{XY} = \overrightarrow{XP} + \overrightarrow{PY}$ so $\overrightarrow{XY} = \mathbf{a} + \mathbf{b}$

As X is the midpoint of QP then $\overrightarrow{QX} = \overrightarrow{XP}$.

Similarly as Y is the midpoint of PR then $\overrightarrow{PY} = \overrightarrow{YR}$.

From the diagram:

$\overrightarrow{QR} = \overrightarrow{QP} + \overrightarrow{PR}$

$\overrightarrow{QR} = 2\mathbf{a} + 2\mathbf{b}$ As $\overrightarrow{QP} = \overrightarrow{QX} + \overrightarrow{XP} = 2\mathbf{a}$ and $\overrightarrow{PR} = \overrightarrow{PY} + \overrightarrow{YR} = 2\mathbf{b}$

$\overrightarrow{QR} = 2(\mathbf{a} + \mathbf{b}) = 2\overrightarrow{XY}$ As $\overrightarrow{XY} = \mathbf{a} + \mathbf{b}$ above.

This tells you that the magnitude of $\overrightarrow{QR}$ is twice the magnitude of $\overrightarrow{XY}$ so that QR = 2XY. Since $\overrightarrow{QR}$ is a multiple of $\overrightarrow{XY}$ then QR and XY are in the same direction and must, therefore be parallel.

Check yourself

QUESTIONS

Q1 The diagram shows a series of parallel lines with $\overrightarrow{AB} = \mathbf{a}$ and $\overrightarrow{AE} = \mathbf{b}$.

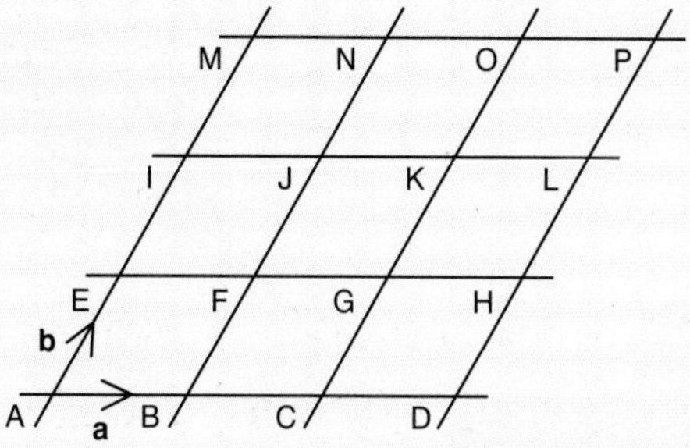

Write down the following in terms of **a** and **b**.

(a) $\overrightarrow{AC}$ (b) $\overrightarrow{AM}$ (c) $\overrightarrow{AF}$
(d) $\overrightarrow{AK}$ (e) $\overrightarrow{AL}$ (f) $\overrightarrow{FK}$
(g) $\overrightarrow{BA}$ (h) $\overrightarrow{GA}$ (i) $\overrightarrow{NE}$
(j) $\overrightarrow{PE}$

Q2 A quadrilateral ABCD has diagonals which cross at X. Write the following as single vectors.

(a) $\overrightarrow{AB} + \overrightarrow{BC}$ (b) $\overrightarrow{AC} + \overrightarrow{CD}$
(c) $\overrightarrow{AX} + \overrightarrow{XB}$ (d) $\overrightarrow{CD} - \overrightarrow{AD}$
(e) $\overrightarrow{CX} - \overrightarrow{BX}$ (f) $\overrightarrow{AB} + \overrightarrow{BC} + \overrightarrow{CX}$

Q3 $\overrightarrow{AB} = \begin{pmatrix} 3 \\ 4 \end{pmatrix}$ and $\overrightarrow{BC} = \begin{pmatrix} 5 \\ -1 \end{pmatrix}$.

Find:

(a) $|\overrightarrow{AB}|$ (b) $|\overrightarrow{BC}|$ (c) $|\overrightarrow{AC}|$.

Q4 For the regular hexagon ABCDEF with centre O, $\overrightarrow{AB} = \mathbf{a}$ and $\overrightarrow{BC} = \mathbf{b}$.

Find: (a) $\overrightarrow{AC}$ (b) $\overrightarrow{AO}$
(c) $\overrightarrow{OB}$ (d) $\overrightarrow{AD}$.

What can you say about the quadrilateral ACDF? Give reasons for your answer.

REMEMBER! Cover the answers if you want to.

ANSWERS

A1
(a) $\overrightarrow{AC} = 2\mathbf{a}$
(b) $\overrightarrow{AM} = 3\mathbf{b}$
(c) $\overrightarrow{AF} = \mathbf{a} + \mathbf{b}$
(d) $\overrightarrow{AK} = 2\mathbf{a} + 2\mathbf{b}$
(e) $\overrightarrow{AL} = 3\mathbf{a} + 2\mathbf{b}$
(f) $\overrightarrow{FK} = \mathbf{a} + \mathbf{b}$
(g) $\overrightarrow{BA} = {}^{-}\mathbf{a}$
(h) $\overrightarrow{GA} = {}^{-}2\mathbf{a} - \mathbf{b}$
(i) $\overrightarrow{NE} = {}^{-}\mathbf{a} - 2\mathbf{b}$
(j) $\overrightarrow{PE} = {}^{-}3\mathbf{a} - 2\mathbf{b}$

TUTORIAL

T1 *Remember that* $\mathbf{a} + \mathbf{b} = \mathbf{b} + \mathbf{a}$ *and if* $\overrightarrow{AB} = \mathbf{a}$ *then* $\overrightarrow{BA} = {}^{-}\mathbf{a}$.

ANSWERS

A2

(a) $\overrightarrow{AB} + \overrightarrow{BC} = \overrightarrow{AC}$

(b) $\overrightarrow{AC} + \overrightarrow{CD} = \overrightarrow{AD}$

(c) $\overrightarrow{AX} + \overrightarrow{XB} = \overrightarrow{AB}$

(d) $\overrightarrow{CD} - \overrightarrow{AD} = \overrightarrow{CA}$

(e) $\overrightarrow{CX} - \overrightarrow{BX} = \overrightarrow{CB}$

(f) $\overrightarrow{AB} + \overrightarrow{BC} + \overrightarrow{CX} = \overrightarrow{AX}$

A3

(a) $|\overrightarrow{AB}| = 5$ units

(b) $|\overrightarrow{BC}| = \sqrt{26}$ units

(c) $|\overrightarrow{AC}| = \sqrt{73}$ units

A4

(a) $\overrightarrow{AC} = \mathbf{a} + \mathbf{b}$

(b) $\overrightarrow{AO} = \mathbf{b}$

(c) $\overrightarrow{OB} = \mathbf{a} - \mathbf{b}$

(d) $\overrightarrow{AD} = 2\mathbf{b}$

ACDF is a parallelogram.

TUTORIAL

T2

A diagram might help.

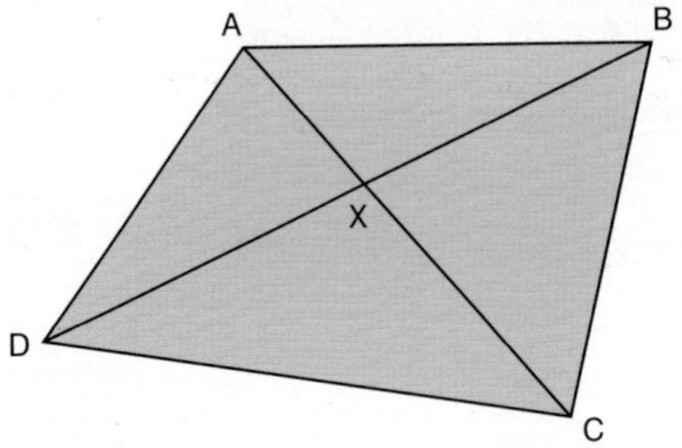

T3

Using the fact that $|\overrightarrow{AB}| = \sqrt{x^2 + y^2}$

(*a*) $|\overrightarrow{AB}| = \left|\begin{pmatrix}3\\4\end{pmatrix}\right| = \sqrt{3^2 + 4^2} = \sqrt{25} = 5$

(*b*) $|\overrightarrow{BC}| = \left|\begin{pmatrix}5\\^-1\end{pmatrix}\right| = \sqrt{5^2 + (^-1)^2} = \sqrt{26}$

(*c*) $|\overrightarrow{AC}| = \left|\begin{pmatrix}8\\3\end{pmatrix}\right| = \sqrt{8^2 + 3^2} = \sqrt{73}$

T4

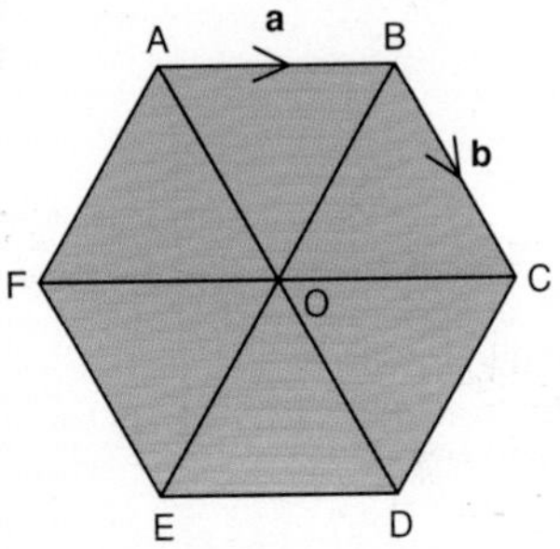

(*a*) $\overrightarrow{AC} = \overrightarrow{AB} + \overrightarrow{BC} = \boldsymbol{a} + \boldsymbol{b}$

(*b*) $\overrightarrow{AO} = \boldsymbol{b}$

(*c*) $\overrightarrow{OB} = \overrightarrow{OA} + \overrightarrow{AB} = {}^-\boldsymbol{b} + \boldsymbol{a} = \boldsymbol{a} - \boldsymbol{b}$

(*d*) $\overrightarrow{AD} = 2\boldsymbol{b}$

$\overrightarrow{AC} = \boldsymbol{a} + \boldsymbol{b}$ *and* $\overrightarrow{FD} = \boldsymbol{a} + \boldsymbol{b}$

$\overrightarrow{AF} = \boldsymbol{b} - \boldsymbol{a}$ *and* $\overrightarrow{CD} = \boldsymbol{b} - \boldsymbol{a}$

so AC *is parallel and equal to* FD

AF *is parallel and equal to* CD.

Therefore ACDF *is a parallelogram.*

SHAPE, SPACE AND MEASURES

CHAPTER 78

EXAMINATION QUESTIONS 15

EXAM PRACTICE

Sample Student's Answers & Examiner's Comments

1 AB is the arc of a sector of a circle with radius 12 cm. The angle in the sector is 250°.

(a) Calculate the length of the arc AB. Give your answer to the nearest millimetre.

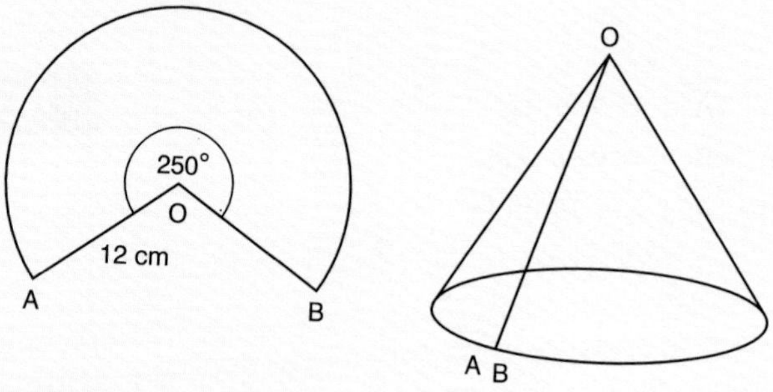

OA is then joined to OB to form a cone.

(b) Calculate the radius of the base of this cone.

(MEG syllabus B, specimen paper 1998)

(a) Length of arc $= \frac{250}{360} \times 2 \times \pi \times 12$

$= 52.359\,878$ cm

$= 523.598\,78$ mm

$= 524$ mm (to the nearest mm)

(b) Circumference of cone $= 2 \times \pi \times r$

$= 523.598\,78$ mm

So $2 \times \pi \times r = 523.598\,78$ mm

$r = \frac{523.598\,78}{2 \times \pi}$ mm

$r = 83.333\,333$ mm

Radius $= 83.3$ mm (3 s.f.)

EXAMINER'S COMMENTS

1 **a)**

The candidate has correctly identified the necessary formula for arc length and obtained a solution which is converted to millimetres so that it can be rounded to the nearest mm.

b)

He has correctly equated the arc length to the circumference of the base of the cone and has used this information to find the radius of the base of the cone.

He has remembered to include the units which is essential to gain full marks in this question.

2 Two chords of a circle, AB and CD, intersect at a point P.

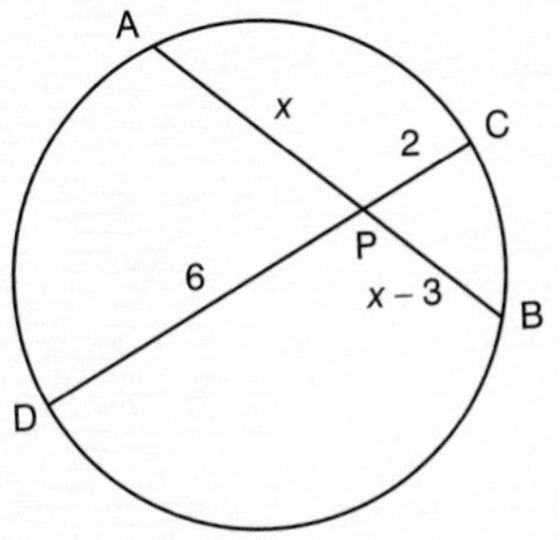

CP = 2 cm and PD = 6 cm.

AP = x cm and PB is 3 cm shorter than AP.

(a) Using the fact that AP × PB = CP × PD, show that $x^2 - 3x - 12 = 0$.

(b) By solving the equation find the length of AP, correct to one decimal place.

(SEG modular syllabus, specimen paper 1998)

(a) AP × PB = CP × PD

$x(x - 3) = 2 \times 6$

$x^2 - 3x = 12$

$x^2 - 3x - 12 = 0$

(b) $x^2 - 3x - 12 = 0$

$(x \quad)(x \quad) = 0$ Does not factorise.

Use formula.

$$x = \frac{-b \pm \sqrt{b^2 - 4ac}}{2a}$$

$$x = \frac{3 \pm \sqrt{9 - 4 \times -12}}{2 \times 1}$$

$$x = \frac{3 \pm \sqrt{57}}{2}$$

$$x = \frac{3 \pm 7.5498344}{2}$$

$x = -2.2749172$ or 5.2749172

$x = -2.27$ or 5.27 (3 s.f.)

EXAMINER'S COMMENTS

2 a)
The candidate has used the diagram to form an equation which is rearranged to the required form.

b)
She has attempted to factorise the given quadratic and then used the formula (provided on the examination paper) to solve the equation.

She should reject the negative solution as a negative length is not possible and the final answer should be given to ***one decimal place*** *as requested in the question. A final answer of* AP = 5.3 *cm would gain full marks.*

EXAMINER'S COMMENTS

3 *The candidate has provided no explanation for the solution although the question does not specifically ask for reasons.*

A suggestion that $\angle TQN = \angle NPQ = 35°$ by the alternate segment theorem would allow some marks for method if the final answer were incorrect.

He has marked $\angle PKQ$ on the diagram as 54°, allowing the possibility of some credit if the final answer were incorrect.

$$\angle PKQ = \tfrac{1}{2} \times \angle POQ$$
$$= \tfrac{1}{2} \times 108°$$
$$= 54°$$

$\angle PKQ + \angle QNP = 180°$ *as opposite angles of a cyclic quadrilateral add up to 180° and*

$$\angle QNP = 180° - 54°$$
$$= 126° \text{ so } n = 126°.$$

4 *She has expressed $\overrightarrow{PQ}$ and $\overrightarrow{BC}$ in terms of $\vec{a}$ and $\vec{b}$.*

She has rewritten the contents of the brackets to highlight the relationship with $\overrightarrow{PQ}$.

Unfortunately, this does not describe fully the relationship between the lines PQ and BC. The response should mention that BC and PQ are parallel and that the length of BC is twice the length of PQ.

3 O is the centre of the circle. P, K, Q and N are points on the circumference. QT is the tangent to the circle at Q. Angle POQ = 108°. Angle NPQ = 35°. Calculate the values of m and n.

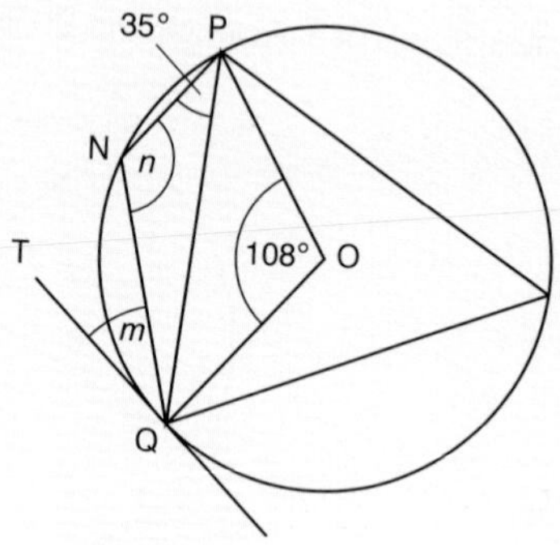

(London specimen paper 1998, part question)

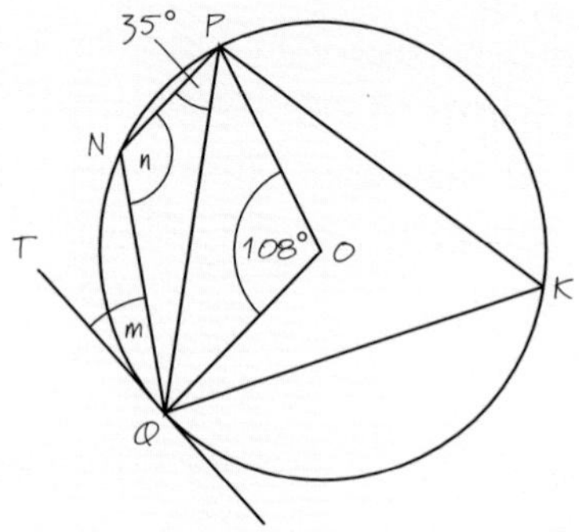

$m = 35°$

$n = 126°$

4 The diagram shows a triangle ABC. P is a point on AC such that **AP** = **a** and AP = $\tfrac{1}{2}$AC. Q is the midpoint of AB and **AQ** = **b**.

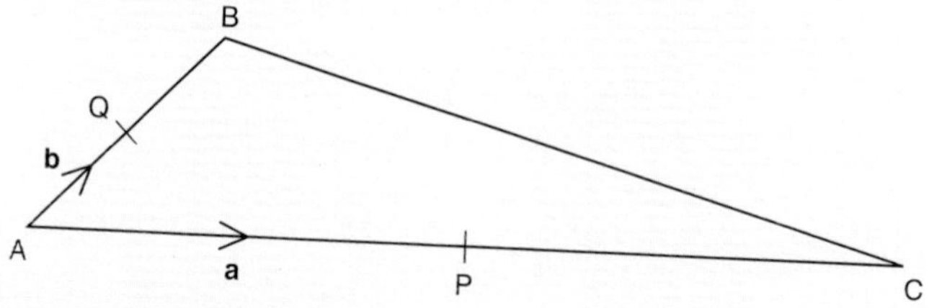

By expressing **PQ** and **BC** in terms of **a** and **b**, describe fully the relationship between the lines PQ and BC.

(SEG modular syllabus, specimen paper 1998)

$\overrightarrow{PQ} = \overrightarrow{PA} + \overrightarrow{AQ}$

$\overrightarrow{PQ} = -\underline{a} + \underline{b}$

$\overrightarrow{BC} = \overrightarrow{BA} + \overrightarrow{AC}$

$\overrightarrow{BC} = -2\underline{b} + 2\underline{a}$

$\overrightarrow{BC} = -2(\underline{b} - \underline{a})$

$\overrightarrow{BC} = -2(-\underline{a} + \underline{b})$

So $\overrightarrow{BC} = -2\overrightarrow{PQ}$

Questions to Answer

1 Find the area and perimeter of the shaded area in the diagram.

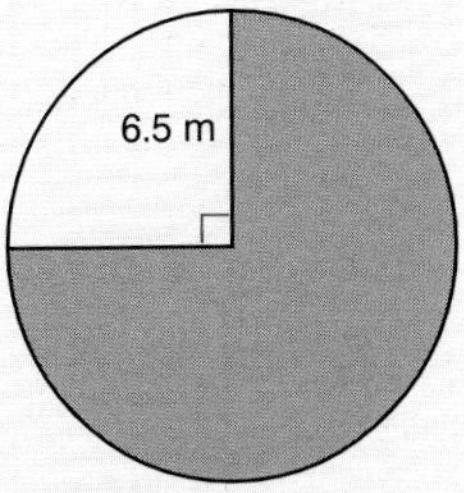

2 A windscreen wiper of length 25 cm sweeps out an angle of 110° as illustrated in the diagram. What is the area of the screen covered?

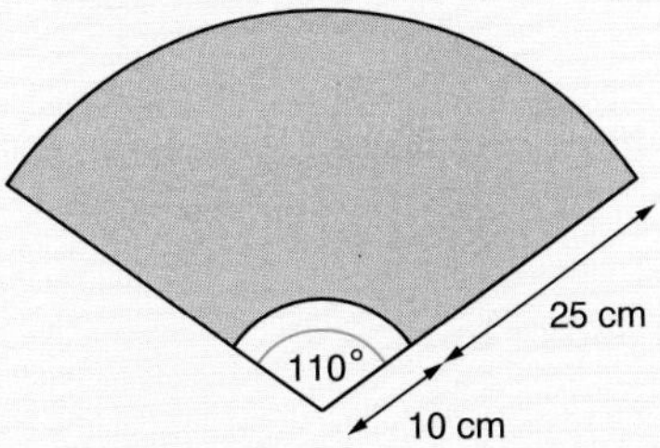

3 In the diagram, RT and PT are tangents to the circle, calculate:

(a) ∠ROP (b) ∠RSP (c) ∠RQP.

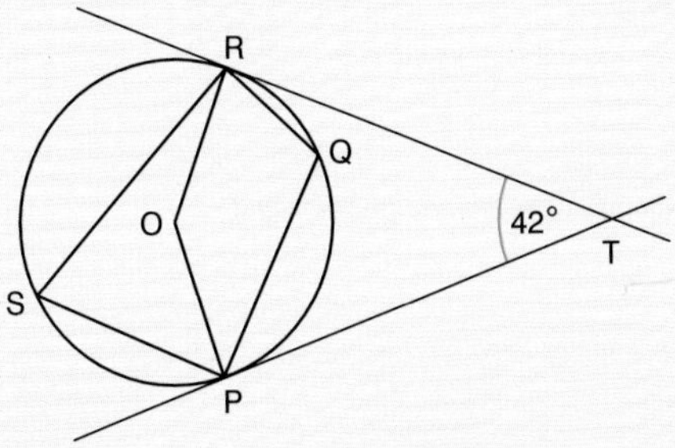

4 ABC and ADE are straight lines. CE is a diameter. Angle DCE = x° and angle BCD = $2x$°. Find, in terms of x, the sizes of the angles:

(a) ABD (b) DBE (c) BAD.

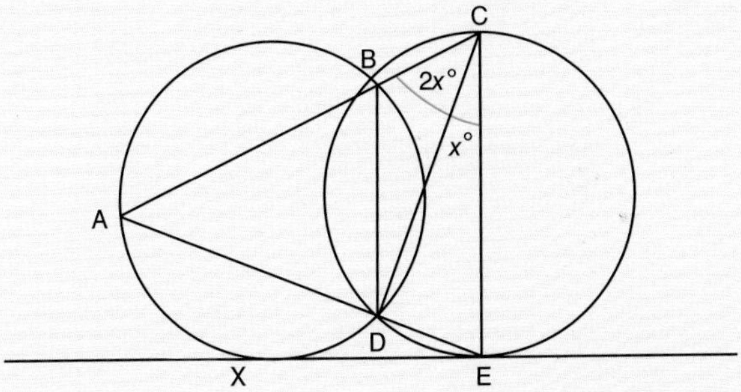

(London specimen paper 1998, part question)

5 PQRS and PSTU are parallelograms. $\overrightarrow{PQ}$ is **a**, $\overrightarrow{PS}$ is **b**, $\overrightarrow{ST}$ is **c**.

Find in terms of **a**, **b** and **c** expressions in their simplest forms for:

(a) $\overrightarrow{PT}$ (b) $\overrightarrow{US}$ (c) $\overrightarrow{PX}$ where X is the midpoint of QT.

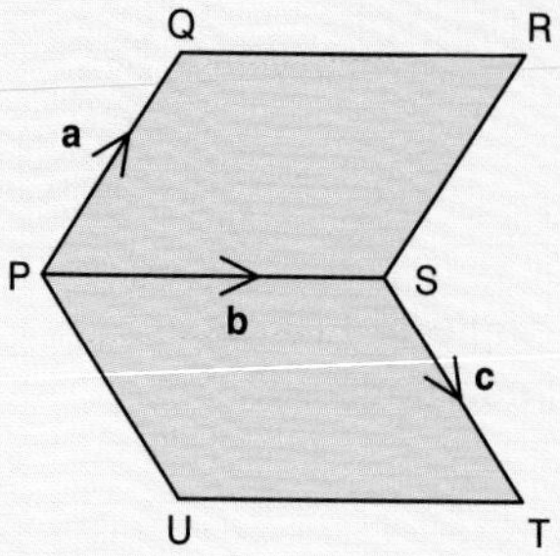

(London specimen paper 1998, part question)

6 In the diagram X is the point on AB such that AX = 3XB.
Given that $\overrightarrow{OA} = 8\mathbf{a}$ and $\overrightarrow{OB} = 4\mathbf{b}$, express in terms of **a** and/or **b**.

(a) $\overrightarrow{AB}$ (b) $\overrightarrow{AX}$ (c) $\overrightarrow{OX}$.

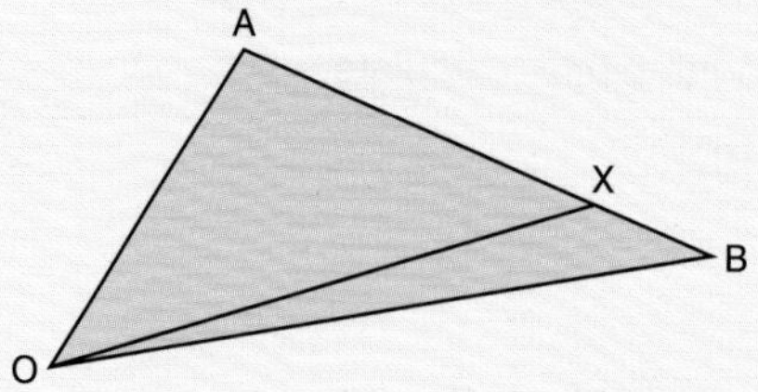

(MEG syllabus A, specimen paper 1998)

CHAPTER 79

COLLECTING AND ORGANISING DATA

You need to know the following definitions as a basis for your work on statistics and data handling.

RAW DATA AND TALLY CHARTS

Raw data is information that has been collected but has not yet been organised in any way. A **tally chart** is often used to collect data. A tally chart consists of a series of tallies which are grouped into fives as shown below.

Tallies

||||	=	4
~~||||~~	=	5
~~||||~~ |	=	6
~~||||~~ ||	=	7
~~||||~~ ~~||||~~	=	10 etc.

Notice that to construct a tally chart, one stroke is made for every item of data counted, and every fifth stroke is drawn through the preceding four.

FREQUENCY DISTRIBUTIONS

NOTE

The frequency is how often a value occurs.

A **frequency distribution** can easily be obtained from a tally chart by totalling the tallies to find the frequencies. In some circumstances, it is helpful to group the data and produce a grouped frequency distribution.

Worked example

This raw data gives the times (in minutes) for 20 pupils to complete a test.

23	23	26	22	19	23	22	24	20	21
25	15	22	24	20	17	21	22	24	18

Construct a grouped frequency distribution using class intervals 15–17, 18–20, 21–23 and 24–26.

The grouped frequency distribution, with the given class intervals, is like this.

Times	Tallies	Frequency
15–17	||	2
18–20	||||	4
21–23	~~||||~~ ||||	9
24–26	~~||||~~	5

CLASSIFYING DATA

Data that can take any values within a given range is called **continuous** data. This includes heights, temperatures, lengths and mass. Data that can only take particular values (such as whole or half numbers) is called **discrete** data. This includes numbers of children, separate colours and shoe sizes.

Quantitative data can only take numerical values such as length, mass, capacity or temperature. **Qualitative** (or **categorical**) data measures qualities such as colour, taste, shade or touch.

PRIMARY AND SECONDARY DATA

Primary data is collected as part of a statistical investigation such as a census or survey, whereas **secondary** data is data which already exists. Once primary data is collected and processed then it becomes secondary data. Examples of secondary data include information provided by government departments, businesses, market research companies, etc.

Data can be collected in a number of ways although observation, interviewing and the use of questionnaires are the most popular.

OBSERVATION

This method involves collecting information and using some means of recording it, such as observation sheets, tape recorders or video recorders. This form of data collection can include **systematic** observation, where the observer tries to be as unobtrusive as possible, or else **participant** observation, where the observer participates in the activity.

INTERVIEWING

This method involves asking questions of individuals, or groups of individuals, using some set format. Interviewing can be **formal**, where the questions will follow a strict format, or else **informal**, where the questions will follow some general format.

QUESTIONNAIRES

This is the most popular method of collecting data. It usually involves postal questionnaires or else questionnaires that are left for the respondents to complete in their own time. A good questionnaire should be simple, short, clear and precise and the questions should be unambiguous, written in appropriate language, avoid personal or offensive questions and be free from bias.

PILOT SURVEY

The pilot survey is a preliminary survey carried out on a small number of people. The pilot survey is useful to check for likely problems and highlight areas requiring further clarification before the actual survey is undertaken.

Check yourself

QUESTIONS

Q1 Give two advantages of using a pilot survey.

Q2 The following three questions were found on a questionnaire.

(a) 'What do you think of our improved magazine?'
(b) 'How many hours of television do you watch?'
(c) 'Do you or do you not listen to the radio?'

What is wrong with these questions?

REMEMBER! Cover the answers if you want to.

ANSWERS

A1 Allows you to identify problems and to improve the design of the actual survey.

A2
(a) The question suffers from bias in that it suggests the magazine is 'improved'.
(b) The question is ambiguous and lacks clarity, as no mention is given to the time scale.
(c) The question is not very clear and would be confusing for the respondent.

TUTORIAL

T1 *Several answers are possible here including the opportunity to find problems with the misinterpretation or misunderstanding of questions, which can then be altered before the actual survey – identifying problems early will save time and money by making sure that the survey does not cause any unexpected problems.*

T2 *Remember that a good questionnaire should be simple, short, clear and precise and the questions should be unambiguous, written in appropriate language, avoid personal or offensive questions and be free from bias.*

CHAPTER 80

REPRESENTING DATA

Representing data is an important aspect of statistics and data handling, but you should ensure that the representation chosen is appropriate to the data to be represented. The following information details a variety of different representations with which you will need to be familiar.

PICTOGRAMS

A **pictogram** (or **pictograph** or **ideograph**) is a simple way of representing data where the frequency is indicated by a number of identical pictures. When using a pictogram, you must remember to include a key to explain what the individual pictures represent as well as giving the diagram an overall title. You may also need to use a symbol which can easily be divided into halves, quarters, tenths and so on.

Drink	Frequency
Tea	10
Coffee	13
Soup	4
Chocolate	8
Other	3

Worked example

The following frequency distribution shows the number of different drinks purchased from a vending machine

Show this information as a pictogram.

The pictogram looks like this.

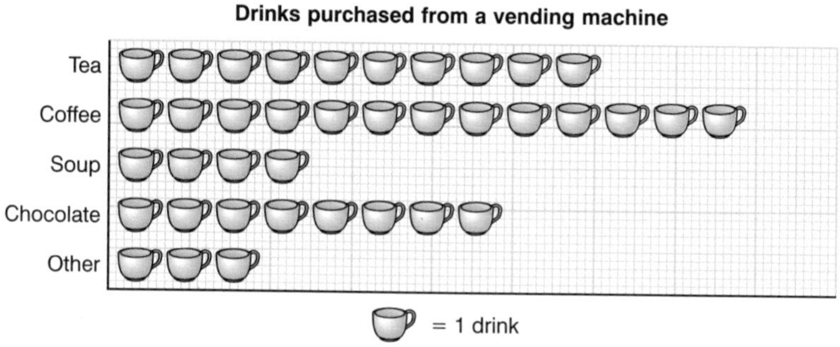

Alternatively, you could allow one cup to stand for two drinks.

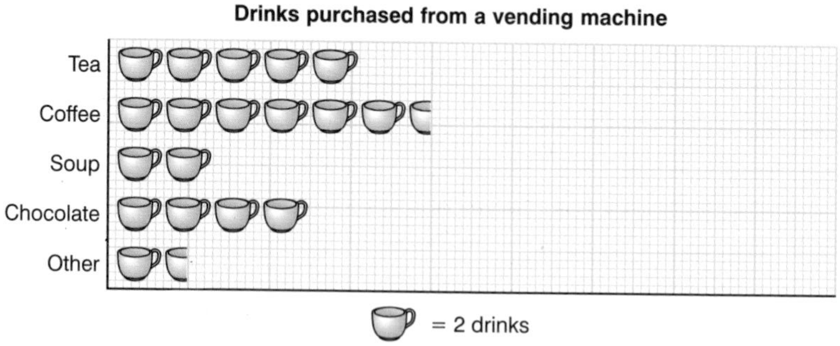

BAR CHARTS

A bar chart is a common way of representing data where the frequency is indicated by a number of vertical or horizontal bars, all with the same width. When using a bar chart, you must remember to label the axes clearly and give the diagram a title to explain what it represents.

Worked example

The bar chart at the top of page 253, shows the average price for cars in a number of trade magazines.

NOTE

A bar chart is sometimes called a histogram, but this is not strictly correct, since histograms can have bars of unequal width, and the data is represented by the area of the bar, not just the height.

a) What was the average price for a car in the *What Car?* magazine?

b) Which magazine had the lowest average price for a car?

c) In which two magazines were the average prices the closest?

d) What is the biggest difference between average prices in the trade magazines?

From the graph:

(a) The average price for a car in the *What Car?* magazine is £3900.

(b) The *Car Weekly* had the lowest average price for a car.

(c) The two magazines where the average prices were the closest are *What Car?* and *Car Buyer*.

(d) The biggest difference between average prices is £3400 (£6700 – £3300).

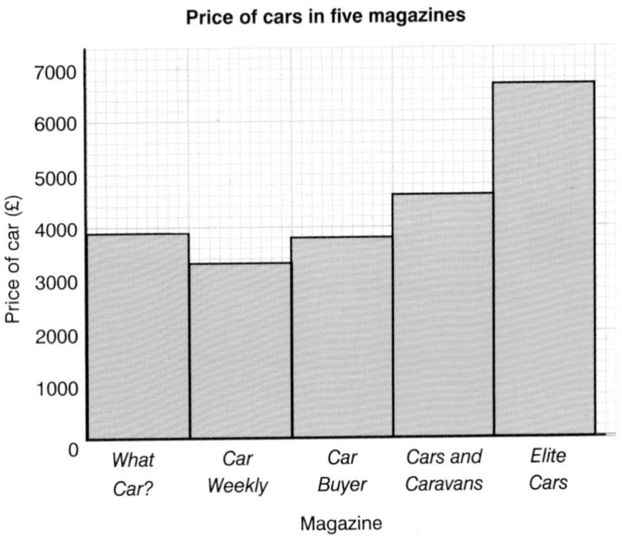

COMPOUND AND COMPONENT BAR CHARTS

Compound and **component** bar charts are similar to bar charts except that they show how individual items are split into different groups. When using a compound or a component bar chart, you must remember to label the axes clearly and give the diagram a title to explain what it represents.

NOTE

The compound bar chart is sometimes called a multiple bar chart and the component bar chart is sometimes called a composite or sectional bar chart.

Worked example

The information in the table shows the favourite crisp flavours of pupils in a class.

Show this information as a:

a) a compound bar chart

b) a component bar chart.

Favourite flavour	Girls	Boys
Plain	19	36
Salt & vinegar	12	14
Cheese & onion	13	12
Smokey bacon	6	8
Others	2	1

(a)

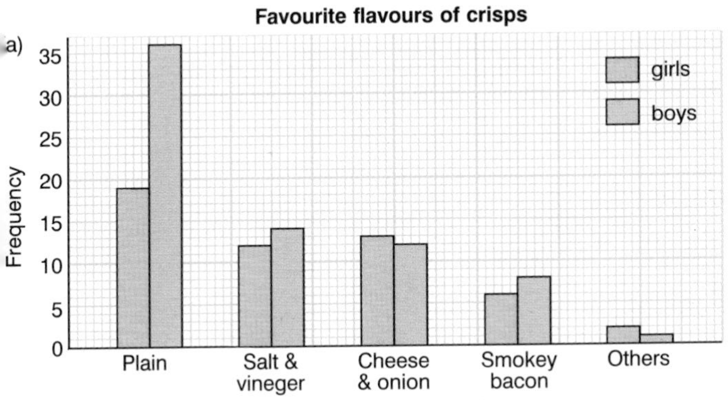

(b)

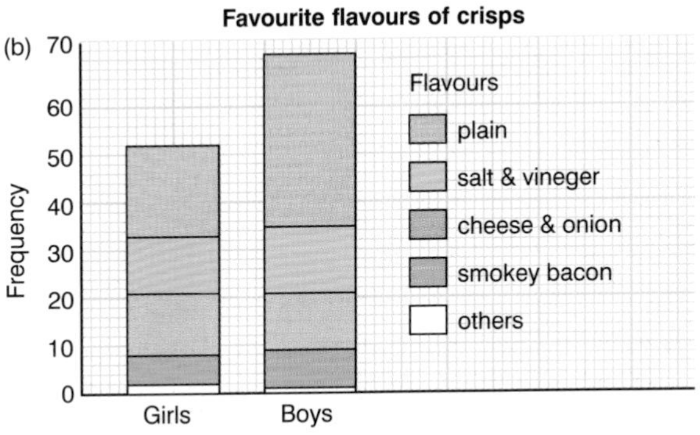

Check yourself

QUESTIONS

Q1 The following distribution shows the number of different flavoured sorbets sold in a restaurant one weekend.

Flavour	Frequency
Lemon	22
Strawberry	17
Raspberry	14
Other	24

Show this information as a pictogram.

Q2 The following information shows the sales of different coloured socks in a department store.

Colour	Frequency
White	18
Black	13
Blue	10
Brown	6
Other	7

Show this information as:

(a) a pictogram (b) a bar chart.

Q3 The following information shows the length of time (in minutes) which it takes to get served in a shop.

Time	Frequency
0–1	4
1–2	3
2–3	11
3–4	8
4–5	3
5–6	1

Draw a bar chart to illustrate this information.

Q4 The following information shows the income (in thousands of £) taken in different departments over three consecutive years.

Year	Sweets	Stationery	Books
1995	47	22	34
1996	32	41	26
1997	36	52	28

Show this information as:

(a) a compound bar chart
(b) a component bar chart.

Q5 The following chart shows the sales in a music shop over five days.

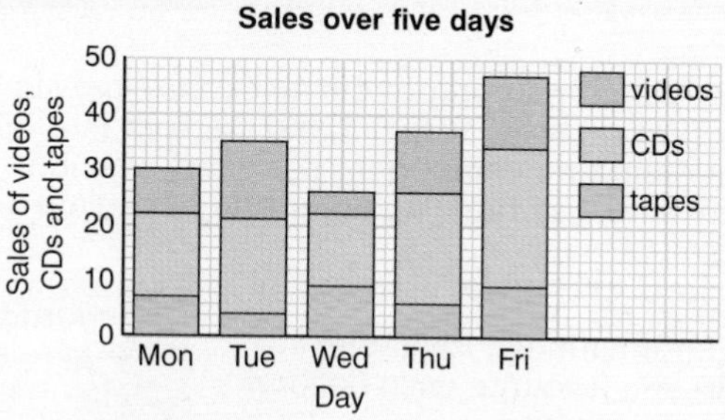

(a) On which day were the most videos sold?
(b) On which day were the most tapes sold?
(c) How many CDs were sold on Wednesday?
(d) How many tapes were sold altogether?
(e) How many sales were made on Thursday?

Use the above information to draw a percentage component bar chart.

REMEMBER! Cover the answers if you want to.

ANSWERS AND TUTORIAL

A1

Sorbets sold in one weekend

A2

(a) Sales of coloured socks

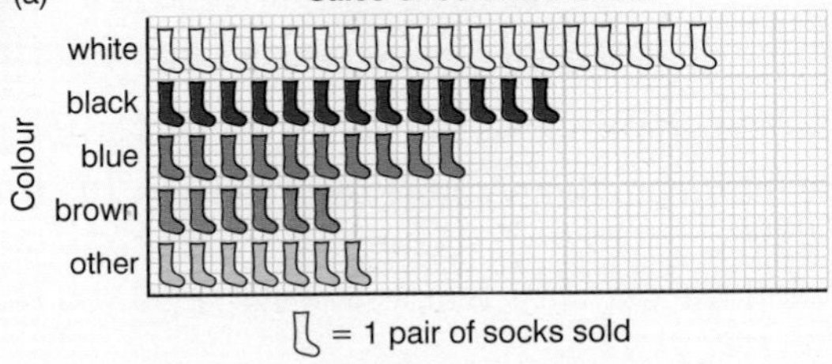

(b) Sales of coloured socks

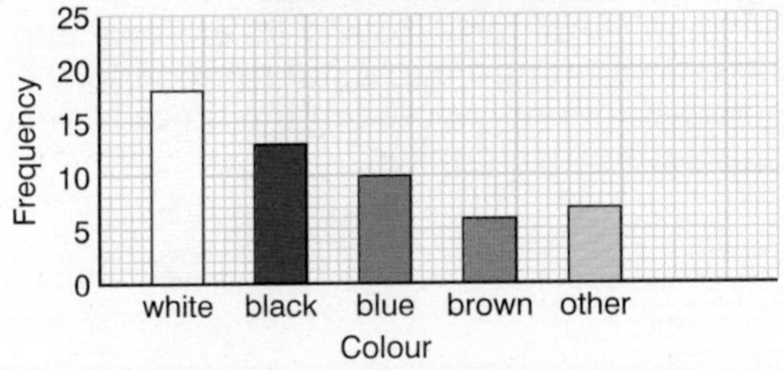

A3

Time taken to be served

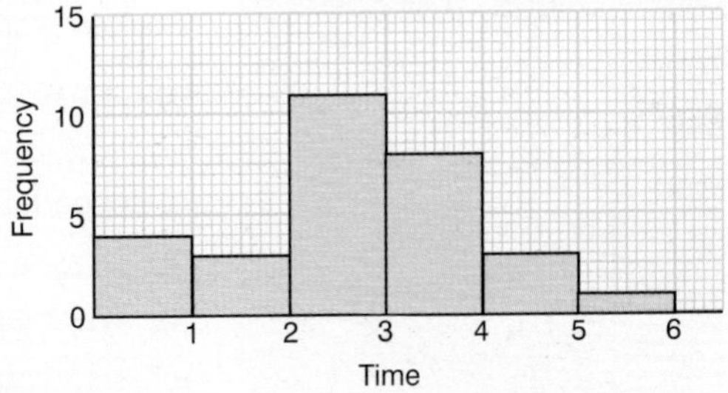

A4

(a) Income in three departments

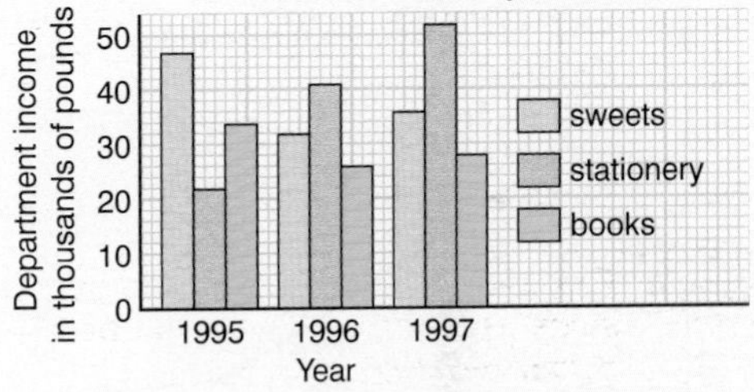

(b) Income in three departments

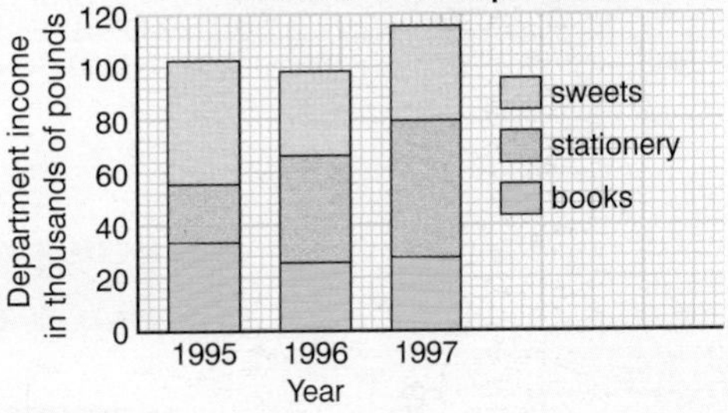

A5

(a) Tuesday
(b) Wednesday and Friday
(c) 13
(d) 35
(e) 37

Sales over five days

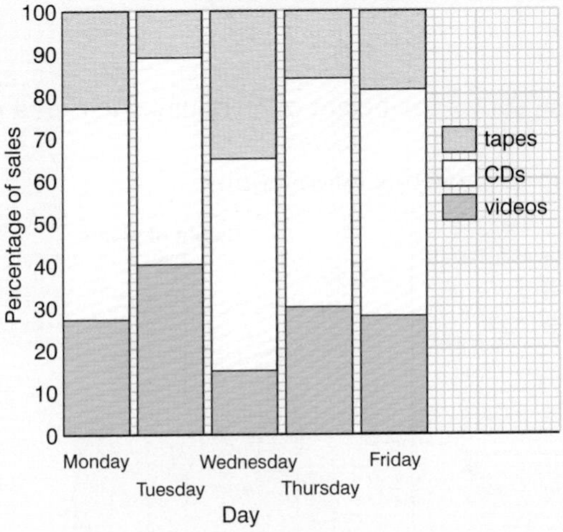

TUTORIAL

T1–T4 *Remember to give your representations a title and keep the symbols simple.*

T5

(a) *Tuesday (from the graph)*
(b) *Wednesday and Friday (from the graph)*
(c) *13 (from the graph)*
(d) $7 + 4 + 9 + 6 + 9 = 35$
(e) $11 + 20 + 6 = 37$

To draw the percentage component bar chart the following information needs to be converted to percentages.

Day	*Videos*	*CDs*	*Tapes*	*Total*
Monday	8 (27%)	15 (50%)	7 (23%)	30 (100%)
Tuesday	14 (40%)	17 (49%)	4 (11%)	35 (100%)
Wednesday	4 (15%)	13 (50%)	9 (35%)	26 (100%)
Thursday	11 (30%)	20 (54%)	6 (16%)	37 (100%)
Friday	13 (28%)	25 (53%)	9 (19%)	47 (100%)
	50	90	35	

All figures rounded to nearest integer.

LINE GRAPHS

Time	Temperature (°F)
0600	102.5
0700	102.8
0800	101.5
0900	100.2
1000	99.0
1100	98.8
1200	98.6

A **line graph** is another way of representing data. The frequencies are plotted at suitable points and joined by a series of straight lines. Once again, you must remember to label the axes clearly and give the diagram a title to explain what it represents.

Worked example

The information on the left shows the temperatures of a patient over a period of seven hours.

(a) Draw a line graph to show this data.

(b) What was the maximum recorded temperature?

(c) Use your graph to find an estimate of the patient's temperature at 0930.

(d) Explain why your answers in parts (b) and (c) are only approximate.

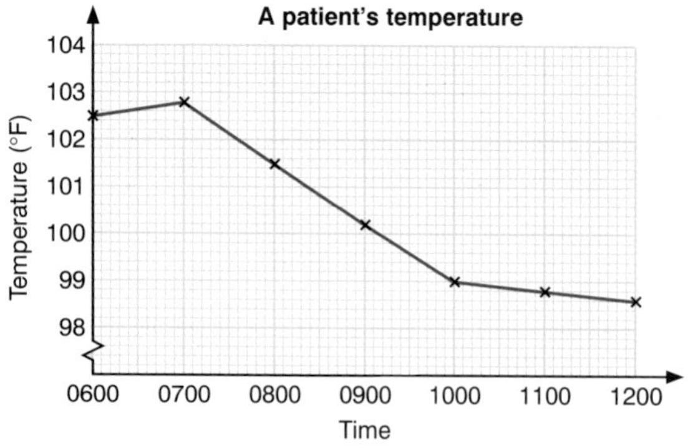

(a) This line graph shows the data.

(b) The maximum temperature recorded was 102.8 °F at 0700.

(c) To find the patient's temperature at 0930 you read off the value at 0930.
The patient's temperature at 0930 was 99.6 °F.

(d) The answers in parts (b) and (c) are only approximate as the temperature might not necessarily increase or decrease uniformly between the times at which the temperatures were taken.

FREQUENCY POLYGONS

Height (cm)	Frequency
6–10	1
11–15	3
16–20	7
21–25	9
26–30	7
31–35	10
36–40	7
41–45	5
46–50	1

A **frequency polygon** can be drawn from a bar chart (or histogram), by joining the midpoints of the tops of consecutive bars, with straight lines, to form a polygon. The lines should be extended to the horizontal axis on both sides, so that the area under the frequency polygon is the same as the area under the bar chart (or histogram).

Worked example

The frequency distribution shows the height of 50 plants, measured to the nearest centimetre.

Draw a frequency polygon to show this information.

The length is continuous and a bar chart (or histogram) of the information can be drawn as usual. The frequency polygon can be obtained by joining up the midpoints of the tops of the bars, in order, with straight lines.

The lines are extended to the horizontal axis on each side as shown.

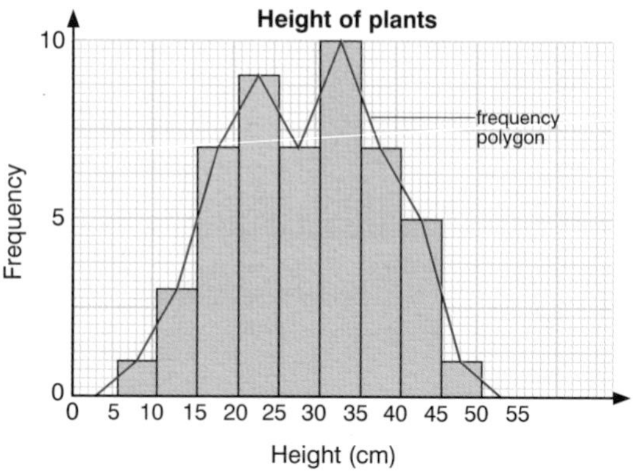

NOTE

The frequency polygon can be drawn without the bar chart, by plotting the frequencies at the midpoints of each interval.

Check yourself

QUESTIONS

Q1 The weight of a child is recorded at birth and at the end of each month as follows.

Age (months)	0	1	2	3	4	5
Weight (pounds)	8	9.5	10.8	12.4	13.8	15.2

Draw a line graph to represent this information.

Q2 The number of inches of rainfall at a holiday resort is recorded as follows.

Day	Friday	Saturday	Sunday	Monday	Tuesday	Wednesday	Thursday
Inches of rain	0.5	0.8	1.1	0.1	0.0	0.3	0.2

Draw a line graph to illustrate this information.

Q3 The lengths of 100 bolts are measured as follows.

Length (mm)	20–25	25–30	30–35	35–40	40–45
Frequency	3	8	15	7	2

Draw a frequency polygon to show this information.

REMEMBER! Cover the answers if you want to.

ANSWERS AND TUTORIAL

A1

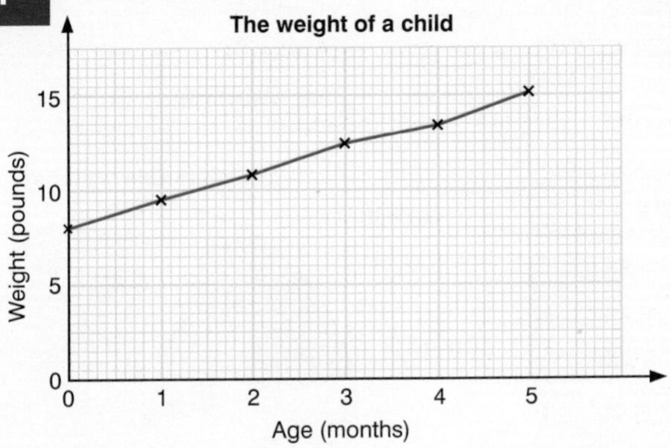

A2

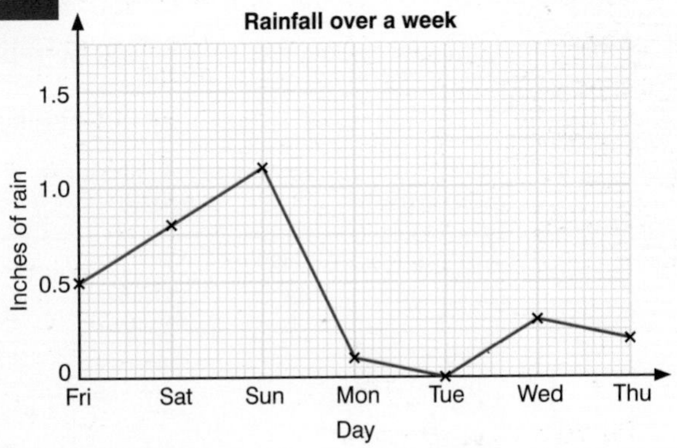

ANSWERS

A3

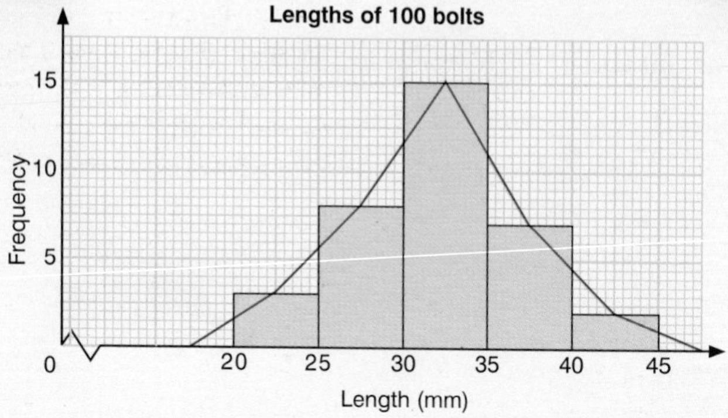

TUTORIAL

T3

Remember that the lines at each end should be extended to the horizontal axis as shown.

PIE CHARTS

A **pie chart** is another common way of representing data where the frequency is represented by the angles (or areas) of the sectors of a circle. When using a pie chart, you must remember to label each of the sectors clearly and give the diagram a title to explain what it represents.

The following worked examples are given to illustrate the construction of a pie chart.

Worked example

In a survey, 180 people were asked which TV channel they watched the most the previous evening. The answers to the survey are given in the table.

Channel	Frequency
BBC1	58
BBC2	20
ITV	42
C4	21
Channel 5	11
Other	18
Not watching	10

Construct a pie chart to show this information.

The pie chart needs to be drawn to represent 180 people. There are 360° in a full circle so each person will be represented by $\frac{360°}{180} = 2°$ of the pie chart.

Channel	Number	Angle
BBC1	58	$58 \times 2° = 116°$
BBC2	20	$20 \times 2° = 40°$
ITV	42	$42 \times 2° = 84°$
C4	21	$21 \times 2° = 42°$
Channel 5	11	$11 \times 2° = 22°$
Other	18	$18 \times 2° = 36°$
Not watching	10	$10 \times 2° = 20°$
		360°

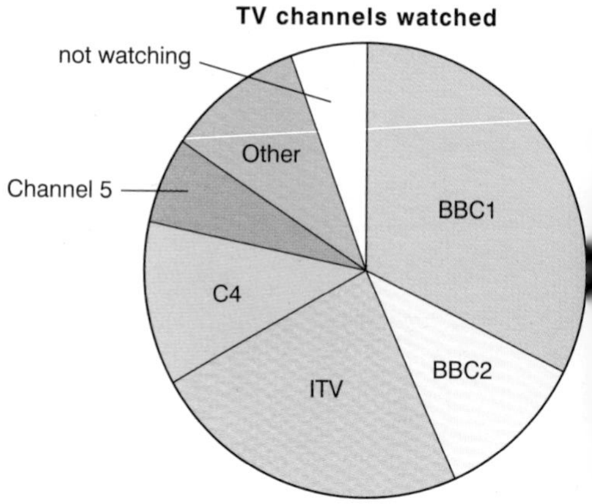

NOTE

It is always a good idea to check that your angles do add up to 360° before proceeding with the construction of the pie chart. Sometimes, rounding means that the total is not exact. In this case, you need to adjust the sizes of the angles to give a total of 360°.

Check yourself

QUESTIONS

Q1 250 students at a college were asked about the courses which they are following. Their responses were as follows.

A levels	106
GCSEs	42
GNVQs	86
Others	16

Construct the pie chart to show the different courses.

Q2 The following pie chart shows how 100 stockbrokers travelled to work one day in winter.

Travelling in winter

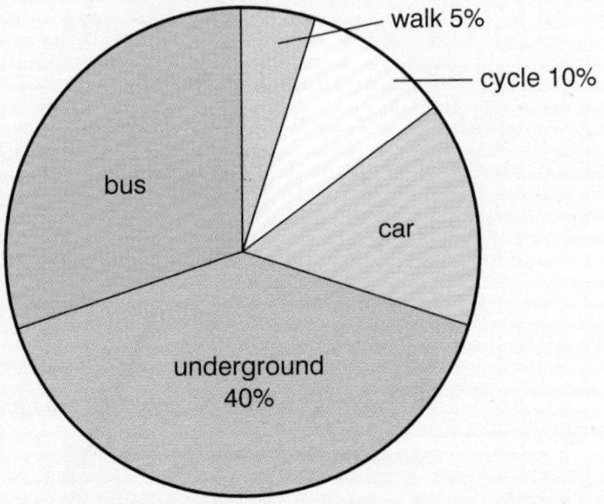

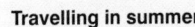

(a) Which method of transport was the most popular in winter?
(b) Which method of transport was the least popular in winter?
(c) What angle is represented by the 'cycle' sector?

The number who travelled to work by bus is twice as many as the number who travelled by car.

(d) How many stockbrokers travelled to work by bus in winter?
(e) How many stockbrokers travelled to work by car in winter?

The following pie chart shows how stockbrokers travelled to work one day in summer.

Travelling in summer

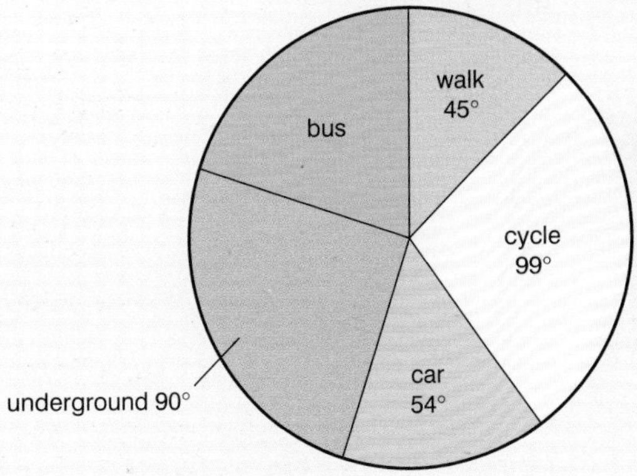

(f) Which method of transport was the most popular in summer?

During the summer 18 stockbrokers travelled to work by car.

(g) How many stockbrokers were surveyed in the summer?
(h) How many stockbrokers travelled to work by bus in the summer?

ANSWERS

A1

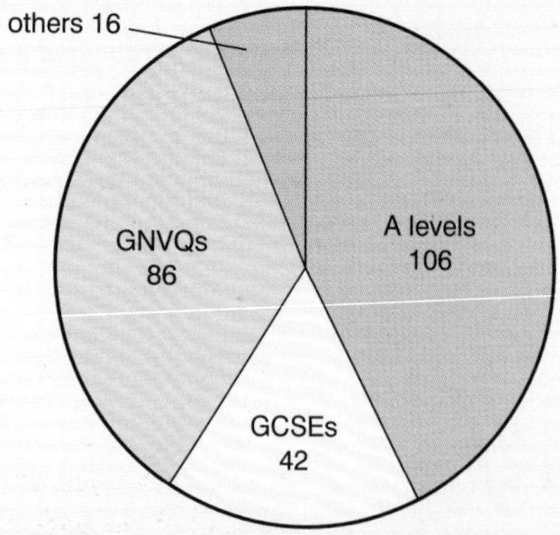

NOTE

The angles do not work out exactly, so they have to be rounded to the nearest whole number.

A2

(a) underground
(b) walking
(c) 36°
(d) 30
(e) 15
(f) cycling
(g) 120
(h) 24

TUTORIAL

T1

The pie chart needs to be drawn to represent 250 people. There are 360° in a full circle so each person is represented by 360° ÷ 250 = 1.44° of the pie chart.

Course	Number	Angle
A levels	106	106 × 1.44° = 153°
GCSEs	42	42 × 1.44° = 60°
GNVQs	86	86 × 1.44° = 124°
Others	16	16 × 1.44° = 23°
		360°

All angles are rounded to the nearest degree.

T2

(a) Underground (from the pie chart)
(b) Walking (from the pie chart)
(c) 10% × 360° = 36°

The number travelling by bus or by car = 45% × 100 = 45 people

Number who travelled by bus is twice the number who travelled by car.

(d) Travelled by bus: $\frac{2}{3} \times 45 = 30$

(e) Travelled by car: $\frac{1}{3} \times 45 = 15$

(f) Cycling (from the pie chart)
(g) 54° represents 18 stockbrokers.

1° represents $\frac{18}{54}$ *stockbrokers.* Dividing both sides by 54.

360° represents $360 \times \frac{18}{54}$ *stockbrokers.*

Multiplying both sides by 360.

360° represents 120 stockbrokers.
120 stockbrokers are surveyed altogether.

(h) Sector angle = 72°

1° represents $\frac{18}{54}$ *stockbrokers.* As shown in part (g).

72° represents $72 \times \frac{18}{54}$ *stockbrokers.*

Multiplying both sides by 72.

72° represents 24 stockbrokers.
24 stockbrokers travel by bus.

CHAPTER 81

MEASURES OF CENTRAL TENDENCY

Measures of central tendency are more often referred to as measures of average. You will need an understanding of the mode, median and mean for the examination.

MODE OF A DISTRIBUTION

The **mode** of a distribution is the value that occurs most frequently. If there are two modes then the distribution is called **bimodal**. If there are more than two modes then the distribution is called **multimodal**.

Worked example

Find the mode of the following distribution.

8, 6, 7, 4, 9, 8, 8, 6, 7, 6, 8

The number 4 occurs 1 time. The number 7 occurs 2 times.

The number 5 occurs 0 times. The number 8 occurs 4 times.

The number 6 occurs 3 times. The number 9 occurs 1 time.

The number 8 occurs the most frequently so the mode is 8.

NOTE

Make sure that you write down the value *of the mode and not the* frequency *in this work.*

MODE OF A FREQUENCY DISTRIBUTION

The mode of frequency distribution is the value that has the highest frequency.

Worked example

Find the mode of this frequency distribution.

Value	Frequency
4	1
5	0
6	3
7	2
8	4
9	1

Mode of frequency distribution. → 8

The mode of the frequency distribution is 8.

MODE OF A GROUP FREQUENCY DISTRIBUTION

The mode of a group frequency distribution has little meaning, although it is possible to identify a modal group.

Worked example

Find the modal group of this grouped frequency distribution.

Weight (grams)	Frequency
15–25	11
25–35	17
35–45	23
45–55	16
55–65	10

Modal group. → 35–45

The modal group of the grouped frequency distribution is 35–45.

Check yourself

QUESTIONS

Q1 Find the mode of the following pictorial representations.

(a)

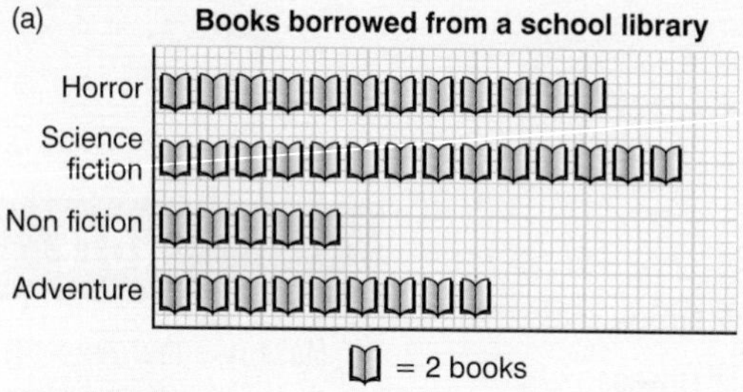

(b)

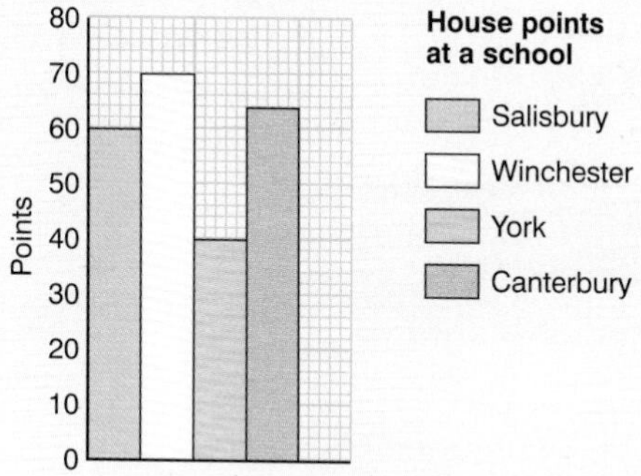

(c)

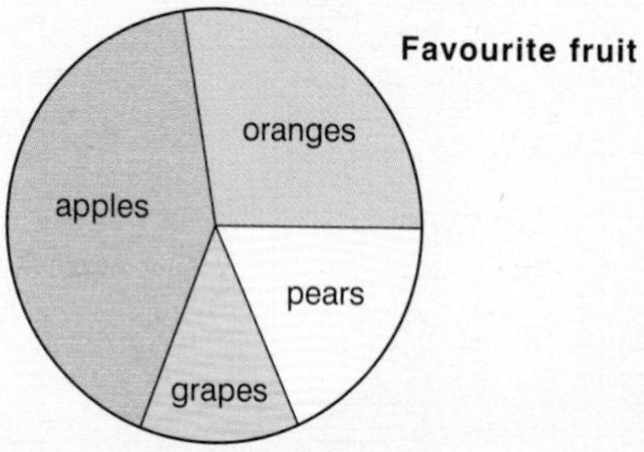

Q2 Find the mode of the following frequency distribution which shows the sizes of shoes sold in a shop.

Shoe size	$4\frac{1}{2}$	5	$5\frac{1}{2}$	6	$6\frac{1}{2}$	7	$7\frac{1}{2}$
Frequency	8	14	23	25	14	8	6

Q3 Find the modal group of the following grouped frequency distribution.

Weight (g)	**Frequency**
0.0–0.5	3
0.5–1.0	9
1.0–1.5	12
1.5–2.0	19
2.0–2.5	13
2.5–3.0	8

REMEMBER! Cover the answers if you want to.

ANSWERS

A1
(a) Science fiction books
(b) Winchester House
(c) Apples

A2 Size 6

A3 Group 1.5–2.0

TUTORIAL

T1
(a) Science Fiction books – greatest number of pictures
(b) Winchester House – longest bar
(c) Apples – largest angle/sector area

T2 *Size 6 – from the frequency*

T3 *Group 1.5–2.0 – from the frequency*

MEDIAN OF A DISTRIBUTION

The **median** of a distribution is the middle value when the values are arranged in order. Where there are two middle values (i.e. for an even number of values) then you add the two numbers and divide by 2.

NOTE

If there are n values in the distribution then the median position is given by $\frac{1}{2}(n + 1)$.

Worked example

Find the median of each of the following distributions.

(a) 8, 6, 7, 4, 9, 8, 8, 6, 7, 6, 8

(b) 8, 6, 7, 4, 9, 8, 8, 6, 7, 6, 8, 10

(a) For the distribution: 8, 6, 7, 4, 9, 8, 8, 6, 7, 6, 8

Rearrange in order: 4, 6, 6, 6, 7, 7, 8, 8, 8, 8, 9

The median position is given by $\frac{1}{2}(n + 1) = \frac{1}{2}(11 + 1) = 6$th value.

4, 6, 6, 6, 7, (7) 8, 8, 8, 8, 9
↑

So the median is 7.

(b) For the distribution: 8, 6, 7, 4, 9, 8, 8, 6, 7, 6, 8, 10

Rearrange in order: 4, 6, 6, 6, 7, 7, 8, 8, 8, 8, 9, 10

The median position is given by $\frac{1}{2}(n + 1) = \frac{1}{2}(12 + 1) = 6\frac{1}{2}$th value (i.e. between the 6th and 7th values).

4, 6, 6, 6, 7, (7, 8) 8, 8, 8, 9, 10
↑

So the median is $\frac{1}{2}(7 + 8) = 7\frac{1}{2}$.

MEDIAN OF A FREQUENCY DISTRIBUTION

To find the median of a frequency distribution you should work out the cumulative frequency distribution, as shown in the next example.

Worked example

Find the median of the frequency distribution in the table on the right.

Value	Frequency
4	1
5	0
6	3
7	2
8	4
9	1

To find the median of a of a frequency distribution you should work out the cumulative frequency like this.

From the cumulative frequency you can see that the middle value (i.e. the 6th value) occurs at 7.

Value	Frequency	Cumulative frequency
4	1	1
5	0	1
6	3	4
7	2	6
8	4	10
9	1	11

NOTE

To find the median of a group frequency distribution you need to draw up a cumulative frequency diagram. Cumulative frequency diagrams are considered fully in Chapter 83, Measures of spread and cumulative frequency diagrams.

Check yourself

QUESTIONS

Q1 Find the median of each of the following sets of numbers.

(a) 4, 1, 7, 8, 3
(b) 8, 1, 7, 11, 4, 3

Q2 Find the median of the following frequency distribution which shows the number of goals scored in 34 premier division matches.

Number of goals	Frequency
0	4
1	8
2	11
3	7
4	3
5	0
6	1

REMEMBER! Cover the answers if you want to.

ANSWERS

A1 (a) Median = 4
(b) Median = $5\frac{1}{2}$

A2 Median = 2

TUTORIAL

T1 (*a*) *Arranging in order* 1, 3, 4, 7, 8
Middle value = $\frac{1}{2}(5 + 1)$ *value* = *3rd value* = 4

(*b*) *Arranging in order* 1, 3, 4, 7, 8, 11
Middle value = $\frac{1}{2}(6 + 1)$ *value* = $3\frac{1}{2}$*th value*
= $\frac{1}{2}(4 + 7) = 5\frac{1}{2}$

T2 *To find the median of a frequency distribution you need to find the cumulative frequency.*

Number of goals	Frequency	Cumulative frequency
0	4	4
1	8	12
2	11	23
3	7	30
4	3	33
5	0	33
6	1	34

Median value = $\frac{1}{2}(34 + 1) = 17\frac{1}{2}$*th value which is* 2.

MEAN OF A DISTRIBUTION

The **mean** (or **arithmetic mean**) of a distribution is found by summing the values of the distribution and dividing by the number of values.

Worked example

Find the mean of the following distribution.

8, 6, 7, 4, 9, 8, 8, 6, 7, 6, 8

The mean is found by adding up the values of the distribution and dividing by the number of values.

$$\text{Mean} = \frac{8+6+7+4+9+8+8+6+7+6+8}{11}$$

$$= \frac{77}{11} = 7$$

The definition for the mean is often written as:

$$\text{mean} = \frac{\text{sum of the values}}{\text{number of values}} \quad \text{or} \quad \text{mean} = \frac{\Sigma \text{ values}}{\text{number of values}}$$

NOTE

Σ means 'the sum or total of'.

MEAN OF A FREQUENCY DISTRIBUTION

The mean of a frequency distribution is found by summing the values of the distribution and dividing by the number of values.

$$\text{Mean} = \frac{\text{sum of the values}}{\text{number of values}}$$

For a frequency distribution the sum of the values is equal to the sum of the products (frequency × value) or Σfx.

The number of values is the sum of the frequencies Σf. So the mean is $\frac{\Sigma fx}{\Sigma f}$.

Worked example

Find the mean of this frequency distribution.

Value x	Frequency f	Frequency × value fx
4	1	$1 \times 4 = 4$
5	0	$0 \times 5 = 0$
6	3	$3 \times 6 = 18$
7	2	$2 \times 7 = 14$
8	4	$4 \times 8 = 32$
9	1	$1 \times 9 = 9$
	$\Sigma f = 11$	$\Sigma fx = 77$

$$\text{Mean for the frequency distribution} = \frac{\Sigma fx}{\Sigma f}$$

$$= \frac{77}{11}$$

$$= 7 \quad \text{(as before)}$$

Check yourself

QUESTIONS

Q1 The diameters of five washers are 1.523 cm, 1.541 cm, 1.498 cm, 1.505 cm and 1.518 cm. Find the mean diameter of the five washers.

Q2 Find the mean of the following data.

Age (years)	17	18	19	20	21
Frequency	23	13	4	0	1

REMEMBER! Cover the answers if you want to.

ANSWERS

A1 1.517 cm

A2 17.6 years (3 s.f.)

TUTORIAL

T1

$$\textit{Total of diameters} = 1.523 + 1.541 + 1.498 + 1.505 + 1.518$$
$$= 7.585\ \textit{cm}$$
$$\textit{Mean diameter} = \frac{\textit{sum of the diameters}}{\textit{number of diameters}}$$
$$= \frac{7.585}{5}$$
$$= 1.517\ \textit{cm}$$

T2

Age (years) x	Frequency f	Frequency × age fx
17	23	391
18	13	234
19	4	76
20	0	0
21	1	21
	$\Sigma f = 41$	$\Sigma fx = 722$

$$\textit{Mean} = \frac{\Sigma fx}{\Sigma f}$$
$$= \frac{722}{41}$$
$$= 17.609\,756$$
$$= 17.6\ \textit{years}\ (3\ \textit{s.f.})$$

NOTE

The mid-interval values are used as an estimate of the particular interval so that the final answer will not be exact but will be an 'estimate of the mean'. The mid-interval value is found by taking the mean of the upper and lower class boundaries – see Chapter 83, Measures of spread and cumulative frequency diagrams, for definitions.

MEAN OF A GROUP FREQUENCY DISTRIBUTION

The mean of a group frequency distribution is found in the same way as for a frequency distribution, using the mid-interval values (or midpoints) as representative of the interval.

Worked example

The following table shows the heights of trees growing in a nursery. Calculate an estimate of the mean height of the trees.

Height (cm)	15–20	20–30	30–40	40–50	50–60	60–70	70–80
Frequency	8	4	5	11	17	2	1

Height	**Mid-interval value** x	**Frequency** f	**Frequency × mid-interval value** fx
15–20	17.5	8	$8 \times 17.5 = 140$
20–30	25	4	$4 \times 25 = 100$
30–40	35	5	$5 \times 35 = 175$
40–50	45	11	$11 \times 45 = 495$
50–60	55	17	$17 \times 55 = 935$
60–70	65	2	$2 \times 65 = 130$
70–80	75	1	$1 \times 75 = 75$
		$\Sigma f = 48$	$\Sigma fx = 2050$

For the group frequency distribution:

$$\text{mean} = \frac{\Sigma fx}{\Sigma f} = \frac{2050}{48} = 42.708\,333\,3 = 43\text{ cm to an appropriate degree of accuracy.}$$

NOTE

An answer of 43 cm is appropriate bearing in mind the accuracy of the original data and the inaccuracies resulting from the use of the mid-interval values as an estimate of the particular interval.

Check yourself

QUESTIONS

Q1 Calculate an estimate of the mean for this distribution.

Weight (kg)	0–10	10–20	20–30	30–40	40–50	50–60
Frequency	11	18	16	11	5	2

Q2 The amount of money spent by customers in a restaurant is shown in this table.

Use the information to calculate an estimate of the mean.

Amount (£)	**Frequency**
0 and less than 5	12
5 and less than 10	15
10 and less than 15	8
15 and less than 20	7
20 and less than 25	3

REMEMBER! Cover the answers if you want to.

ANSWERS

A1 23 kg to an appropriate degree of accuracy.

A2 £9.61 to an appropriate degree of accuracy.

TUTORIAL

T1

Weight (kg)	Mid-interval value x	Frequency f	Frequency × mid-interval value fx
0–10	5	11	55
10–20	15	18	270
20–30	25	16	400
30–40	35	11	385
40–50	45	5	225
50–60	55	2	110
		$\Sigma f = 63$	$\Sigma fx = 1445$

For the group frequency distribution:

$$\textit{mean} = \frac{\Sigma fx}{\Sigma f}$$

$$= \frac{1445}{63}$$

$$= 22.936\,508$$

= 23 *kg to an appropriate degree of accuracy.*

T2

Amount (£)	Mid-interval value x	Frequency f	Frequency × mid-interval value fx
0 and less than 5	2.5	12	30
5 and less than 10	7.5	15	112.5
10 and less than 15	12.5	8	100
15 and less than 20	17.5	7	122.5
20 and less than 25	22.5	3	67.5
		$\Sigma f = 45$	$\Sigma fx = 432.5$

For the group frequency distribution:

$$\textit{mean} = \frac{\Sigma fx}{\Sigma f}$$

$$= \frac{432.5}{45}$$

$$= 9.611\,111\,1$$

= £9.61 *to an appropriate degree of accuracy.*

EXAM PRACTICE

Sample Student's Answers & Examiner's Comments

1 The following pie chart shows how nurses travel to work at one hospital.
You are given that 39 nurses walk to work.

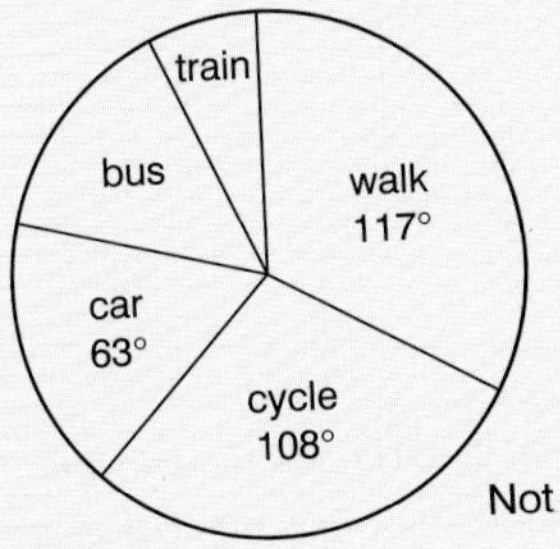

Not to scale

(a) How many nurses cycle?

(b) How many nurses travel by car?

(c) How many nurses are there altogether?

Three times as many nurses travel by bus as travel by train.

(d) How many nurses travel by bus?

(e) How many nurses travel by train?

39 nurses = 117°

1 nurse $= \frac{117°}{39} = 3°$

(a) Cycle $= \frac{108}{3} = 36$

36 nurses cycle

(b) Car $= \frac{63}{3} = 21$

21 nurses travel by car

(c) Altogether $= \frac{360}{3} = 120$ nurses

(d) 24 nurses left

18 nurses travel by bus

(e) 6 nurses travel by train

EXAMINER'S COMMENTS

1 **a), b), c)**
The candidate has correctly found the number of nurses cycling and travelling by car, using the angles given on the diagram. He has appreciated that the total for the nurses is represented by 360° and has used this to find how many nurses there are altogether.

d), e)
The candidate has used the information from the question rather than attempting to measure the angles (especially as the diagram is labelled that it is not drawn to scale).

120 – (39 + 36 + 21) = 24

He has divided the number of nurses left into 3 + 1 = 4 parts according to the work on ratio – see Chapter 13, Ratios and proportional division.

EXAMINER'S COMMENTS

2 Ian looked at a passage from a book. He recorded the number of words in each sentence in a frequency table using class intervals of 1–5, 6–10, 11–15, etc.

Class interval	Frequency f
1–5	16
6–10	28
11–15	26
16–20	14
21–25	10
26–30	3
31–35	1
36–40	0
41–45	2

(a) Write down:

(i) the modal class interval
(ii) the class interval in which the median lies.

(b) Work out an estimate of the mean number of words in a sentence.

(London specimen paper 1998)

(a) (i) Modal class interval = 6–10
(ii) Median $= \frac{1}{2}(n + 1)$

$= \frac{1}{2} \times 101$

$= 50\frac{1}{2}$

$= 11–15$

Class interval	Frequency f	Mid-interval x	fx
1–5	16	3	48
6–10	28	8	224
11–15	26	13	338
16–20	14	18	252
21–25	10	23	230
26–30	3	28	84
31–35	1	33	33
36–40	0	38	0
41–45	2	43	86
	$\Sigma f = 100$		$\Sigma fx = 1295$

(b) Mean $= \frac{1295}{100}$

$= 12.95$

2 **a)**
The candidate would be advised to lay the work out more clearly here to avoid possible errors. The median value is the $50\frac{1}{2}$th value (although for large numbers $\frac{1}{2}n$ is just as acceptable). This value occurs in the 11–15 interval.

b)
She has correctly calculated an estimate for the mean although an answer to 3 s.f. or 2 s.f. might be more appropriate in view of the approximating involved.

Questions to Answer

1 A questionnaire for schoolchildren includes the following question.

'How much pocket money do you get?'

Less than £1 ☐ Between £2 and £5 ☐ More than £5 ☐

Write down two criticisms of this question.

(SEG Modular syllabus, specimen paper 1998)

2 The following frequency distribution shows the different types of books borrowed from a library one weekend.

Type of book	Frequency
Sport	12
Crime	31
Horror	29
Romance	34
Other	14

Show this information as:

(a) a pictogram

(b) a bar chart

(c) a pie chart.

3 A teacher records the marks of students in the record book as follows.

16 18 17 15 20 16 15 17 16 19

(a) Calculate:

(i) the mean mark (ii) the median mark (iii) the modal mark?

The teacher realises that one of the marks recorded as 15 should have been 14.

(b) What effect will this have on:

(i) the mean mark (ii) the median mark (iii) the modal mark?

4 The following information shows the marks awarded to students in an end of term exam.

Mark	14	15	16	17	18	19	20
Frequency	2	1	6	12	11	10	8

What is the mean mark?

5 The length of time which it takes to get through to information services is given in the table.

Calculate an estimate of the mean length of time which it takes to get through to information services.

Length of time (minutes)	Frequency
0 and less than 1	107
1 and less than 2	89
2 and less than 3	36
3 and less than 4	21
4 and less than 5	11
5 and less than 6	6

CHAPTER 83

MEASURES OF SPREAD AND CUMULATIVE FREQUENCY DIAGRAMS

The measure of spread gives a measure of how spread out the data are. For the purposes of the examination you will need to be familiar with the range, interquartile range and standard deviation (although the work on standard deviation is confined to the Higher level).

THE RANGE

The **range** of a distribution is found by working out the difference between the greatest value and least value. The range should therefore always be given as a single value.

Worked example

The following information details the insurance premiums paid by eleven households.

£340 £355 £400 £320 £380 £320 £632 £365 £340 £380 £370

Calculate the mean and the range.

$$\text{Mean} = \frac{£340 + £355 + £400 + £320 + £380 + £320 + £632 + £365 + £340 + £380 + £370}{11}$$

$$\text{Mean} = \frac{4202}{11}$$

$$= £382$$

Greatest value = £632

Least value = £320

Range = greatest value – least value = £632 – £320 = £312

> **NOTE**
>
> The mean of £382 is deceptive as a measure of central tendency because it is affected by the value of £632 – this type of value is sometimes called an extreme value. Similary, the range of £312 is deceptive as a measure of spread because again, it is also affected by the value of £632.

INTERQUARTILE RANGE

Although the range is affected by extreme values, the **interquartile range** considers only the middle 50% of the distribution. The interquartile range is found by dividing the data into four parts or **quartiles** and working out the difference between the upper quartile and the lower quartile, as shown in the following worked example.

Worked example

The following information details the insurance premiums paid by eleven households.

£340 £355 £400 £320 £380 £320 £632 £365 £340 £380 £370

Calculate the interquartile range.

Arranging the data in order and considering the middle 50%:

£320 £320 £340 £340 £355 £365 £370 £380 £380 £400 £632

(↑ LQ at the third value £340; ↑ Median at £365; ↑ UQ at the ninth value £380)

Upper quartile = £380

Lower quartile = £340

Interquartile range = upper quartile – lower quartile

= £380 – £340

= £40

> **NOTE**
>
> You have already seen that if there are n values in the distribution then the median position is given by $\frac{1}{2}(n + 1)$. Similarly, the lower quartile position is given by $\frac{1}{4}(n + 1)$ and the upper quartile position is given by $\frac{3}{4}(n + 1)$.

CUMULATIVE FREQUENCY DIAGRAMS

Cumulative frequency diagrams can be used to find the median and the quartiles of a variety of distributions, including grouped frequency distributions. To find the cumulative frequency you find the accumulated totals which are then plotted on the cumulative frequency diagram (or **ogive**) by joining them up with a smooth curve.

NOTE

For this work it might be helpful to remind yourself of these definitions.

CLASS LIMITS

Class limits are the values given in each of the individual groups (or class intervals). For the class interval 1–3, in the example below, the class limits are 1 and 3 (where the lower class limit is 1 and the upper class limit is 3).

CLASS BOUNDARIES

As the times are given to the nearest minute then the interval 1–3 will actually include times from 0.5 minutes to 3.5 minutes. The class boundaries are 0.5 and 3.5 (where the lower class boundary is 0.5 and the upper class boundary is 3.5).

CLASS WIDTH

The class width, class length or class size is the difference between the upper and lower class boundaries.

For the class interval 1–3 with lower class boundary 0.5 and upper class boundary 3.5 then the class width equals $3.5 - 0.5 = 3$ minutes.

Worked example

The following information shows the time (given to the nearest minute) which customers have to wait in a checkout queue.

Waiting time (minutes)	Frequency
1–3	8
4–6	19
7–9	11
10–12	6
13–15	2
16–18	1

Draw the cumulative frequency diagram for the information and use it to find:

(a) how many customers waited less than 5 minutes

(b) how many customers waited more than 10 minutes

(c) the median and the interquartile range.

First complete the table to include the cumulative frequencies. Then draw the cumulative frequency diagram.

Waiting time (minutes)	Frequency	Cumulative frequency
1–3	8	8
4–6	19	27
7–9	11	38
10–12	6	44
13–15	2	46
16–18	1	47

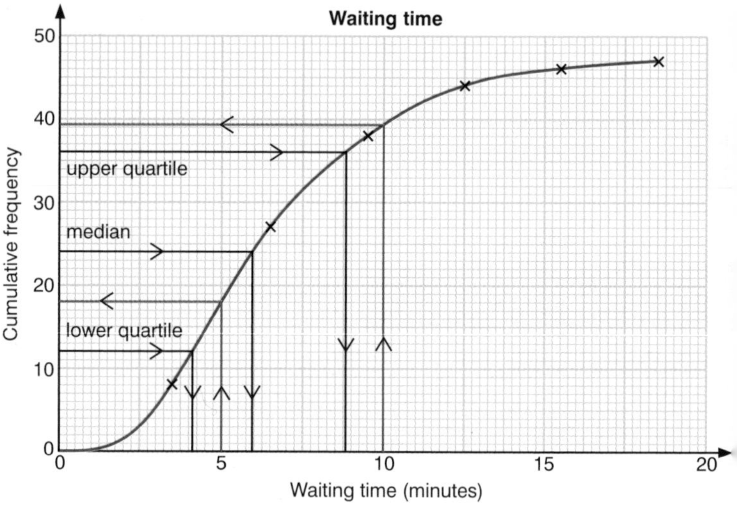

NOTE

The final cumulative frequency should, of course, equal the sum of the frequencies.

NOTE

The cumulative frequencies must be plotted at the upper class boundaries (i.e. 3.5, 6.5, 9.5, 12.5, 15.5 and 18.5).

(a) To find out how many customers waited less than 5 minutes, the information should be read off against a waiting time of 5 minutes.

From the graph, the number of customers who waited less than 5 minutes is 18.

(b) To find out how many customers waited more than 10 minutes then the information should be read off against a waiting time of 10 minutes.

From the graph, the number of customers who waited less than 10 minutes is 39 so that the number of customers who waited more than 10 minutes is $47 - 39 = 8$.

(c) To find the median waiting time you need to read off the median value on the cumulative frequency.

The median position is given by $\frac{1}{2}(n + 1) = \frac{1}{2}(47 + 1)$ $= 24$th value.

To find the interquartile range you need to find the lower quartile and the upper quartile.

The lower quartile position is given by

$\frac{1}{4}(n + 1) = \frac{1}{4}(47 + 1) = 12$th value and the upper quartile position is given by $\frac{3}{4}(n + 1) = \frac{3}{4}(47 + 1) = 36$th value.

From the graph:

median = 6

Similarly, from the graph:

upper quartile = 8.8

lower quartile = 4.1

interquartile range = upper quartile – lower quartile

= 8.8 – 4.1

= 4.7

Check yourself

QUESTIONS

Q1 The following information gives the heights of 15 plants.

13 cm, 36 cm, 15 cm, 13 cm, 21 cm, 18 cm, 37 cm, 18 cm, 12 cm, 21 cm, 18 cm, 20 cm, 6 cm, 37 cm, 39 cm

Find the range and interquartile range.

Q2 The frequency distribution for the time taken to obtain clearance through customs is given in the following table.

Time (t minutes)	Frequency
$20 \leq t < 25$	3
$25 \leq t < 30$	7
$30 \leq t < 35$	15
$35 \leq t < 40$	18
$40 \leq t < 45$	22
$45 \leq t < 50$	17
$50 \leq t < 55$	8
$55 \leq t < 60$	2

Draw a cumulative frequency curve and use it to find an estimate of the median and the interquartile range.

Q3 The following table shows the number of words per paragraph in a children's book.

Number of words per paragraph	Number of paragraphs
1–10	17
11–20	33
21–30	51
31–40	21
41–50	18

Draw a cumulative frequency curve to illustrate this information and use your graph to estimate:

(a) the median and interquartile range
(b) the percentage of paragraphs over 35 words in length.

REMEMBER! Cover the answers if you want to.

ANSWERS

A1 Range = 33 cm
Interquartile range = 23 cm

TUTORIAL

T1 *Arranging the information in order:*

6 cm, 12 cm, 13 cm, 13 cm, 15 cm, 18 cm, 18 cm, 18 cm, 20 cm, 21 cm, 21 cm, 36 cm, 37 cm, 37 cm, 39 cm

The range of the heights is 39 – 6 = 33 cm.

To find the interquartile range you need to find the lower quartile and the upper quartile. The lower quartile is the $\frac{1}{4}(15 + 1) = 4$*th value and the upper quartile is the* $\frac{3}{4}(15 + 1) = 12$*th value.*

From the data: 6 cm, 12 cm, 13 cm, 13 cm (↑ LQ), 15 cm, 18 cm, 18 cm, 18 cm (↑ Median), 20 cm, 21 cm, 21 cm, 36 cm (↑ UQ), 37 cm, 37 cm, 39 cm

Interquartile range = upper quartile – lower quartile
= 36 – 13 = 23 cm

ANSWERS

A2

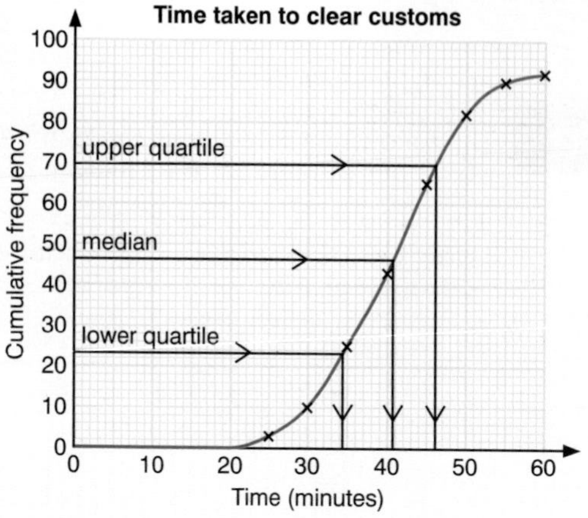

Median = 41
Interquartile range = $11\frac{1}{2}$

NOTE

The cumulative frequencies must be plotted at the upper class boundaries (i.e. 25, 30, 35, 40, etc.)

A3

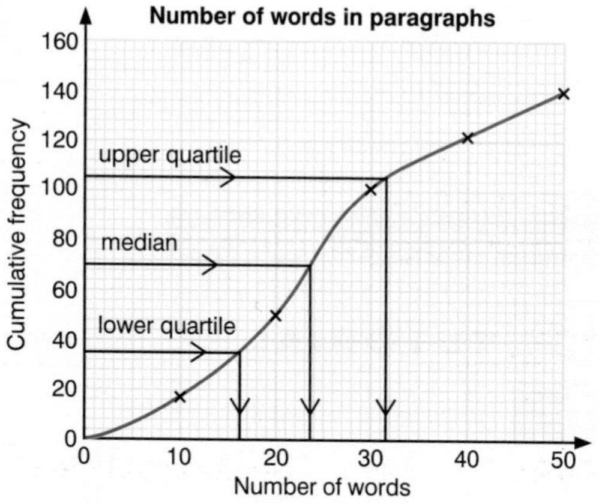

(a) Median = 24, interquartile range = 16
(b) Percentage of paragraphs over 35 words = 19%

NOTE

The cumulative frequencies must be plotted at the upper class boundaries (i.e. 10, 20, 30, etc.) as the number of words is discrete in this instance.

TUTORIAL

T2

Time (t minutes)	Frequency	Cumulative frequency
$20 \leqslant t < 25$	3	3
$25 \leqslant t < 30$	7	10
$30 \leqslant t < 35$	15	25
$35 \leqslant t < 40$	18	43
$40 \leqslant t < 45$	22	65
$45 \leqslant t < 50$	17	82
$50 \leqslant t < 55$	8	90
$55 \leqslant t < 60$	2	92

From the graph:
median = 41
upper quartile = 46
lower quartile = $34\frac{1}{2}$
Interquartile range = *upper quartile* – *lower quartile*
= $46 - 34\frac{1}{2}$
= $11\frac{1}{2}$

T3

Number of words per paragraph	Number of paragraphs	Cumulative frequency
1–10	17	17
11–20	33	50
21–30	51	101
31–40	21	122
41–50	18	140

(*a*) *From the graph:*
median = 24
upper quartile = 32
lower quartile = 16
Interquartile range
= *upper quartile* – *lower quartile* = 32 – 16
= 16

(*b*) *Number of paragraphs under 35 words in length*
= 114

Number of paragraphs over 35 words in length
= 140 – 114 = 26

Percentage of paragraphs over 35 words in length
$= \frac{26}{140} \times 100 = 19\%$

CHAPTER 84

SCATTER DIAGRAMS AND LINES OF BEST FIT

Scatter diagrams are representations that are used to show the relationship between two variables. Each of the two variables is assigned to a different axis and the information is plotted as different coordinates on the scatter diagram.

SCATTER DIAGRAMS

Worked example

The following table shows the heights and shoe sizes of 10 pupils.

Shoe size	3	2	5	$6\frac{1}{2}$	4	3	6	1	$3\frac{1}{2}$	$7\frac{1}{2}$
Height (cm)	133	126	150	158	135	128	152	118	142	101

Draw a scatter diagram of the information.

Considering each pair of values as a different pair of coordinates then the following scatter diagram is produced.

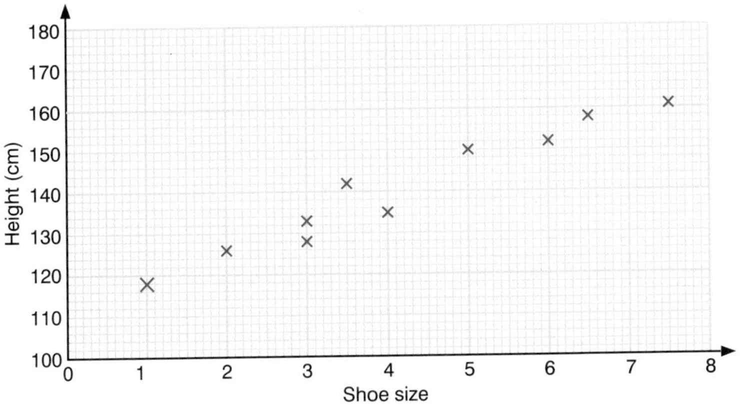

CORRELATION

Scatter graphs can be used to show whether there is any relationship or correlation between the two variables. You need to know the following descriptions of such relationships or correlation.

Little or no apparent correlation

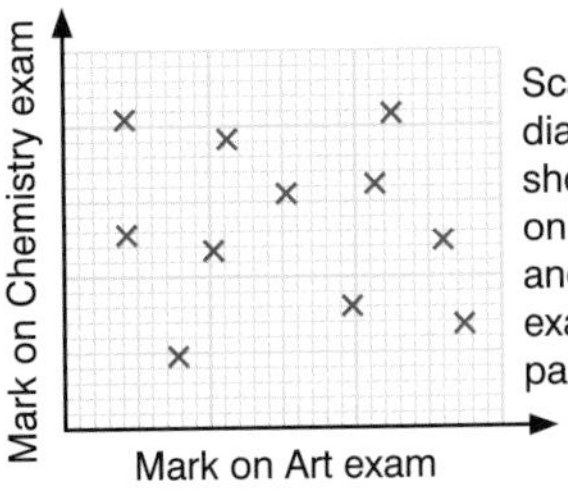

Scatter diagram to show marks on Chemistry and Art examination papers.

The points are scattered randomly over the graph indicating little or no correlation between the two variables.

Moderate correlation

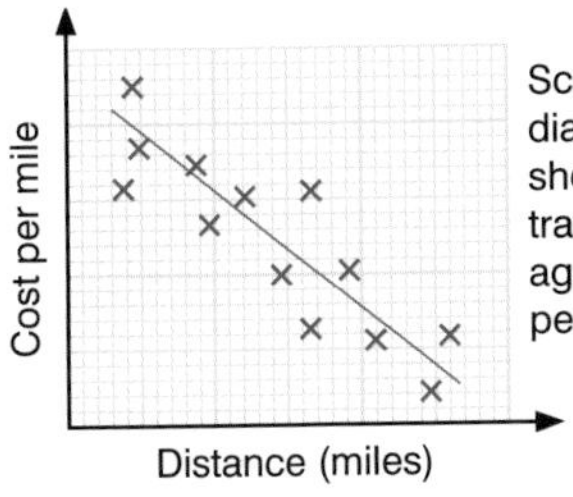

Scatter diagram to show distance travelled against cost per mile.

The points lie close to a straight line indicating moderate correlation between the two variables (the closer the points are to the line then the stronger the correlation).

Strong correlation

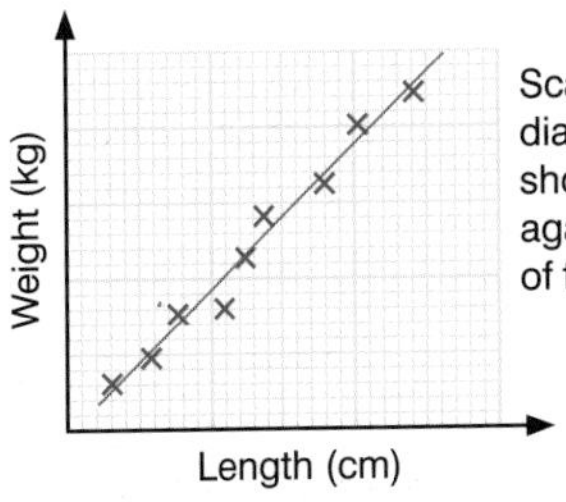

Scatter diagram to show length against weight of fish.

The points lie along a straight line indicating a strong correlation between the two variables.

Correlation can also be described in terms of **positive** and **negative** correlation.

Positive correlation

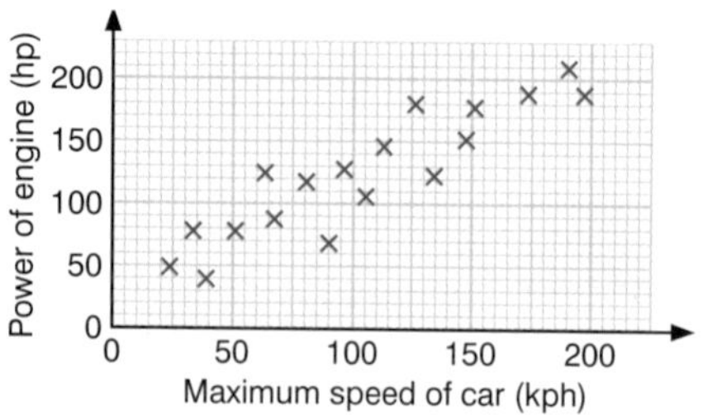

Where an increase in one variable is associated with an increase in the other variable then the correlation is said to be positive or direct.

Negative correlation

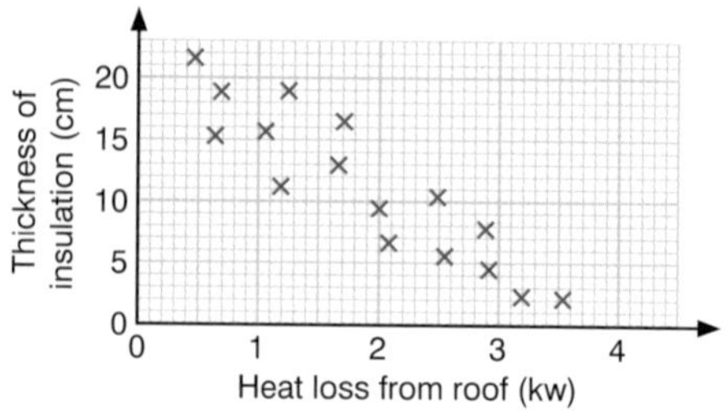

Where an increase in one variable is associated with a decrease in the other variable then the correlation is said to be negative or inverse.

Check yourself

QUESTIONS

Q1 Describe fully the relationship between the two variables in each of the following scatter diagrams.

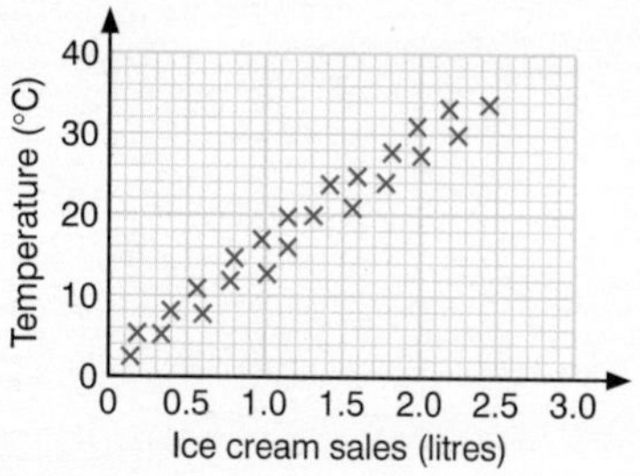

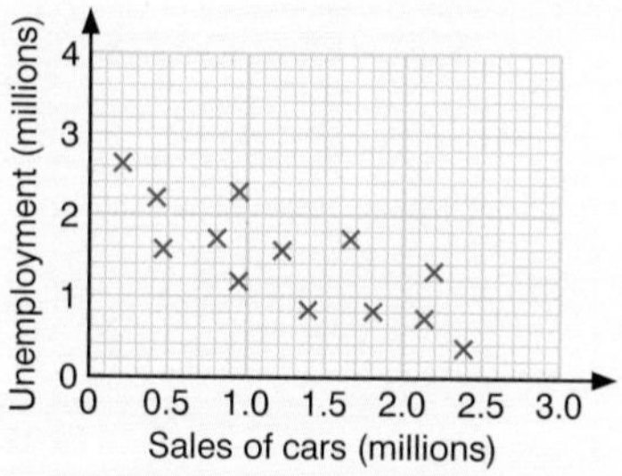

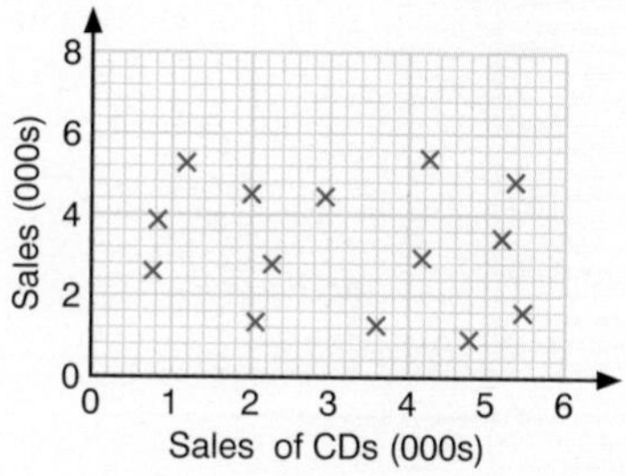

Q2 For each of the following sets of data, draw a scatter diagram and describe fully the relationship between the two variables.

(a)

Variable P	19	17	5	14	26	5	24	8
Variable Q	13	19	37	21	4	33	10	29

(b)

Variable R	35	10	50	18	39	24
Variable S	0.35	0.05	0.58	0.13	0.47	0.27

(c)

Variable M	30	10	20	37	16	45	55	25	46
Variable N	68	20	57	48	40	47	21	57	30

Q3 The following information shows the marks awarded to students on two examination papers. Plot the points on a graph and comment on your findings.

Paper 1	30	56	40	68	14	85	64	28	79	48	44	59
Paper 2	30	39	20	73	16	83	45	15	62	44	32	61

REMEMBER! Cover the answers if you want to.

ANSWERS AND TUTORIAL

A1
(a) Strong positive correlation
(b) Moderate negative correlation
(c) No correlation

A2
(a)

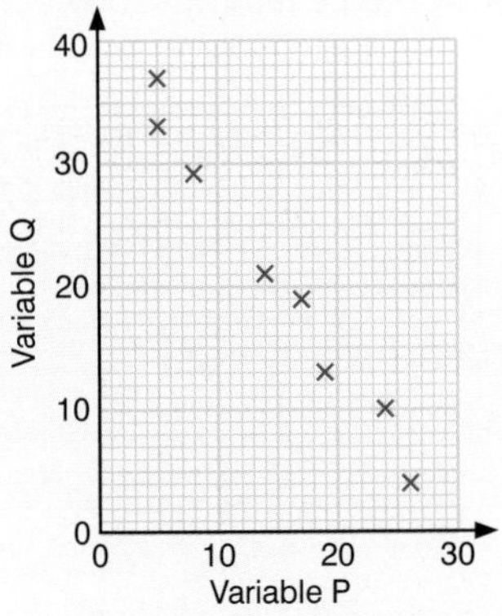

Strong negative correlation

(b)

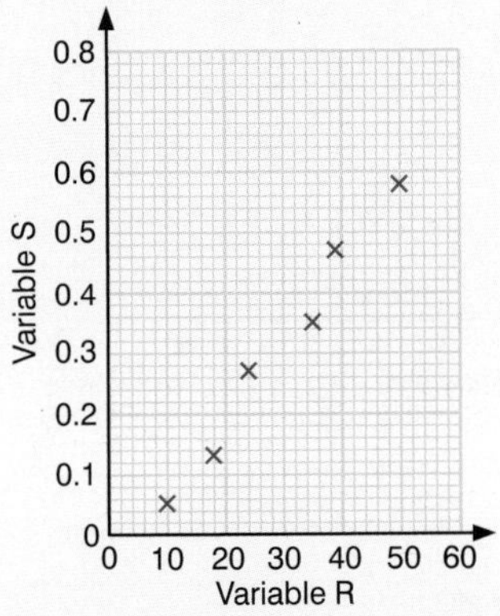

Strong positive correlation

(c)

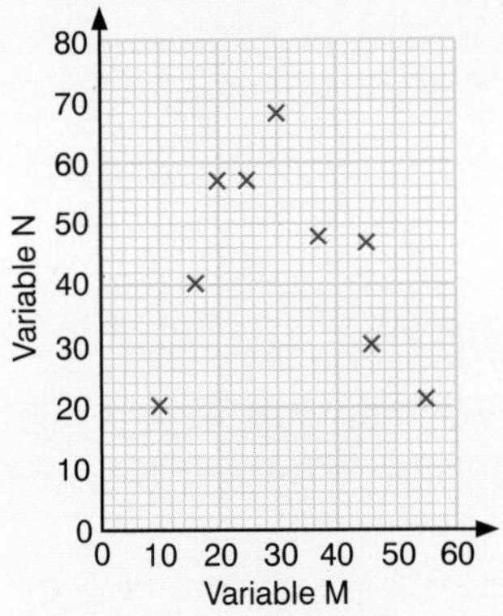

There is no definable correlation between the two variables although there is some correlation here as demonstrated by the U-shaped curve.

A3

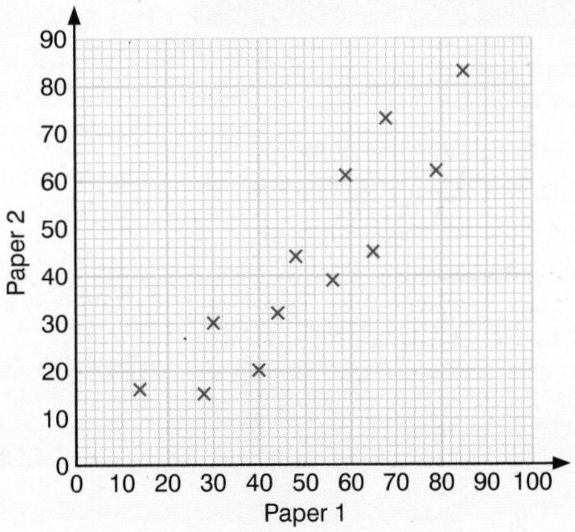

The diagram shows moderate positive correlation.

NOTE

Refer to the section on correlation for further information.

LINE OF BEST FIT

NOTE

In most cases a line of best fit can be drawn 'by eye'. For more accurate work the line of best fit should pass through $(\bar{x}, \bar{y})$ where $\bar{x}$ and $\bar{y}$ are the mean values of x and y respectively.

Where the points on a scatter diagram show moderate or strong correlation then a line can be drawn to approximate the relationship. This line is called the **line of best fit** (or **regression line**) and this line can be used to predict other values from the given data.

Worked example

The following table (from the previous worked example) shows the heights and shoe sizes of 10 pupils.

Shoe size	3	2	5	$6\frac{1}{2}$	4	3	6	1	$3\frac{1}{2}$	$7\frac{1}{2}$
Height (cm)	133	126	150	158	135	128	152	118	142	161

Draw a scatter diagram and the line of best fit.

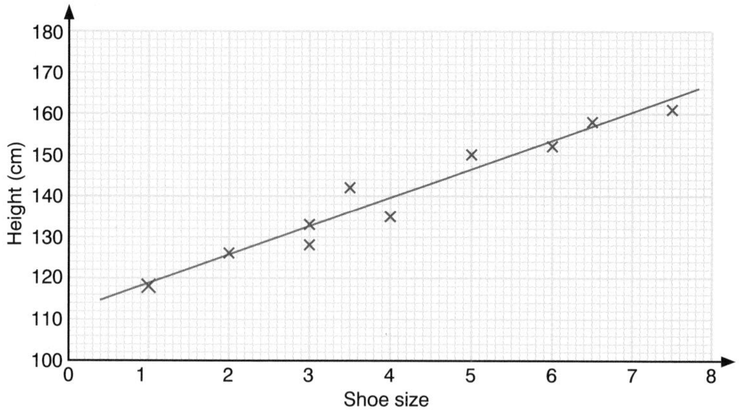

Check yourself

QUESTIONS

Q1 The following information shows the temperature and the production rate for a factory.

Temperature (°C)	23.4	21.4	25.1	23.4	24.8	21.7	20.9	22.7	21.7	23.8	22.2
Production rate (units)	47	53	35	53	41	58	60	59	64	40	50

Draw a scatter diagram and show the line of best fit.

Use your line of best fit to calculate:

(a) the production rate for a temperature of 22.8°C
(b) the temperature for a production rate of 50 units.

QUESTIONS

Q2 The following information shows the marks awarded to students on two examination papers.

Paper 1	30	56	40	68	14	85	64	28	79	48	44	59
Paper 2	30	39	20	73	16	83	45	15	62	44	32	61

Draw the line of best fit and use it to calculate:

(a) the mark of someone who gets 65 on paper 1
(b) the mark of someone who gets 75 on paper 2.

REMEMBER! Cover the answers if you want to.

ANSWERS

A1

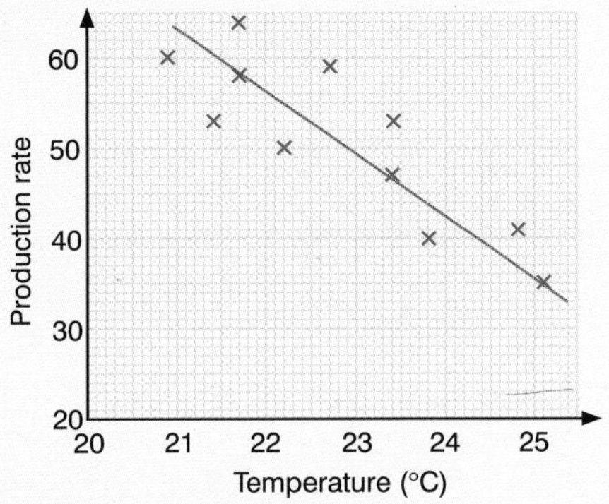

From the graph:

(a) production rate = 51 units
(b) temperature = 22.9°C

A2

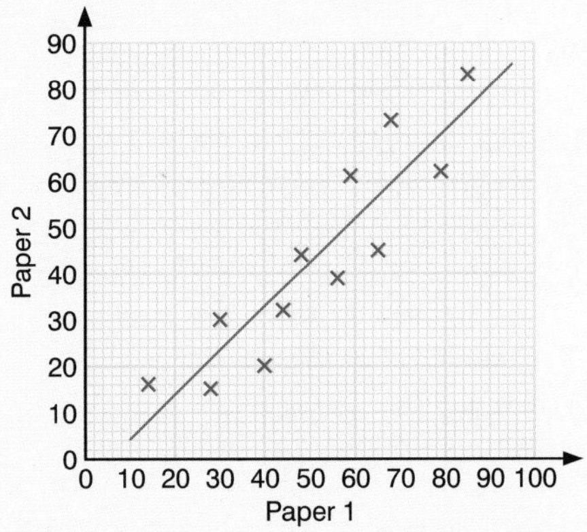

(a) 57 marks (b) 84 marks

TUTORIAL

T1 *From the graph:*

(*a*) *production rate* = 51 *units, by reading off the data when the temperature equals* 22.8°C
(*b*) *temperature* = 22.9°C, *by reading off the data when the production rate equals* 50 *units.*

T2 *From the graph:*

(*a*) 57 *marks, by reading off the data when paper* 1 *equals* 65
(*b*) 84 *marks, by reading off the data when paper* 2 *equals* 75.

CHAPTER 85

EXAMINATION QUESTIONS 17

EXAM PRACTICE

Sample Student's Answers & Examiner's Comments

1 The speeds in miles per hour (mph) of 200 cars travelling on the A320 road were measured. The results are shown in the table.

Speed (mph)	Cumulative frequency
not exceeding 20	1
not exceeding 25	5
not exceeding 30	14
not exceeding 35	28
not exceeding 40	66
not exceeding 45	113
not exceeding 50	164
not exceeding 55	196
not exceeding 60	200
TOTAL	200

(a) Draw a cumulative frequency graph to show these figures.

(b) Use your graph to find an estimate for:

(i) the median speed (in mph)
(ii) the interquartile range (in mph)
(iii) the percentage of cars travelling at less than 48 miles per hour.

(London specimen paper 1998)

(a)

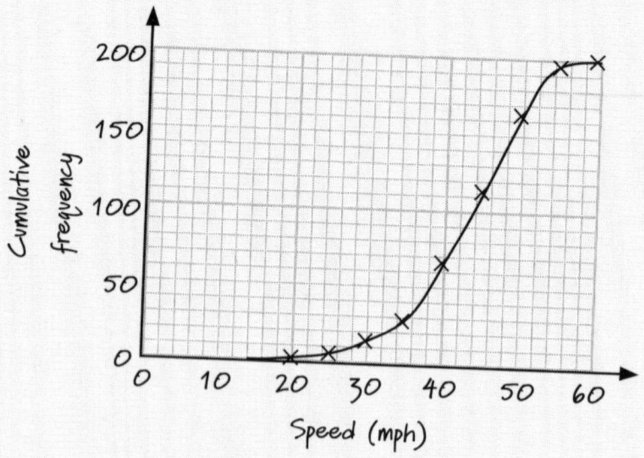

(b) (i) Median = 44 mph

(ii) LQ = 38

UQ = 49

interquartile range = 49 − 38 = 11 mph

(iii) Number of cars travelling at less than 48 miles per hour = 140

Percentage of cars travelling at less than 48 miles per hour = $\frac{140}{200} \times 100\% = 70\%$

EXAMINER'S COMMENTS

1 **a)**
The candidate has drawn the cumulative frequency graph and remembered to plot the points at the upper boundaries. The accuracy of the curve needs to be ± half a square in most instances.

b)(i)
An answer of 44 ± 0.5 was deemed to be acceptable.

(ii)
She has remembered that the interquartile range is a single value found by subtracting the lower quartile from the upper quartile.

(iii)
An answer of 70 ± 1% was deemed to be acceptable.

2 Van Winkel sells tulip bulbs. The cumulative frequency graph shows the sizes of the bulbs in a random sample from his stock. (The size is the largest circumference in centimetres.)

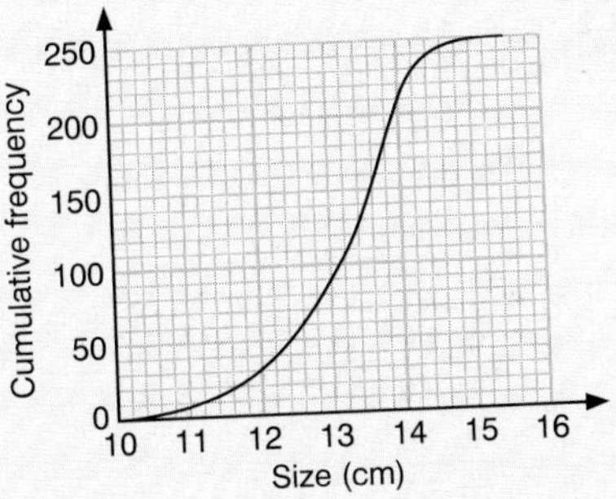

(a) How many bulbs were in the sample?

(b) Estimate (i) the median size bulb and (ii) the interquartile range.

Frans Gretel also sells tulip bulbs. This is a frequency distribution for a sample from his stock.

Size (x cm)	Frequency
10.5 ≤ x < 11.5	53
11.5 ≤ x < 12.5	90
12.5 ≤ x < 13.5	57
13.5 ≤ x < 14.5	30
14.5 ≤ x < 15.5	20

(c) Draw up a cumulative frequency table. Draw the cumulative frequency graph on the same grid as that used for Van Winkel's bulbs.

(d) Compare the two samples.

(MEG syllabus B, specimen paper 1998)

(a) 250 bulbs

(b) (i) Median = 13.35 cm

(ii) LQ = 12.65

UQ = 13.8

interquartile range = 13.8 − 12.65 = 1.15 cm

(c)

Size (x cm)	Frequency	Cumulative frequency
10.5 ≤ x < 11.5	53	53
11.5 ≤ x < 12.5	90	143
12.5 ≤ x < 13.5	57	200
13.5 ≤ x < 14.5	30	230
14.5 ≤ x < 15.5	20	250

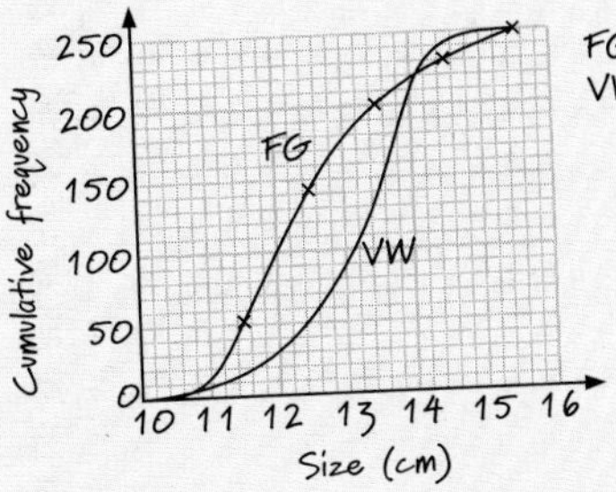

FG Frans Gretal's bulbs
VW Van Winkel's bulbs

(d) The first set of bulbs has the bigger median and the smaller interquartile range.

EXAMINER'S COMMENTS

2 a), b)
The candidate has obtained this information from the cumlative frequency graph provided.

c)
He has worked out the cumulative frequency for these bulbs.

He has remembered to plot the cumulative frequencies at the boundaries (i.e. 11.5, 12.5,13.5, etc.).

d)
Comparisons of measures of central tendency (median) and spread (interquartile range) are always the most appropriate measurements to include in such an explanation.

EXAMINER'S COMMENTS

3 The following distribution shows the weight (in kilograms to the nearest kilogram) of 65 farmyard animals.

Weight (kg)	Frequency
26–30	9
31–35	13
36–40	20
41–45	15
46–50	6
51–55	2

Draw the cumulative frequency diagram for this information and use this to find:

(a) the median and the interquartile range

(b) the percentage of animals greater than 42 kg in weight.

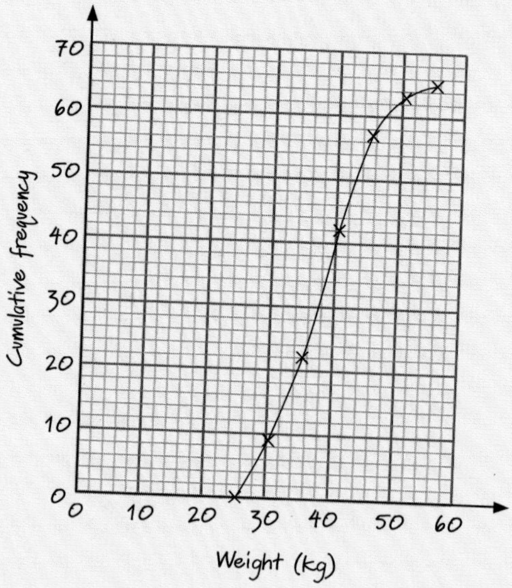

(a) median = 38 kg

LQ = 33

UQ = 42

interquartile range = 42 − 33 = 9 kg

(b) The number of animals less than 42 kg in weight = 50

The number of animals greater than 42 kg in weight = 65 − 50 = 15

The percentage of animals greater than 42 kg in weight = $\frac{15}{65}$ X 100% = 23%

3 a)
The candidate has correctly drawn the cumulative frequency diagram but neglected to show where this has come from – this is important in the case where any errors are made.

She has remembered to plot the cumulative frequencies at the upper boundaries (i.e. 30.5, 35.5, 40.5, etc.).

b)
She has correctly remembered that to find the number of animals weighing more than 42 kg it is necesssary first to find the number of animals weighing less than 42 kg and subtract from 65 – as there are 65 animals in the survey.

4 This table gives you the marks scored by pupils in a French test and in a German test.

German	20	37	35	25	33	30	39	36	27	20	33	35	27	32	28
French	15	35	34	23	35	27	36	34	23	24	30	40	25	35	20

(a) Find the range of the pupils' marks in French.

(b) Draw a scatter graph of the marks scored in the French and German tests.

(c) Draw the line of best fit on the diagram.

(d) Use your line of best fit to estimate a pupil's mark in French when their mark in German was 23.

(e) Describe the relationship between the marks scored in the two tests.

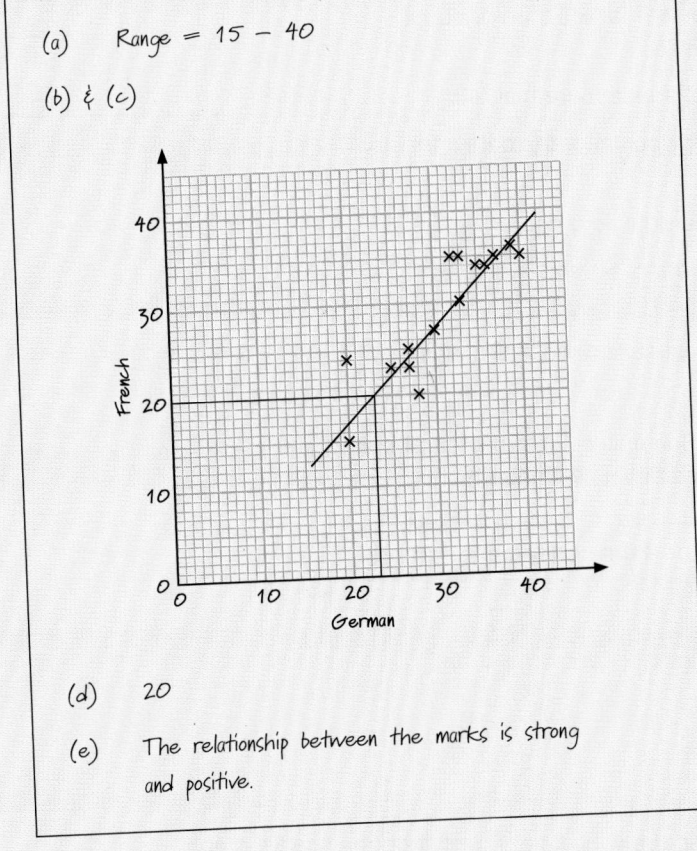

(a) Range = 15 – 40

(b) & (c)

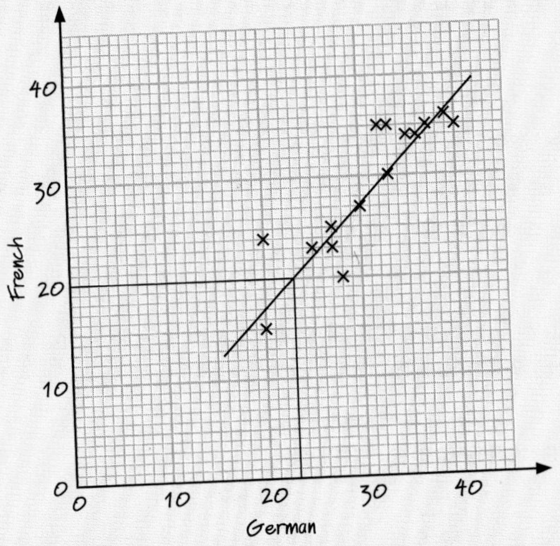

(d) 20

(e) The relationship between the marks is strong and positive.

EXAMINER'S COMMENTS

4 **a)**
The candidate has answered the question incorrectly as the range (in mathematics) should be a single value. An answer of 25 (i.e. 40 – 15) will gain full marks.

e)
All correlation should be described in terms of the strength of the correlation and whether it is positive or negative.

Questions to Answer

1 Find the range and interquartile range for the following set of data.

29 27 38 35 50 38 41 38 28 32 39

2 This table shows the lifetimes of batteries given to the nearest hour.

Lifetime (hours)	Frequency
100–199	32
200–299	98
300–399	65
400–499	14
500–599	3
600–699	2

Draw a cumulative frequency curve to illustrate this information and use your graph to find:

(a) how many batteries lasted less than 250 hours

(b) how many batteries lasted more than 350 hours

(c) the median and the interquartile range.

3 This table shows the cumulative frequency for the test results of 72 students.

Marks	Cumulative frequency
$\leqslant 10$	6
$\leqslant 20$	18
$\leqslant 30$	33
$\leqslant 40$	54
$\leqslant 50$	72

From the table calculate:

(a) how many students got marks less than or equal to 20

(b) how many students got marks more than 40.

Draw the cumulative frequency curve for the data and use it to calculate:

(c) the median and the interquartile range.

In the next test, also marked out of 50, the interquartile range was 30.

(d) Comment on the two tests using the interquartile ranges.

4 Dr Malik wants to buy a car. She collects information about engine capacity and fuel consumption as shown in the table.

Engine capacity (litres)	Fuel consumption (mpg)
2.6	20.0
2.0	27.5
1.2	34.5
1.6	32.0
3.0	19.5
1.8	28.5
1.4	33.5

(a) Plot this information and draw the line of best fit on the scatter graph.

Dr Malik decides to use her line to estimate the fuel consumption for other engine capacities.

(b) Use the line of best fit to estimate the fuel consumption of a car with an engine capacity of:

(i) 2.3 litres
(ii) 3.5 litres.

(c) Explain why one of these two estimates is more reliable than the other.

(SEG Modular syllabus, specimen paper 1998 – modified)

CHAPTER 86

PROBABILITY

Probability is the branch of statistics which allows you to work out how likely or unlikely an outcome or result might be.

In probability, an event which is certain to happen will have a probability of 1 whereas an event which is impossible will have a probability of 0. Probabilities greater than 1 or less than 0 do not have any meaning.

EVENTS AND OUTCOMES

n **event** is something that happens, such as throwing a die, or tossing a oin, or picking a card from a pack.

n **outcome** is the result of an event, such as scoring a 3 or a 6 on throwing die.

the outcome is the required result, such as throwing a 6 to start a game, ien the outcome is a **success**.

he **probability** is a measure of how likely an outcome is to happen.

ı general:

$$\text{probability of success} = \frac{\text{number of 'successful' outcomes}}{\text{number of 'possible' outcomes}}$$

nd you can use p(success) as shorthand for the probability of success.

$$\text{p(success)} = \frac{\text{number of 'successful' outcomes}}{\text{number of 'possible' outcomes}}$$

Vorked example

. box contains 25 coloured balls, where seven balls are red, ten balls are blue nd eight balls are yellow. A ball is selected from the box at random. alculate the probability of selecting:

a) a red ball (b) a blue ball (c) a red or a yellow ball

d) a red or a blue or a yellow ball (e) a green ball.

Use:

$$\text{p(success)} = \frac{\text{number of 'successful' outcomes}}{\text{number of 'possible' outcomes}}$$

(a) $\text{p(red ball)} = \frac{\text{number of red balls}}{\text{number of balls}} = \frac{7}{25}$

(b) $\text{p(blue ball)} = \frac{\text{number of blue balls}}{\text{number of balls}} = \frac{10}{25} = \frac{2}{5}$ Cancelling down to its lowest terms.

(c) $\text{p(red or yellow ball)} = \frac{\text{number of red or yellow balls}}{\text{number of balls}}$

$= \frac{15}{25} = \frac{3}{5}$ Cancelling down to its lowest terms.

(d) $\text{p(red or yellow or blue ball)} = \frac{\text{number of red or yellow or blue balls}}{\text{number of balls}}$

$= \frac{25}{25} = 1$ Meaning that this outcome is certain to happen.

(e) $\text{p(green ball)} = \frac{\text{number of green balls}}{\text{number of balls}} = \frac{0}{25}$

$= 0$ Meaning that this outcome is impossible, it cannot happen.

NOTE

The answer could also be given as a decimal or a percentage.

So p(red ball) = 0.28
or p(red ball) = 28%

TOTAL PROBABILITY

The probability of an event happening is equal to 1 minus the probability of the event not happening.

Worked example

The probability that it will rain tomorrow is $\frac{1}{5}$. What is the probability that it will not rain tomorrow?

p(rain) $= \frac{1}{5}$

p(not rain) $= 1 - \frac{1}{5} = \frac{4}{5}$ p(not rain) = 1 – p(rain)

Check yourself

QUESTIONS

Q1 A box contains 50 balls coloured blue, red and green. The probability of getting a blue ball is 32% and the probability of getting a red ball is 0.46.

(a) How many blue balls are there in the box?
(b) How many red balls are there in the box?
(c) How many green balls are there in the box?

Q2 A manufacturer produces microchips and in a sample of 100 it is found that 17 are faulty. What is the probability that a microchip chosen at random is not faulty?

Q3 The probability that a train arrives early is 0.2 and the probability that the train arrives on time is 0.45. What is the probability that the train arrives late?

REMEMBER! Cover the answers if you want to.

ANSWERS

A1
(a) Number of blue balls = 16
(b) Number of red balls = 23
(c) Number of green balls = 11

A2 p(not faulty) $= \frac{83}{100} = 0.83 = 83\%$

A3 p(late) = 0.35

TUTORIAL

T1
(a) Number of blue balls $= 32\% \times 50 = 16$
(b) Number of red balls $= 0.46 \times 50 = 23$
(c) Number of green balls $= 50 - (16 + 23)$
$= 11$ *As the remaining balls are green.*

T2 *p(faulty)* $= \frac{17}{100}$

p(not faulty) = 1 – *p(faulty)*

$= 1 - \frac{17}{100}$

$= \frac{83}{100}$

T3 *p(late)* = 1 – *p(not late)*

$= 1 - (0.2 + 0.45)$ *Total probability = 1.*

$= 0.35$

CHAPTER 87

THEORETICAL AND EXPERIMENTAL PROBABILITY

Theoretical probability is based on equally likely outcomes and is used to tell you how an event should perform in theory, whereas experimental probability (or relative frequency) is used to tell you how an event performs in an experiment.

NOTE

You may see the word 'dice' used instead of 'die' in these questions. Die is the singular form, dice is the plural.

Worked example

A die is thrown 100 times.

a) How many times would you expect to throw a six?

The following frequency distribution is obtained.

Score	1	2	3	4	5	6
Frequency	18	15	19	17	16	15

b) What is the relative frequency of a score of 6?

c) What is the relative frequency of getting an even number?

d) For which score are the theoretical probability and relative frequency the closest?

(a) When throwing a die 100 times:
expected number of sixes $= 100 \times \frac{1}{6} = 16.666\,66\ldots$
$= 17$ (to the nearest whole number)

(b) The relative frequency of a score of 6 is $\frac{15}{100} = \frac{3}{20}$.

(c) The frequency of getting an even number is $15 + 17 + 15 = 47$.
The relative frequency of getting an even number is $\frac{47}{100}$.

(d) The theoretical and experimental probabilities are closest for a score of 4.

Check yourself

QUESTIONS

Q1 A die is thrown 120 times. What is the expected frequency of a number greater than 4?

Q2 The probability that a new car will develop a fault in the first month is 0.062%. A garage sells 1037 new cars in one year. How many of these cars will be expected to develop a fault in the first month?

REMEMBER! Cover the answers if you want to.

ANSWERS

A1 40

A2 1 car (rounding to an appropriate degree of accuracy)

TUTORIAL

T1 *p(number greater than 4)* $= \frac{2}{6} = \frac{1}{3}$

Expected frequency $= 120 \times \frac{1}{3} = 40$

T2 *Expected number of cars*

$= 1037 \times 0.062\% = 1037 \times \frac{0.062}{100}$

$= 0.64294$

= 1 *car (rounding to an appropriate degree of accuracy)*

CHAPTER 88

DIAGRAMMATIC REPRESENTATIONS AND TREE DIAGRAMS

One way of showing possible outcomes in probability is in a diagram, called a possibility space, or a sample space. Tree diagrams are another method of showing the probabilities involved, especially where more than one event is taking place.

POSSIBILITY SPACES

		Second die					
		1	2	3	4	5	6
First die	1	2	3	4	5	6	7
	2	3	4	5	6	7	8
	3	4	5	6	7	8	9
	4	5	6	7	8	9	10
	5	6	7	8	9	10	11
	6	7	8	9	10	11	12

A **possibility space** is a diagram which can be used to show the outcomes of various events.

Worked example

Two fair dice are thrown and the sum of the scores on the faces is noted. What is the probability that the sum is 8?

Draw a diagram to illustrate the possible outcomes.

The diagram to illustrate the possible outcomes is shown on the left. There are 36 possible outcomes. There are 5 outcomes that give a total of 8.

Probability that the sum is 8 $= \frac{5}{36}$.

TREE DIAGRAMS

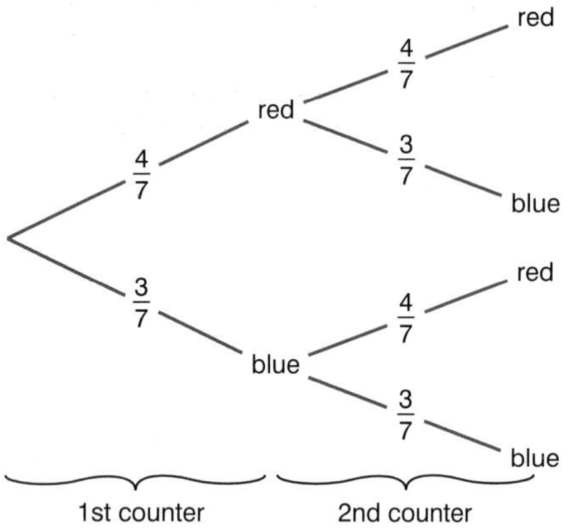

In a tree diagram the probabilities of the outcomes of different events are written on different branches of the diagram.

Worked example

A bag contains four red and three blue counters. A counter is drawn from the bag, replaced and then a second counter is drawn from the bag. Draw a tree diagram to show the various possibilities that can occur.

The tree diagram on the left shows the various possibilities.

NOTE

Further work on tree diagrams will be found in later chapters.

Check yourself

QUESTIONS

Q1 An experiment consists of throwing a die and tossing a coin. Draw a possibility space for the two events and use this to calculate the probability of scoring:

(a) a head and a 1
(b) a tail and an odd number.

Q2 Two tetrahedral dice, numbered 1 to 4 are thrown simultaneously. Draw a possibility space for the total of the two dice and use this information to calculate the probability of scoring a total of :

(a) 2 (b) 6 (c) 9.

What is the most likely outcome?

Q3 A counter is selected from a box containing three red, four green and five blue counters and a second counter is selected from a different box containing five red and four green counters. Draw a tree diagram to show the various possibilities when a counter is drawn from each bag.

REMEMBER! Cover the answers if you want to.

ANSWERS

A1

		Die					
		1	2	3	4	5	6
Coin	H	H1	H2	H3	H4	H5	H6
	T	T1	T2	T3	T4	T5	T6

(a) $\frac{1}{12}$

(b) $\frac{1}{4}$

A2

		Second die			
		1	2	3	4
First die	1	2	3	4	5
	2	3	4	5	6
	3	4	5	6	7
	4	5	6	7	8

(a) $\frac{1}{16}$

(b) $\frac{3}{16}$

(c) 0

The most likely outcome is a 5.

A3

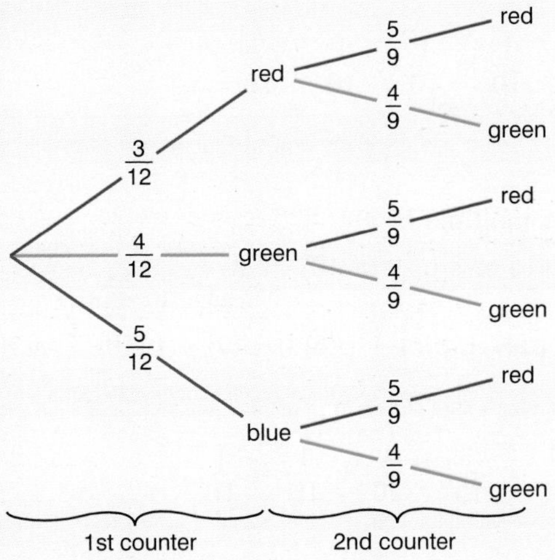

TUTORIAL

T1 *From the possibility space diagram, there are 12 possible outcomes.*

(a) *p(a head and a one)* $= \frac{1}{12}$ *As only one outcome gives this result.*

(b) *p(a tail and an odd number)*

$= \frac{3}{12}$ *As three outcomes give this result.*

$= \frac{1}{4}$ *Cancelling down to its lowest terms.*

T2 *From the possibility space diagram, there are 16 possible outcomes.*

(a) *p(total of 2)* $= \frac{1}{16}$ *As only one outcome gives this result.*

(b) *p(total of 6)* $= \frac{3}{16}$ *As three outcomes give this result.*

(c) *p(total of 9)* $= 0$ *As there is no possibility of getting a total of 9.*

The most likely outcome is a 5.

T3 *See the tree diagram.*

CHAPTER 89

ADDITION RULE – MUTUALLY EXCLUSIVE EVENTS

Two or more events are mutually exclusive if they cannot happen at the same time.

MUTUALLY EXCLUSIVE EVENTS

In the case of mutually exclusive events you can apply the **addition rule** (also called the **or rule**) which states that:

$$p(A \text{ or } B) = p(A) + p(B)$$

and for more than two mutually exclusive events:

$$p(A \text{ or } B \text{ or } C \text{ or } ...) = p(A) + p(B) + p(C) + \dots .$$

Worked example

A spinner with ten sides numbered 1 to 10 is spun. What is the probability of getting:

(a) a five

(b) a five or a six

(c) a multiple of 3 or a multiple of 4

(d) a multiple of 2 or a multiple of 3?

(a) $p(5) = \frac{1}{10}$

(b) $p(5 \text{ or } 6) = p(5) + p(6)$ As the events are mutually exclusive.

$= \frac{1}{10} + \frac{1}{10}$

$= \frac{2}{10}$

$= \frac{1}{5}$ Cancelling down to the lowest terms.

(c) p(multiple of 3 or multiple of 4)

$= p(3) + p(6) + p(9) + p(4) + p(8)$ As the events are mutually exclusive.

$= \frac{1}{10} + \frac{1}{10} + \frac{1}{10} + \frac{1}{10} + \frac{1}{10} = \frac{5}{10}$

$= \frac{1}{2}$ Cancelling down to the lowest terms.

(d) p(multiple of 2 or multiple of 3)

$= p(2 \text{ or } 3 \text{ or } 4 \text{ or } 6 \text{ or } 8 \text{ or } 9 \text{ or } 10)$ As the events are not mutually exclusive.

$= p(2) + p(3) + p(4) + p(6) + p(8) + p(9) + p(10)$ As these events are now mutually exclusive.

$= \frac{1}{10} + \frac{1}{10} + \frac{1}{10} + \frac{1}{10} + \frac{1}{10} + \frac{1}{10} + \frac{1}{10}$

$= \frac{7}{10}$

NOTE

p(multiple of 3) + p(multiple of 4) = p(3 or 6 or 9) + p (4 or 8) As the events are mutually exclusive.

NOTE

The events are not mutually exclusive as:

p(multiple of 2) = p(2 or 4 or 6 or 8 or 10)
p(multiple of 3) = p(3 or 6 or 9)
The number 6 is common to both events and if the probabilities are added then this probability will be added twice.

Check yourself

QUESTIONS

Q1 A die with faces numbered 1 to 6 is rolled and the face uppermost noted. Find the probability that the result will be:

(a) a 5 or a 6 (b) an even number
(c) a factor of 8.

Q2 Letters are chosen from the word:

PROBABILITY

Find the probability that the chosen letter is:

(a) the letter P (b) the letter B
(c) the letter B or the letter I.

Q3 A box of chocolates contains seven milk chocolates, five plain chocolates and four mint chocolates. Calculate the probability of picking:

(a) a milk chocolate
(b) a milk chocolate or a plain chocolate
(c) a milk chocolate or a plain chocolate or a mint chocolate.

REMEMBER! Cover the answers if you want to.

ANSWERS

A1 (a) $\frac{1}{3}$

(b) $\frac{1}{2}$

(c) $\frac{1}{2}$

A2 (a) $\frac{1}{11}$

(b) $\frac{2}{11}$

(c) $\frac{4}{11}$

A3 (a) $\frac{7}{16}$

(b) $\frac{3}{4}$

(c) 1

TUTORIAL

T1 *(a)* $p(5 \text{ or } 6) = p(5) + p(6)$ *As the events are mutually exclusive.*

$= \frac{1}{6} + \frac{1}{6} = \frac{2}{6} = \frac{1}{3}$

(b) *p(even number)*

$= p(2 \text{ or } 4 \text{ or } 6)$ *As the events are mutually exclusive.*

$= p(2) + p(4) + p(6)$

$= \frac{1}{6} + \frac{1}{6} + \frac{1}{6} = \frac{3}{6} = \frac{1}{2}$

(c) $p(\text{factor of } 8) = p(1 \text{ or } 2 \text{ or } 4)$

As the factors of 8 are 1, 2 4 and 8.

$= p(1) + p(2) + p(4)$ *As the events are mutually exclusive.*

$= \frac{1}{6} + \frac{1}{6} + \frac{1}{6} = \frac{3}{6} = \frac{1}{2}$

T2 *(a)* $p(\text{letter P}) = \frac{1}{11}$

(b) $p(\text{letter B}) = \frac{2}{11}$

(c) $p(\text{letter B or letter I}) = p(\text{letter B}) + p(\text{letter I})$

$= \frac{2}{11} + \frac{2}{11} = \frac{4}{11}$ *As the events are mutually exclusive.*

T3 *(a)* $p(\text{milk chocolate}) = \frac{7}{16}$

(b) *p(milk chocolate or plain chocolate)*

$= p(\text{milk chocolate}) + p(\text{plain chocolate})$

$= \frac{7}{16} + \frac{5}{16} = \frac{12}{16} = \frac{3}{4}$ *As the events are mutually exclusive.*

(c) *p(milk or plain or mint chocolate)*

$= p(\text{milk}) + p(\text{plain}) + p(\text{mint})$

$= \frac{7}{16} + \frac{5}{16} + \frac{4}{16} = \frac{16}{16} = 1$

The probability is therefore a certainty.

CHAPTER 90

MULTIPLICATION RULE – INDEPENDENT EVENTS

Events are independent if the outcome of one event does not affect the other event.

INDEPENDENT EVENTS

For independent events you can use the **multiplication rule** (also called the **and rule**) which states that:

p(A and B) = p(A) × p(B)

Similarly for more than two independent events:

p(A and B and C and ...) = p(A) × p(B) × p(C) × ...

Worked example

A bag contains four red and three blue counters. A counter is drawn from the bag, replaced and then a second counter is drawn from the bag. Draw a tree diagram and use it to calculate the probability that:

(a) both counters will be red (b) both counters will be blue

(c) the first counter will be red and the second counter blue

(d) one counter will be red and one counter will be blue.

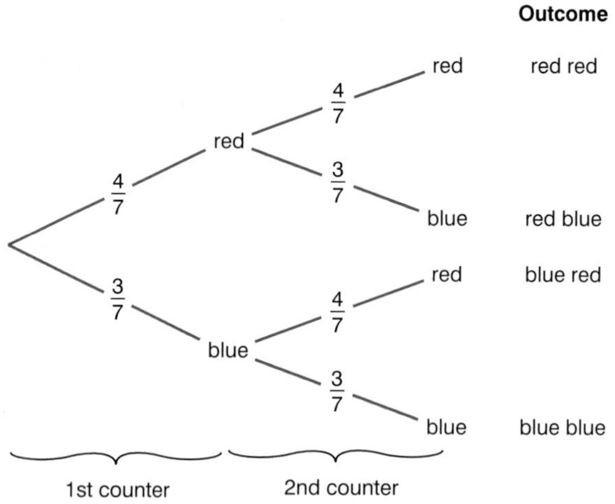

(a) p(red and red) = p(red) × p(red) As the events are independent.

$$= \frac{4}{7} \times \frac{4}{7} = \frac{16}{49}$$

(b) p(blue and blue) = p(blue) × p(blue) As the events are independent.

$$= \frac{3}{7} \times \frac{3}{7} = \frac{9}{49}$$

(c) p(red and blue) = p(red) × p(blue) As the events are independent.

$$= \frac{4}{7} \times \frac{3}{7} = \frac{12}{49}$$

(d) p(one counter will be red and one counter will be blue) is the same as p(red and blue or blue and red).

= p (red and blue) + p(blue and red). As the events are mutually exclusive.

= p(red) × p(blue) + p(blue) × p(red) As the events are independent.

$$= \frac{4}{7} \times \frac{3}{7} + \frac{3}{7} \times \frac{4}{7} = \frac{12}{49} + \frac{12}{49} = \frac{24}{49}$$

NOTE

Both of these outcomes give one red and one blue

Check yourself

QUESTIONS

Q1 A die is thrown and a card drawn from a pack of 52 cards. Calculate the probability that the outcome will be:

(a) a 4 and a heart
(b) a 6 and a king
(c) an even number and a club
(d) an odd number and the ace of hearts.

Q2 The probability that a car will fail its MOT because of the lights is 0.32 and the probability that a car will fail its MOT because of the brakes is 0.55. Calculate the probability that the car fails because of:

(a) its lights and its brakes
(b) its lights only.

Q3 For a biased coin, the probability of getting a head is 0.4. Calculate the probability of getting:

(a) a tail
(b) two heads when the coin is tossed twice
(c) exactly one head when the coin is tossed twice.

Q4 The probability that a particular component will fail is 0.015. Draw and label a tree diagram to show the possible outcomes when two such components are chosen at random. Calculate the probability that:

(a) both components will fail
(b) exactly one component will fail.

REMEMBER! Cover the answers if you want to.

ANSWERS

A1

(a) $\frac{1}{24}$

(b) $\frac{1}{78}$

(c) $\frac{1}{8}$

(d) $\frac{1}{104}$

A2

(a) 0.176

TUTORIAL

T1

(a) *p (4 and a heart)*
$= p(4) \times p(heart)$ *As the events are independent.*
$= \frac{1}{6} \times \frac{1}{4} = \frac{1}{24}$

(b) *p(6 and a king)*
$= p(6) \times p(king)$ *As the events are independent.*
$= \frac{1}{6} \times \frac{1}{13} = \frac{1}{78}$

(c) *p(even number and a club)*
$= p(even\ number) \times p(club)$ *As the events are independent.*
$= \frac{1}{2} \times \frac{1}{4} = \frac{1}{8}$

(d) *p(odd number and ace of hearts)*
$= p(odd) \times p(ace\ of\ hearts)$ *As the events are independent.*
$= \frac{1}{2} \times \frac{1}{52} = \frac{1}{104}$

T2

(a) *p(lights and brakes)*
$= p(lights) \times p(brakes)$ *As the events are independent.*
$= 0.32 \times 0.55 = 0.176$

ANSWERS

A2 (b) 0.144

A3
(a) 0.6
(b) 0.16
(c) 0.48

A4
(a) 0.000 225 or 2.25×10^{-4}
(b) 0.029 55

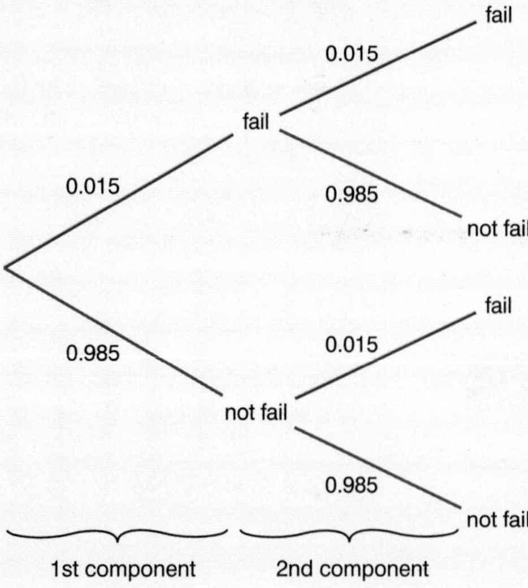

TUTORIAL

T2

(*b*) *The probability that the car fails because of its lights only is equal to the probability that the car fails because of its lights and does not fail because of the brakes.*

The probability that a car will fail its MOT because of the brakes is 0.55.

The probability that a car will not fail its MOT because of the brakes is 0.45. (1 – 0.55)

p(car fails because of its lights only)

$= 0.32 \times 0.45 = 0.144$

T3

(*a*) $p(tail) = 1 - p(head) = 1 - 0.4 = 0.6$

(*b*) *p(two heads)*

$= p(\text{head and head})$

$= p(head) \times p(head)$ *As the events are independent.*

$= 0.4 \times 0.4 = 0.16$

(*c*) *The probability of exactly one head when the coin is tossed twice is equivalent to the probability that the first is a head and the second is a tail or the first is a tail and the second is a head*

$= p(\text{head and tail or tail and head})$

$= p(\text{head and tail}) + p(\text{tail and head})$
As the events are mutually exclusive.

$= p(head) \times p(tail) + p(tail) \times p(head)$
As the events are independent.

$= 0.4 \times 0.6 + 0.6 \times 0.4 = 0.48$

T4

(*a*) $p(\text{both components will fail}) = p(\text{fail and fail})$

$= p\,(fail) \times p(fail)$ *As the events are independent.*

$= 0.015 \times 0.015$

$= 0.000\,225$ *or* 2.25×10^{-4}

(*b*) *The probability that exactly one component will fail is equivalent to the probability that the first component fails and the second doesn't or the first component doesn't and the second fails.*

$= p(\text{fails and doesn't fail or doesn't fail and fails})$

$= p(\text{fails and doesn't fail}) + p(\text{doesn't fail and fails})$ *As the events are mutually exclusive.*

$= p(\text{fails}) \times p(\text{doesn't fail}) + p(\text{doesn't fail}) \times p(\text{fails})$
As the events are independent.

$= 0.015 \times 0.985 + 0.985 \times 0.015$
As $p(\text{doesn't fail}) = 1 - p(\text{fail})$

$= 0.029\,55$

EXAM PRACTICE

Sample Student's Answers & Examiner's Comments

1 A manufacturer makes switches for electric circuits. The probability that a switch is faulty is 0.1. William buys 200 of these switches.

(a) (i) How many of the 200 switches can be expected to be faulty?

(ii) Explain why your answer is only an approximation.

(b) Draw and label a probability tree diagram to show the possible outcomes when two switches are chosen at random.

(c) Calculate the probability that:

(i) both switches are not faulty

(ii) exactly one switch is faulty.

(SEG Modular syllabus, specimen paper 1998)

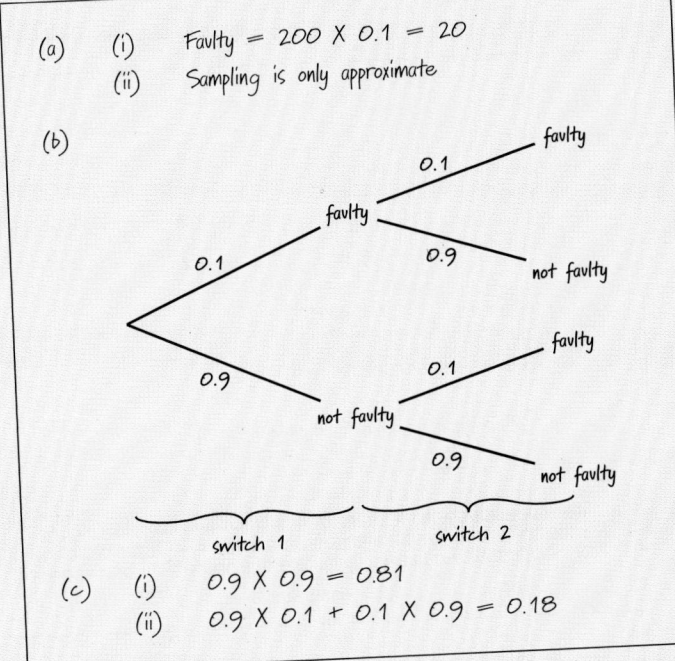

(a) (i) Faulty = 200 X 0.1 = 20

(ii) Sampling is only approximate

(b)

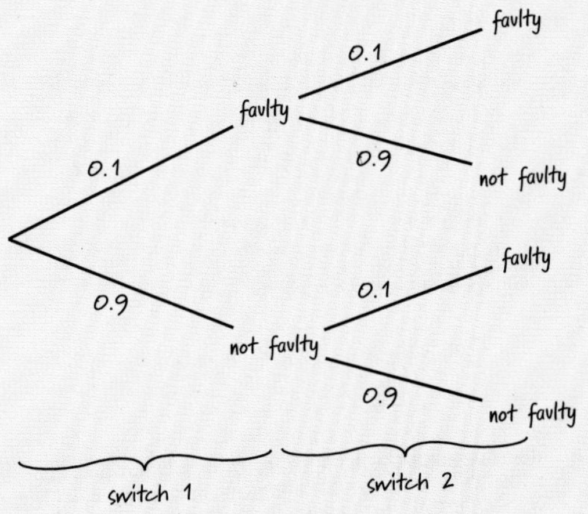

(c) (i) 0.9 X 0.9 = 0.81

(ii) 0.9 X 0.1 + 0.1 X 0.9 = 0.18

EXAMINER'S COMMENTS

1 **a)**
The candidate has worked out the answer correctly and appreciated that the answer is approximate because sampling is approximate. Alternatively, he might respond that it depends upon the batch.

b)
He has labelled the tree diagram correctly and included the probabilities on the branches.

c)
The candidate has used the tree diagram to calculate the probabilities. The probability that exactly one switch is faulty must take account of the fact that the first switch is faulty and the second switch is not faulty or the first switch is not faulty and the second switch is faulty.

EXAMINER'S COMMENTS

2 In a survey about family size, Trevor asks his workmates how many brothers and how many sisters they each have. None of Trevor's workmates are related. His results are shown in the table.

Number of brothers						
5						
4	1	1				
3	3	4				
2	3		1			1
1	4	7	1		1	
0	4	5	2	1	1	
	0	1	2	3	4	5
	Number of sisters					

(a) How many of Trevor's workmates have no sisters?

(b) How many workmates did Trevor survey?

(c) Find the modal number of sisters.

(d) How many sisters did Trevor's workmates have altogether?

(e) Calculate the mean number of sisters.

(f) What is the probability that a workmate chosen at random has one brother?

(SEG Modular syllabus, specimen paper 1998)

(a) 15

(b) 40

(c) 1

(d) $(1 \times 17) + (2 \times 4) + (3 \times 1) + (4 \times 2) + (5 \times 1)$
$= 17 + 8 + 3 + 8 + 5 = 41$

(e) Mean $= \frac{41}{40} = 1.025$

(f) Number of workmates with one brother $= 13$

Probability $= \frac{13}{40} = 0.325$

2 **a), b), c)**
The candidate has correctly identified the required information from the given table.

d)
She has multiplied the number of sisters by the totals of each column. The working allows method marks to be given if the final answer was wrong.

e)
An answer of 1.025 or 1.03 is acceptable as an answer for the mean. An answer rounded to an integer would lose marks.

f)
The answer $\frac{13}{40}$ is quite acceptable as an answer to a probability question. 0.325 and 32.5% are also acceptable answers in this case.

3 On a biased six-sided die, a score of 6 is twice as likely as a score of 5. A score of 5 is twice as likely as a score of 4. The probabilities of a score of 1, 2, 3 or 4 are all the same. Calculate the probability of a score of:

(a) 6 (b) 5 (c) less than 4.

Let p be the probability of a score of 1.

$p(1) = p$

$p(2) = p$

$p(3) = p$

$p(4) = p$

$p(5) = 2p$

$p(6) = 4p$

(a) a score of $6 = 4p$

(b) a score of $5 = 2p$

(c) a score less than $4 = p + p + p = 3p$

3 *Although some credit will be gained for this work, the candidate has not 'calculated' the probability. He should have used the idea that the total probability adds up to 1 so*

$p + p + p + p + 2p + 4p = 1$

$10p = 1 \Rightarrow p = \frac{1}{10}$

Using the fact that $p = \frac{1}{10}$ *then:*

(a) a score of $6 = 4p = \frac{4}{10} = \frac{2}{5}$

(b) a score of $5 = 2p = \frac{2}{10} = \frac{1}{5}$

(c) a score less than $4 = 3p = \frac{3}{10}$

Questions to Answer

1 Q, R and S are three mutually exclusive events where p(event Q) = 30%, p(event R) = $\frac{1}{4}$ and p(event S) = 0.2. Calculate the probability of:

(a) either Q or R occurring

(b) either R or S occurring

(c) either Q or R or S occurring.

2 Vijay and Baljit are playing a game with two fair five-sided spinners, one red and one blue. The blue spinner is numbered 5, 6, 7, 8, 9 and the red spinner is numbered 1, 2, 3, 4, 5. The final score is calculated by multiplying the two spinner scores together.

(a) Complete a grid to show all the possible final scores.

(b) Find the probability that the final score is a square number.

(c) Find the probability that the final score is less than 30. *(MEG syllabus B, specimen paper 1998)*

3 A bag contains a number of counters. Each counter is coloured red, blue, yellow or green. Each counter is numbered 1, 2 or 3.

The table shows the probability of colour and number for these counters.

		Colour of counter			
		Red	**Blue**	**Yellow**	**Green**
No. of counter	**1**	0.2	0	0.1	0
	2	0.2	0.1	0.1	0
	3	0.1	0.1	0	0.1

(a) A counter is taken from the bag at random.

(i) What is the probability that it is red and numbered 2?

(ii) What is the probability that it is green or numbered 2?

(iii) What is the probability that it is red or numbered 2?

(b) There are two green counters in the bag. How many counters are in the bag altogether? *(SEG specimen paper 1998)*

4 The probability of different numbers of births per week in a village are shown in this table.

Number of births	0	1	2	3
Probability	0.4	0.3	0.2	0.1

The probability of different numbers of deaths per week in a village are shown in this table.

Number of deaths	0	1	2	3
Probability	0.1	0.3	0.4	0.2

Assume that the number of births per week and the number of deaths per week are independent.

(a) Find the probability that during any particular week, there are exactly two births and two deaths.

(b) Find the probability that during any particular week, the numbers of births and deaths are the same. *(SEG specimen paper 1998)*

CHAPTER 92

SAMPLING METHODS

Sampling should be representative of the population as a whole and sample sizes should be as large as possible. Sampling techniques considered here include:

- **convenience sampling**
- **random sampling**
- **systematic sampling**
- **stratified sampling**
- **quota sampling.**

CONVENIENCE SAMPLING

In **convenience sampling** or **opportunity sampling**, you choose the first people who come along. Convenience sampling might mean sampling friends and members of your own family and is therefore likely to involve some element of bias.

RANDOM SAMPLING

In **random sampling**, each member of the population has an equally likely chance of being selected. Random sampling might involve giving each member of the population a number and then choosing the numbers at random, using some appropriate means to generate random numbers.

SYSTEMATIC SAMPLING

Systematic sampling involves random sampling, using some system to choose the members of the population to be sampled. Systematic sampling might include numbering each member of the population according to some criterion, such as by name, age, height, etc.

STRATIFIED SAMPLING

Stratified sampling involves dividing the population into groups or strata. From each stratum you choose a random or systematic sample so that the sample size is proportional to the size of the group in the population as a whole.

For example, in a class where there are twice as many girls as boys, then the sample should include twice as many girls as boys.

QUOTA SAMPLING

Quota sampling involves choosing population members who have specific characteristics which are selected beforehand. Quota sampling is popular in market research where interviewers are told that there should be equal numbers of men and women, or twice as many adults as teenagers, or that the sample should include ten shoppers and five commuters etc.

Check yourself

QUESTIONS

Q1 A bus company attempted to estimate the number of people who travel on local buses in a certain town. They telephoned 100 people in the town one evening and asked, 'Have you travelled by bus in the last week?'. Nineteen people said 'Yes'.

The bus company concluded that 19% of the town's population travel on local buses. Give three criticisms of this method of estimation.

(MEG *syllabus* A, *specimen paper* 1998)

Q2 The numbers of people living in three villages are given in the table.

Village	Population
Atford	2500
Beeham	4100
Calbridge	5900

A sample of 240 is taken by Mr James. He selects, at random, 80 people from each village.

(a) Explain why this might be an inappropriate sampling method.

(b) Explain how Mr James could select a more representative sample of 240 people, and write down the number of people selected from each village.

REMEMBER! Cover the answers if you want to.

ANSWERS

A1 Choose three from:

(i) Only people with access to a telephone will be contacted.

or (ii) The survey is limited to only one town.

or (iii) Some of the respondents may not have used the bus 'last week'.

or (iv) Some of the respondents may be at work during the evening.

A2 (a) The villages have different sized populations and stratified sampling would be most appropriate.

(b) Sample size

Atford	48
Beeham	79
Calbridge	113

TUTORIAL

T1 *Responses to such questions need to take account of the people surveyed and the question asked.*

In this case, the words 'travel on local buses', 'in a certain town', 'telephoned 100 people' and 'one evening' are all relevant.

T2 *Population of all villages* = 12 500

For stratified sampling, the following sample size are appropriate.

Atford $= \frac{2500}{12\,500} \times 240 = 48$

Beeham $= \frac{4100}{12\,500} \times 240 = 78.72 = 79$

(to the nearest integer)

Calbridge $= \frac{5900}{12\,500} \times 240 = 113.28 = 113$

(to the nearest integer)

Answers for Beeham and Calbridge are rounded to the nearest integer to give sensible numbers.

CHAPTER 93

HISTOGRAM – UNEQUAL CLASS INTERVALS

Histograms are like bar charts except that it is the area of each 'bar' which represents the frequency and not the length or height.

HISTOGRAMS

The bars on the horizontal axis should be drawn at the class boundaries and the area of the bars should be proportional to the frequency, i.e.

$$\text{class width} \times \text{height} = \text{frequency}$$

$$\text{so height} = \frac{\text{frequency}}{\text{class width}}$$

and the height is referred to as the **frequency density**. This means that the vertical axis of a histogram should be labelled frequency density where:

$$\text{frequency density} = \frac{\text{frequency}}{\text{class width}}$$

Worked example

The frequency distribution on the left shows the height of 50 bushes, measured to the nearest centimetre.

Draw a histogram to represent this information.

Height (cm)	Frequency
10–14	3
15–19	6
20–24	7
25–29	9
30–39	12
40–49	8
50–74	4
75–99	1

NOTE

Since the height is measured to the nearest centimetre then the 10–14 interval extends from 9.5 – 14.5, giving it a width of 5 cm, etc.

First draw up a table to calculate the respective frequency densities, and then draw the histogram.

Height (cm)	Frequency	Class width	Frequency density
10–14	3	5	3 ÷ 5 = 0.6
15–19	6	5	6 ÷ 5 = 1.2
20–24	7	5	7 ÷ 5 = 1.4
25–29	9	5	9 ÷ 5 = 1.8
30–39	12	10	12 ÷ 10 = 1.2
40–49	8	10	8 ÷ 10 = 0.8
50–74	4	25	4 ÷ 25 = 0.16
75–99	1	25	1 ÷ 25 = 0.04

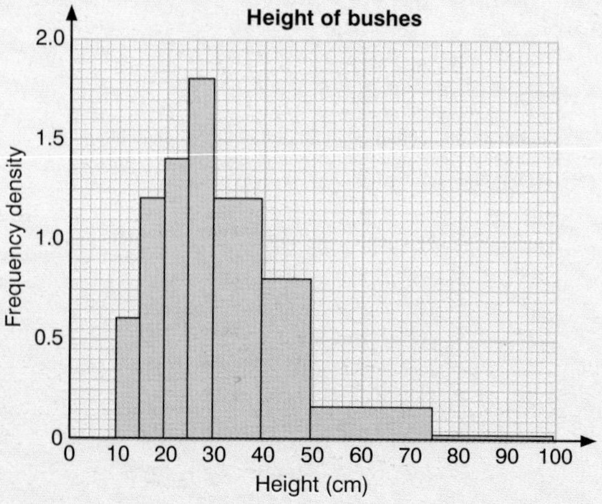

FREQUENCY POLYGONS

ғrequency polygons were first discussed in Chapter 80, Representing data. A requency polygon can be drawn from a histogram by joining the midpoints of he tops of the bars and extending the lines to the horizontal axis on either side. The area under the frequency polygon should be the same as the area under the histogram.

Worked example

The heights of people queuing for a fairground ride were as follows.

Height (inches)	Frequency
35 up to 45	8
45 up to 55	13
55 up to 60	20
60 up to 65	29
65 up to 70	23
70 up to 90	11

Draw a frequency polygon to represent this information.

First draw up a table to calculate the respective frequency densities.

Height (inches)	Frequency	Class width	Frequency density
35 up to 45	8	10	0.8
45 up to 55	13	10	1.3
55 up to 60	20	5	4
60 up to 65	29	5	5.8
65 up to 70	23	5	4.6
70 up to 90	11	20	0.55

Then draw the histogram and frequency polygon.

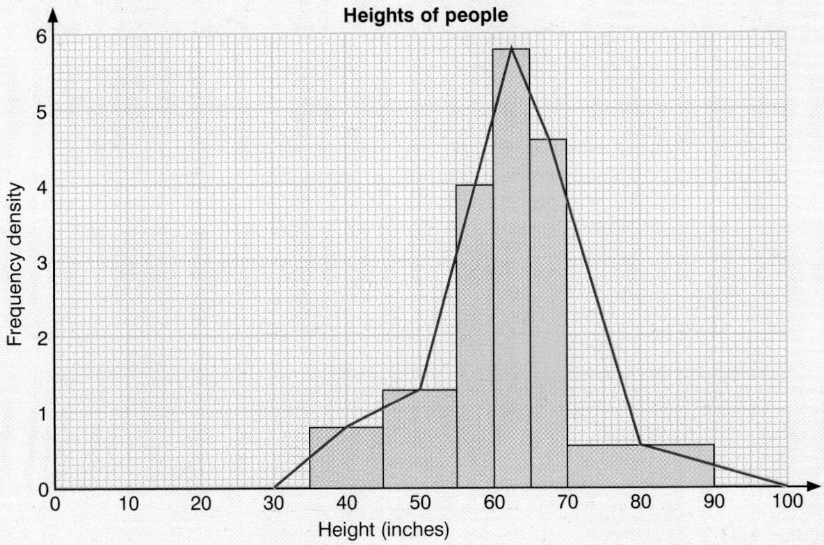

Check yourself

QUESTIONS

Q1 The following table shows the length of time for 120 workers to travel home one evening.

Time (*t* minutes)	Frequency
$0 \leqslant t < 10$	8
$10 \leqslant t < 20$	17
$20 \leqslant t < 30$	23
$30 \leqslant t < 60$	42
$60 \leqslant t < 90$	18
$90 \leqslant t < 120$	9
$120 \leqslant t < 240$	3

Draw a histogram to represent these data.

Q2 The following information shows the distance travelled by 100 salespeople one week. The distance is measured in miles to the nearest mile.

Distance (miles)	Frequency
0–500	3
501–1000	19
1001–2000	27
2001–4000	36
4001–6000	15

Use this information to construct:

(a) a histogram
(b) a frequency polygon.

Q3 The distance travelled by 50 lecturers to work is shown in the histogram below.

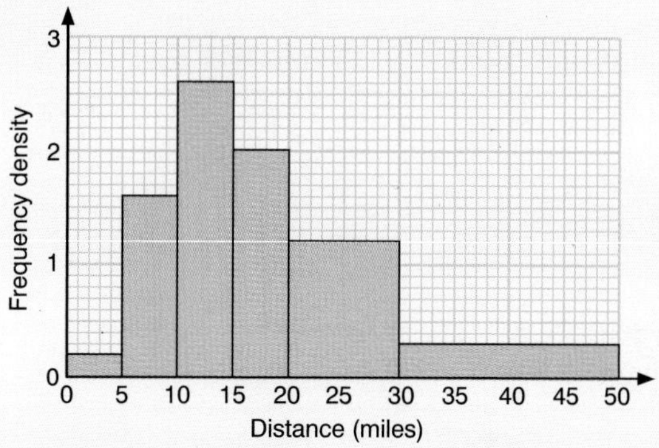

Use the information in the histogram to complete the following table.

Distance (miles)	0–	5–	10–	15–	20–	30–50
Number of lecturers						

ANSWERS

A1

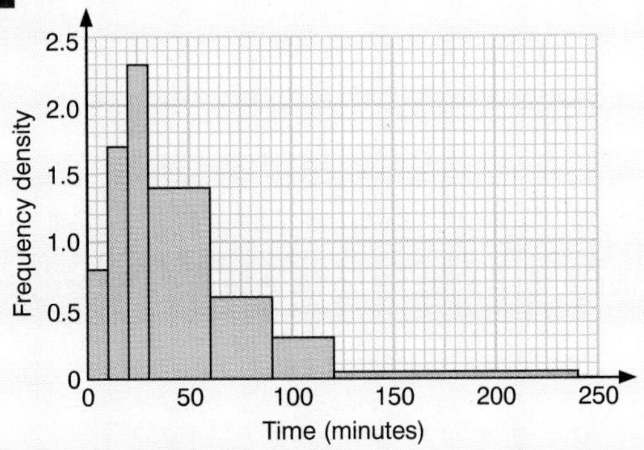

A2

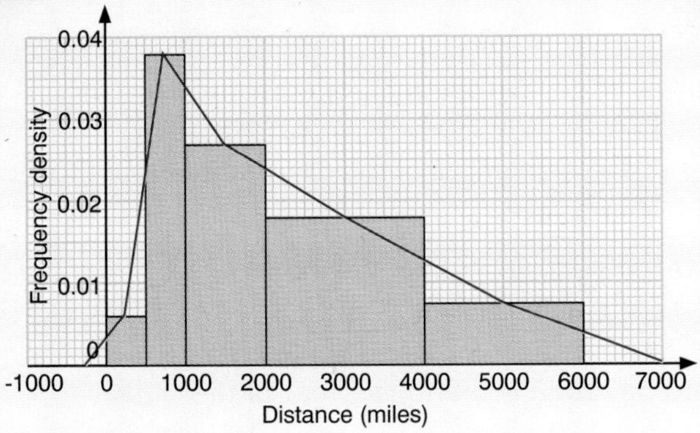

A3

Distance (miles)	0–	5–	10–	15–	20–	30–50
Number of lecturers	1	8	13	10	12	6

NOTE

You should check that the total number of lectures is 50.

TUTORIAL

T1

Completing the table:

Time (t minutes)	Frequency	Class width	Frequency density
$0 \leq t < 10$	8	10	0.8
$10 \leq t < 20$	17	10	1.7
$20 \leq t < 30$	23	10	2.3
$30 \leq t < 60$	42	30	1.4
$60 \leq t < 90$	18	30	0.6
$90 \leq t < 120$	9	30	0.3
$120 \leq t < 240$	3	120	0.025

The histogram is shown in the answer.

T2

Completing the table:

Distance (miles)	Frequency	Class width	Frequency density
0–500	3	500.5	0.005 994
501–1000	19	500	0.038
1001–2000	27	1000	0.027
2001–4000	36	2000	0.018
4001–6000	15	2000	0.0075

The histogram and frequency polygon are shown in the answer.

NOTE

The boundaries on the first interval are 0 and 500.5 so the class width = 500.5.

T3

Reading from the histogram and working backwards:

Distance (miles)	0–	5–	10–	15–	20–	30–50
Class width	5	5	5	5	10	20
Frequency density	0.2	1.6	2.6	2.0	1.2	0.3
Number of lecturers	1	8	13	10	12	6

where number of lectures = class width × frequency density.

CHAPTER 94

STANDARD DEVIATION

The standard deviation is a measure of spread which, unlike the range or interquartile range, takes account of all of the values of a distribution.

The following formulae for standard deviation are given on the examination paper.

Standard deviation, s, for a set of numbers $x_1, x_2, \ldots, x_n$, having a mean of $\bar{x}$ is given by:

$$s = \sqrt{\frac{\Sigma(x-\bar{x})^2}{n}} \text{ or } s = \sqrt{\frac{\Sigma x^2}{n} - \left(\frac{\Sigma x}{n}\right)^2}$$

STANDARD DEVIATION

Worked example

Find the standard deviation of 1, 2, 3, 4, 5 using the formula:

(a) $s = \sqrt{\frac{\Sigma(x-\bar{x})^2}{n}}$ (b) $s = \sqrt{\frac{\Sigma x^2}{n} - \left(\frac{\Sigma x}{n}\right)^2}$

(a) The arithmetic mean $\bar{x} = \frac{1+2+3+4+5}{5} = \frac{15}{5}$

$= 3$ The arithmetic mean also equals $\frac{\Sigma x}{n}$.

The differences of the distribution from the mean $(x - \bar{x})$ are ⁻2, ⁻1, 0, 1, 2.

The squares of the differences of the distribution from the mean $(x - \bar{x})^2$ are 4, 1, 0, 1, 4.

The mean of these squares of the differences

$$\frac{\Sigma(x-\bar{x})^2}{n} = \frac{4+1+0+1+4}{5} = \frac{10}{5} = 2$$

The standard deviation is the square root of the variance.

$$\sqrt{\frac{\Sigma(x-\bar{x})^2}{n}} = \sqrt{2} = 1.4142136 = 1.41 \text{ (3 s.f.)}$$

(b) The arithmetic mean $\frac{\Sigma x}{n} = \frac{1+2+3+4+5}{5} = \frac{15}{5}$

$= 3$ The arithmetic mean also equals $\bar{x}$.

The squares of the distribution are 1, 4, 9, 16, 25.

The mean of these squares $\frac{\Sigma x^2}{n} = \frac{1+4+9+16+25}{5} = \frac{55}{5} = 11$

The standard deviation is $\sqrt{\frac{\Sigma x^2}{n} - \left(\frac{\Sigma x}{n}\right)^2}$

$= \sqrt{11-(3)^2} = \sqrt{11-9} = \sqrt{2} = 1.4142136 = 1.41\text{(3 s.f.)}$

NOTE

The value $\frac{\Sigma(x-\bar{x})^2}{n}$ is called the **variance** and the variance is equal to the square of the standard deviation.

NOTE

The second formula is usually the easier to use and can easily be calculated in tabular form as shown in the next worked example.

Worked example

The weekly pay for five workers is given as follows.

£200 £220 £220 £260 £280 £320

(a) Find the mean and the standard deviation of the pay.

(b) If, one week, the workers are each given a £50 bonus how will this affect the mean and the standard deviation of the pay?

(c) If the workers are each given a 5% pay rise how will this affect the mean and the standard deviation of the pay?

(a) Mean $= \frac{\Sigma x}{n} = \frac{1500}{6}$

$= £250$

Standard deviation $= \sqrt{\frac{\Sigma x^2}{n} - \left(\frac{\Sigma x}{n}\right)^2} = \sqrt{\frac{385\,200}{6} - (250)^2}$

$= \sqrt{1700}$

$= 41.231\,056$

$= £41.23$ (to the nearest penny).

x	x^2
200	40 000
220	48 400
220	48 400
260	67 600
280	78 400
320	102 400
$\Sigma x = 1500$	$\Sigma x^2 = 385\,200$

(b) After a £50 bonus the weekly pay for the five workers will be:

£250 £270 £270 £310 £330 £370.

Expressing this in tabular form:

Mean $= \frac{\Sigma x}{n} = \frac{1800}{6} = £300$

Standard deviation $= \sqrt{\frac{\Sigma x^2}{n} - \left(\frac{\Sigma x}{n}\right)^2} = \sqrt{\frac{550\,200}{6} - (300)^2}$

$= \sqrt{1700}$

$= 41.231\,056$

$= £41.23$ (to the nearest penny).

x	x^2
250	62 500
270	72 900
270	72 900
310	96 100
330	108 900
370	136 900
$\Sigma x = 1800$	$\Sigma x^2 = 550\,200$

(c) After a 5% pay rise the weekly pay for the five workers will be:

£210 £231 £231 £273 £294 £336

Expressing this in tabular form:

Mean $= \frac{\Sigma x}{n} = \frac{1575}{6}$

$= £262.50$ Writing the calculator display of 262.5 as £262.50.

Standard deviation $= \sqrt{\frac{\Sigma x^2}{n} - \left(\frac{\Sigma x}{n}\right)^2} = \sqrt{\frac{424\,683}{6} - (262.50)^2}$

$= \sqrt{1874.25}$

$= 43.292\,609$

$= £43.29$ (to the nearest penny).

x	x^2
210	44 100
231	53 361
231	53 361
373	74 529
294	86 436
336	112 896
$\Sigma x = 1575$	$\Sigma x^2 = 424\,683$

From these answers, you can see that in general:

If you increase (or decrease) all of the values of a distribution by an equal amount then the mean will increase (or decrease) by that amount and the standard deviation (spread) will not be affected.

If you multiply (or divide) all of the values of a distribution by an equal amount then the mean and the standard deviation will also be multiplied (or divided) by that same amount.

Worked example

The following five numbers have a mean of 2 and a standard deviation of 4.

¯4 ¯1 3 5 7

Use this information to find the mean and standard deviation of the following sets of numbers.

(a)	¯3	0	4	6	8
(b)	¯10	¯7	¯3	¯1	1
(c)	¯12	¯3	9	15	21
(d)	¯2	¯0.5	1.5	2.5	3.5
(e)	¯7	¯1	7	11	15

(a) New series = original series + 1

New mean = mean + 1 = 3 — Standard deviation unaffected = 4

As the addition of 1 has no effect on the spread.

(b) New series = original series – 6

New mean = mean – 6 = ¯4 — Standard deviation unaffected = 4

As subtracting 6 has no effect on the spread.

(c) New series = original series × 3

New mean = mean × 3 = 6 — New standard deviation = SD × 3 = 12

(d) New series = original series ÷ 2

New mean = mean ÷ 2 = 1 — New standard deviation = SD ÷ 2 = 2

(e) New series = original series × 2 then + 1

New mean = mean × 2 + 1 = 5 — New standard deviation = SD × 2 = 8

INTERPRETING THE STANDARD DEVIATION

The standard deviation shows the **spread** (or **dispersion**) of values about the arithmetic mean. The greater the standard deviation, then the greater the spread.

Under normal circumstances:

68% of the distribution lies within one standard deviation of the mean

95% of the distribution lies within two standard deviations of the mean

99% of the distribution lies within three standard deviations of the mean.

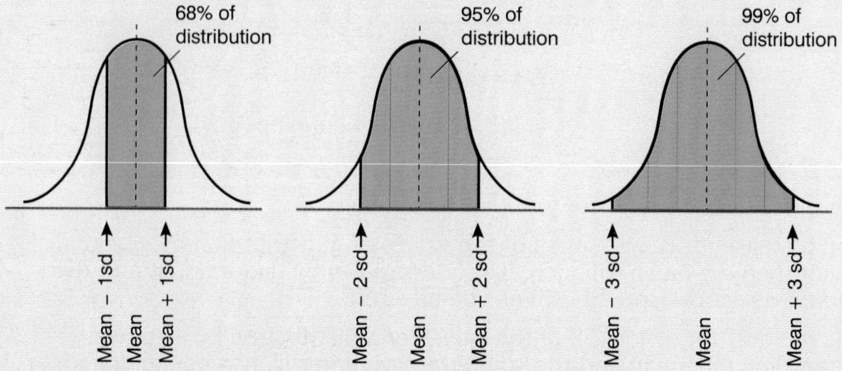

Check yourself

QUESTIONS

Q1 Find the mean and standard deviation of the following sets of data.

(a) $^-2, ^-1, 0, 1, 2$
(b) 5, 5, 5, 5, 5
(c) 2, 3, 4
(d) 2, 2, 3, 3, 4, 4

Q2 The mathematics and history marks for ten pupils are shown in the following table. Calculate the mean and standard deviation and make a comparison of the two sets of data.

History	8	9	10	6	8	7	9	9	10	8
Mathematics	9	5	7	6	8	8	7	10	5	9

Q3 For the numbers 1, 2, 3, 4 and 5, the mean is 2 and the standard deviation is $\sqrt{2}$. Find the mean and standard deviation of the following sets of numbers.

(a) 3, 6, 9, 12, 15
(b) $^-1, ^-2, ^-3, ^-4, ^-5$
(c) 101, 102, 103, 104, 105
(d) 11, 21, 31, 41, 51

Q4 Calculate the mean and standard deviation of the following distribution which shows the mass of 40 suitcases.

Mass (kg)	Frequency
6–10	2
11–15	6
16–20	17
21–25	11
26–30	4

REMEMBER! Cover the answers if you want to.

ANSWERS

A1 (a) Mean = 0
Standard deviation = 1.41 (3 s.f.)

(b) Mean = 5
Sandard deviation = 0

TUTORIAL

T1 (*a*) Mean $\frac{\Sigma x}{n} = \frac{^-2 + ^-1 + 0 + 1 + 2}{5} = \frac{0}{5} = 0$

Mean of the squares $\frac{\Sigma x^2}{n} = \frac{4 + 1 + 0 + 1 + 4}{5}$

$= \frac{10}{5} = 2$

Standard deviation is $\sqrt{\frac{\Sigma x^2}{n} - \left(\frac{\Sigma x}{n}\right)^2}$

$= \sqrt{2 - (0)^2} = \sqrt{2}$

$= 1.414\,213\,6 = 1.41$ (3 s.f.)

(*b*) Mean $\frac{\Sigma x}{n} = \frac{5 + 5 + 5 + 5 + 5}{5} = \frac{25}{5} = 5$

Mean of the squares $\frac{\Sigma x^2}{n}$

$= \frac{25 + 25 + 25 + 25 + 25}{5} = \frac{125}{5} = 25$

Standard deviation is $\sqrt{\frac{\Sigma x^2}{n} - \left(\frac{\Sigma x}{n}\right)^2}$

$= \sqrt{25 - (5)^2} = \sqrt{0} = 0$

ANSWERS

A1 (c) Mean = 3
Standard deviation = 0.816 (3 s.f.)

(d) Mean = 3
Standard deviation = 0.816 (3 s.f.)

A2 History: mean = 8.4,
standard deviation = 1.2

Mathematics: mean = 7.4,
standard deviation = 1.62 (3 s.f.)

The mean of the mathematics marks is lower than the mean of the history marks and the mathematics marks are more spread out than the history marks.

TUTORIAL

T1 (c) 2, 3, 4

Mean $\frac{\Sigma x}{n} = \frac{2+3+4}{3} = \frac{9}{3} = 3$

Mean of the squares $\frac{\Sigma x^2}{n} = \frac{4+9+16}{3} = \frac{29}{3}$

$= 9.67$ (3 s.f.)

Standard deviation is $\sqrt{\frac{\Sigma x^2}{n} - \left(\frac{\Sigma x}{n}\right)^2}$

$= \sqrt{\frac{29}{3} - (3)^2} = \sqrt{\frac{2}{3}} = 0.816\,496\,5$

$= 0.816$ (3 s.f.)

(d) 2, 2, 3, 3, 4, 4

Mean $\frac{\Sigma x}{n} = \frac{2+2+3+3+4+4}{6} = \frac{18}{6} = 3$

Mean of the squares $\frac{\Sigma x^2}{n}$

$= \frac{4+4+9+9+16+16}{6} = \frac{58}{6} = 9.67$ (3 s.f.)

Standard deviation is $\sqrt{\frac{\Sigma x^2}{n} - \left(\frac{\Sigma x}{n}\right)^2}$

$= \sqrt{\frac{58}{6} - (3)^2} = \sqrt{\frac{2}{3}}$

$= 0.816\,496\,5 = 0.816$ (3 s.f.)

T2 History

x	x^2
8	64
9	81
10	100
6	36
8	64
7	49
9	81
9	81
10	100
8	64
$\Sigma x = 84$	$\Sigma x^2 = 720$

Mean $= \frac{\Sigma x}{n} = \frac{84}{10} = 8.4$

Standard deviation $= \sqrt{\frac{\Sigma x^2}{n} - \left(\frac{\Sigma x}{n}\right)^2}$

$= \sqrt{\frac{720}{10} - (8.4)^2} = \sqrt{1.44} = 1.2$

ANSWERS

TUTORIALS

T2 *Mathematics*

x	x^2
9	81
5	25
7	49
6	36
8	64
8	64
7	49
10	100
5	25
9	81
$\Sigma x = 74$	$\Sigma x^2 = 574$

$$\text{Mean} = \frac{\Sigma x}{n} = \frac{74}{10} = 7.4$$

$$\text{Standard deviation} = \sqrt{\frac{\Sigma x^2}{n} - \left(\frac{\Sigma x}{n}\right)^2}$$

$$= \sqrt{\frac{574}{10} - (7.4)^2} = \sqrt{2.64} = 1.624\,807\,7$$

$= 1.62$ (3 s.f.)

The mean of the mathematics marks is lower than the mean of the history marks (suggesting that the history paper was an easier paper or else the pupils were better prepared for the test). The mathematics marks are more spread out than the history marks.

NOTE

In these questions you should always make reference to the mean (measure of central tendency) and the standard deviation (measure of spread).

A3

(a) Mean = 6, standard deviation = $3\sqrt{2}$
(b) Mean = $^-2$, standard deviation = $\sqrt{2}$
(c) Mean = 102, standard deviation = $\sqrt{2}$
(d) Mean = 21, standard deviation = $10\sqrt{2}$

NOTE

You can ignore the – sign here, as it is a deviation from a number.

T3

(a) *New series = original series* × 3
New mean = mean × 3 = 6
New standard deviation = SD × 3
= $3\sqrt{2}$

(b) *New series = original series* × $^-1$
New mean = mean × $^-1$ = $^-2$
New standard deviation = SD × $^-1$
= $\sqrt{2}$

(c) *New series = original series* + 100
New mean = mean +100 = 102
Standard deviation unaffected = $\sqrt{2}$

(d) *New series = original series* × 10 *then* + 1
New mean = mean × 10 + 1
= 21
New standard deviation = SD × 10
= $10\sqrt{2}$
As the addition of 1 *has no effect on the spread here.*

ANSWERS

A4 Mean = 19.1 kg (3 s.f.)
Standard deviation = 4.94 kg (3 s.f.)

TUTORIALS

T4 *To calculate the mean and standard deviation you need to find the mid-interval values.*

Mass (kg)	Frequency f	Mid-interval value x	x^2	fx	fx^2
6–10	2	8	64	16	128
11–15	6	13	169	78	1014
16–20	17	18	324	306	5508
21–25	11	23	529	253	5819
26–30	4	28	784	112	3136
	$\Sigma f = 40$			$\Sigma fx = 765$	$\Sigma fx^2 = 15\,605$

NOTE

The value of $n = \Sigma f$

The formula needs to be modified to take account of the fact that all of the values have to be multiplied by the frequencies.

$$\text{Mean} = \frac{\Sigma fx}{n} = \frac{765}{40}$$

$$= 19.125 = 19.1\ kg\ (3\ s.f.)$$

$$\text{Standard deviation} = \sqrt{\frac{\Sigma fx^2}{n} - \left(\frac{\Sigma fx}{n}\right)^2}$$

$$= \sqrt{\frac{15\,605}{40} - (19.125)^2}$$

$$= \sqrt{24.359\,375}$$

$$= 4.935\,521\,8$$

$$= 4.94\ kg\ (3\ s.f.)$$

CHAPTER 95

MULTIPLICATION RULE – DEPENDENT EVENTS

Two or more events are dependent if one event affects the probability of the other event.

DEPENDENT EVENTS

Worked example

A bag contains four red and three blue counters. A counter is drawn from the bag and then a second counter is drawn from the bag. Draw a tree diagram to show the various possibilities that can occur and use the diagram to find the probability that both counters are blue.

The question does not make it clear whether the first counter is replaced before the second counter is drawn. This gives rise to two possibilities as shown in the following tree diagrams.

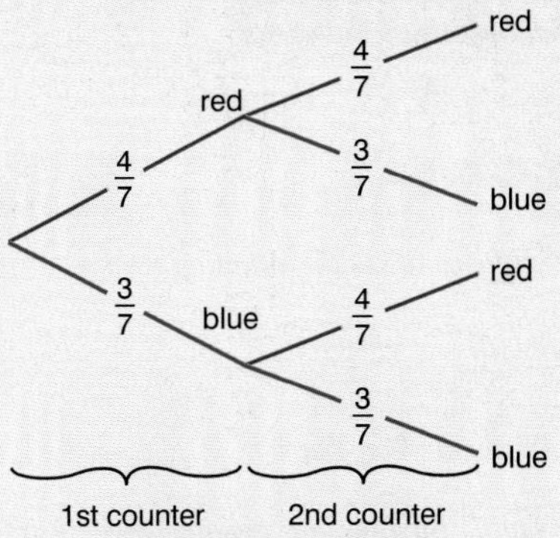

With replacement
Independent events

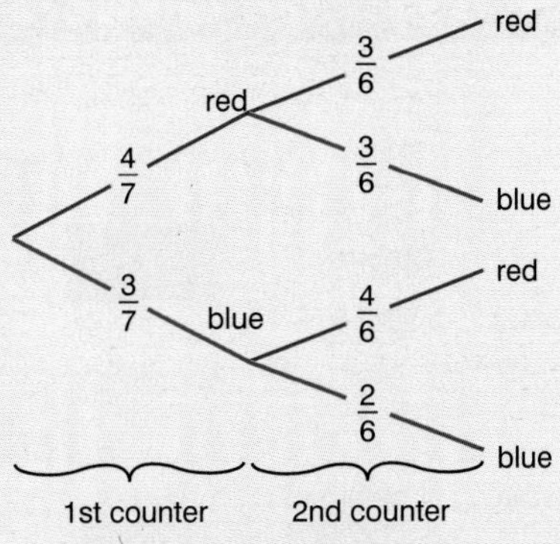

Without replacement
Dependent events

With replacement (independent events)

If the first counter is replaced before the second counter is drawn, then the two events are independent and the probabilities for each event are the same.

From the diagram, the probability that both counters are blue

= p(blue counter drawn first and blue counter drawn second)

= p(blue counter drawn first) × p(blue counter drawn second)

$= \frac{3}{7} \times \frac{3}{7} = \frac{9}{49}$

Without replacement (dependent events)

If the first counter is not replaced before the second counter is drawn, then the two events are not independent (i.e. they are dependent) and the probabilities on the second event will be affected by the outcomes on the first event.

For the second counter:

- if the first counter was blue, there are now two blue counters and six counters altogether
- if the first counter was not blue, there are still three blue counters and six counters altogether.

From the diagram you can see that:

the probability that both counters are blue

= p(blue counter drawn first and blue counter drawn second)

= p(blue counter drawn first) × p(blue counter drawn second)

$= \frac{3}{7} \times \frac{2}{6} = \frac{6}{42} = \frac{1}{7}$

Check yourself

QUESTIONS

Q1 Jack has ten black and six brown socks in his drawer. If two socks are removed from the drawer, one after the other, calculate the probability that:

(a) both socks are black
(b) both socks are brown
(c) the socks are different colours.

Q2 The probability that Rebecca passes the driving theory test on her first attempt is $\frac{6}{7}$. If she fails then the probability that she passes on any future attempt is $\frac{7}{8}$.

Draw a tree diagram to represent this situation and use it to calculate the probability that Rebecca passes the driving test on her third attempt.

REMEMBER! Cover the answers if you want to.

ANSWERS

A1

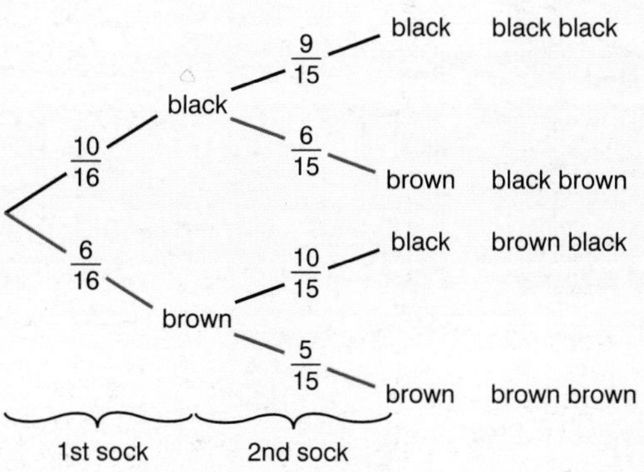

(a) $\frac{3}{8}$

(b) $\frac{1}{8}$

(c) $\frac{1}{2}$

A2 Probability $= \frac{1}{64}$

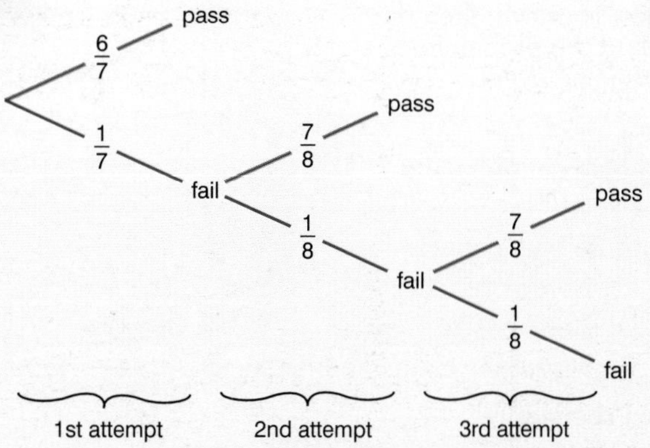

TUTORIAL

T1 *A diagram helps to see the situation clearly.*

(*a*) *p(black and black)* $= \frac{10}{16} \times \frac{9}{15} = \frac{3}{8}$

(*b*) *p(brown and brown)* $= \frac{6}{16} \times \frac{5}{15} = \frac{1}{8}$

(*c*) *Reading from the tree diagram p(socks are different colours)*
= *p(black and brown or brown and black)*
$= \frac{10}{16} \times \frac{6}{15} + \frac{6}{16} \times \frac{10}{15} = \frac{1}{2}$

T2 *p(passing on third attempt)* = *p(fail and fail and pass)*

$= \frac{1}{7} \times \frac{1}{8} \times \frac{7}{8} = \frac{1}{64}$

CHAPTER 96

EXAMINATION QUESTIONS 19

EXAM PRACTICE

Sample Student's Answers & Examiner's Comments

1 (a) The histogram shows the distribution of marks obtained by St Wilfred's Year 11 pupils in their mock examinations.

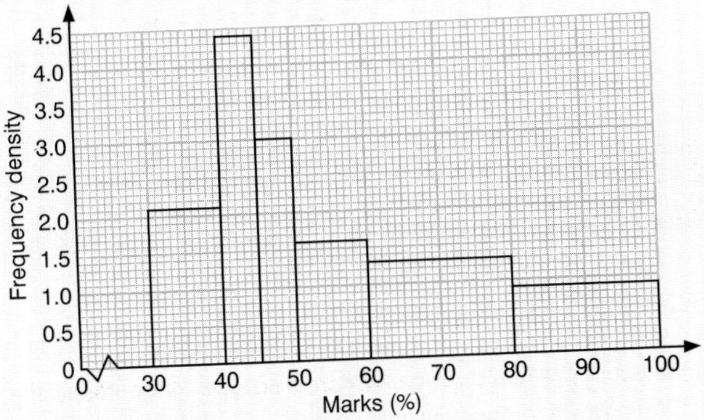

Use the histogram to:

(i) estimate how many pupils gained between 50 and 60%

(ii) find which mark range illustrated is the modal range.

(b) A holiday insurance company carries out a survey on the ages of travellers using their service in one particular week.

The information is given in this table.

Draw a histogram to represent this information

Age (x years)	Frequency (*f*)
$0 < x \leqslant 5$	20
$5 < x \leqslant 10$	54
$10 < x \leqslant 20$	106
$20 < x \leqslant 30$	223
$30 < x \leqslant 40$	180
$40 < x \leqslant 60$	252
$60 < x \leqslant 90$	54

(NEAB specimen paper 1998)

(a) From the histogram:

(i) frequency $= 1.6 \times 10 = 16$

(ii) Modal range $= 60 - 80$

Age (*x* years)	Frequency (*f*)	Class width	Frequency density
$0 < x \leqslant 5$	20	5	4
$5 < x \leqslant 10$	54	5	10.8
$10 < x \leqslant 20$	106	10	10.6
$20 < x \leqslant 30$	223	10	22.3
$30 < x \leqslant 40$	180	10	18
$40 < x \leqslant 60$	252	20	12.6
$60 < x \leqslant 90$	54	30	1.8

EXAMINER'S COMMENTS

1 **a)** *The candidate has correctly identified the frequency by multiplying the class width by the frequency density and has correctly identified the modal group (which has the largest area).*

EXAMINER'S COMMENTS

b)
She has identified the respective frequency densities and drawn the histogram.

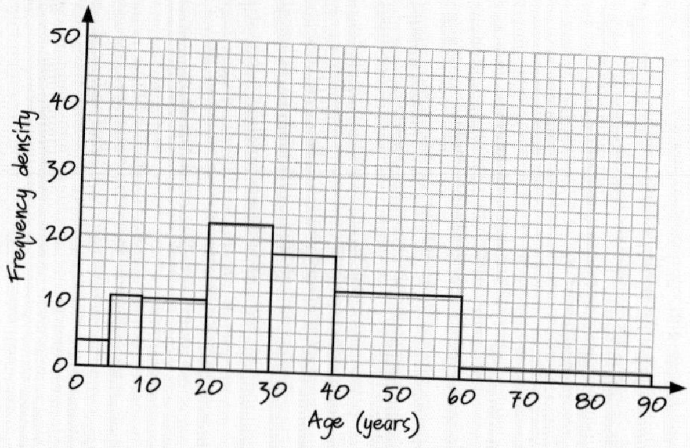

2 This set of marks was obtained by 10 pupils in an English GCSE examination.

73, 70, 62, 67, 69, 76, 55, 65, 61, 82

The mean of this set of marks is 68.

(a) Calculate the standard deviation of these marks.

All the ten pupils had their marks increased by 3 for good spelling.

(b) Write down for the new set of marks:

(i) the mean
(ii) the standard deviation.

(London specimen paper 1998)

(a) Standard deviation $= \sqrt{\frac{\Sigma x^2}{n} - \left(\frac{\Sigma x}{n}\right)^2}$

$\Sigma x^2 = 73^2 + 70^2 + 62^2 + 67^2 + 69^2 + 76^2 + 55^2 + 65^2 + 61^2 + 82^2$

$= 5329 + 4900 + 3844 + 4489 + 4761 + 5776 + 3025 + 4225 + 3721 + 6724$

$= 46\,794$

$\sqrt{\frac{\Sigma x^2}{n} - \left(\frac{\Sigma x}{n}\right)^2} = \sqrt{\frac{46\,794}{10} - (68)^2}$

$= \sqrt{55.4}$

$= 7.443\,1176$

$= 7.44$ (3 s.f.)

(b) (i) Mean $= 68 + 3 = 71$

(ii) Standard deviation $= 7.44$ (3 s.f.)

2 a)
As the mean is given then the other formula provided on the examination paper would have been easier to use. The candidate is wasting time by writing out each step of Σx^2 when much of this can be done on the calculator – sufficient working should be shown to show the method only.

b)
He has, fortunately, realised that when all marks are increased by 3 the mean is increased by 3 and the standard deviation remains unchanged.

3 Sam was making a survey of pupils in his school. He wanted to find out their opinions on noise pollution by motorbikes. The size of each year group is shown below.

Year	Boys	Girls	Total
8	85	65	150
9	72	75	147
10	74	78	152
11	77	72	149
6th Form	93	107	200
			798

Sam took a sample of 80 pupils.

(a) Explain whether or not he should have sampled equal numbers of boys and girls in Year 8.

(b) Calculate the number of pupils he should have sampled in Year 8.

(London, specimen paper 1998)

(a) There are more boys than girls in Year 8.

(b) $\frac{150}{798} \times 80 = 15.037\,594$

4 On each day that I go to work, the probability that I leave home before 0800 is 0.2. The probability that I leave home after 0810 is 0.05.

(a) What is the probability that I leave home between 0800 and 0810 inclusive?

(b) The probability that I am late for work depends upon the time I leave home. The probabilities are given in the table below.

Time of leaving home	Before 0800	Between 0800 and 0810	After 0810
Probability of being late	0.01	0.1	0.2

I work 230 days each year. Estimate how many times I would expect to be late in a year.

(NEAB specimen paper 1998)

(a) Probability $= 1 - (0.2 + 0.05) = 0.75$

(b) $0.2 \times 0.01 + 0.75 \times 0.1 + 0.05 \times 0.2$

$= 0.087$

On 230 days $= 230 \times 0.087$

$= 20.01$

EXAMINER'S COMMENTS

3 a)
The candidate might benefit from being more descriptive by stating, 'He should not sample equal numbers of boys and girls as there are more boys than girls in Year 8.'

b)
The working is correct but the candidate loses marks for not rounding to the nearest integer – you cannot sample 15.037 594 people. An answer of 15 gains full credit.

4 a)
Using the idea that total probability = 1.
Using p(late)
= p(leaving before 0800 and being late)
+ p(leaving 0800–0810 and being late)
+ p(leaving after 0800 and being late)

b)
The candidate might round this to 20 days.

Questions to Answer

1 The age of each person in a coach party is recorded. The table shows the number of people in each age category.

Age group (years)	10–	20–	30–	45–	50–	70–100
Frequency	2	0	6	4	22	12

Draw a histogram to represent the data.

(SEG specimen paper 1998)

2 A small building company employs ten people. The weekly pay of each is as follows.

£150 £150 £200 £220 £220 £240 £260 £290 £290 £350

The mean weekly pay for these employees is £237.

(a) Calculate the standard deviation of the weekly pay.

A building project is completed early and each person in the company receives a bonus on their weekly pay.

(b) The manager suggests that everyone should receive a bonus of £24.50 added to their usual weekly pay. What effect would this have on the mean and standard deviation of the weekly pay?

(c) The foreman suggests that everyone should receive a bonus of 10% added to their usual weekly pay. What effect would this have on the mean and standard deviation of the weekly pay?

(d) Which of these proposals should be implemented to give the greater benefit to the greater number of employees?
Explain your answer.

(SEG specimen paper 1998)

3 This list shows the maximum daily temperature, in °F, throughout the month of April.

56.1	49.4	63.7	56.7	55.3	53.5	52.4	57.6	59.8	52.1
45.8	55.1	42.6	61.0	61.9	60.2	57.1	48.9	63.2	68.4
55.5	65.2	47.3	59.1	53.6	52.3	46.9	51.3	56.7	64.3

(a) Complete the grouped frequency table below.

Temperature, T	**Frequency**
$40 < T \leqslant 50$	
$50 < T \leqslant 54$	
$54 < T \leqslant 58$	
$58 < T \leqslant 62$	
$62 < T \leqslant 70$	

3 (b) Use your table of values in part (a) to calculate an estimate of the mean and the standard deviation of the distribution. You must show your working clearly.

(c) Draw a histogram to represent your distribution in part (a).

(d) For the month of June, the mean maximum daily temperature was 58.9 °F and the standard deviation was 5.3 °F.

Comment on the difference between these figures and the corresponding figures for April.

(MEG syllabus A, specimen paper 1998)

4 The South Midshires Pullman train is timetabled to arrive at Central Station at midday. Over a period of 42 days Mrs Eva Grumpy noted how late this train was each day and produced the following histogram.

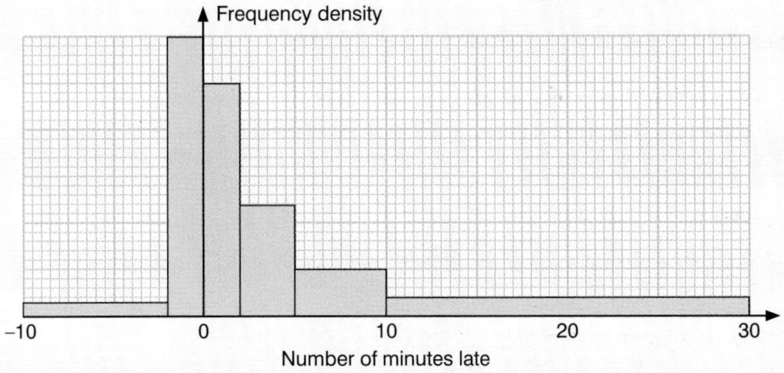

(a) Complete the first two columns of this table:

Number of minutes late (t)	Frequency
$^{-}10 \leqslant t < {}^{-}2$	2
$^{-}2 \leqslant t < 0$	
$0 \leqslant t < 2$	7
$2 \leqslant t < 5$	
....	
....	

(b) Calculate, to 1 decimal place, estimates for the mean and the standard deviation of these data.

(MEG syllabus B, specimen paper 1998 – modified)

5 A bag contains four red and six blue balls. One ball is chosen and its colour noted. It is not put back into the bag. A second ball is chosen and its colour noted.

(a) Draw a tree diagram to represent this situation.

(b) (i) Find the probability of obtaining two red balls.
(ii) Find the probability of obtaining one ball of each colour.

(SEG specimen paper 1998)

CHAPTER 97

ANSWERS TO EXAM PRACTICE

Answers	Examiner's comments
7 EXAMINATION QUESTIONS 1	**(Page 17)**
1 **$3960 = 2 \times 2 \times 2 \times 3 \times 3 \times 5 \times 11$ or $2^3 \times 3^2 \times 5 \times 11$**	You find the prime factors by successively dividing by prime numbers in order i.e. 2, 3, 5, 7, 11, 13, 17, 19, 23, … . Check that all the numbers are prime numbers. (See Chapter 1, The number system for further information.)
2 **40 inches ≈ 1 metre**	Using the fact that 1 inch ≈ 2.5 cm and 1 metre = 100 centimetres.
3 **160 000 cm ≈ 1 mile**	Using 1 foot = 30 cm, 3 feet = 1 yard and 1760 yards = 1 mile. The calculated answer is 158 400 cm but this needs to be rounded to an appropriate degree of accuracy.
4 **$\sqrt{81} = 9$** **$\sqrt{50} = 7.071\,067\,8\ldots$**	You should use the √ key on your calculator. 81 is a square number so the square root is exact but 50 is not a square number so the square root is not exact.
5 **$\sqrt[3]{512} = 8$** **$\sqrt[3]{50} = 3.684\,031\,5\ldots$**	You should use the ∛ key on your calculator. 512 is a cube number so the cube root is exact but 50 is not a cube number so the cube root is not exact.
6 **$\frac{4}{1}$ or 4**	To find the reciprocal of a number you need to convert the number to a fraction. The number $0.25 = \frac{1}{4}$ and turning the fraction upside-down gives the answer $\frac{4}{1}$ or 4.
7 **12^{18}**	It is possible to multiply these out but the intended method involves using the rules of indices so that $12^7 \times 12^{11} = 12^{7+11} = 12^{18}$
8 **$9^{-2} = \frac{1}{9^2} = \frac{1}{81}$**	Again, using the rules of indices, $9^3 \div 9^5 = 9^{3-5} = 9^{-2}$ which is written as $\frac{1}{9^2}$ or $\frac{1}{81}$ (as $9^2 = 9 \times 9 = 81$).
9 **7^{14}**	Again, using the rules of indices, $7^4 \times 7^2 \times 7^8 = 7^{4+2+8} = 7^{14}$.
10 **2304**	You cannot use the rules of indices here as the base numbers are not the same. The numbers have to be evaluated separately and then multiplied. $4^3 = 64$ and $6^2 = 36$ so $4^3 \times 6^2 = 64 \times 36 = 2304$.
12 EXAMINATION QUESTIONS 2	**(page 36)**
1 **1.25×10^7**	Using the fact that 1 million = 10^6, 12.5 million = $12.5 \times 10^6 = 1.25 \times 10^7$ (writing the number in standard form).
2 **2.67×10^{-3} (3 s.f.)**	$(4 \times 10^5) \div (1.5 \times 10^8) = (4 \div 1.5) \times (10^5 \div 10^8) = 2.6666\ldots \times 10^{5-8}$ Using the rules of indices. $= 2.6666\ldots \times 10^{-3} = 2.67 \times 10^{-3}$ (3 s.f.).
3 **8 minutes**	Use time = distance ÷ speed = $1.496 \times 10^8 \div 299\,800$ seconds = 498.999 33 seconds = 8.316 655 5 minutes — Dividing by 60 to convert to minutes as required by the question. = 8 minutes (to the nearest minute) — As required by the question.
4 **Order is $\frac{6}{11}$, 59%, $\frac{3}{5}$, 62%, 0.65, 0.666, $\frac{2}{3}$**	Converting each to a percentage: $0.65 = 65\%$, $\frac{3}{5} = 60\%$, $\frac{6}{11} = 54.54\ldots\%$, $\frac{2}{3} = 66.6666\ldots\%$, $0.666 = 66.6\%$. Read the question carefully to see which way around the answer should be given.
5 **$3\frac{8}{15}$**	$5\frac{1}{5} - 1\frac{2}{3} = \frac{26}{5} - \frac{5}{3}$ — Converting to improper fractions. $= \frac{78}{15} - \frac{25}{15}$ — Writing as equivalent fractions with a common denominator. $= \frac{53}{15} = 3\frac{8}{15}$ — Rewriting as a mixed number.
6 **16 bottles**	$12 \div \frac{3}{4} = 12 \times \frac{4}{3} = \frac{48}{3} = 16$

Answers	Examiner's comments
£5.50	$2\frac{1}{2}\%$ of £220 $= \frac{2\frac{1}{2}}{100}\% \times £220 = £5.50$
(a) 560 (b) 627	(a) An increase of 12% is equivalent to a multiplier of $\frac{112}{100}$. $\frac{112}{100} \times 500 = 560$ (b) $\frac{112}{100} \times 560$ Representing an increase of 12% on the previous year's figure of 560. $= 627.2$ Which is then rounded off to give a realistic answer.
29.2% or 29%	Fraction $= \frac{£0.70}{£2.40}$ and to convert a fraction to a percentage you multiply by 100 giving 29.166 667%. This is then rounded off to an appropriate degree of accuracy.
0 2100	20% = 420, 1% = 21 and 100% = 2100
1 £534 000	After an increase of 8%, £576 720 represents 108% (100% + 8%) of the original value so 108% = £576 720, 1% = £5340 (dividing by 108) and 100% = £534 000.

17 EXAMINATION QUESTIONS 3 (Page 47)

Answers	Examiner's comments
24 mph and 36 mph	Ratio = 2 : 3 : 5 = 24 : 36 : 60, as an equivalent ratio found by multiplying by 12. So the speed in the other two lanes is 24 miles per hour and 36 miles per hour.
16 g	There are 9 parts (8 + 1) and each part is worth 2 g. The copper has $8 \times 2\,\text{g} = 16\,\text{g}$.
£889, £1244.60 and £1422.40	There are 20 parts (5 + 7 + 8) and each part is worth £177.80 (£3556 ÷ 20). The three friends get 5 × £177.80, 7 × 177.80 and 8 × £177.80.
0.067	Rounding as $\frac{30 \times 0.1}{15 \times 3} = 0.066\,666\,6...$
$s = 8.75$ m The actual value is $s = 8.625$ m.	Using $v = 20$, $u = 15$ and $a = 10$ and remembering to work out the numerator before dividing by the denominator.
(a) 1.082 132 8 (b) $\frac{5 \times 10^2}{500}$ (c) 1	Rounding off to 1 s.f.

21 EXAMINATION QUESTIONS 4 (Page 59)

Answers	Examiner's comments
27.8 m/s (3 s.f.)	Using the fact that 1 km = 1000 m and 1 hour = 3600 seconds, 100 km/h = 100 000 m/h $= \frac{100\,000}{3600}$ m/s = 27.8 m/s (3 s.f.)
$8 \times 10^1\,\text{km}^2$ (square kilometres)	Area per head of population $= \frac{4 \times 10^{11}}{5 \times 10^9} = 0.8 \times 10^2 = 8 \times 10^1\,\text{km}^2$, leaving the answer in standard form.
6.71 gallons (3 s.f.)	In 80 miles on the motorway, the car uses $\frac{80}{35} = 2.285\,714\,3$ gallons. In 115 miles on the other road, the car uses $\frac{115}{26} = 4.423\,076\,9$ gallons. Total number of gallons used = 6.708 791 2 gallons which is then rounded to an appropriate degree of accuracy (3 s.f. or 2 s.f. would be most appropriate here).
£131.73 (to the nearest penny)	Using the formula for simple interest and remembering that 10 weeks needs to be expressed in years as $\frac{10}{52}$ years, $I = \frac{10\,000 \times 6.85 \times \frac{10}{52}}{100} = £131.73$ Rounding the answer to the nearest penny.

Answers | **Examiner's comments**

5 29 months or 2 years 5 months

The interest is found from $A = P + I$ and substituted into the simple interest formula.

$$72.50 = \frac{600 \times 5 \times T}{100} \text{ so } T = \frac{72.50}{30} = 2.416\,666\,667 \text{ years.}$$

It is important that the value for T is properly interpreted. The time is converted to months by remembering that there are 12 months in 1 year.
$2.416\,666\,667 \times 12 = 29$ months $= 2$ years 5 months.

6 £5376.85 (to the nearest penny)

In this question, the compound interest is best worked out by repeatedly applying the simple interest formula where $P = £5000$, $R = 7.4\%$ and $T = 6$ months $= \frac{1}{2}$ year.

$$\text{After 6 months } A = P + \frac{PRT}{100} = 5000 + \frac{5000 \times 7.4 \times \frac{1}{2}}{100} = £5185$$

$$\text{After 12 months } A = £5185 + \frac{5185 \times 7.4 \times \frac{1}{2}}{100} = £5376.845.$$

27 EXAMINATION QUESTIONS 5

(Page 71)

1 (a) $3\sqrt{3}$
(b) $6\sqrt{2}$

(a) The expressions can be simplified using the fact that

$$\sqrt{a} \times \sqrt{b} = \sqrt{a \times b} \text{ and } \frac{\sqrt{a}}{\sqrt{b}} = \sqrt{\frac{a}{b}}.$$

$$\sqrt{12} + \sqrt{3} = \sqrt{4 \times 3} + \sqrt{3} = 2\sqrt{3} + \sqrt{3} = \mathbf{3\sqrt{3}}$$

(b) $\sqrt{12} \times \sqrt{6} = \sqrt{12 \times 6} = \sqrt{72} = \sqrt{36 \times 2} = \mathbf{6\sqrt{2}}$

2 $0.\dot{2}3\dot{4} = \frac{26}{111}$

A rational number can be expressed in the form $\frac{p}{q}$ where p and q are integers.

$$1000 \times 0.\dot{2}3\dot{4} = 234.234\,234...$$
$$1 \times 0.\dot{2}3\dot{4} = 0.234\,234...$$
$$999 \times 0.\dot{2}3\dot{4} = 234$$

$0.\dot{2}3\dot{4} = \frac{234}{999} = \frac{26}{111}$ Cancelling down.

As $0.\dot{2}3\dot{4} = \frac{26}{111}$ then $0.\dot{2}3\dot{4}$ must be a rational number.

3 $\frac{37}{1100}$

Let $0.033\,636\,36... = x$.
Then $100 \times x = 3.363\,636\,3...$
$1 \times x = 0.033\,636\,3...$
$99 \times x = 3.33$

$x = \frac{3.33}{99}$

$= \frac{333}{9900}$ As a proper fraction.

$= \frac{37}{1100}$ Cancelling down.

You should always write the fraction as a proper fraction, using equivalent fractions, and cancel down to the lowest terms where possible.

4 Order is $(^-2)^3, 5^{-1}, 2^{-2}, 20^0, 32^{\frac{4}{5}}, 243^{\frac{3}{5}}$

With the smallest first …

$243^{\frac{3}{5}} = (243^{\frac{1}{5}})^3 = 3^3 = 27$ You might use your x^y key on your calculator to help you with these. The power $\frac{1}{5}$ can be keyed in as 0.2.

$32^{\frac{4}{5}} = (32^{\frac{1}{5}})^4 = 2^4 = 16$

$5^{-1} = \frac{1}{5}$ As $a^{-n} = \frac{1}{a^n}$

$20^0 = 1$ As $a^0 = 1$

$(^-2)^3 = {^-2} \times {^-2} \times {^-2} = {^-8}$

$2^{-2} = \frac{1}{2^2} = \frac{1}{4}$

Answers	Examiner's comments
5 (a) **£2000** (b) **2 years 10 months (to nearest month)**	(a) Using $v \propto \frac{1}{a}$ to write $v = \frac{k}{a}$, the value of k is found by substituting $v = 7000$ when $a = 1$. The constant of proportionality = 7000 and $v = \frac{7000}{a}$. When $a = 3\frac{1}{2}$, $v = £2000$. (b) When $v = 2500$, $a = 2.8$ years. It is important that this is not interpreted as 2 years 8 months as it is closer to 2 years 10 months (to the nearest month).
6 **$R = \frac{2}{3}$ when $T = 9$** **$R = \frac{2}{\sqrt{5}}$ when $T = 5$**	Using $R \propto \frac{1}{\sqrt{T}}$ to write $R = \frac{k}{\sqrt{T}}$, the value of k is found by substituting $R = 1$ when $T = 4$. The constant of proportionality = 2 and $R = \frac{2}{\sqrt{T}}$. When $T = 9$, $R = \frac{2}{3}$ and when $T = 5$, $R = \frac{2}{\sqrt{5}}$.
7 (a) **251 kg** (b) **2 kg**	(a) Weight is 12.6 kg to nearest 0.1 kg. Weight $_{min}$ = 12.55 kg. Weight $_{max}$ = 12.65 kg. Lower bound for 20 videos = 20 × 12.55 = 251 kg (b) Upper bound for 20 videos = 20 × 12.65 = 253 kg Difference = 253 − 251 = 2 kg
8 **Maximum = 73.9 mph (3 s.f.)** **Minimum = 70.1 mph (3 s.f.)**	Distance = 42 miles. Distance_{min} = 41.5 miles. Distance_{max} = 42.5 miles. Time = 35 minutes. Time_{min} = 34.5 minutes. Time_{max} = 35.5 minutes. Using speed = $\frac{\text{distance}}{\text{time}}$, and expressing time in hours: $\text{Speed}_{max} = \frac{42.5}{\frac{34.5}{60}} = 73.9$ mph (3 s.f.) $\text{Speed}_{min} = \frac{41.5}{\frac{35.5}{60}} = 70.1$ mph (3 s.f.)
9 **6.7%**	Minimum volume = 0.145 × 829.5 × 634.5 = 76 316.074 mm^3 Maximum volume = 0.155 × 830.5 × 635.5 = 81 806.326 mm^3 Saving = 81 806.326 − 76 316.074 = 5490.252 mm^3 Percentage saving = $\frac{5490.252}{81\,806.326} \times 100\%$ = 6.711 2805% = 6.7% (to an appropriate degree of accuracy) It is not necessary to work to such a degree of accuracy especially since the dimensions are approximated to the nearest mm/nearest 0.01 mm.

32 EXAMINATION QUESTIONS 6

(Page 87)

1 **48, 63 and 80** **nth term = $n^2 + 2n$** **Difference = 53**	3, 8, 15, 24, 35 ... Differences: +5, +7, +9, +11 Differences of differences: +2, +2, +2 The differences of differences are constant so a quadratic rule can be applied to the terms to find the nth term. (See Chapter 28, Patterns and sequences for further examples.) The 25th term = $25^2 + 2 \times 25 = 675$ and the 26th term = $26^2 + 2 \times 26 = 728$. The difference between the two terms is 53.
2 (a) **72** (b) **nth term = $2n^2$**	(a) 2, 8, 18, 32, 50 ... Differences: +6, +10, +14, +18 Differences of differences: +4, +4, +4 (b) The differences of differences are constant so a quadratic rule can be applied to the terms to find the nth term. (See Chapter 28, Patterns and sequences for further examples.) (c) 50th term = $2 \times 50^2 = 5000$

Answers / Examiner's comments

(c) 50th term = 5000

3 $d = 19.8$ cm (1 d.p.)

The perimeter of the triangle $= 4x + 8$ cm, so the perimeter of the square $= 4x + 8$ cm and each side is $(4x + 8) \div 4 = x + 2$ cm.
Area of triangle $= 84\,\text{cm}^2$.

Area of triangle $= \frac{1}{2} \times \text{base} \times \text{height}$

$= \frac{1}{2} \times 2x \times 7 = 7x$

and $7x = 84$ so $x = 12$ cm. Each side of the square is $x + 2$ cm $= 14$ cm.

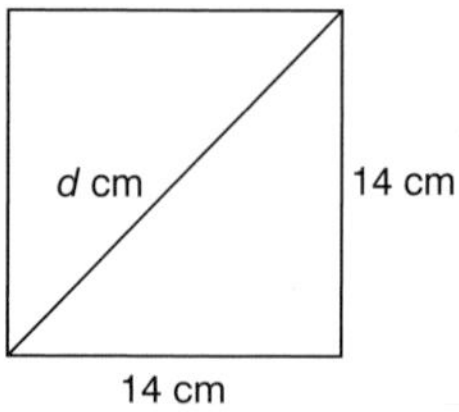

$d^2 = 14^2 + 14^2 = 392$ Using Pythagoras' theorem.

$d = 19.798\,99$ Taking square roots on both sides.

You can do all of this question using Pythagoras' theorem if you prefer.

4 (a) 3
(b) 15

(a) $2^2 - 3 \times 2 + 5 = 4 - 6 + 5 = 3$
(b) $(^-2)^2 - 3 \times {}^-2 + 5 = 4 + 6 + 5 = 15$

5 $2\pi r(h + r)$

$2\pi r$ is a common factor and is taken outside the brackets.

6 $a = 5, b = {}^-3$

$(2x - 1)(x + 3) = 2x^2 + 5x - 3$ and comparing with $2x^2 + ax + b$ you can find a and b.

7 $20x$

$(x + 5)^2 - (x - 5)^2 = \{x^2 + 10x + 25\} - \{x^2 - 10x + 25\}$
$= x^2 + 10x + 25 - x^2 + 10x - 25$
$= 20x$

8 $(x - 3)(x - 11)$

$x^2 - 14x + 33 = (x \quad)(x \quad)$
$= (x - 3)(x - 11)$

Looking for numbers which multiply together to give $^+33$.
$^+1 \times {}^+33 \quad {}^-1 \times {}^-33$
$^+3 \times {}^+11 \quad {}^-3 \times {}^-11$

36 EXAMINATION QUESTIONS 7

(Page 95)

1 26, 28 and 30

Let the three consecutive even numbers be x, $x + 2$ and $x + 4$. Then $x + (x + 2) + (x + 4) = 84$ and $x = 26$ so the numbers are 26, 28 and 30.

2 20 metres

Let the length of the room be x metres. Then the width is $(x - 4)$ metres, as the room is 4 metres longer than its width.
The perimeter $= l + w + l + w$ so $(x - 4) + x + (x - 4) + x = 72$ and $x = 20$.

3 $a = \dfrac{2(s - ut)}{t^2}$

$s = ut + \frac{1}{2}at^2$

$s - ut = \frac{1}{2}at^2$ Subtracting ut to isolate the term containing a on one side.

$at^2 = 2(s - ut)$ Multiplying both sides of the equation by 2 and turning the equation around.

$a = \dfrac{2(s - ut)}{t^2}$ Dividing both sides by t^2.

4 (a) $A = 5030\,\text{cm}^2$ (3 s.f.)
(b) $r = \sqrt{\dfrac{A}{4\pi}}$
(c) $r = 2.82$ cm (3 s.f.)

(a) $A = 4 \times \pi \times r^2 = 5026.5482\,\text{cm}^2$

(b) $A = 4\pi r^2$

$\dfrac{A}{4\pi} = r^2$ Dividing both sides by 4π.

$r^2 = \dfrac{A}{4\pi}$ Turning the equation around.

$r = \sqrt{\dfrac{A}{4\pi}}$ Taking square roots on both sides.

(c) $r = \sqrt{\dfrac{A}{4\pi}} = \sqrt{\dfrac{100}{4\pi}} = 2.820\,947\,9$ cm

5 (a) $x = 2$
(b) $y = 0$
(c) $x = 3$

(a) $5^4 = 625$ So $2x = 4$ and $x = 2$.
(b) $10^0 = 1$ So $y = 0$.
(c) $2^8 = 256$ So $3x - 1 = 8$ and $x = 3$.

Answers

Examiner's comments

41 EXAMINATION QUESTIONS 8

(Page 116)

1 (a) £12.30
(b) £16.40
(c) 48 therms

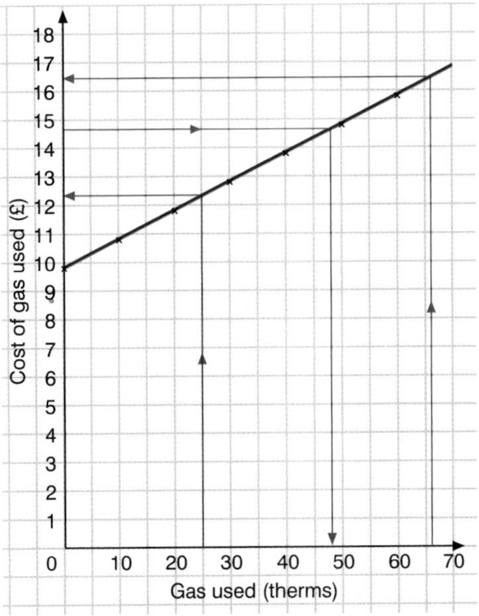

The information should be obtained from the graph. It is important to show sufficient working to demonstrate that you have used the graph. It is quite acceptable for you to check your work by calculating the answers.

2 (a) 60 kph
(b) 65.5 kph (3 s.f.)

(a) For speed in kilometres per hour, the time must be expressed in hours.
40 minutes $= \frac{40}{60}$ hours

Between A and B distance travelled = 40 km and time taken = 40 minutes.
So speed = distance ÷ time $= 40 \div \frac{40}{60} = 60$ kph

(b) Again 110 minutes $= \frac{110}{60}$ hours.

Between A and C distance travelled = 120 km and time taken = 110 minutes.
So speed = distance ÷ time $= 120 \div \frac{110}{60} = 65.454\,545 = 65.5$ kph (3 s.f.)

3

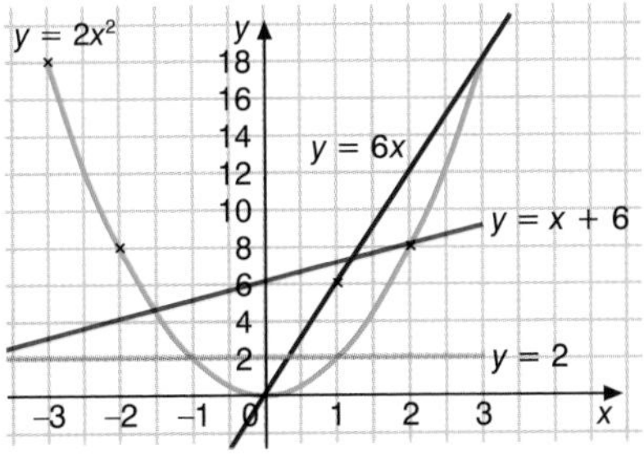

(a) $x = {}^{-}1$ and $x = 1$
(b) $x = 0$ and $x = 3$
(c) $x = {}^{-}1.5$ and $x = 2$

(a) Using the intersection of $y = 2x^2$ and $y = 2$.
(b) Using the intersection of $y = 2x^2$ and $y = 6x$.
(c) Using the intersection of $y = 2x^2$ and $y = x + 6$ so that $2x^2 = x + 6$ or $2x^2 - x - 6 = 0$.

4

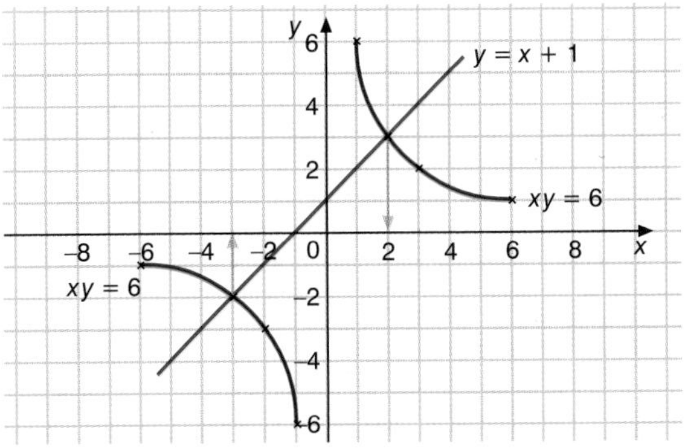

The solutions are $x = {}^{-}3$ and $x = 2$.

It is important to use the graph to find the solution of $x^2 + x - 6 = 0$.

Answers	**Examiner's comments**
5 $x > {}^{-}\frac{1}{2}$	$2(3 + 2x) > 4$ $6 + 4x > 4$ Expanding the brackets. $4x > {}^{-}2$ Subtracting 6 from both sides. $x > {}^{-}\frac{1}{2}$ Dividing both sides by 4.

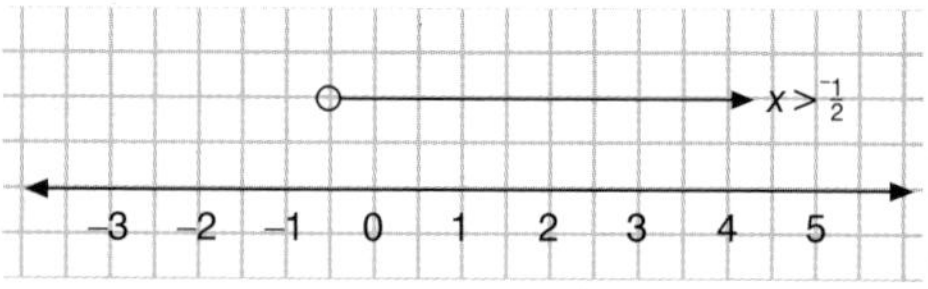

Answers	**Examiner's comments**
6 **Least value of x is ${}^{-}2$.**	$8 - 2x \leqslant 12$ ${}^{-}2x \leqslant 4$ Subtracting 8 from both sides. $x \geqslant {}^{-}2$ Dividing both sides by ${}^{-}2$ and changing the sign.
7 $x \leqslant 3$ $x + y \geqslant 4$ $y \leqslant 2x + 1$	The inequalities < and > would be acceptable in defining the region, if the lines were dotted.

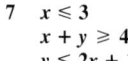

45 EXAMINATION QUESTIONS 9

(Page 129)

1 **$x = 4$ and $y = 3$**	The process of elimination is applied to the two equations which are added to produce: $7x = 28$ and $x = 4$ This value is then substituted into the first equation to give $y = 3$.
2 $y = 5x - 7$	Substituting the points (2, 3) and (1, ${}^{-}2$) in the equation $y = mx + c$: $3 = 2m + c$ ${}^{-}2 = 1m + c$ The process of elimination is applied to the two equations which are subtracted to produce: $5 = m$ This value is then substituted into the first equation to give $c = {}^{-}7$.
3 **7 and 5**	Let the two numbers be x and y (say) so that: $x + y = 12$ $x - y = 2$ The process of elimination is applied to the two equations which are added to produce the answers.
4 **(a) $s = 11$ or $s = {}^{-}5$** **(b) $n = 2$ or $n = 5$**	(a) $s^2 - 6s - 55 = 0$ can be factorised as $(s - 11)(s + 5) = 0$. (b) $n^2 + 10 = 7n$ is rewritten as $n^2 - 7n + 10 = 0$ which factorises as $(n - 2)(n - 5) = 0$.
5 $(y - 5)(y + 5)$ **$y = 5$ or $y = {}^{-}5$**	$y^2 - 25$ should be recognised as $y^2 - 5^2$ which is the difference of two squares. as $a^2 - b^2 = (a - b)(a + b)$ so $y^2 - 5^2 = (y - 5)(y + 5)$ $y^2 - 25 = 0$ factorises as $(y - 5)(y + 5) = 0$.
6 **3 or ${}^{-}4$**	$x + x^2 = 12$ is rewritten as $x^2 + x - 12 = 0$ which factorises as $(x - 3)(x + 4) = 0$.
7 **$x = 2.1$ (1 d.p.)**	Using trial and improvement to home in on the solution – See Chapter 44, Trial and improvement methods, for further help.
8 **(a) 69.153** **(b) $x = 3.49$ (2 d.p.)**	(a) $(3.7)^2 + 5(3.7) = 69.153$ (b) See Chapter 44, Trial and improvement methods, for further help.

51 EXAMINATION QUESTIONS 10

(Page 149)

1 $v = \dfrac{p}{t - 1}$

$t = \dfrac{v + p}{v}$

$tv = v + p$ Multiplying both sides by v.

$tv - v = p$ Subtracting v from both sides.

$v(t - 1) = p$ Factorising the left-hand side.

$v = \dfrac{p}{t - 1}$ Dividing both sides by $(t - 1)$.

2 **(a) $t = 10(v - u)$**
(b) $s = 5(v - u)(v + u)$
$s = 5(v^2 - u^2)$
$\frac{1}{5}s = v^2 - u^2$
$v^2 = u^2 + \frac{1}{5}s$
$v^2 = u^2 + 0.2s$

(a) $v = u + 0.1t$
$v - u = 0.1t$
$10(v - u) = t$
$t = 10(v - u)$

(b) $s = ut + 0.05t^2$
$s = u \times 10(v - u) + 0.05[10(v - u)]^2$ Substituting $t = 10(v - u)$.
$s = 10u(v - u) + 5(v - u)^2$ Expanding the brackets.
$s = 5(v - u)\{2u + (v - u)\}$ Removing $5(v - u)$ as a common factor.
$s = 5(v - u)(v + u)$ Simplifying.

Answers / Examiner's comments

3 **19**

Let the consecutive numbers be x and $(x + 1)$ so that the sum of the squares is $x^2 + (x + 1)^2$.

$x^2 + (x + 1)^2 = 685$

$x^2 + x^2 + 2x + 1 = 685$ — Expanding the brackets.

$2x^2 + 2x - 684 = 0$ — Writing in the form $ax^2 + bx + c = 0$.

$x^2 + x - 342 = 0$ — Simplifying the equation by dividing by 2.

This can be solved by the formula or else by factorisation to give $x = 18$ and $x = {}^{-}19$. The largest number is $(18 + 1)$ or 19. The other solution will not give any larger values than this.

4 **$x = 0.290$ or $^{-}0.690$ (3 s.f.)**

Using the formula with $a = 5$, $b = 2$ and $c = {}^{-}1$, leads to $x = 0.289\,897\,9$ and $x = {}^{-}0.689\,897\,9$.

5 $\mathbf{\dfrac{2x - 1}{(x + 3)(x - 4)}}$

A common denominator is $(x + 3) \times (x - 4)$ so writing each part as an equivalent fraction:

$$\frac{1}{x+3} + \frac{1}{x-4} = \frac{x-4}{(x+3)(x-4)} + \frac{x+3}{(x+3)(x-4)}$$

$$= \frac{x - 4 + x + 3}{(x+3)(x-4)} \quad \text{where } \frac{1}{x+3} = \frac{x-4}{(x+3)(x-4)} \text{ and } \frac{1}{x-4} = \frac{x+3}{(x+3)(x-4)}$$

$$= \frac{2x - 1}{(x+3)(x-4)}$$

6 **(a) Time = $\dfrac{15}{x} + \dfrac{11}{x - 2}$**

(b) $\dfrac{15}{x} + \dfrac{11}{x - 2} = 4$

(c) $2x^2 - 17x + 15 = 0$

(d) Average speed = 7.5 mph

(a) Using time $= \dfrac{\text{distance}}{\text{speed}}$.

(b) $\dfrac{15}{x} + \dfrac{11}{x - 2} = 4$

(c) $\dfrac{15(x - 2) + 11x}{x(x - 2)} = 4$ — Using a common denominator.

$15(x^2 - 30) + 11x = 4x(x - 2)$ — Multiplying both sides by $x(x - 2)$.

$15x - 30 + 11x = 4x^2 - 8x$ — Expanding the brackets on both sides.

$4x^2 - 34x + 30 = 0$ — Collecting like terms and rewriting with the expression on the LHS.

$2x^2 - 17x + 15 = 0$ — Dividing both sides of the equation by 2.

(d) Solving the equation by factorising:

$2x^2 - 17x + 15 = 0$

$(2x - 15)(x - 1) = 0$

$2x - 15 = 0$ gives $x = 7.5$.

$x - 1 = 0$ gives $x = 1$.

The value $x = 1$ is rejected as this would give a negative speed for the last 11 miles.

7

(a)

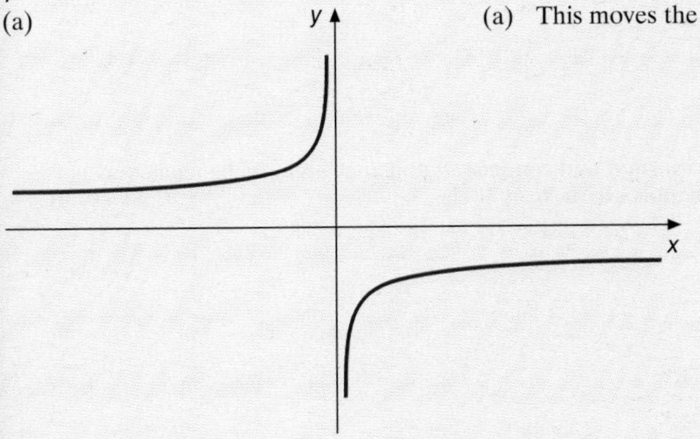

(a) This moves the graph along the x-axis, one unit to the right.

(b)

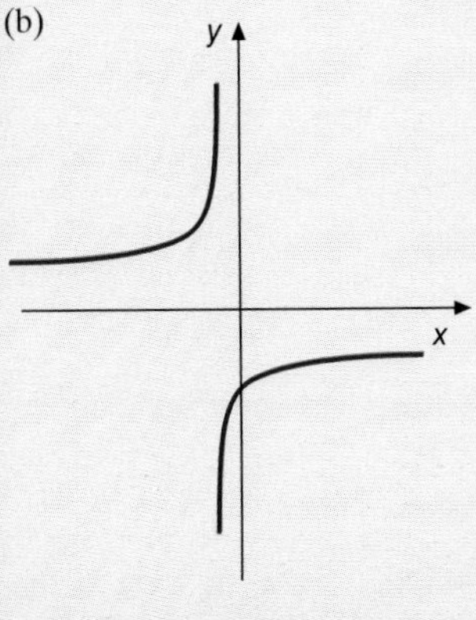

(b) This shrinks the graph along the x-axis.

Answers

Examiner's comments

57 EXAMINATION QUESTIONS 11

(Page 170)

1 **The perimeter is 1880 m.**

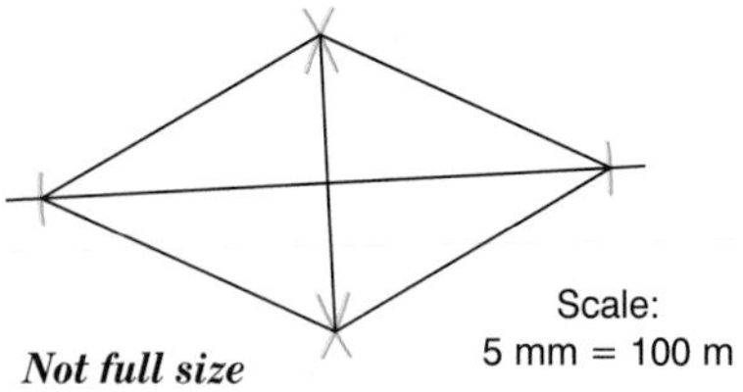

Not full size

It is important to choose a suitable scale, such as 1 cm to 100 m.
The fact that the diagonals of the rhombus bisect each other, at right angles, makes the construction easier.
From your scale drawing the length of each side should be 4.7 cm so the perimeter is $4 \times 4.7 = 18.8$ cm which corresponds to 1880 m on the actual playground.

2

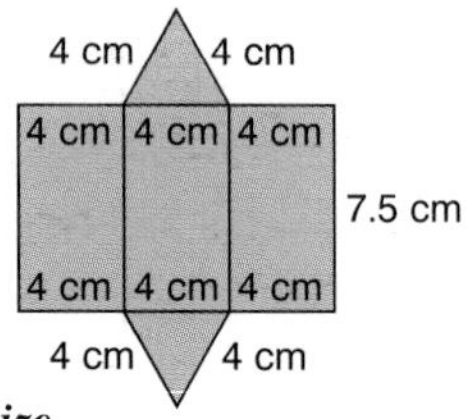

Not full size

The net should be drawn, using the fact that the cross-section is an equilateral triangle. The lengths should be drawn to the required accuracy of ±1 mm and the angles are drawn to the required accuracy of ±1°.

3

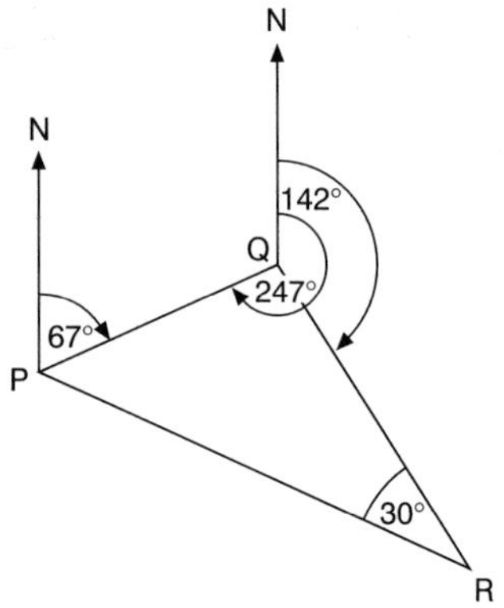

(a) The bearing of Q from P = 067°
(b) The bearing of P from Q = 247°
(c) The bearing of R from P = 112°

(a) Bearings are always described in three figures.
(b) This is found from the diagram, $67° + 180° = 247°$.
(c) $\angle PQR = 247° - 142° = 105°$
$\angle QPR + 105° + 30° = 180°$ Sum of angles of a triangle = 180°.
$\angle QPR = 45°$
The bearing of R from P = $67° + 45° = 112°$.
It is important to explain your working and explain how you arrive at the answer.

4

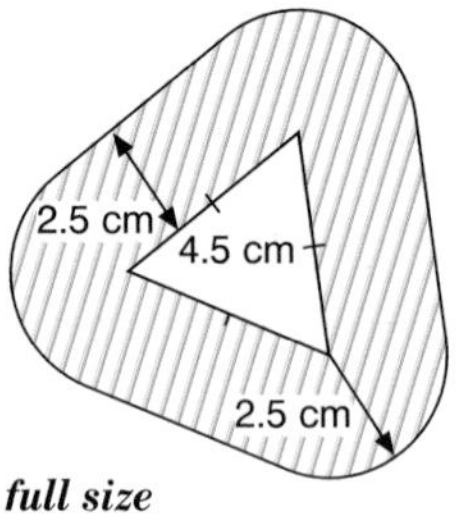

Not full size

The locus should be constructed with ruler and compasses, and angles should be measured accurately. The curved areas at the ends are parts of circles.

5

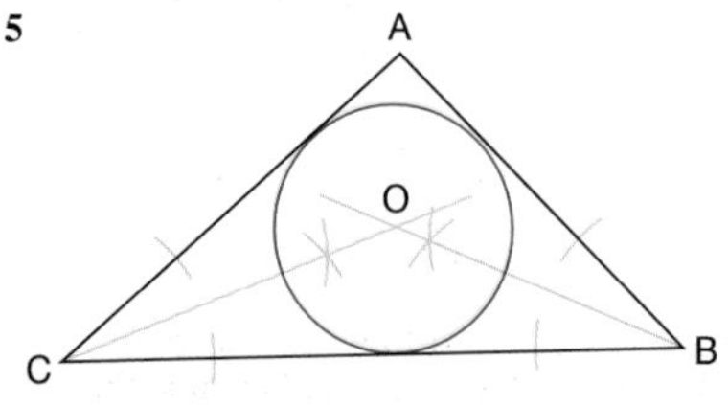

Accurate construction results in the circle inscribed in the triangle ABC.

Answers

Examiner's comments

64 EXAMINATION QUESTIONS 12

(Page 191)

1 (a) B, C, D, E, H, I, each have a horizontal line of symmetry.
(b) A, H, I each have a vertical line of symmetry.
(c) H, I have both horizontal and vertical lines of symmetry.
(d) F, G, J, K have no line symmetry.
(e) H, I have rotational symmetry of order 2.

It is always important to read the question carefully. The correct types of symmetry must be identified, for maximum marks.

2 (a) ΔAGF onto ΔBFE is a translation of 3 cm to the right.
(b) ΔBCE onto ΔDCE is a reflection about the line CE.
(c) ΔAGF onto ΔFBA is a rotation through 180° about the midpoint of AF.
(d) ΔAFB onto ΔEBF is a rotation through 180° about the midpoint of BF.

It would be easy to lose marks on this question if the transformations are not described fully. Remember that:

- a reflection is defined by giving the position of the line of symmetry,
- a rotation is defined by giving the position of the centre of rotation along with the angle and direction of the rotation,
- an enlargement is defined by giving the position of the centre of enlargement and the factor, a translation is defined by giving the distance and direction of the translation.

3

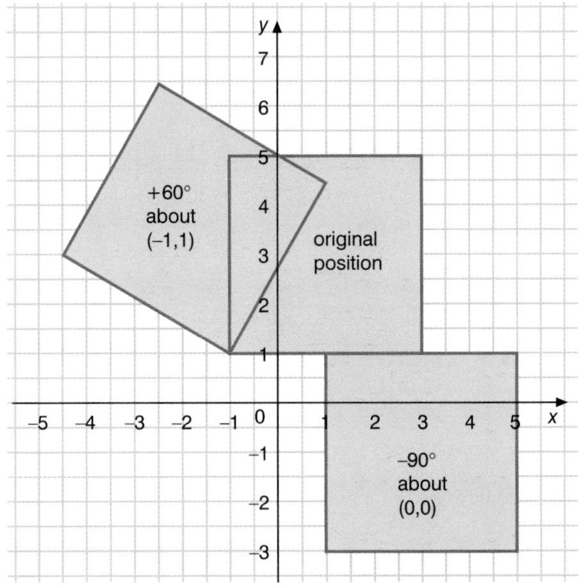

Remember that a negative rotation is clockwise and a positive rotation is anticlockwise. A typical error here would be to use the same centre of rotation for both parts of the question.

4 x = 7 cm or 2.29 cm (to 2 d.p.)

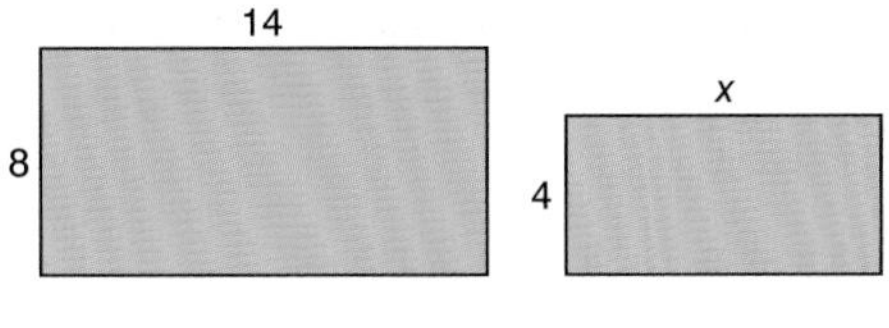

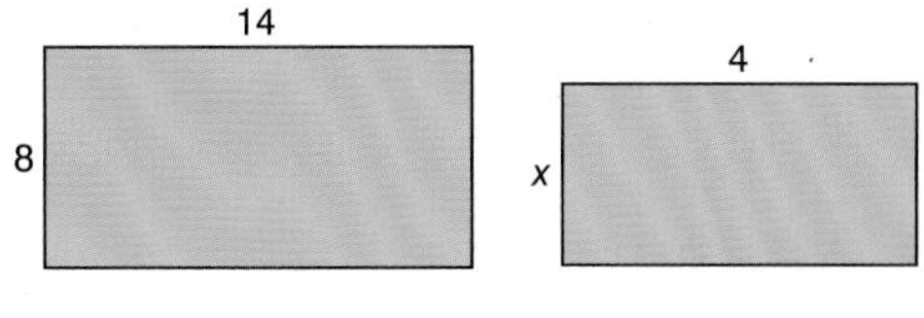

$$\frac{14}{x} = \frac{8}{4}$$

$$\frac{x}{14} = \frac{4}{8}$$ Turning the expressions upside-down.

$$x = \frac{4}{8} \times 14$$

$$x = 7$$

$$\frac{14}{4} = \frac{8}{x}$$

$$\frac{4}{14} = \frac{x}{8}$$ Turning the expressions upside-down.

$$x = \frac{4}{14} \times 8$$

$$x = 2.285\,714\,3$$

It is important to round off the answer to a reasonable degree of accuracy and include the units so that x = 2.29 cm (3 s.f.).

The way that the question is asked does suggest the possibility of more than one solution as it says 'possible values'. Always take care to check whether a second solution exists.

Answers / **Examiner's comments**

5 **∠PBC = 72°** **(exterior angle of regular pentagon)**
∠PCB = 72° **(exterior angle of regular pentagon)**
∠BPC = 36° **(angles of a triangle add up to 180°)**

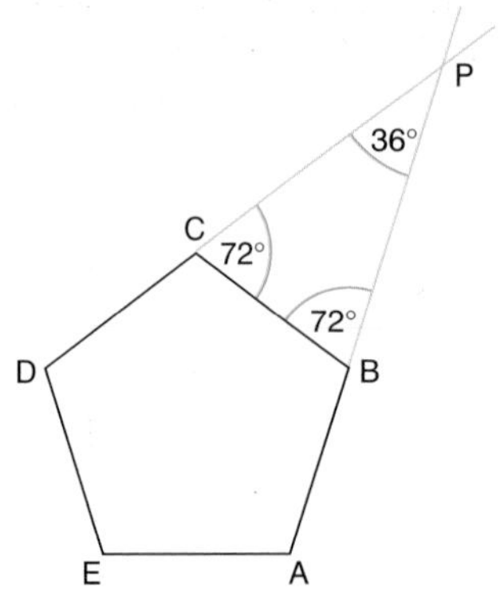

It is always a good idea to give reasons, so that the examiner can give credit for the methods used, even if the answers are wrong.

External angle of regular pentagon $= \frac{360°}{5} = 72°$.

∠PBC = ∠PCB = 72° as these are exterior angles.
∠BPC = 36° as the angles of triangle BPC add up to 180°.

6 **100° and 150°**

Angle sum of a hexagon = (6 – 2) × 180° = 720°
Sum of angles = 470°
Total of remaining angles = 250°
In the ratio 2 : 3 there is a total of 5 parts.
Each part is $\frac{250°}{5} = 50°$
2 parts are 2 × 50° = 100° and 3 parts are 3 × 50° = 150°.
Angles are 100° and 150°.

7 **5**

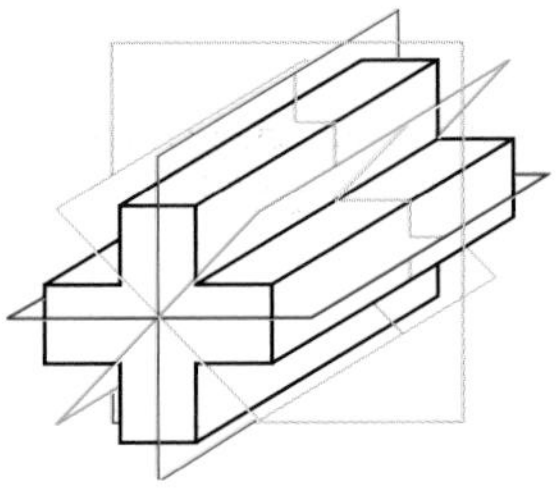

A rough sketch of the different planes helps to check the answer.

68 EXAMINATION QUESTIONS 13

(Page 207)

1 **PT = 16.4 cm (3 s.f.)**

Using Pythagoras' theorem on the triangle PQR:
$PR^2 = PQ^2 + QR^2$
$QR^2 = PR^2 - PQ^2 = 400 - 225 = 175$
QR = 13.228 757 cm
Then in triangle PQT:
$PT^2 = PQ^2 + QT^2$ As $QT = \frac{1}{2}QR$.
$= 225 + 43.75 = 268.75$
PT = 16.393 596 cm

2 **Area of the kite = 270 cm²**
Angle BAD = 73.7° (3 s.f.)

It is a good idea to use or draw a diagram, adding to it all the information that is either given or calculated.
By Pythagoras' theorem:
$OB^2 = AB^2 - AO^2 = 225 - 144 = 81$
OB = 9 cm
BD = 2 × OB = 18 cm
Area of kite
$= \frac{1}{2}AC \times BD = \frac{1}{2} \times 18 \times 30 = 270\text{ cm}^2$

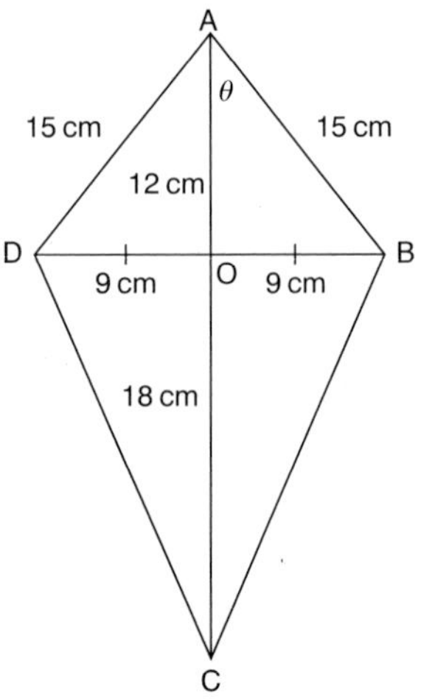

By the symmetry of the kite: BAO is marked θ in the diagram.
∠BAD = 2 × ∠BAO
$\cos \angle BAO = \frac{12}{15} = 0.8$
∠BAO = 36.869 898°
∠BAD = 73.739 796°

Answers

Examiner's comments

3 Distance = 1330 m (3 s.f.)

Start by drawing a sketch of the situation and completing the given details on the diagram.
Since the angle of depression is 20° the top angle in the triangle is 70° (as angles forming a right angle add up to 90°).

From the triangle:

$$\tan 70° = \frac{x}{485}$$

$$x = 485 \times \tan 70°$$
$$= 1332.5265\text{ m}$$

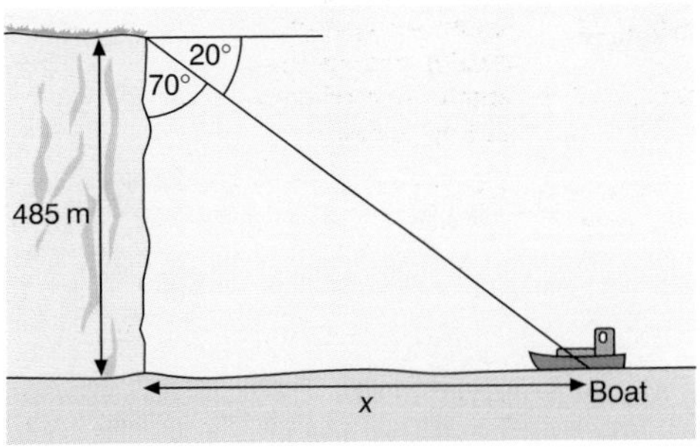

4 Height = 2.8 m (2 s.f.)

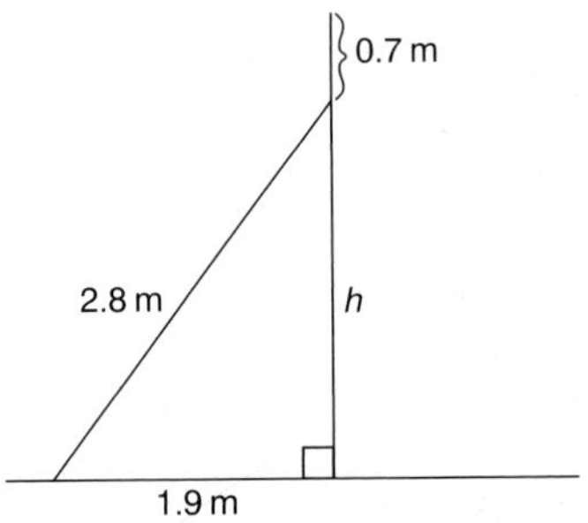

Use Pythagoras' theorem to find $h = 2.056\,696\,4$ m and then add 0.7 m to the answer to allow for the fact that the ropes are secured 0.7 metres from the top of the pole. The final answer is given to 2 s.f. to reflect the accuracy of the information given in the question.

5 Area of the washer = 84.8 mm^2 (3 s.f.)

The cross-sectional area is calculated by subtracting the area of the smaller circle from the area of the larger circle, remembering to halve the diameter to give the radius each time.
Area $= \pi \times 6^2 - \pi \times 3^2 = 84.823\,002\text{ mm}^2$

6 Volume of packet = 384 cm^3
Volume of smaller packet = 307 cm^3 (3 s.f.)

Volume $= 8 \times 4 \times 12 = 384\text{ cm}^3$
Using the fact that 125 grams gives a volume of 384 cm^3:
1 gram gives a volume of $\frac{384}{125}$ cm^3.

100 grams give a volume of $100 \times \frac{384}{125}\text{ cm}^3 = 307.2\text{ cm}^3$.

7 Volume = 1140 mm^3 (3 s.f.)

Consider the coin to be a cylinder so that:
Volume $= \pi r^2 h$
$= \pi \times 11^2 \times 3$ Where $r = \frac{1}{2} \times$ diameter
$= 1140.3981\text{ mm}^3$

8 Volume = 3120 cm^3 (3 s.f.)

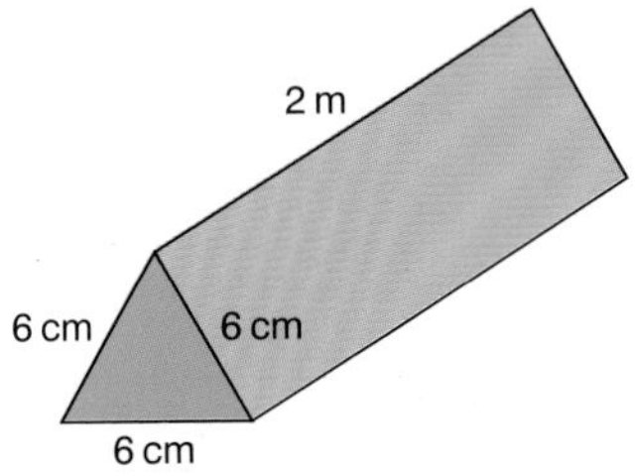

Use Pythagoras' theorem on the given triangle to find the height and thus the area of the equilateral triangle.
Height = 5.196 152 4 cm
Area = 15.588 457 cm^2
Volume of the prism = area of cross-section × length
= 15.588 457 × 200 Converting 2 m to 200 cm.
= 3117.6915 cm^3

9 The expression $\frac{4}{3}\pi$ is a constant and r^2 gives units of length × length = area so the expression cannot represent the volume of a sphere.

Answers | Examiner's comments

73 EXAMINATION QUESTIONS 14 (Page 227)

1 **464 spheres**

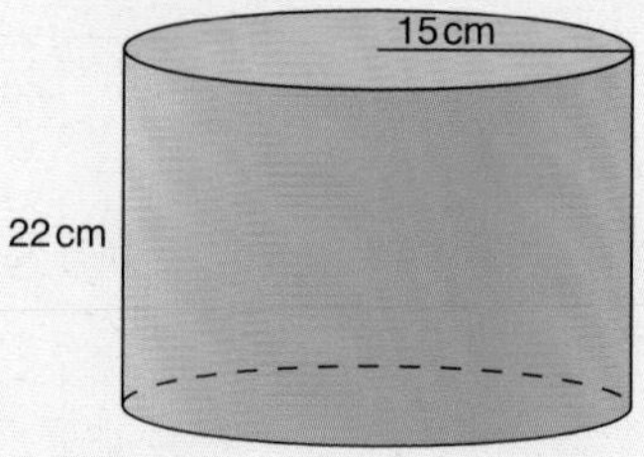

Volume of cylinder $= \pi r^2 h = \pi \times 15^2 \times 22$
$= 15\,550.884\,\text{cm}^3$

Volume of sphere $= \frac{4}{3}\pi r^3 = \frac{4}{3}\pi \times 2^3 = 33.510\,322\,\text{cm}^3$

Number of spheres $= 15\,550.884 \div 33.510\,322 = 464.062\,51$

The answer has to be a whole number, rounded down.

2 **(a) Volume = 497 cm³ (3 s.f.)**
(b) Surface area = 319 cm² (3 s.f.)

(a) Volume of top = volume of cone + volume of hemisphere

$= \frac{1}{3}\pi r^2 h + \frac{1}{2} \times \frac{4}{3}\pi r^3 = 235.619\,45 + 261.799\,39$

$= 497.418\,84\,\text{cm}^3$

(b) Surface area of top = surface area of cone + surface area of hemisphere

$= \pi r l + \frac{1}{2} \times 4\pi r^2$

Where l is the slant height of the cone.

$l^2 = 9^2 + 5^2$ By Pythagoras' theorem.

$l^2 = 106$

$l = 10.295\,63$

Surface area $= 161.723\,38 + 157.079\,63$
$= 318.803\,01\,\text{cm}^2$

3 **Volume of sand = 11.9 cm³**
h = 1 cm

Volume of sand = volume of cone + volume of cylinder

$= \frac{1}{3}\pi r^2 h_{\text{cone}} + \pi r^2 h_{\text{cylinder}} = \frac{1}{3}\pi \times 1.4^2 \times 1.3 + \pi \times 1.4^2 \times 1.5$

$= 2.668\,259\,4 + 9.236\,282\,4$

$= 11.904\,542\,\text{cm}^3$

Volume of hemisphere $= \frac{1}{2}(\frac{4}{3}\pi r^3) = 5.747\,020\,2\,\text{cm}^3$

Remaining sand $= 11.904\,542 - 5.747\,020\,2 = 6.157\,521\,8\,\text{cm}^3$

Volume of cylinder $= \pi r^2 h = 6.157\,521\,8\,\text{cm}^3$

$h = \frac{6.157\,521\,8}{\pi r^2} = \frac{6.157\,521\,8}{\pi \times 1.4^2} = 1\,\text{cm}$

4 **(a) The angle edge CP makes with the base = 73.1° (3 s.f.)**
(b) The angle face APD makes with the base = 77.9° (3 s.f.)
(c) The area of one of the triangular faces = 10.7 cm³ (3 s.f.)

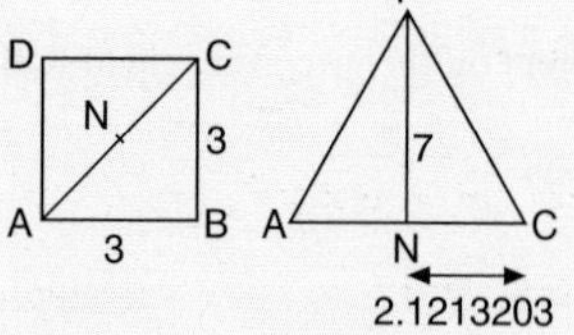

(a) By Pythagoras' theorem:

$AC = \sqrt{3^2 + 3^2} = 4.242\,640\,7$

$NC = 2.121\,320\,3$ Where N is the centre of the base.

$\tan NCP = \frac{7}{2.121\,320}$

$\angle NCP = 73.140\,79°$

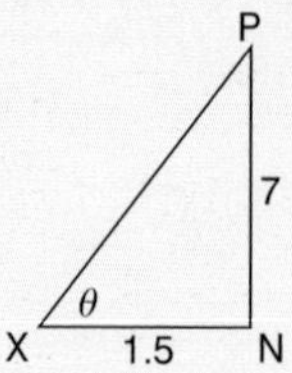

(b) $\tan\theta = \frac{PN}{NX} = \frac{7}{1.5}$ Where X is the midpoint of AD.

$\theta = 77.905\,243°$

(c) Using Pythagoras' theorem in triangle PXN:

$PX^2 = PN^2 + XN^2$

$PX = \sqrt{7^2 + 1.5^2} = 7.158\,910\,5$

Area of triangle $= \frac{1}{2} \times$ base $\times$ perpendicular height

$= \frac{1}{2} \times 3 \times 7.158\,910\,5$

$= 10.738\,366\,\text{cm}^3$

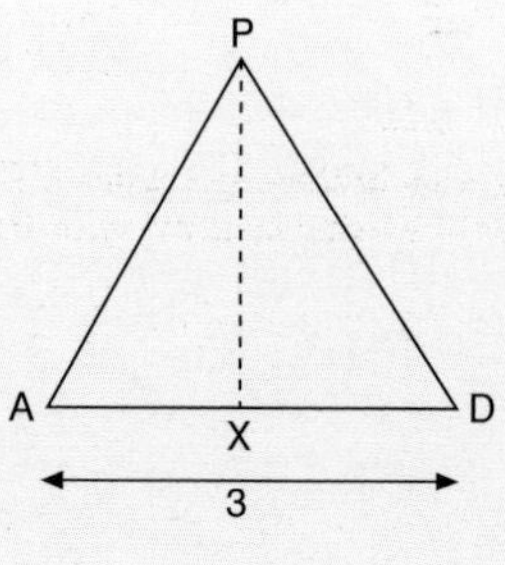

Answers | **Examiner's comments**

5 θ **= 0°, 45°, 135° and 180°**

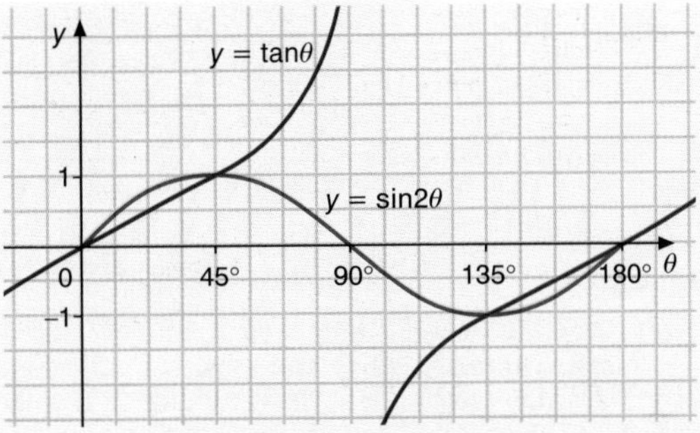

From the sketch, the values of θ which are common to both graphs can clearly be seen.

6

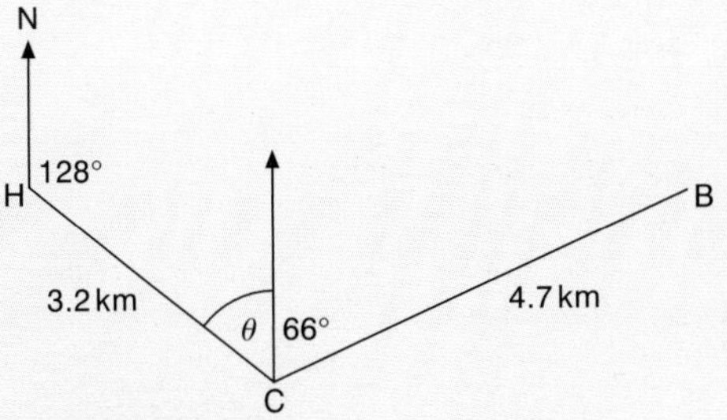

(a) Angle HCB = 118°
(b) Distance HB = 6.8 km (2 s.f.) or 6.82 km (3 s.f.)

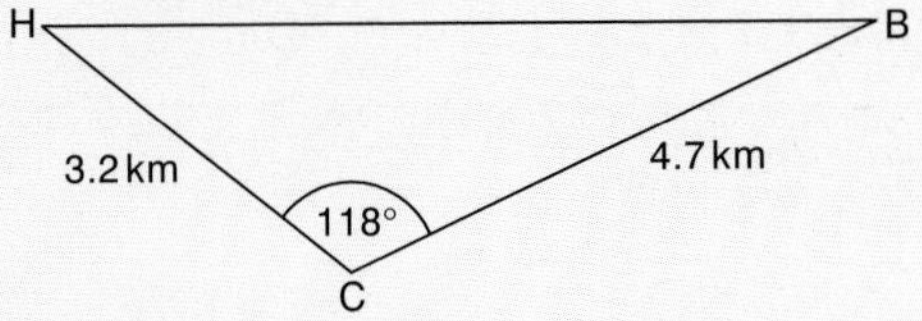

(a) $\theta = 52°$ The interior angles between parallel lines add up to 180°.
$\angle HCB = 52° + 66° = 118°$ (as required)

(b) The question is set so that the answer in part (a) should be used in part (b) even if you are unable to prove that angle HCB = 118°.
Using the cosine rule on the triangle HCB:
$a^2 = b^2 + c^2 - 2bc \cos A$
$HB^2 = 3.2^2 + 4.7^2 - (2 \times 3.2 \times 4.7 \times \cos 118°)$
$HB^2 = 10.24 + 22.09 - (^{-}14.121\,705)$
$HB^2 = 46.451\,705$
$HB = 6.815\,548\,7$
HB = 6.8 km (2 s.f.) or 6.82 km (3 s.f.) Reflecting the accuracy of the original question.

78 EXAMINATION QUESTIONS 15

(Page 247)

1 **Area = 99.5 cm² (3 s.f.)**
Perimeter = 43.6 cm (3 s.f.)

Area $= \frac{270}{360} \times \pi r^2 = 99.549\,217\,\text{cm}^2$

Perimeter = arc + radius + radius As the straight lines are also part of the perimeter.

$= \frac{270}{360} \times 2\pi r + r + r$

$= 43.630\,528$ cm

2 **Area = 1080 cm² (3 s.f.)**

Area $= \frac{110}{360} \times \pi \times 35^2 - \frac{110}{360} \times \pi \times 10^2$

$= \frac{110}{360} \times \pi \times (35^2 - 10^2)$

$= 1079.9225$

3 **(a) ∠ROP = 138°**
(b) ∠RSP = 69°
(c) ∠RQP = 111°

(a) $\angle OPT = \angle ORT = 90°$ As the tangents to a circle are perpendicular to the radius at the point of contact.

$\angle ROP = 360° - (42° + 90° + 90°)$ As the angles of quadrilateral RTPO add up to 360°.
$\angle ROP = 138°$

(b) $\angle RSP = \frac{1}{2} \times \angle ROP$ The angle subtended at the circumference of a circle equals half of the angle subtended at the centre.

(c) $\angle RQP = 180° - \angle RSP$ Opposite angles of cyclic quadrilateral RQPS add up to 180°.

$\angle RQP = 180° - 69° = 111°$

Remember that the diagrams in these questions will not be drawn accurately so you should not attempt to reach solutions by using measuring instruments.

Answers

Examiner's comments

4 **(a)** $\angle ABD = 90° - x$
(b) $\angle DBE = x$
(c) $\angle BAD = 90° - 2x$

(a) $\angle CDE = 90°$ — CE is a diameter and the angle in a semicircle = 90°.
$\angle DEC = 90° - x$ — As the angles of the triangle DEC add up to 180°.
$\angle DBC = 90° + x$ — As opposite angles of a cyclic quadrilateral add up to 180°. $180° - (90° - x) = 90° + x$
$\angle ABD = 90° - x$ — As the angles on a straight line ABC add up to 180°. $180° - (90° + x) = 90° - x$

(b) $\angle DBE = DCE$ — As the angles subtended by the same chord DE at the circumference are equal (it would be helpful to draw in the line BE on the diagram).

(c) $\angle ACE = 3x$ and $\angle AEC = 90° - x$
$\angle BAD = 180° - \{3x + (90° - x)\}$ — As the angles of Δ ACE add up to 180°.
$\angle BAD = 180° - (3x + 90° - x) = 180° - (2x + 90°) = 180° - 2x - 90° = 90° - 2x$

5 **(a)** $\overrightarrow{PT} = \mathbf{b} + \mathbf{c}$
(b) $\overrightarrow{US} = \mathbf{b} - \mathbf{c}$
(c) $\overrightarrow{PX} = \frac{1}{2}(\mathbf{a} + \mathbf{b} + \mathbf{c})$

(a) $\overrightarrow{PT} = \overrightarrow{PS} + \overrightarrow{ST} = \mathbf{b} + \mathbf{c}$

(b) $\overrightarrow{US} = \overrightarrow{UT} + \overrightarrow{TS} = \mathbf{b} + {}^-\mathbf{c} = \mathbf{b} - \mathbf{c}$

(c) $\overrightarrow{PX} = \overrightarrow{PQ} + \overrightarrow{QX}$

$\overrightarrow{PX} = \overrightarrow{PQ} + \frac{1}{2}\overrightarrow{QT}$ $\quad\overrightarrow{QT} = \overrightarrow{QR} + \overrightarrow{RS} + \overrightarrow{ST}$

$\overrightarrow{QT} = \mathbf{b} + {}^-\mathbf{a} + \mathbf{c}$

$\overrightarrow{QT} = {}^-\mathbf{a} + \mathbf{b} + \mathbf{c}$

$\overrightarrow{PX} = \mathbf{a} + \frac{1}{2}({}^-\mathbf{a} + \mathbf{b} + \mathbf{c})$

$\overrightarrow{PX} = \frac{1}{2}(\mathbf{a} + \mathbf{b} + \mathbf{c})$

6 **(a)** $\overrightarrow{AB} = {}^-8\mathbf{a} + 4\mathbf{b}$ or $4\mathbf{b} - 8\mathbf{a}$
(b) $\overrightarrow{AX} = {}^-6\mathbf{a} + 3\mathbf{b}$ or $3\mathbf{b} - 6\mathbf{a}$
(c) $\overrightarrow{OX} = 2\mathbf{a} + 3\mathbf{b}$

(a) $\overrightarrow{AB} = \overrightarrow{AO} + \overrightarrow{OB} = {}^-8\mathbf{a} + 4\mathbf{b}$

(b) $\overrightarrow{AX} = \frac{3}{4}\overrightarrow{AB} = \frac{3}{4}({}^-8\mathbf{a} + 4\mathbf{b}) = {}^-6\mathbf{a} + 3\mathbf{b}$

(c) $\overrightarrow{OX} = \overrightarrow{OA} + \overrightarrow{AX} = 8\mathbf{a} + ({}^-6\mathbf{a} + 3\mathbf{b}) = 2\mathbf{a} + 3\mathbf{b}$

82 EXAMINATION QUESTIONS 16

(Page 271)

1 **Not all amounts of money are covered.**
No time period is specified for the answer.

Full answers are essential for maximum marks on this type of question. Similar answers also gain full credit but it is important to be clear exactly what is meant. Other acceptable answers include:
'The classes are too wide for further analysis.'
'At least four classes should be provided in the questionnaire.'
'The respondents may not get the same amount each week.'

2 (a)

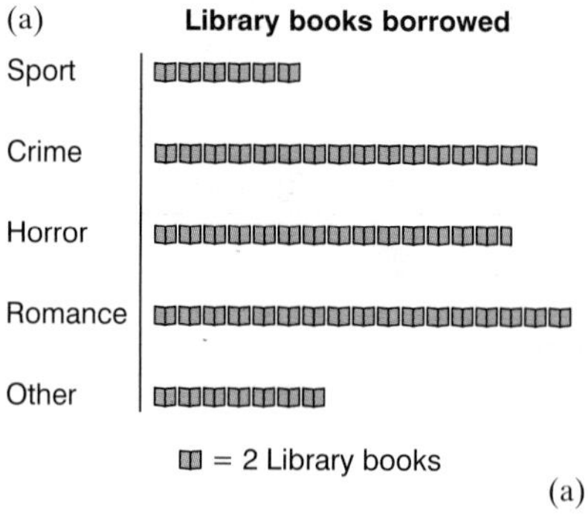

(b)

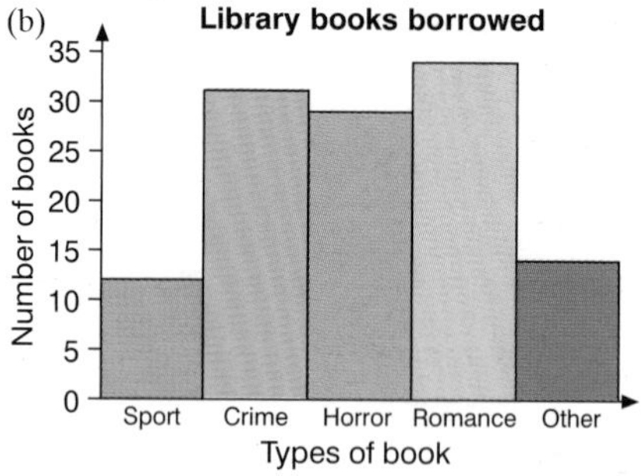

(c)

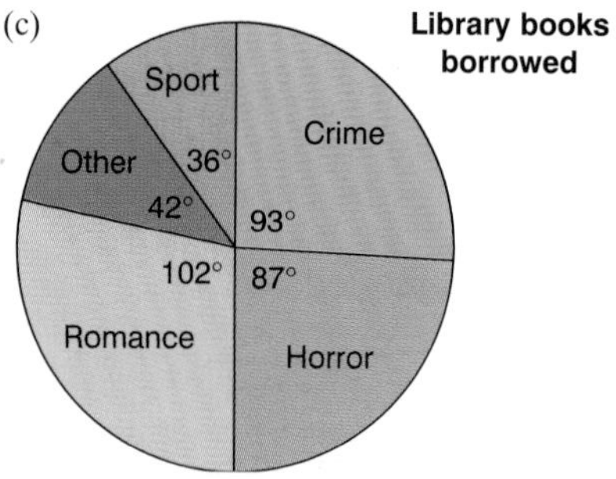

(a) Notice the inclusion of a key to support the pictures used in the pictogram.

(b) Always remember to label the axes appropriately and give the bar chart a title to explain its purpose.

(c) Always label each sector and include the angles at the centre for further information.
Sum of frequencies = 120 so each book gets 360° ÷ 120 = 3°.
Angles of pie chart:

Type of book	Frequency	Angle
Sport	12	12 × 3° = 36°
Crime	31	31 × 3° = 93°
Horror	29	29 × 3° = 87°
Romance	34	34 × 3° = 102°
Other	14	14 × 3° = 42°
		Total 360°

It is always a good idea to check that the angles add up to 360°.

Answers

3 (a) (i) Mean = 16.9
(ii) Median = 16.5
(iii) Mode = 16
(b) (i) The mean mark will decrease to 16.8.
(ii) The median mark stays the same.
(iii) The modal mark stays the same.

Examiner's comments

(a) (i) The mean = $\frac{\text{total of all the values}}{\text{number of values}}$

$$= \frac{16+18+17+15+20+16+15+17+16+19}{10} = \frac{169}{10}$$

(ii) Rearrange the numbers in order to find the median.
15, 15, 16, 16, 16, 17, 17, 18, 19, 20
For 10 values the median is the $5\frac{1}{2}$ th value i.e. $\frac{1}{2}(16 + 17)$.

(iii) Once the numbers are in order, it is easy to see which one occurs the most.

(b) (i) The question only asks, 'What effect will this have,' so it is not necessary to work out the new value of the mean.
(ii) The order is not affected by decreasing the 15 to 14.
(iii) The modal mark is not changed.

4 Mean = 17.8 (3 s.f.)

Drawing up a table to find Σf and Σfx is often helpful.

Mark x	Frequency f	Mark × frequency fx
14	2	28
15	1	15
16	6	96
17	12	204
18	11	198
19	10	190
20	8	160
	$\Sigma f = 50$	$\Sigma fx = 891$

Mean mark $\frac{\Sigma fx}{\Sigma f} = \frac{891}{50} = 17.82 = 17.8$ (3 s.f.)

The answer is rounded off to an appropriate degree of accuracy.

5 Mean = 1.60 minutes (3 s.f.)

Drawing up the table to find Σf and Σfx is helpful. Mid-interval values are used to represent the grouped data.

Length of time (minutes)	Frequency f	Mid-interval value x	Frequency × mid-interval value fx
0 and less than 1	107	0.5	53.5
1 and less than 2	89	1.5	133.5
2 and less than 3	36	2.5	90
3 and less than 4	21	3.5	73.5
4 and less than 5	11	4.5	49.5
5 and less than 6	6	5.5	33
	$\Sigma f = 270$		$\Sigma fx = 433$

Mean time = $\frac{\Sigma fx}{\Sigma f} = \frac{433}{270} = 1.603\,703\,7 = 1.60$ minutes (3 s.f.)

The answer is rounded off to an appropriate degree of accuracy.

85 EXAMINATION QUESTIONS 17

(Page 285)

1 Range = 23
Interquartile range = 10

Start the question by arranging the data in order.

27 28 29 32 35 38 38 38 39 41 50

Remember that the range is the difference between the highest and lowest values, giving a single number.

Range = 50 – 27 = 23

The lower quartile is one-quarter of the way from the bottom value, which makes it the 3rd value.

The upper quartile is three-quarters of the way up, which makes it the 9th value.

Interquartile range = UQ – LQ = 39 – 29 = 10

Answers

2 (a) **78 batteries lasted less than 250 hours**
(b) **36 batteries lasted more than 350 hours**
(c) **Median = 270 and interquartile range = 100**

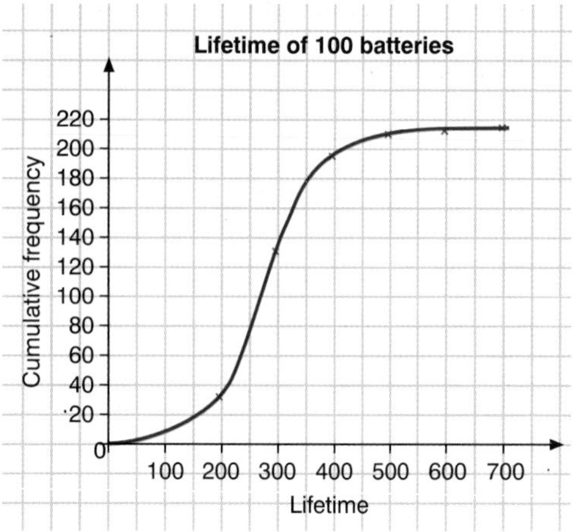

Examiner's comments

First, you need to produce the cumulative frequency chart. Then you can draw a cumulative frequency curve to illustrate this information. The cumulative frequencies should be plotted at the upper boundaries of 199.5, 299.5, 399.5, etc.

Lifetime (hours)	Frequency	Cumulative frequency
100–199	32	32
200–299	98	130
300–399	65	195
400–499	14	209
500–599	3	212
600–699	2	214

The values should be read from the graph.

(a) 78 batteries lasted less than 250 hours

(b) 178 batteries lasted less than 350 hours
36 batteries lasted more than 350 hours

(c) Median = 270
Upper quartile = 325
Lower quartile = 225
Interquartile range = upper quartile – lower quartile
= 325 – 225
= 100

3 (a) **18** (b) **18**

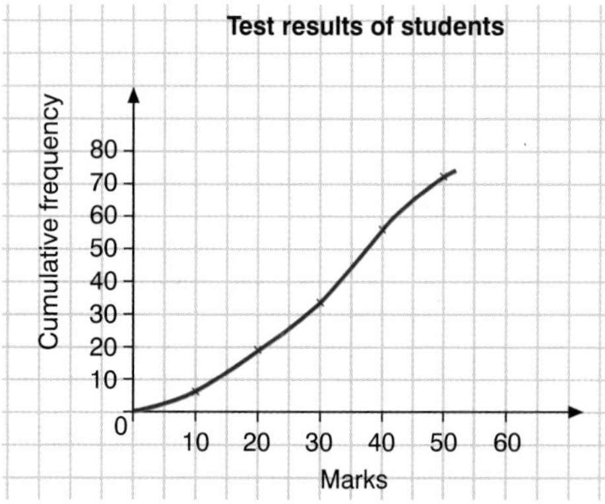

(c) **Median = 33 and interquartile range = 21**
(d) **The interquartile range is smaller so the results are less spread out in the second test.**

(a) 18 (from table)

(b) 18 (72 – 54)

When plotting the cumulative frequency diagram, remember to plot the points at the upper class boundaries (i.e. 10, 20, 30, 40 and 50).

Marks	Cumulative frequency	Frequency
⩽ 10	6	6
⩽ 20	19	13
⩽ 30	33	14
⩽ 40	56	23
⩽ 50	72	16

(c) Median = 33
Lower quartile = 19
Upper quartile = 40
Interquartile range = 40 – 19 = 21

It is a good idea to show your working, so that the examiner can see where your answers come from.

(d) The conclusions should be stated clearly.

4 (a)

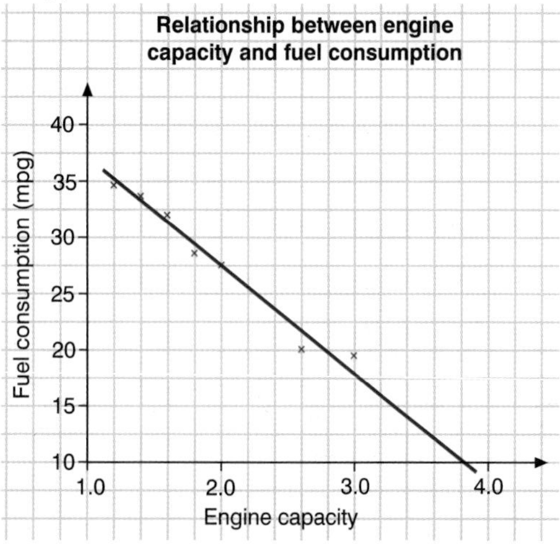

(b) **(i) 24 mpg (ii) 13mpg**
(c) **The first value is better as it lies within the plotted points whereas the second value lies at the extremity of the plotted points and is therefore more prone to error.**

You should use a line of best fit to find estimates for the given questions. More marks are awarded if you provide a thorough explanation for the choice. This should always be borne in mind when undertaking work of this nature.

Answers

Examiner's comments

91 EXAMINATION QUESTIONS 18

(Page 299)

1 (a) 0.55
(b) 0.45
(c) 0.75

Converting to decimals or percentages helps with calculations and comparisons.
You should use the fact that the events Q, R and S are mutually exclusive to solve the question.
Answers in fractions, decimals or percentages are acceptable.

(a) $p(Q \text{ or } R) = p(Q) + p(R) = 30\% + \frac{1}{4} = 0.30 + 0.25 = 0.55$

(b) $p(R \text{ or } S) = p(R) + p(S) = \frac{1}{4} + 0.2 = 0.25 + 0.2 = 0.45$

(c) $p(Q \text{ or } R \text{ or } S) = p(Q) + p(R) + p(S) = 30\% + \frac{1}{4} + 0.2 = 0.30 + 0.25 + 0.2 = 0.75$

2 (a)

		red				
		1	2	3	4	5
blue	5	5	10	15	20	25
	6	6	12	18	24	30
	7	7	14	21	28	35
	8	8	16	24	32	40
	9	9	18	27	36	45

(b) p(square number) $= \frac{4}{25}$

(c) p(final score is less than 30) $= \frac{19}{25}$

(a) Using an appropriate grid helps to show the final scores.
(b) Results can be read from the table.
(c) Results can be read from the table.

3 (a) (i) 0.2 (ii) 0.5 (iii) 0.7
(b) 20

(a) (iii) Remember to add red **and** 2 only once, not twice.
(b) The probability of a green counter = 0.1, so $\frac{2}{\text{total}} = 0.1$ and total = 20.

4 (a) 0.08
(b) 0.23

(a) The information can be read from the two tables. The probabilities are multiplied, giving probability $0.2 \times 0.4 = 0.08$.
(b) The probability that the numbers of births and deaths are the same is equivalent to the probability of
exactly one birth and one death
or exactly two births and two deaths
or exactly three births and three deaths
or exactly four births and four deaths.
Probability $= 0.4 \times 0.1 + 0.3 \times 0.3 + 0.2 \times 0.4 + 0.1 \times 0.2 = 0.23$

96 EXAMINATION QUESTIONS 19

(Page 318)

1

Age group (years)	10–	20–	30–	45–	50–	70–100
Frequency	2	0	6	4	22	12
Class width	10	10	15	5	20	30
Frequency density	0.2	0	0.4	0.8	1.1	0.4

You need to work out the class width and the frequency density before drawing the histogram.

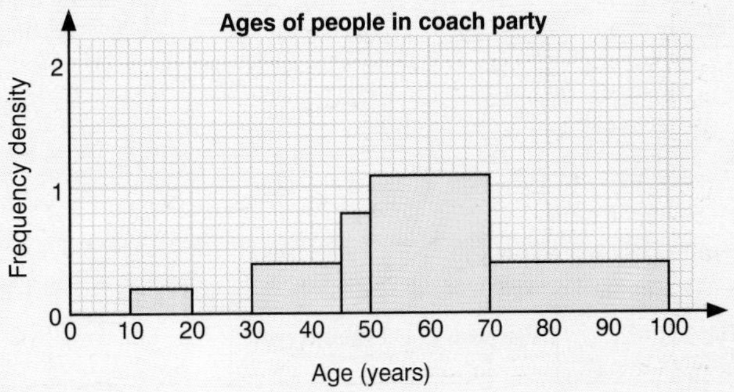

Answers

Examiner's comments

2 (a) £60.01
(b) The mean increased by £24.50 to £261.50.
The standard deviation was unchanged.
(c) The mean increased by 10% to £260.70.
The standard deviation increased by 10% to £66.01.
(d) The manager's scheme is better for the lowest-paid workers.

(a) Standard deviation = $\sqrt{\frac{\Sigma x^2}{n} - \left(\frac{\Sigma x}{n}\right)^2}$ (given on examination paper)

$\Sigma x^2 = 597\,700$

$$\sqrt{\frac{\Sigma x^2}{n} - \left(\frac{\Sigma x}{n}\right)^2} = \sqrt{\frac{597\,700}{10} - (237)^2}$$

$= \sqrt{3601} = 60.008\,333 = £60.01$ (to the nearest penny)

Incorrect rounding to £60.00 would be penalised in the mark scheme.

(b) After a bonus of £24.50:
mean = £237 + £24.50 = £261.50
standard deviation = £60.01 (unchanged)

(c) After a bonus of 10%:
mean = £237 + 10% × £237 = £260.70
standard deviation = £60.01 + 10% = £66.01

(d) Substantiated answers showing some understanding of the work will gain credit here. Clea the manager's scheme is better for the six lowest-paid workers whereas the foreman's sche is better for the four highest-paid workers.

3 (a)

Temperature, T	Tally	Frequency
$40 < T \leqslant 50$	𝍸 I	6
$50 < T \leqslant 54$	𝍸 I	6
$54 < T \leqslant 58$	𝍸 III	8
$58 < T \leqslant 62$	𝍸	5
$62 < T \leqslant 70$	𝍸	5

(b) Mean = 55.3 °F (3 s.f.)
Standard deviation = 6.80 °F (3 s.f.)

(c)

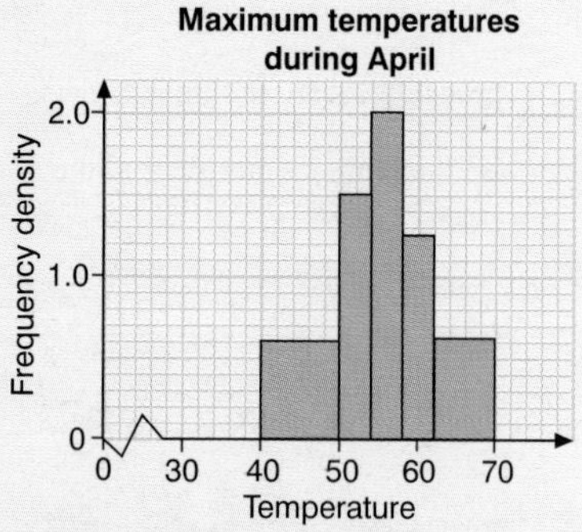

(d) The weather in June is generally hotter and less spread out.

(a) It is often helpful to use a tally chart to calculate the frequencies, as shown.

(b) Mean $= \frac{\Sigma x}{n} = \frac{1660}{30} = 55.333\,333... = 55.3$ °F (3 s.f.)

Standard deviation $= \sqrt{\frac{\Sigma x^2}{n} - \left(\frac{\Sigma x}{n}\right)^2} = \sqrt{\frac{93\,242}{30} - (55.333\,333)^2}$

$= \sqrt{46.288\,889} = 6.803\,593\,8 = 6.80$ °F (3 s.f.)

Final answers to 2 s.f. or 3 s.f. take account of the rounding incurred by taking mid-interval values to represent the values although original values might have been used to find the mean and standard deviation.

(c) Having identified the respective frequency densities, you can draw the histogram, as shown.

Temperature, T	Frequency	Class width	Frequency density
$40 < T \leqslant 50$	6	10	0.6
$50 < T \leqslant 54$	6	4	1.5
$54 < T \leqslant 58$	8	4	2.0
$58 < T \leqslant 62$	5	4	1.25
$62 < T \leqslant 70$	5	8	0.625

(d) April: mean = 55.3 °F
standard deviation = 6.80 °F
June: mean = 58.9 °F
standard deviation = 5.3°F
The word 'variable' might be better instead of 'spread out'.

4 (a)

Number of minutes late (t)	Frequency
$^-10 \leqslant t < {}^-2$	2
$^-2 \leqslant t < 0$	12
$0 \leqslant t < 2$	10
$2 \leqslant t < 5$	7
$5 \leqslant t < 10$	5
$10 \leqslant t < 30$	6

(a) In this type of question, you can use the idea of equivalent areas to calculate the remaining frequencies.
It is useful to check the sum of frequencies to make sure no simple mistakes are made.

(b) Mean = 4.0 (1 d.p.)
Standard deviation = 7.2 (1 d.p.)

Number of minutes late (t)	Frequency f	Mid-interval x	fx	fx^2
$^-10 \leqslant t < {}^-2$	2	$^-6$	$^-12$	72
$^-2 \leqslant t < 0$	12	$^-1$	$^-12$	12
$0 \leqslant t < 2$	10	1	10	10
$2 \leqslant t < 5$	7	3.5	24.5	85.75
$5 \leqslant t < 10$	5	7.5	37.5	281.25
$10 \leqslant t < 30$	6	20	120	2400
	$\Sigma f = 42$		$\Sigma fx = 168$	$\Sigma fx^2 = 2861$

Particular care must be taken when dealing with the negative signs.

(b) Mean $= \frac{\Sigma x}{n} = \frac{168}{42} = 4 = 4.0$ (1 d.p.)

Standard deviation $= \sqrt{\frac{\Sigma x^2}{n} - \left(\frac{\Sigma x}{n}\right)^2}$

$= \sqrt{\frac{2861}{42} - (4)^2}$

$= \sqrt{52.119\,048}$
$= 7.219\,352\,3$
$= 7.2$ (1 d.p.)

The question asks for answers correct to 1 d.p. Marks can be lost if this is not done.

Answers / **Examiner's comments**

5 (a)

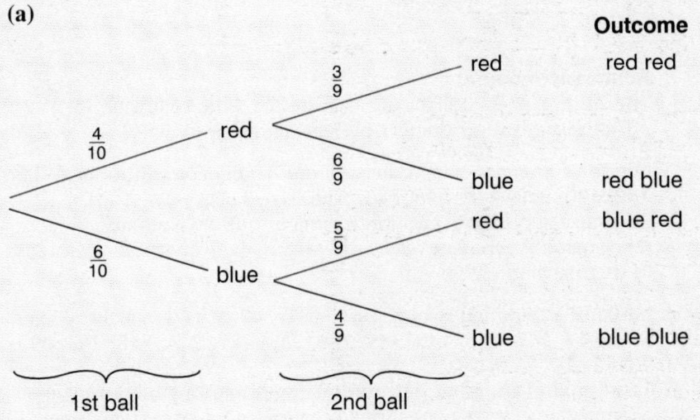

(b) (i) $\frac{2}{15}$

(ii) $\frac{3}{5}$

(b) (i) The probability of obtaining two red balls = p(red and red)

$$= \frac{4}{10} \times \frac{3}{9} = \frac{2}{15}$$

(ii) The probability of obtaining one ball of each colour = p(red and blue or blue and red)

$$= \frac{4}{10} \times \frac{6}{9} + \frac{6}{10} \times \frac{5}{9} = \frac{54}{90} = \frac{9}{15} = \frac{3}{5}$$

ACKNOWLEDGEMENTS

Published by HarperCollins*Publishers* Ltd
77–85 Fulham Palace Road
London W6 8JB

www.**Collins**Education.com
On-line support for schools and colleges

© HarperCollins*Publishers* Ltd 2001

First published 2001
Reprinted 2001
Reprinted 2002

ISBN 0 00 711202 5

Paul Metcalf asserts the moral right to be identified as the author of this work.

All rights reserved. No part of this publication may be reproduced, stored in a retrieval system, or transmitted in any form or by any means, electronic, mechanical, photocopying, recording or otherwise, without either the prior permission of the Publisher or a licence permitting restricted copying in the United Kingdom issued by the Copyright Licensing Agency Ltd, 90 Tottenham Court Road, London W1P 0LP. This book is sold subject to the condition that it shall not by way of trade or otherwise be lent, hired out or otherwise circulated without the Publisher's prior consent.

British Library Cataloguing in Publication Data

A catalogue record for this book is available from the British Library.

Edited by Joan Miller

Production by Kathryn Botterill

Cover design by Susi Martin-Taylor

Book design by Rupert Purcell and produced by Gecko Limited

Index compiled by Laurence Errington

Printed and bound by Bath Press

Acknowledgements
The Author and Publishers are grateful to the following for permission to reproduce copyright material:

London Examinations, a division of Edexcel Foundation (pp. 86, 189, 190, 223, 246, 248, 270, 282, 316, 317)
Edexcel Foundation, London Examinations accepts no responsibility whatsoever for the accuracy or method of working in the answers given.

Midland Examining Group (pp. 21, 31, 47, 85, 91, 94, 115, 117, 129, 144, 145, 146, 148, 149, 206, 225, 227, 244, 248, 283, 299, 301, 319)
The Midland Examining Group bears no responsibility for the example answers to questions taken from its past question papers which are contained in this publication.

Northern Examinations and Assessment Board (pp.62, 127, 129, 149, 205, 227, 315, 317)
The author is responsible for the possible answers/solutions and the commentaries on the past questions from the Northern Exminations and Assessment Board. They may not constitute the only possible solutions.

Southern Examining Group (pp. 21, 33, 54, 67, 69, 71, 87, 147, 227, 245, 246, 271, 286, 297, 298, 299, 318, 319)
Answers to questions taken from past examination papers are entirely the responsibility of the author and have neither been provided nor approved by the Southern Examining Group.

Illustrations
Roger Bastow, Harvey Collins, Gecko Ltd and Tony Warne

Every effort has been made to contact the holders of copyright material, but if any have been inadvertently overlooked, the Publishers will be pleased to make the necessary arrangements at the first opportunity.

You might also like to visit:
www.**fire**and**water**.com
The book lover's website

Study Notes